AF598320

38
Topics in Organometallic Chemistry

Topics in Organometallic Chemistry

Recently Published Volumes

Bifunctional Molecular Catalysis
Volume Editors: T. Ikariya, M. Shibasaki
Vol. 37, 2011

Asymmetric Catalysis from a Chinese Perspective
Volume Editor: Shengming Ma
Vol. 36, 2011

Higher Oxidation State Organopalladium and Platinum Chemistry
Volume Editor: A. J. Canty
Vol. 35, 2011

Iridium Catalysis
Volume Editor: P. G. Andersson
Vol. 34, 2011

Iron Catalysis – Fundamentals and Applications
Volume Editor: B. Plietker
Vol. 33, 2011

Medicinal Organometallic Chemistry
Volume Editors: G. Jaouen, N. Metzler-Nolte
Vol. 32, 2010

C-X Bond Formation
Volume Editor: A. Vigalok
Vol. 31, 2010

Transition Metal Complexes of Neutral η^1-Carbon Ligands
Volume Editors: R. Chauvin, Y. Canac
Vol. 30, 2010

Photophysics of Organometallics
Volume Editor: A. J. Lees
Vol. 29, 2010

Molecular Organometallic Materials for Optics
Volume Editors: H. Le Bozec, V. Guerchais
Vol. 28, 2010

Conducting and Magnetic Organometallic Molecular Materials
Volume Editors: M. Fourmigué, L. Ouahab
Vol. 27, 2009

Metal Catalysts in Olefin Polymerization
Volume Editor: Z. Guan
Vol. 26, 2009

Bio-inspired Catalysts
Volume Editor: T. R. Ward
Vol. 25, 2009

Directed Metallation
Volume Editor: N. Chatani
Vol. 24, 2007

Regulated Systems for Multiphase Catalysis
Volume Editors: W. Leitner, M. Hölscher
Vol. 23, 2008

Organometallic Oxidation Catalysis
Volume Editors: F. Meyer, C. Limberg
Vol. 22, 2007

N-Heterocyclic Carbenes in Transition Metal Catalysis
Volume Editor: F. Glorius
Vol. 21, 2006

Dendrimer Catalysis
Volume Editor: L. H. Gade
Vol. 20, 2006

Metal Catalyzed Cascade Reactions
Volume Editor: T. J. J. Müller
Vol. 19, 2006

Catalytic Carbonylation Reactions
Volume Editor: M. Beller
Vol. 18, 2006

Bioorganometallic Chemistry
Volume Editor: G. Simonneaux
Vol. 17, 2006

Surface and Interfacial Organometallic Chemistry and Catalysis
Volume Editors: C. Copéret, B. Chaudret
Vol. 16, 2005

Transition Metal Catalyzed Enantioselective Allylic Substitution in Organic Synthesis

Volume Editor: Uli Kazmaier

With Contributions by

A. Alexakis · J.-M. Begouin · M.L. Crawley · P.J. Guiry ·
C. Kammerer-Pentier · J. Kleimark · J.E.M.N. Klein ·
J.-B. Langlois · F. Liron · W.-B. Liu · L. Milhau ·
C. Moberg · P.-O. Norrby · B. Plietker · G. Poli ·
G. Prestat · B.M. Trost · D. Weickmann ·
J.-B. Xia · S.-L. You

Editor
Prof. Dr. Uli Kazmaier
Institut für Organische Chemie
Universität des Saarlandes
66123 Saarbrücken
Germany
u.kazmaier@mx.uni-saarland.de

ISBN 978-3-642-22748-6 e-ISBN 978-3-642-22749-3
DOI 10.1007/978-3-642-22749-3
Springer Heidelberg Dordrecht London New York

Library of Congress Control Number: 2011940292

Printed on acid-free paper

Springer is part of Springer Science+Business Media (www.springer.com)

Volume Editor

Prof. Dr. Uli Kazmaier

Institut für Organische Chemie
Universität des Saarlandes
66123 Saarbrücken
Germany
u.kazmaier@mx.uni-saarland.de

Editorial Board

Topics in Organometallic Chemistry Also Available Electronically

Topics in Organometallic Chemistry is included in Springer's eBook package *Chemistry and Materials Science*. If a library does not opt for the whole package the book series may be bought on a subscription basis. Also, all back volumes are available electronically.

For all customers who have a standing order to the print version of *Topics in Organometallic Chemistry*, we offer free access to the electronic volumes of the Series published in the current year via SpringerLink.

If you do not have access, you can still view the table of contents of each volume and the abstract of each article by going to the SpringerLink homepage, clicking on "Chemistry and Materials Science," under Subject Collection, then "Book Series," under Content Type and finally by selecting *Topics in Organometallic Chemistry*.

You will find information about the

- Editorial Board
- Aims and Scope
- Instructions for Authors
- Sample Contribution

at springer.com using the search function by typing in *Topics in Organometallic Chemistry*.
Color figures are published in full color in the electronic version on SpringerLink.

Aims and Scope

The series *Topics in Organometallic Chemistry* presents critical overviews of research results in organometallic chemistry. As our understanding of organometallic structures, properties and mechanisms grows, new paths are opened for the design of organometallic compounds and reactions tailored to the needs of such diverse areas as organic synthesis, medical research, biology and materials science. Thus the scope of coverage includes a broad range of topics of pure and applied organometallic chemistry, where new breakthroughs are being made that are of significance to a larger scientific audience.

The individual volumes of *Topics in Organometallic Chemistry* are thematic. Review articles are generally invited by the volume editors.

In references *Topics in Organometallic Chemistry* is abbreviated Top Organomet Chem and is cited as a journal. From volume 29 onwards this series is listed with ISI/Web of Knowledge and in coming years it will acquire an impact factor.

Preface

Organometallic chemistry is one of the key tools in modern organic synthesis. Besides stoichiometric reactions of organometallic compounds, especially transition metal-catalyzed reactions play a dominant role, and a wide range of transition metal-catalyzed cross-coupling reactions has been developed during the last decades. Of these C–C coupling reactions, the allylic alkylations became a major player in this field. In 1965, J. Tsuji discovered that C–C bond formation can be achieved by the reaction of π-allylpalladium complexes with C-nucleophiles, typically stabilized carbanions such as malonates. Later on, catalytic and enantioselective versions were developed mainly by B. M. Trost and his group. While in the early years the π-allyl chemistry was clearly dominated by the palladium complexes, in the meanwhile a wide range of other transition metals made their way into the limelight. During the last two decades, complexes of Mo, W, Ir, Rh, Ru and Fe became competitors to the popular Pd catalyst. Each of these transition metals has its own characteristics and reaction behavior.

The aim of this volume of *Topics in Organometallic Chemistry* is to focus on the latest developments of transition metal-catalyzed allylation reactions. Besides mechanistical aspects and the specialities of the different transition metals, applications of this interesting protocol in the asymmetric synthesis of natural products will also be covered.

Saarbrücken, Germany Uli Kazmaier

Contents

Top Organomet Chem (2012) 38: 1–64
DOI: 10.1007/3418_2011_14

Published online: 3 July 2011

Selectivity in Palladium-Catalyzed Allylic Substitution

Giovanni Poli, Guillaume Prestat, Frédéric Liron, and Claire Kammerer-Pentier

Abstract The present chapter introduces the basic fundaments of the palladium-catalyzed allylic substitution reaction. After a brief introduction, the reaction is explored into the different steps of the catalytic cycle in a chronological order. Formation of the crucial η^3-allyl palladium complexes is first commented, followed by a brief description of the static isomerism and dynamic features related to these compounds. Synthetic opportunities to intercept these complexes are then presented. Selectivity is then addressed with a first focus on regioselectivity and memory effects. Finally, selected examples of enantioselective versions are presented and classified according to the position of the enantiodiscriminating step in the catalytic cycle.

Keywords Allylic substitution · Enantioselectivity · Memory effect · Palladium · Regioselectivity · Tsuji–Trost reaction

Contents

G. Poli (✉), G. Prestat, and F. Liron
UPMC Univ Paris 06, Institut Parisien de Chimie Moléculaire, UMR CNRS 7201, FR2769, Case 183, F-75252 Paris Cedex 05, France
e-mail: giovanni.poli@upmc.fr; guillaume.prestat@upmc.fr; frederic.liron@upmc.fr

C. Kammerer-Pentier
Lehrstuhl für Organische Chemie I, Technische Universität München, Lichtenbergstr. 4, 85747 Garching, Germany
e-mail: claire.kammererpentier@mytum.de

1 Introduction

1.1 Foreword

This chapter introduces the book "Transition Metal Catalyzed Enantioselective Allylic Substitution in Organic Synthesis." Undeniably, in view of the considerable efforts spent on this topic over 45 years and the amount of impressive results obtained by hundreds of chemists, times were ripe for the redaction of such a book. The aim of this chapter, which is far from exhaustive, is to introduce this stimulating subject showing its state of the art and giving the reader the opportunity to fully appreciate the more specialized chapters that follow in the book.

1.2 A Touch on the Mechanism

Palladium-mediated allylic substitution was first reported as a stoichiometric reaction by Tsuji in 1965 [1]. In this seminal paper, it was shown that ethyl malonate, acetoacetate as well as an enamine derived from cyclohexanone react smoothly with dimeric π-allyl palladium chloride to afford allylated products. Catalytic version appeared in 1970 [2, 3], and in 1977 Trost reported the first palladium-catalyzed asymmetric allylic alkylation [4]. Since these pioneering works, the Tsuji–Trost reaction has known a remarkable development and numerous transition metals such as molybdenum [5], tungsten [6], iridium [7], rhodium [8], ruthenium [9], platinum [10], nickel [11, 12], copper [13], iron, and cobalt [14] have been recognized as efficient catalysts for this reaction. Transition metal-catalyzed allylic substitution is nowadays a common tool for organic synthesis. Moreover, asymmetric allylic alkylation (AAA) has become a benchmark reaction to test the efficiency of a new chiral ligand [15] and has been widely used as a key step for the preparation of bioactive compounds [16–19].

Scheme 1 Typical mechanism for a generic palladium catalyzed allylic substitution of soft nucleophiles

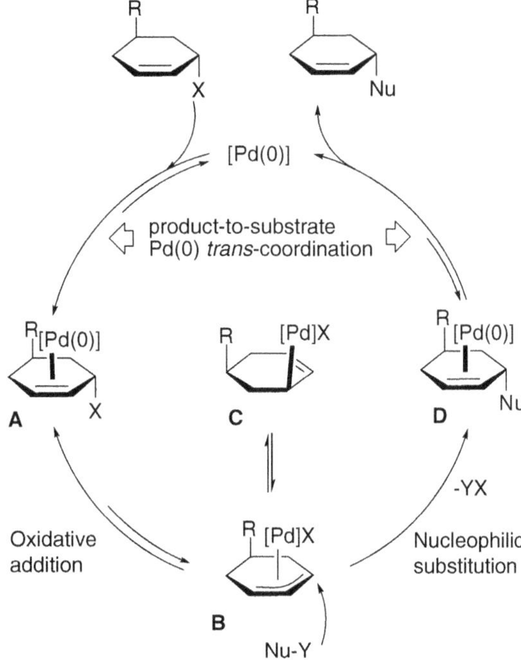

The typical mechanism using a palladium catalyst and "soft" nucleophile (p*K*a < 25) is depicted in Scheme 1.[1]

π-Coordination of the substrate to the electron-rich Pd(0) complex takes place *anti* to its leaving group and generates a Pd(0)-coordinated substrate **A**. *Oxidative addition* then affords the η^3-allyl palladium(II) complex **B**,[2] which can be in equilibrium with the isomeric η^1-allyl palladium(II) complex **C**. *Nucleophilic substitution* on the electrophilic complex **B** *anti* to the metal, affords either (usually irreversibly) the Pd(0)-coordinated product **D**, or gives back reversibly the Pd(0)-coordinated substrate **A**. Finally, product-to-substrate *Pd(0) trans-coordination* releases the product and closes the catalytic cycle.

Some general considerations are worthy. Although the counteranion in **B** lies outside the metal coordination sphere, usually as a tight ion-pair, in **C** it is in the metal coordination sphere. Donor ligands such as phosphines are necessary to enrich palladium atom and thus allow the oxidative addition step to take place. More precisely, moderately donor phosphine ligands are those that allow the fastest turnovers. Indeed, although electron-withdrawing phosphines render the

[1]Throughout this chapter, brackets around palladium atom in the notation of a generic (charged or neutral) allyl complex intend to render implicit the dative ligands. An asterisk next to the brackets indicates the presence of a chiral (usually enantiopure) ligand.

[2]As we will see later, this π-allyl palladium(II) complex can be also generated via interaction between an alkene and a PdX_2 complex.

allyl complex more electrophilic, they also favor ion-pair return (**B** → **A**) more than exogenous nucleophile attack (**B** → **D**) [20, 21]. Further ways of contrasting ion-pair return are the addition of salts of noncoordinating anions, such as $[B((3,5\text{-}(CF_3)_2)C_6H_3)_4]^-$ (BAr'F$^-$) [22], to break the tight ion-pair, or to use more efficient leaving groups (see Sect. 2.1.1).

Formation of the η^3-allyl palladium(II) complex **B** is a reversible process, and in the case of simple allyl derivatives, thermodynamically disfavored. In this case, the Pd(0)-coordinated complex **A** or the η^1-allyl palladium(II) complex **C** may be the resting state of the catalytic cycle. However, this seems to not be the case for diphenyl substituted allyl derivatives, which show the more stable η^3-allyl palladium (II) complex **B** as resting state. The rate-limiting step can be either the oxidative addition or the nucleophilic substitution, depending on the relative height of their respective transition states (Fig. 1). In any case, due to the irreversibility of the nucleophilic substitution by the exogenous nucleophile, the global transformation can be usually entirely shifted toward the product formation.

Finally, the above double inversion mechanism rationalizes the observed global retention of configuration in the substitution process [23].

"Hard" nucleophiles, on the other hand, directly attack the metal, leading to an overall inversion of the stereochemistry after reductive elimination and decoordination (Scheme 2) [24].

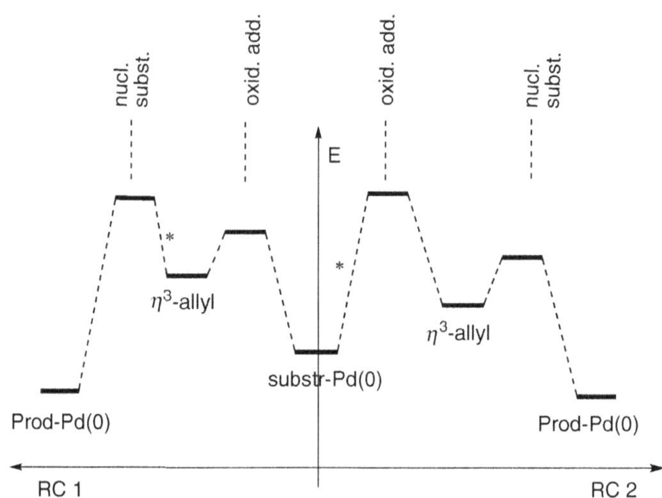

Fig. 1 Qualitative energy profiles for Pd-catalyzed allylic substitutions. The starred paths represent the rate-determining steps

Scheme 2 Mechanism of the palladium catalyzed allylic substitution of hard nucleophiles

2 Generation, Behavior, and Trapping of the π-Allyl Complex

2.1 Generation of the π-Allyl Complex

2.1.1 Via Cleavage of an Allylic C–X Bond

The most common way to generate a π-allyl palladium complex proceeds through the heterolytic cleavage of a C–X allylic bond in the presence of an electron rich palladium(0) complex. This step is an "oxidative addition," as the metal oxidation state increases by 2. However, it is often referred to as "ionization" as the transiently generated reactive η^3-allyl intermediate is normally an ionic species. Numerous allylic substrates have been used in this allylation reaction and some are presented in Fig. 2. As a general rule the better the leaving group, the more ion-pair return is inhibited, the faster the turnover [20].

Oxidative addition of the standard allyl acetate to a Pd(0) complex was demonstrated in 1981 by Yamamoto and coworkers, who isolated and characterized the corresponding η^3-allyl(acetato)palladium intermediate [25]. Formation of the π-allyl Pd(II) complex from allyl acetate and Pd(0) is a reversible process, whose equilibrium lies extensively in favor of the Pd(0) side [26, 27]. On the other hand, submission of an allyl trifluoroacetate to $Pd(dba)_2$ leads to the quantitative formation of the corresponding π-allyl complex [28]. These elements clearly demonstrate that the nature of the leaving group has a dramatic influence on the dislocation of the equilibrium and therefore on the whole substitution process.

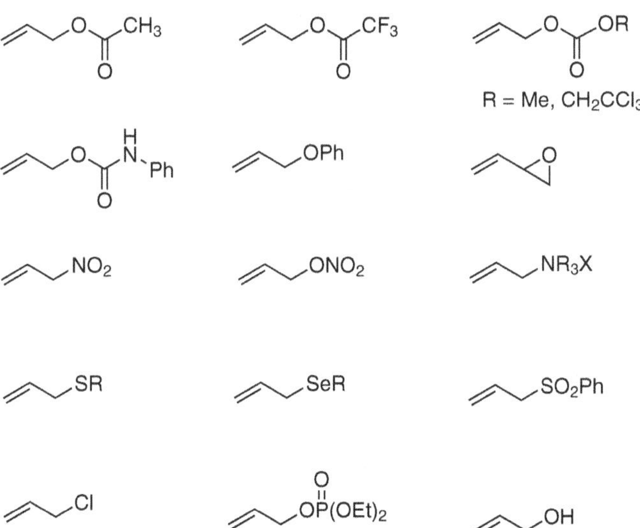

Fig. 2 Classical electrophiles used in the palladium catalyzed allylation reaction

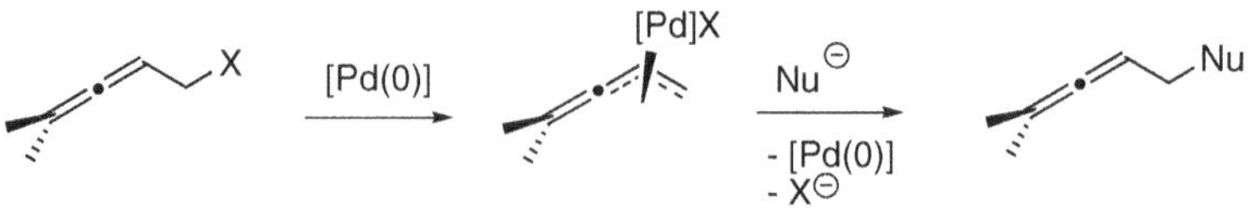

Scheme 3 Palladium catalyzed allylic allenylation

The introduction of carbonate as leaving group, by Tsuji in 1982, was an important breakthrough [29, 30]. Carbonate is a better leaving group than acetate, but its significant advantage over the latter is that it allows the reaction to occur in an almost neutral medium. This behavior was originally accounted for assuming that decarboxylation of the released carbonate generates the corresponding alkoxide, which can in turn deprotonate the pronucleophile. However, such a scenario contrasts with the fact that η^3-allyl palladium alkoxycarbonate complexes can be isolated without showing spontaneous decarboxylation [31] and their reaction with an acidic substrate generates the corresponding hydrogen carbonate complex, which might decarboxylate only at this stage. Recent studies by Amatore, Jutand, and Moreno–Mañas [32] demonstrated that oxidative addition from allylic carbonates is a reversible process, thereby confirming that η^3-allyl palladium alkoxycarbonates do not undergo spontaneous decarboxylation. Allyl carbamates react as allyl carbonates [33]. Following the same principle and introduced one year before allyl carbonates, the less general 1,3-diene monoepoxide reacts also with nucleophiles in the absence of base [34, 35]. Allyl phenoxides [36] are also compatible with neutral conditions, but they are associated with a poor reactivity toward oxidative addition that limits their use.

Apart from these classical substrates, allylic ammonium ions [37], sulfones [38], sulfides and selenides [39], phosphates [40], nitro groups [41, 42], halides [43], nitrates [44] have also been used as π-allyl precursors.

The parent allyl alcohol is an attractive substrate from an atom economical point of view. Although it has been used as early as 1970 by Atkins [2], its development has been hampered by the poor leaving aptitude of the hydroxyl group and is still under active investigation [45, 46].

Although less documented, oxidative addition of a Pd(0) complex on allenyl derivatives generates a vinyl allyl Pd(II) complex that can be trapped by a nucleophile (Scheme 3).

In this context, phosphates [47] and acetates [48] (Scheme 4) have been used successfully as leaving groups.

2.1.2 Via Cleavage of a C–X Benzylic Bond

The synthesis and isolation of benzylic π-allyl complexes, from benzylic chlorides and palladium vapors, have been reported as early as 1977 [49], whereas the first palladium-catalyzed allylic alkylation exploiting such type of benzylic π-complexes was reported by Legros and Fiaud fifteen years later [50]. The disruption

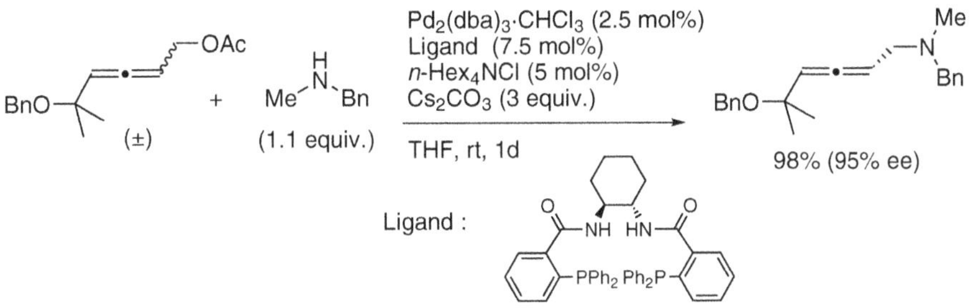

Scheme 4 Palladium catalyzed allylic allenylation of a secondary amine

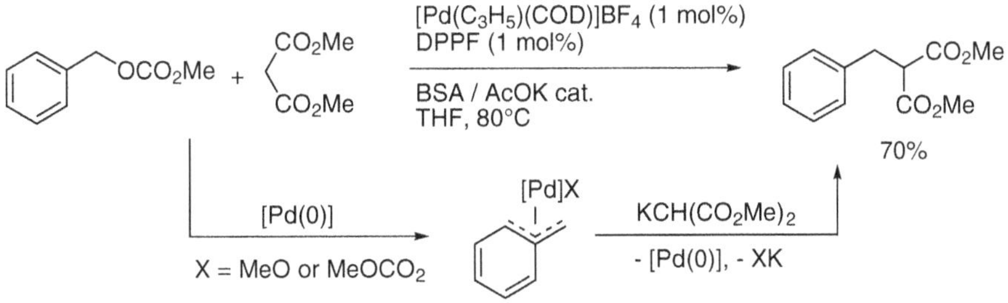

Scheme 5 Palladium catalyzed benzylation of dimethyl malonate

of aromaticity induced by the formation of the η^3-allylic complex was supposed to limit the synthetic application of this reaction to compounds featuring a weak aromaticity such as naphthylmethyl derivatives. Kuwano showed the dramatic influence of the ligand bite angle in this reaction [51]. Indeed, use of large bite angle ligands such as DPPF, DPEphos, or Xantphos allows the palladium-catalyzed substitution of simple benzylic carbonates in high yield (Scheme 5). This new area of allylic alkylation has attracted the attention of several researchers [52, 53].

2.1.3 Via Carbopalladation on a Diene

Use of 1,2 or 1,3 dienes as substrates to induce the formation of a π-allyl palladium complex is also conceivable via a carbopalladation step. Indeed, carbopalladation by an organopalladium complex to such dienes transitorily generates a σ-allyl complex that equilibrates to the more stable π-allyl complex (Scheme 6).

In 1970, Stevens and Shier [54] reported the isolation of a palladium π-allyl complex generated via the carbopalladation of an arylpalladium(II) onto propadiene, thereby giving support to such a strategy (Scheme 7).

In 1984, Tsuji [55] and Goré [56] reported independently the catalytic generation and trapping of π-allyl palladium complexes derived from allene carbopalladation using amine and malonate as nucleophile, respectively (Scheme 8). This route has known considerable development since these pioneering experiments [57].

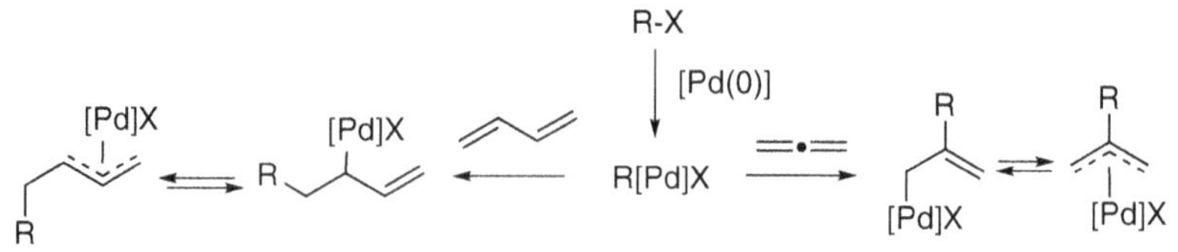

Scheme 6 Generation of π-allyl palladium complexes from 1,2 or 1,3 dienes

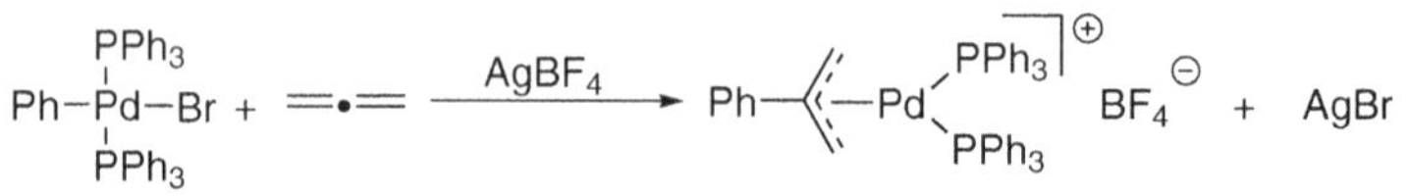

Scheme 7 Isolation of π-allyl palladium complex via carbopalladation of 1,2-propadiene

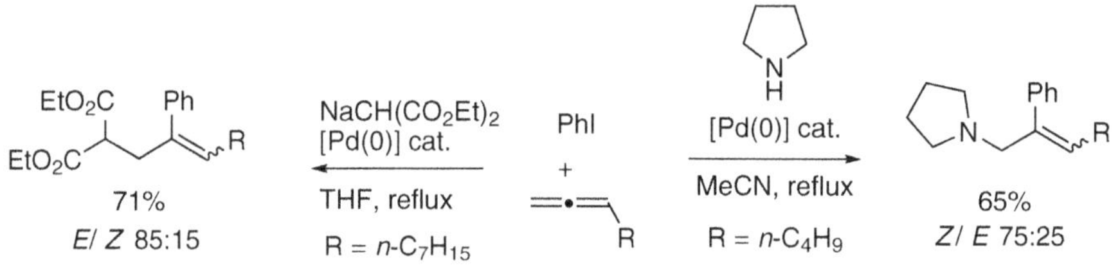

Scheme 8 Generation and interception of π-allyl palladium complexes derived from allene carbopalladation

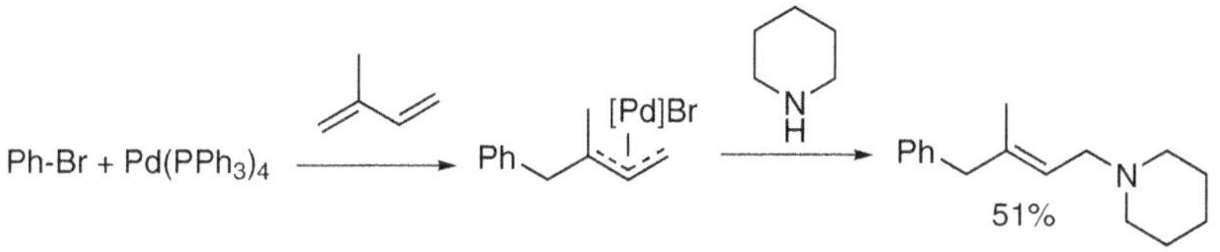

Scheme 9 Generation and interception of a π-allyl palladium complex derived from carbopalladation of a 1,3-diene

Similarly, Heck [58] has reported in 1978 the use of 1,3-dienes with amines as nucleophiles (Scheme 9), whereas the use of malonates was reported by Dieck in 1983 [59].

2.1.4 Via Nucleopalladation of a 1,3-Diene

Attack of a nucleophile to a PdX_2-activated 1,3-diene generates a π-allyl palladium complex [60, 61]. This intermediate can be regioselectively trapped by a second nucleophile thereby releasing the product and a palladium(0) complex. In order to

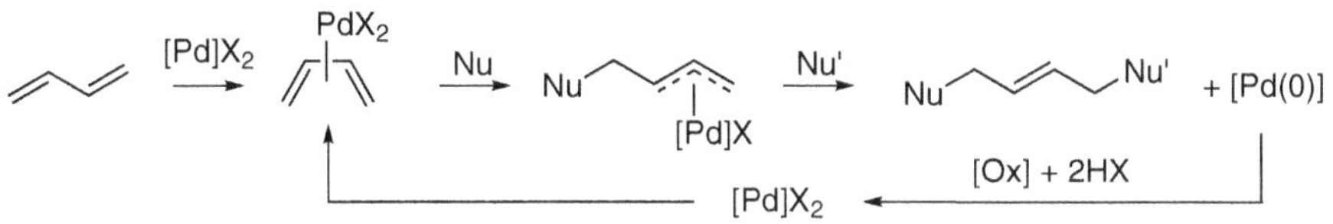

Scheme 10 Generation and interception of a π-allyl palladium complex derived from nucleopalladation of a 1,3-diene

allow a catalytic process, Pd(0) has to be reoxidized to a Pd(II) salt (Scheme 10). Very elegant examples in this field are the Bäckvall's 1,4-diacetoxylation and 1,4-chloroacetoxylation [62, 63]. A similar reactivity is observed on the use of allenes [64, 65].

2.1.5 Via C–H Activation

Allylic C–H bond cleavage is certainly the most straightforward and atom economical way to generate a π-allyl complex. The mechanism of this reaction is believed to proceed through allylic C–H bond cleavage activated by Pd(II)-alkene coordination, in the presence of a suitable ligand and/or additive. The thus generated π-allyl complex is classically intercepted by a nucleophile with release of a Pd(0) species. Reoxidation of the latter to Pd(II) closes the catalytic cycle (Scheme 11).

Parshall and Wilkinson reported in 1962 the synthesis of a π-allyl complex from mesityl oxide and palladium or platinum salts [66]. In 1973, the first stoichiometric allylic alkylation using a nonfunctionalized olefin was reported by Trost and Fullerton [67]. Development of a catalytic version was highly desirable but was hampered by the difficulties to render compatible each step. Eventually, Chen and White reported in 2004 a Pd(II)-catalyzed allylic acetoxylation of terminal olefin [68]. Success was encountered using $Pd(OAc)_2$ in the presence of a disulfoxide ligand and 2,6-dimethylbenzoquinone (DMBQ) as a reoxidant. The same group [69] and Shi and coworkers (see also [70]) showed in 2008 that similar conditions allow the use of "soft" carbon nucleophiles, too (Scheme 12).

Palladium-catalyzed allylic C–H substitution has been further developed and allows C–O, C–N, and C–C bond formation, yet a general asymmetric version remains challenging [71, 72].

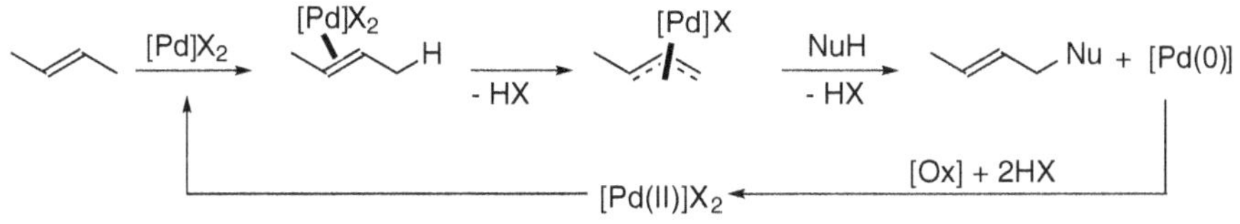

Scheme 11 Generation and interception of a π-allyl palladium complex derived from allylic C–H activation

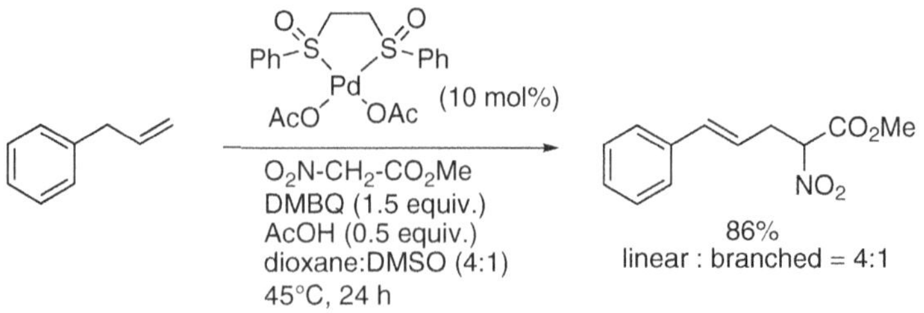

Scheme 12 Generation of a π-allyl palladium complex derived from allylic C–H activation and its interception with methyl nitroacetate

2.2 *Isomerism and Dynamic Equilibria of π-Allyl Palladium Complexes*

2.2.1 Static Stereochemical Analysis of η^3-(Allyl)palladium Complexes

In a cationic η^3-allylpalladium complex of type $[\eta^3\text{-(allyl)PdL}_2]^+X^-$ two coordination sites are occupied by the allyl fragment, whereas the other two are engaged by Lewis-basic ligands such as phosphines, amines or halide ions. Such complexes feature a square planar geometry around the palladium center, and may incorporate zero, one, or two intrinsic stereogenic elements, according to the degree of substitution of the allyl moiety and/or the nature of the complexed ligands. Thus, for example, although the achiral complex **A** incorporates no stereogenic unit, in the chiral complexes **B** and **C** the palladium atom and the allyl plane are stereogenic, respectively. Complex **D**, on the other hand, possesses both the central (Pd atom) as well as the planar (allyl plane) stereogenic units at the same time. Finally, the achiral complex **E** bears an alkene-type stereogenic axis. As a consequence, although complexes **B** and **C** can exist in opposite enantiomeric forms (**B/ent-B**, **C/ent-C**), complexes **D** can be present in four isomeric forms: two diastereomeric forms each one as a pair of enantiomers (**D$_1$/ent-D$_1$**, **D$_2$/ent-D$_2$**) and complex **E** admits two possible diastereoisomeric forms (**E$_{endo}$/E$_{exo}$**) (Fig. 3). As allyl metal complexes of this type exhibit a fluxional behavior, interconversion between these different isomeric forms is possible and, as we will see, can take place via different mechanisms.

2.2.2 Dynamic Stereochemical Analysis of η^3-(Allyl)palladium Complexes

In the absence of a nucleophile, or if the trapping step is slow enough, the π-allylpalladium complex may undergo the following four different equilibria: (a) $\eta^3 - \eta^1$ isomerization, (b) ligand association, (c) ligand dissociation, (d) nucleophilic displacement by a Pd(0)L$_n$ complex (Scheme 13) [73, 74]. Activation of these equilibria depends on the reaction conditions and can trigger exchange of the

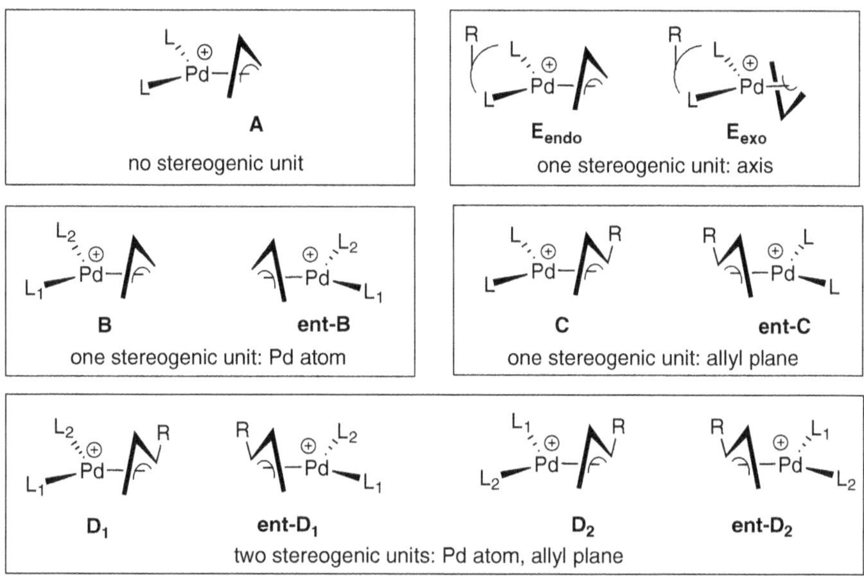

Fig. 3 Number and nature of stereogenic units and possible isomers associated to generic η^3-allylpalladium complexes of type $[\eta^3\text{-(allyl)PdL}_2]^+X^-$

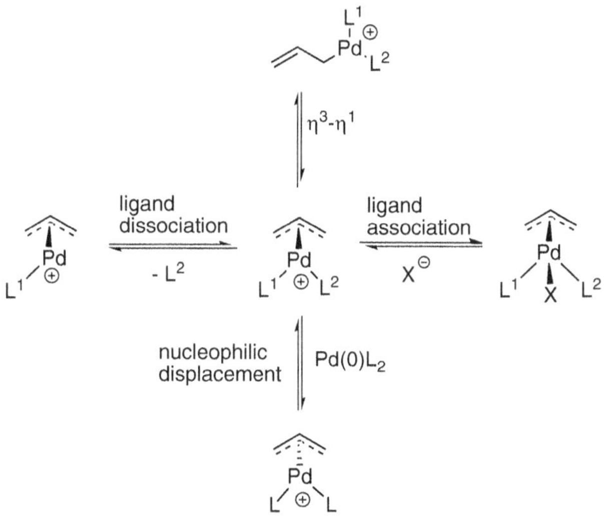

Scheme 13 Possible equilibria associated to a generic π-allylpalladium complex

allyl face complexed by the metal (with or without *syn-anti* switch of the allylic substituents), or the formal rotation of the allyl moiety with respect to the other coordinated ligands (see later).

Syn-Anti Isomerization

$\eta^3 - \eta^1$ Isomerization followed by C–C bond rotation and $\eta^1 - \eta^3$ equilibration leads to a global *syn-anti* exchange of the substituent pair concerned in the rotation, with concomitant exchange of the complexed allyl face (Scheme 14). This equilibrium is very facile when $R^1 = R^2 = H$, and normally displaced toward the *syn* isomer side in the case of monosubstitution.

In the case of the generic complexes $[\eta^3\text{-(allyl)PdL}_2]^+X^-$ **A-E**, such movement may lead to regeneration of the starting molecule, enantiomerization, or diastereomerization (Fig. 4).

Szabó performed a DFT calculation study of *syn-anti* equilibration on a model η^3-allyl palladium complex [75]. Whereas influence of the solvent during the $\eta^3 \rightarrow \eta^1$ process is limited to electrostatic interaction, solvent coordination to the tricoordinated η^1-allylpalladium species induces a stabilization, which was found more efficient in Me_2O than in CH_2Cl_2 by 6.8 kcal mol^{-1}. In the former case, the barrier to C–C bond rotation within the resulting η^1 intermediate was found to be 7.3 kcal mol^{-1}, whereas restoration of the η^3 coordination is almost barrierless. Therefore, this process appears to be very fast in coordinating solvents such as

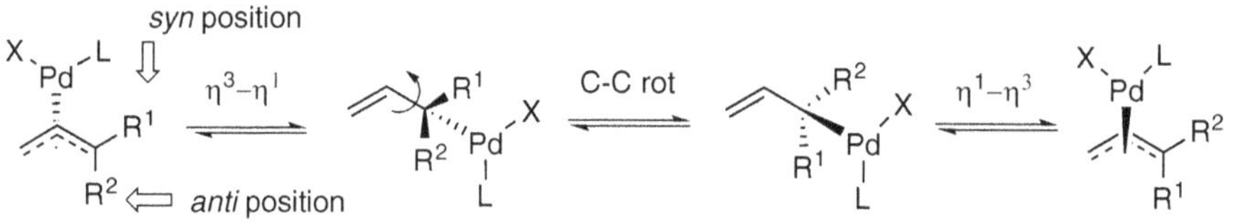

Scheme 14 *Syn-anti* isomerization of a generic π-allylpalladium complex

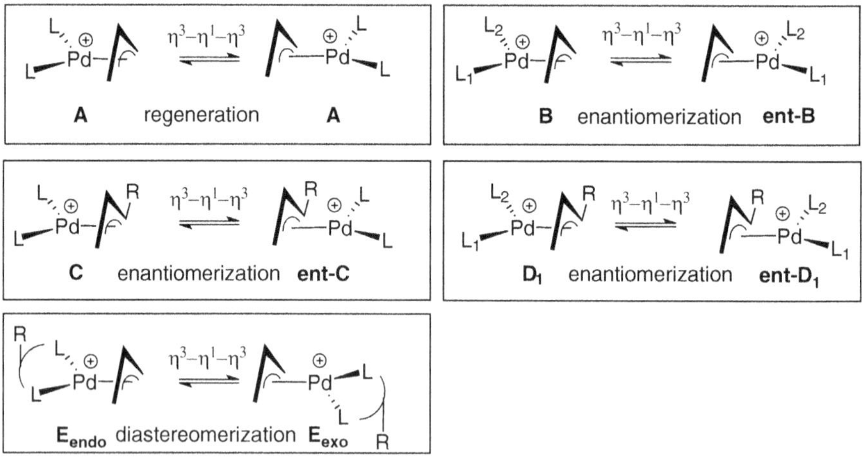

Fig. 4 Results of the *syn-anti* isomerization of differently substituted generic π-allylpalladium complexes of type $[\eta^3\text{-(allyl)PdL}_2]^+X^-$

Me_2O, whereas is less facile in noncoordinating solvents such as CH_2Cl_2 (Scheme 15).

The rate of *syn-anti* equilibration strongly depends on the nature of the ligand. For example, with bis(oxazoline) ligands, fast exchange, indicated by coalescence of the NMR spectra, is obtained at 45 °C, whereas two sets of signals are observed below 0 °C [76]. For *P,P* ligands, lower rate constants of up to 4.3 s^{-1} at 50 °C were measured for the *syn-anti* isomerization [77] and even lower rate constants of 3.5 s^{-1} at 72 °C were observed with *P,S* ligands [78].

Due to the large *trans* influence of the P atom, the Pd–C bond *trans* to phosphorus atom is the bond that is cleaved during *syn-anti* equilibration. Hence, rotation occurs around the allyl terminus originally *cis* to phosphorus (Scheme 16) [79, 80]. The presence of halide ions enhances the rate of *syn-anti* isomerization [81].

At 0 °C, MOP ligands showed a *syn-anti* isomerization rate of 2.2 s^{-1} [82]. This rate is increased with a small bite angle or electron-poor ligands [83], which isomerize 20 times faster than electron-rich ones (Table 1) [81]. Moreover, the

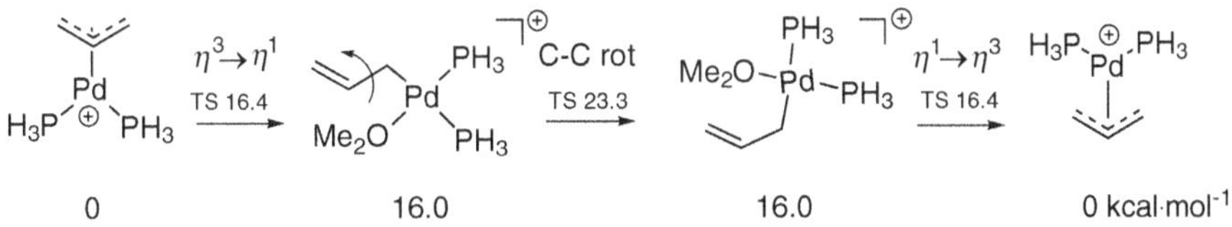

Scheme 15 Computed energy variations in the *syn-anti* equilibration of a model η^3-allyl palladium complex

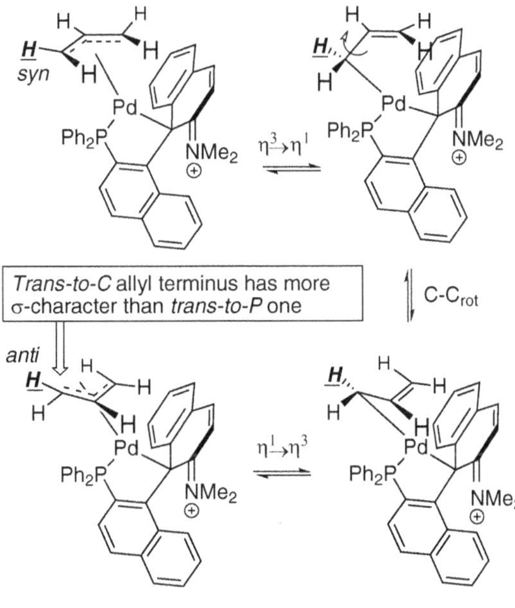

Scheme 16 *Syn-anti* equilibration path in a MAP-coordinated η^3-allyl palladium complex

Table 1 Influence of the electronic density of ligands in the *syn-anti* isomerization of MOP-type coordinated η^3-allyl palladium complexes

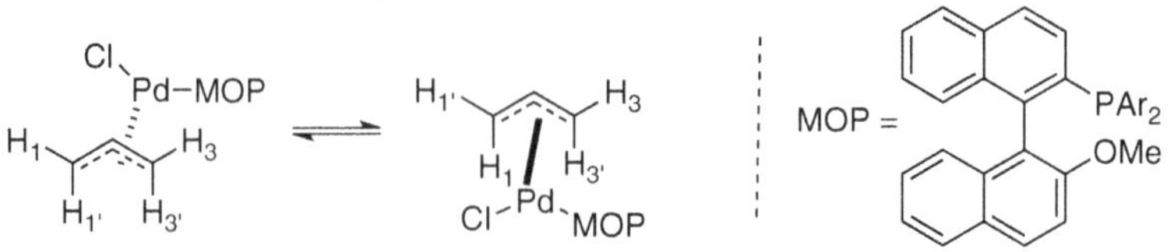

Entry	Ar	k (s^{-1})
1	3-$CF_3C_6H_4$	1.7
2	Ph	0.4
3	4-$MeOC_6H_4$	0.08

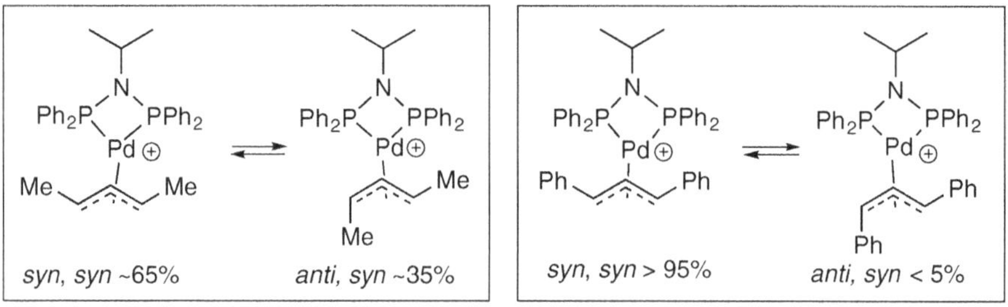

Scheme 17 *Syn-anti* equilibria in 1,3-disubstituted π-allyl palladium complexes

absence of a strong *trans* influence ligand increases the energy of this process. The steric hindrance between the allyl fragment and the ligands also plays a key role.

In the case of 1,3-disubstituted π-allyl complexes, the *syn*, *syn* conformation is usually the preferred one. Thus, for example, if methyl groups are the substituents on the allyl moiety, up to 80% of the π-allyl complex is in the *syn*, *syn* conformation. For bulkier substituents, other conformers are not detected (Scheme 17) [84].

Normally, attack of malonate anion to a π-allyl complex is slower than a possible *anti*-to-*syn* equilibration. Thus, for example, if the nucleophilic substitution to the *syn–syn* η^3-allyl complex **F** is slower than the *anti*-to-*syn* equilibration of the isomeric complexes **B** or **D** derived from *Z* allylic substrates [85], the same selectivity will be obtained independently of the starting material used. In this case, the observed regioselectivity depends exclusively on the differential reactivity of the two allylic termini of the common *syn–syn* intermediate **F**. On the other hand, erosion of the enantiomeric ratio is not a problem. In fact, the Pd(0)-catalyzed racemization via **F**-to-**G** conversion [86] is expected to be slow compared to the rate of nucleophilic substitution (Scheme 18).

The facile equilibration restricted to a single stereochemical set prior to the nucleophilic substitution has been cleverly exploited within the total synthesis of the natural products pyranicin and pyragonicin. In this enantioconvergent step a mixture of (2*E*,4*R*) and (2*Z*,4*S*) allylic isomers, in turn derived from a parallel

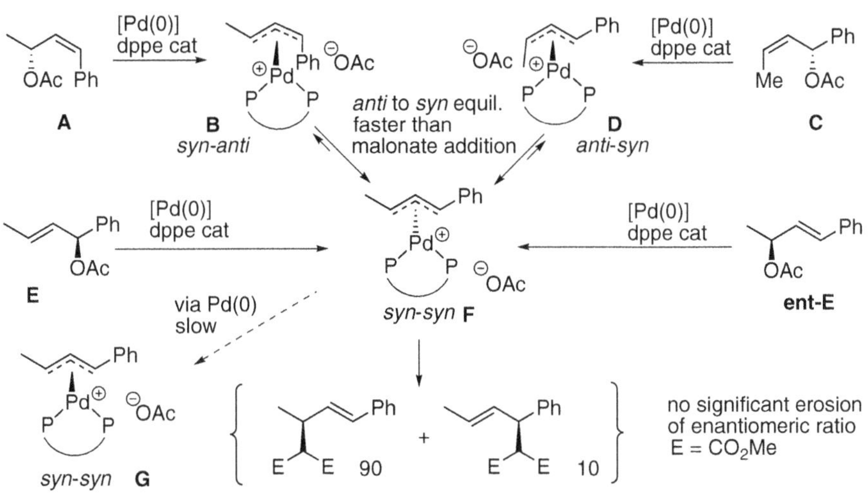

Scheme 18 Generation of a common enantiopure π-allyl palladium complex from different isomeric enantiopure substrates when nucleophilic substitution is the rate determining step

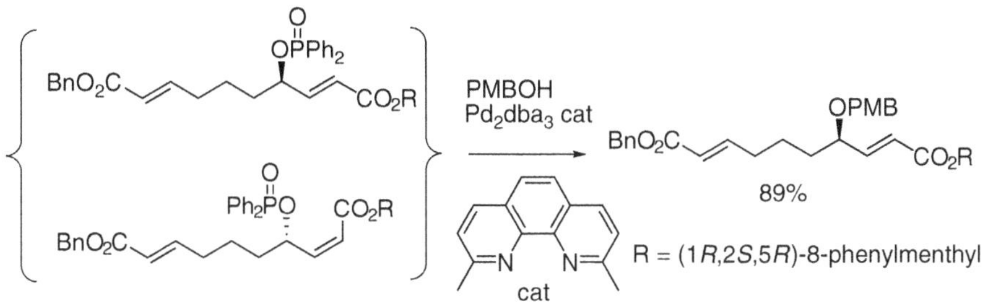

Scheme 19 Enantioconvergent palladium catalyzed allylic substitution

kinetic resolution (PKR) of a racemic precursor, afforded one enantiomer of the desired allyl ether through the involvement of a common η^3-allyl palladium complex (Scheme 19) [87, 88].

Apparent Allyl Rotation

As the name itself reveals, apparent allyl rotation (AAR) implies the formal rotation of the allyl moiety around the imaginary Pd-allyl bond axis. Such a movement can bring about inversion of the stereogenic palladium center, (like in **B** and **D**), or of the stereogenic axis (like in **E**), but never of the stereogenic plane, which sticks complexed to the metal via the same side. As a consequence, apparent allyl rotation can regenerate a structure identical to the starting one, or bring about enantiomerization or diastereomerization, depending on the substitution of the starting complex (Fig. 5). From the practical viewpoint, this movement is only visible when it implies the generation of two diastereoisomers.

Three possible pathways have been postulated for such isomerism, two of them conserving the η^3 coordination throughout the process, either *via* a dissociative or an associative mechanism, and the third one featuring a $\eta^3 - \eta^1$ isomerization/C–Pd rotation/$\eta^1 - \eta^3$ isomerization sequence (Scheme 20).

In complexes containing bidentate ligands with different donor properties, e.g., *P,N-*, *P,S-*, *N,S-*, etc. the mechanism of the apparent allyl rotation is strongly dependent on ligands' *trans* effect, the most donor heteroatom preferring the position *trans* to the more reactive allyl terminus. Donor heteroatoms can be ranked

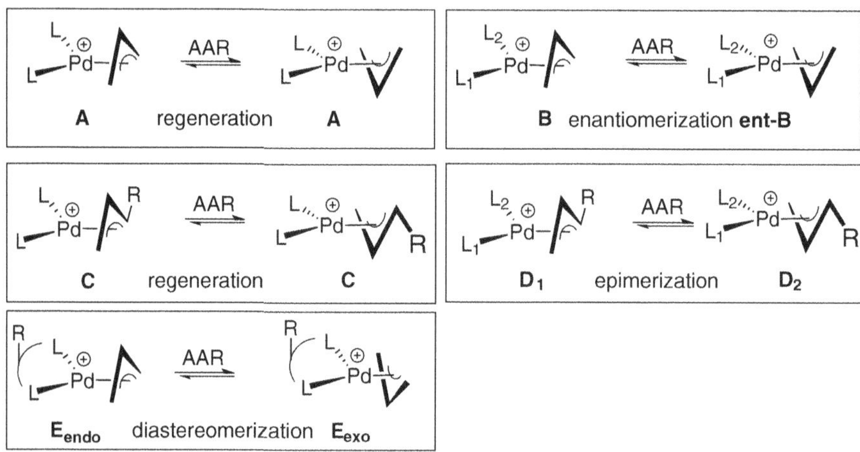

Fig. 5 Results of *apparent allyl rotation* in differently substituted generic π-allylpalladium complexes of type $[\eta^3\text{-(allyl)PdL}_2]^+X^-$

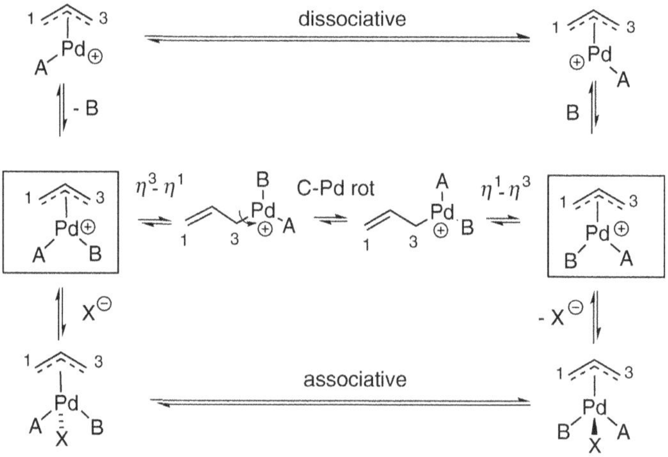

Scheme 20 Possible pathways for the *apparent allyl rotation*

according to their *trans* effect *P* ~ NHC > *S* > *N* (NHC = *N*-heterocyclic carbene) [89, 90].

A preliminary study involving *P,N* ligands concluded that in the absence of excess ligand this isomerization occurs *via* dissociation of one donor atom to lead to a T-shaped intermediate (Scheme 20, top) [91]. Although such an event has a half-life in the order of hours, the presence of excess ligand, coordinating solvents or chloride anions dramatically increases its rate. Apparent allyl rotation occurs faster in DMSO than in $CDCl_3$ [92], whereas in noncoordinating solvents water can increase the rate [93]. Moreover, in the case of *N,N* ligands, ligand dissociation was found to proceed with a higher energy of activation than with *P,N* ligands. These findings led to the hypothesis that pentacoordinated species (Scheme 20, bottom) may be involved, rather than T-shaped, three-coordinated intermediates (Scheme 20, top) [93, 94], and theoretical studies by Norrby confirm such hypothesis [95]. However, the pathway involving the $\eta^3 - \eta^1 - \eta^3$ interconversion has been also postulated (Scheme 20, middle). Use of more basic ligands led to more difficult geometrical apparent rotation.

The lowest energy barrier for apparent allyl rotation was found to be 27 kcal·mol^{-1} [95]. Although this energy barrier is comparable to that for *syn-anti* isomerization, the latter is usually a faster process.

The differential reactivity of *endo*/*exo* isomers is a very important issue. Indeed, nucleophilic attack to one or the other isomer may determine the formation of one or the other enantiomer (Scheme 21).

Although some exceptions exist [96], *endo* isomers are generally the more stable isomers (Table 2) [81].

Endo–exo ratios of up to 10:1 have been observed for other *P,N* ligands with Cl^- counterions, whereas PF_6^- anions led to a somewhat lower selectivity (up to 8.5:1). The nature of the counterion has a great influence on the rate of *endo–exo* equilibration. For example, in $CDCl_3$ the *endo–exo* equilibrium ratio is reached much faster with Cl^- than with PF_6^- counterions [97]. At 25 °C, a rate constant of 0.2 s^{-1} was obtained [97]. However, in an *N,S* ligand series, *endo–exo* isomerization was found to occur within seconds [90]. The authors attributed this effect to a greater flexibility around the sulfur than phosphorus atom, therefore yielding mixtures of *endo* and *exo* isomers (Scheme 22).

Isomerizations via Dative Ligand Flip

A $[\eta^3\text{-(allyl)PdL}_2]^+X^-$ complex can show the presence of slowly interconverting rotamers wherein the η^3-(allyl)PdPP moiety is not directly involved. This is for

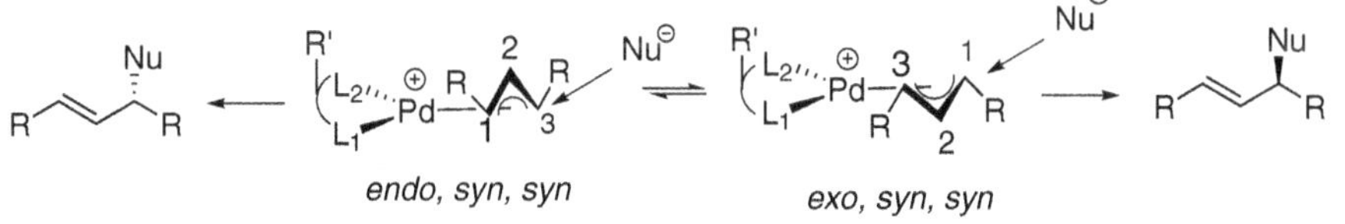

endo, syn, syn   *exo, syn, syn*

Scheme 21 Stereochemical implications associated to *endo*/*exo* isomerism

Table 2 *Endo–exo* ratios for *P,N* ligands

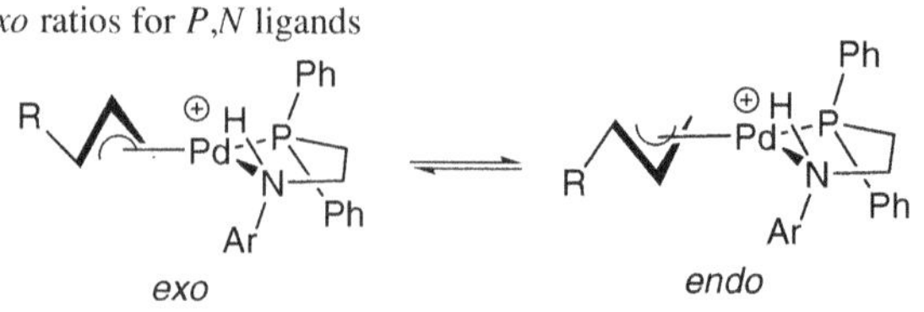

Entry	R	Ar	*Endo/exo*
1	Me	Ph	64/36
2	Me	2,6-$C_6H_3i Pr_2$	89/11
3	Ph	Ph	71/29
4	Ph	2,6-$C_6H_3i Pr_2$	56/44

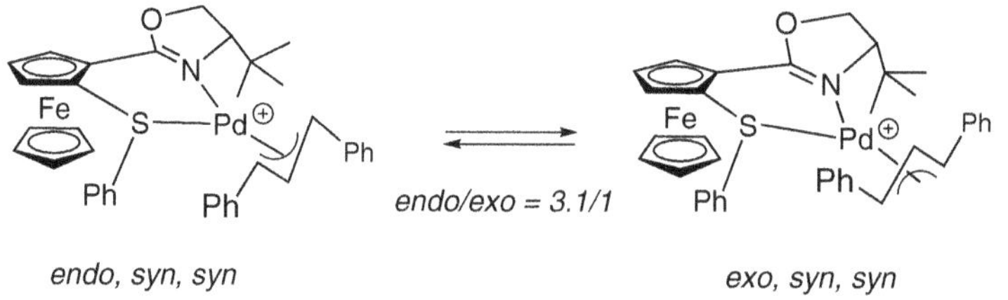

Scheme 22 *Endo–exo* equilibration in a π-allylpalladium complex coordinated to a *N,S* ligand

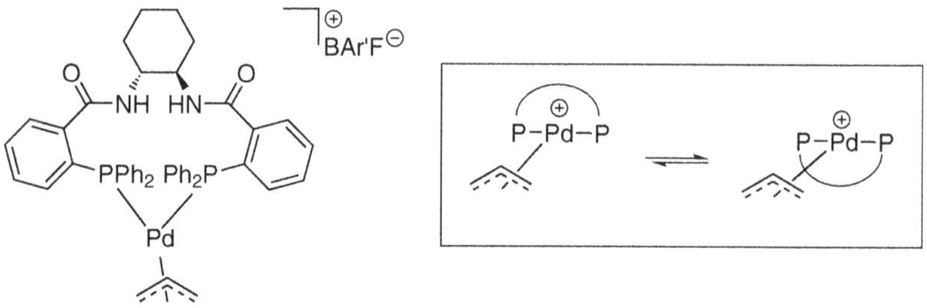

Scheme 23 *Endo–exo* isomerizations not involving the η^3-(allyl)PdPP fragment

example the case of the 13-membered η^3-allyl palladium complex ligated to the Trost Standard Ligand (TSL) [η^3-(allyl)Pd-(TSL)]$^+$BAr'F$^-$ [98] whose $^{31}P\{^1H\}$ NMR spectrum shows two distinct and equally populated rotamers (Scheme 23).

Palladium(0)-Catalyzed Allyl Exchange

As π-allylpalladium(II) complexes are electrophilic and palladium(0) complexes nucleophilic, the two species can react with each other leading to exchange of the complexed π-allyl face (Scheme 24) [32, 99].

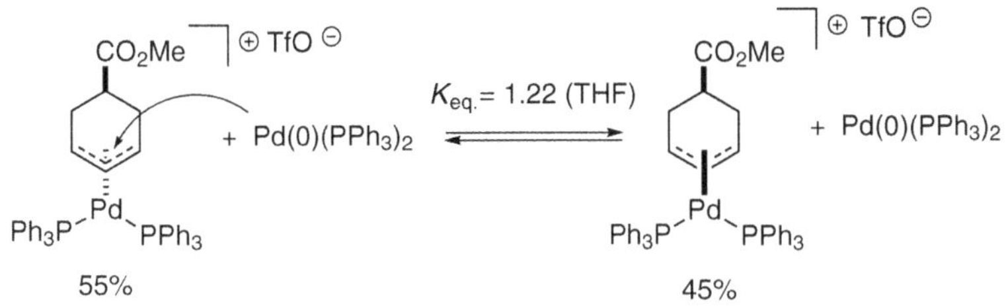

Scheme 24 Equilibration via palladium(0)-catalyzed allyl exchange

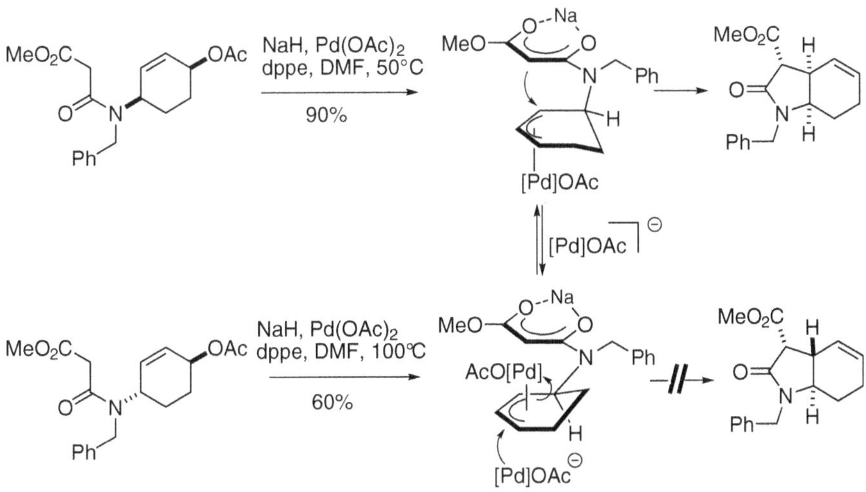

Scheme 25 Example of a stereoconvergent allylation via palladium(0)-catalyzed allyl exchange

As a consequence, if the complexed faces are enantiotopic, final nucleophilic trapping of one or the other complex will generate two opposite enantiomers. This S_N2-type substitution was first reported by Tsuji, who noticed that the higher the palladium(0) concentration, the lower the chirality transfer [100]. In case of high catalyst loading, the amount of palladium not involved in a π-allyl complex increases, making the amount of nucleophilic Pd(0) in solution higher. In contrast to the previously described *syn-anti* exchange and apparent allyl rotation (that are unimolecular) this isomerization is bimolecular. Therefore, as long as the amounts of Pd(0) used are minimized, this isomerization can usually be avoided in catalytic experiments. However, such a process can sometimes become operative, for example when the firstly generated π-allyl complex turns out to be unreactive (Scheme 25) [101].

2.3 Trapping of the π-Allyl Complex

The electrophilic π-allylpalladium intermediate can be trapped by a carbon- or a heteronucleophile, allowing the formation of a new C–C or C–heteroatom bond at one allylic terminus. Palladium(II) is concomitantly reduced to palladium(0), thereby allowing the reaction to be catalytic.

As mentioned above, the mechanism of the nucleophilic substitution depends on the nature of the incoming nucleophile. On the one hand, "soft" nucleophiles, whose conjugate acids have p*K*a < 25, attack directly on the allyl ligand (see Scheme 1). On the other hand, "hard" nucleophiles, whose conjugate acids have p*K*a > 25, coordinate first to the palladium center, then are intramolecularly transferred to the allyl ligand via reductive elimination (see Scheme 2) [102].

2.3.1 Trapping of π-Allyl Intermediates by "Soft" Nucleophiles

Mechanism and Stereochemistry

When the palladium-catalyzed allylic substitution is carried out in the presence of a "soft" nucleophile, the ionization step and the nucleophilic attack are mechanistically similar processes, occurring outside the coordination sphere of the metal. Indeed, the nucleophile approaches the electrophilic π-allyl ligand on the face opposite to palladium and directly attacks the C1 or C3 allylic carbon. This event results in an η^3-to-η^2 reorganization affording a palladium(0)-olefin complex, which readily releases the allylated product and the active catalytic palladium(0) species (see Scheme 1).

The stereochemistry of the allylation of "soft" nucleophiles has been studied on cyclic and acyclic models. Trost and coworkers showed that the reaction of *cis* cyclohexenyl acetate with the sodium salt of dimethylmalonate under palladium(0) catalysis affords the substitution product with a net retention of the relative stereochemistry (Scheme 26) [23]. As the oxidative addition gives rise to the transient π-allyl palladium complex via inversion [103], attack of the nucleophile occurs *anti* to the palladium center.

CO_2Me OAc [Pd(0)] cat. $NaCH(CO_2Me)_2$ *overall retention* CO_2Me CO_2Me CO_2Me

inversion CO_2Me [Pd]OAc *inversion*

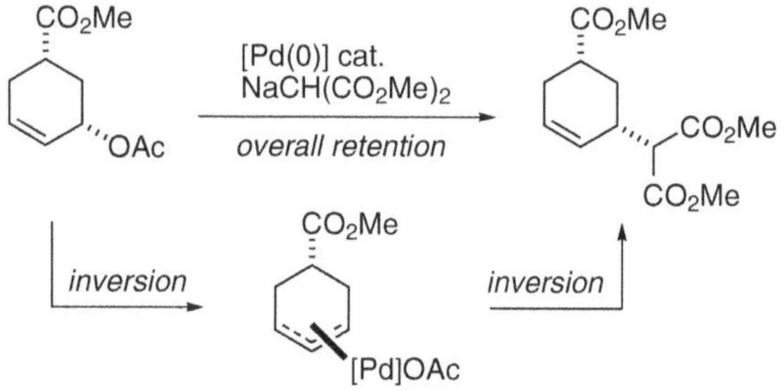

Scheme 26 Proof of the double inversion mechanism in the allylic substitution with "soft" nucleophiles

Range of "Soft" Nucleophiles

Stabilized Carbanions

The allylic substitution reaction involving "soft" carbon nucleophiles is called the Tsuji–Trost reaction [1–3, 104]. In most cases, the nucleophiles are stabilized carbanions, described by the general formula $RXYC^-$, with X and Y being electron-withdrawing groups such as ester, ketone, aldehyde, amide, imine, nitrile, nitro, sulfone, sulfoxide, and phosphonate (Fig. 6).

Usually, the carbanion is preformed by treating the pronucleophile RXYCH with a stoichiometric amount of a strong base, such as sodium hydride. Nevertheless, the Tsuji–Trost reaction is also efficiently performed under neutral conditions via in situ deprotonation of the pronucleophile. In this context, allylic carbonates [30], allylic carbamates [30], vinyl epoxides [34, 35] and aryl allyl ethers [105] are valuable electrophilic partners, as the displaced anion is sufficiently basic to generate the nucleophile (see Sect. 2.1.1).

Poli and coworkers showed that even the common acetate leaving group can perform such an endogenous deprotonation provided the pronucleophile is acidic enough [106]. Further work showed that addition of titanium tetraisopropoxide induces a decrease in the p*K*a value of the coordinated pronucleophile, thereby broadening the range of active methylenes compatible with these mild conditions [107].

Using allylic acetates as electrophilic partners, neutral conditions can also be reached by replacing a strong base with *N*,*O*-bis(trimethylsilyl)acetamide (BSA) in combination with a catalytic amount of acetate anion [108]. This latter initiates the reaction by picking up the trimethylsilyl group from BSA to afford the anion of *N*-trimethylsilylacetamide, which in turn deprotonates the pronucleophile (Scheme 27). The subsequent allylation reaction releases progressively one equivalent of acetate ion, which further reacts with BSA.

In all these procedures, the base is progressively generated in situ in catalytic amounts, thus enabling the application of the Tsuji–Trost reaction to labile compounds. Moreover, these mild conditions are commonly used in asymmetric allylic alkylation.

Owing to the highly variable nature of the nucleophilic and electrophilic partners and to the reaction conditions that tolerate many functional groups, the Tsuji–Trost reaction has been employed as the key step in the synthesis of numerous natural and pharmaceutical compounds (see [109]) [110]. Moreover, performing the Tsuji–Trost reaction intramolecularly allows the preparation of a variety of

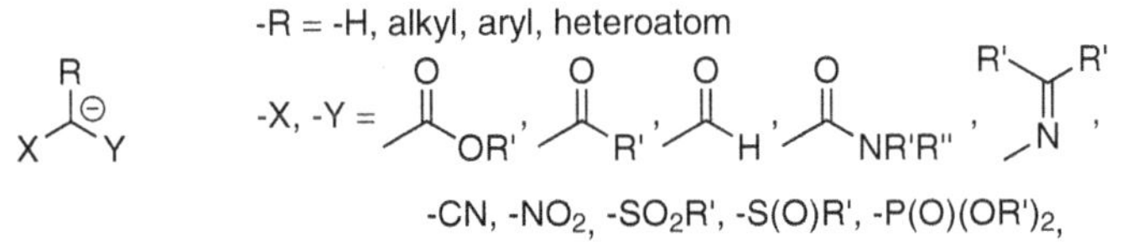

-R = -H, alkyl, aryl, heteroatom

-X, -Y = C(O)OR', C(O)R', C(O)H, C(O)NR'R'', C(=NR')R',

-CN, $-NO_2$, $-SO_2R'$, -S(O)R', $-P(O)(OR')_2$,

Fig. 6 Typical nucleophiles used in the Tsuji–Trost reaction

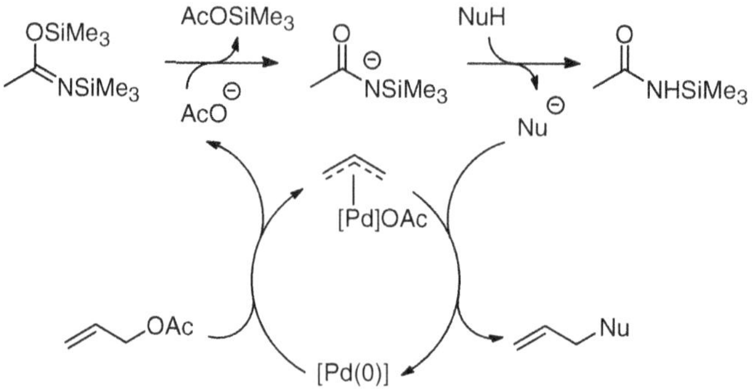

Scheme 27 Mechanism of the (BSA/cat AcO^-) promoted palladium catalyzed allylic substitution

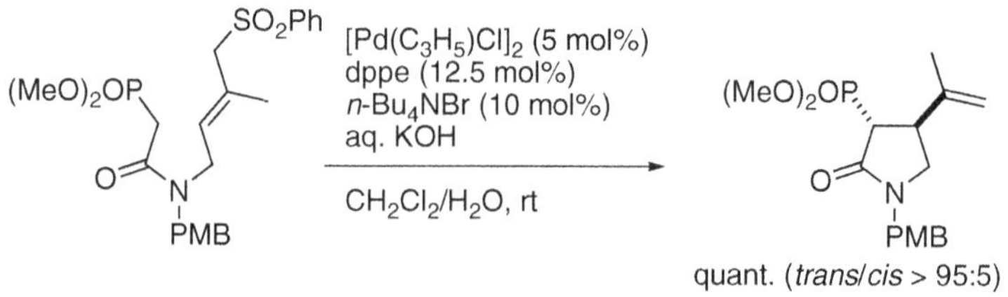

Scheme 28 Pd-catalyzed allylic alkylation reaction, used as key step in the synthesis of α-kainic acid

carbo- and heterocycles [111]. In their synthesis of α-kainic acid, Poli and coworkers used an intramolecular allylic alkylation to generate diastereoselectively the key *trans* pyrrolidinone (Scheme 28) [112]. Interestingly, the authors selected a phosphonoacetamide as nucleophilic partner, instead of the more common malonamide, so that the phosphonate moiety could be exploited in a Horner–Wadsworth–Emmons olefination in the following step.

Nonstabilized Carbanions

Although C–C bond formation via Tsuji–Trost reaction has been mostly carried out with carbanions stabilized by two electron-withdrawing groups, various other carbanions, such as the anions of phenylacetonitrile [30] and nitroalkanes [113], react smoothly to yield the corresponding allylated products.

More synthetically important, the palladium-catalyzed α-allylation of "simple" carbonyl derivatives can be achieved via the corresponding non-stabilized enolates [114, 115]. Indeed, preformed enolates of ketones, aldehydes, and esters bearing various countercations such as Li [116–118], B [119, 120], Sn [121, 122], Si [123], Mg [118] and Zn [120, 124] have been successfully employed as nucleophiles in allylic alkylation. However, their synthetic application has been hampered by several drawbacks such as a poor regioselectivity, polyalkylation, and the strong basicity of the reaction medium.

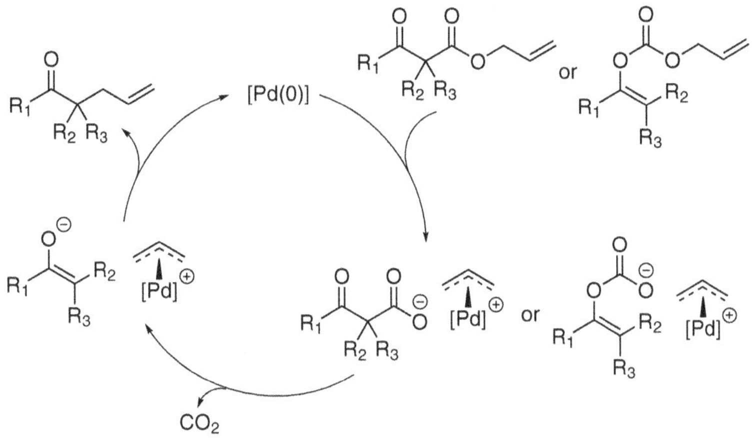

Scheme 29 Mechanism of the palladium catalyzed decarboxylative allylic alkylation

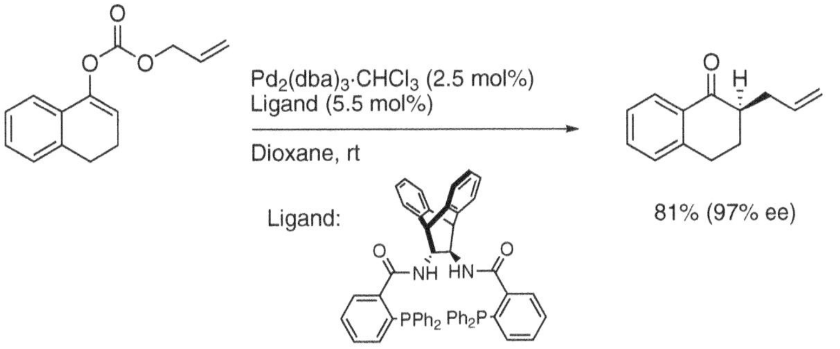

Scheme 30 Example of enantioselective palladium-catalyzed decarboxylative allylation of an allyl enol carbonate

Alternatively, it is possible to generate palladium enolates in situ under neutral conditions when using allyl β-ketoesters [125, 126] and allyl enol carbonates [127] as precursors (Scheme 29) [128, 129]. In this case, oxidative addition of the allyl moiety onto palladium(0) is followed by a decarboxylation, thereby affording a π-allylpalladium enolate. Finally, attack of the carbanion on the allyl ligand yields the α-allylated ketone and releases the active catalytic species [130]. This transformation represents a palladium-catalyzed variation of the thermal Carroll rearrangement.

Trost and coworkers successfully applied the palladium-catalyzed decarboxylative asymmetric allylic alkylation of allyl enol carbonates to the preparation of ketones possessing an α-tertiary center (Scheme 30) [131]. Since the reaction conditions are mild, high yields and ee's were obtained without undesired racemizations or dialkylations of the α-center.

Heteroatom Nucleophiles

Various heteroatom-based nucleophiles have been employed in palladium-catalyzed allylic substitution reactions [132]. Among them, the most studied are *N*-, *O*-, *S*-, and *P*-derivatives.

***N*-Nucleophiles** Ammonia, the simplest *N*-nucleophile, is a very attractive substrate from a cost- and atom-economy viewpoint, but it has long been considered as ineffective in palladium-catalyzed allylic amination. Nevertheless, Kobayashi reported very recently the selective preparation of primary amines employing aqueous ammonia as nitrogen source (Scheme 31) [133].

Various nitrogen compounds such as primary and secondary alkyl amines [2, 3, 36], aryl amines [36], hydroxylamines [134], azides [135], amides [136], cyanamides [137], sulfonamides [138], sulfamides [137], carbamates [139], imides [140], isocyanates [141], and nitrogen heterocycles stabilized by delocalization [142] have been employed as nucleophiles in allylic amination.

***O*-Nucleophiles** Palladium-catalyzed allylic etherification is efficiently performed in the presence of phenols, which are highly reactive substrates [36]. On the contrary, aliphatic alcohols are poor nucleophiles for such reactions [36]. Carboxylates [143] have also been used as *O*-nucleophiles, especially for the asymmetric synthesis of allyl alcohols via dynamic kinetic resolution (Scheme 32) [144]. However, trapping of π-allyl complexes by carboxylates is normally a reversible process, as the fact that allyl esters are ideal π-allyl precursors, reveals. Therefore, in order to avoid racemization of the product, the rate of the reaction with the allylic substrate must outcompete that of the product and the choice of the leaving group is thus crucial.

Additionally, triphenylsilanol [145] and sodium bicarbonate [146] have been used as water surrogates.

Scheme 31 Example of enantioselective palladium catalyzed allylic amination using ammonia

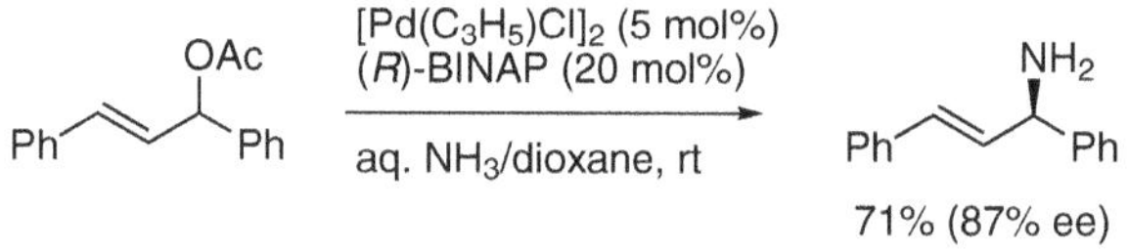

OAc
Ph Ph
$[Pd(C_3H_5)Cl]_2$ (5 mol%)
(*R*)-BINAP (20 mol%)
aq. NH_3/dioxane, rt
NH_2
Ph Ph
71% (87% ee)

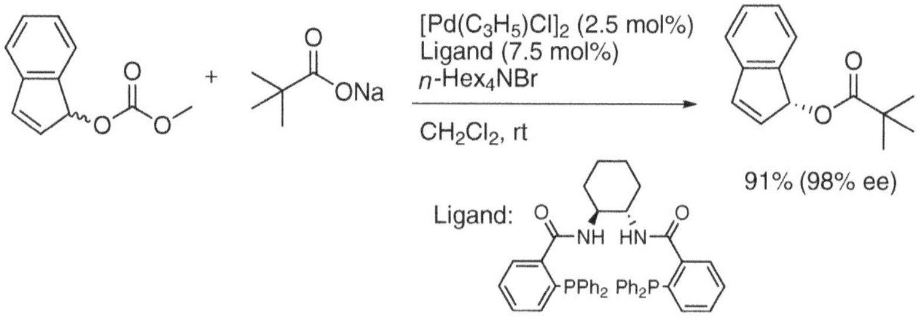

O O
O O
+
O
ONa
$[Pd(C_3H_5)Cl]_2$ (2.5 mol%)
Ligand (7.5 mol%)
n-Hex_4NBr
CH_2Cl_2, rt
O
O
91% (98% ee)
Ligand: O NH HN O
PPh_2 Ph_2P

Scheme 32 Example of enantioselective palladium catalyzed allylic carboxylate allylation

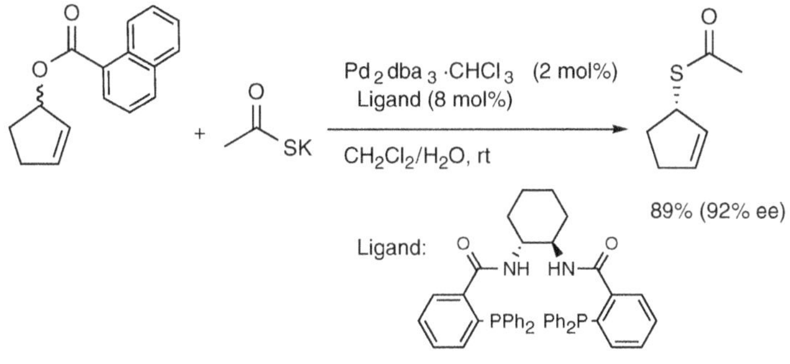

Scheme 33 Example of enantioselective palladium catalyzed thiocarboxylate allylation

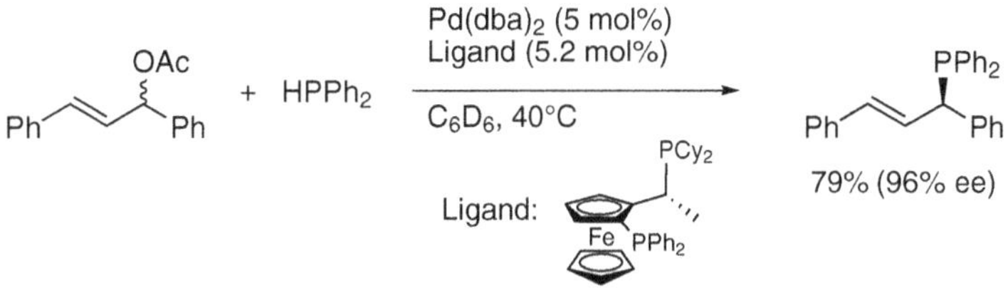

Scheme 34 Example of enantioselective palladium catalyzed allylic phosphination

***S*-Nucleophiles** In spite of the well-known thiophilicity of palladium, which may induce deactivation of the catalytic system, a variety of *S*-nucleophiles such as thiolates [147–150], sulfenates [151], sulfinates [152], and thiocarboxylates [153] have been successfully employed in allylic substitutions. Asymmetric versions of such reactions yield synthetically important sulfur derivatives in high ee (Scheme 33) [154].

Moreover, heterocyclic systems bearing ambident nucleophiles have been selectively allylated at the *S*-atom [155, 156] and various *O*-allylic substrates such as phosphoro- and phosphonothionates [157], sulfinates [158] and sulfites [159] underwent palladium-catalyzed *O*,*S*-rearrangements via transient π-allyl intermediates to afford the corresponding *S*-allylic products.

***P*-Nucleophiles** Few examples of C–P bond formations via palladium-catalyzed allylic substitutions have been reported. Lithium diphenylthiophosphide is smoothly allylated [160] and secondary phosphines have been recently used in the first enantioselective allylic phosphination reaction (Scheme 34) [161].

Additionally, tertiary phosphines have been successfully employed to generate allylphosphonium derivatives, which were trapped in situ by aldehydes in a Wittig-type reaction [162]. Trialkyl phosphites have been used with allyl acetates in a palladium-catalyzed Michaelis–Arbuzov reaction to afford the corresponding dialkyl allylphosphonates [163].

2.3.2 Trapping of π-Allyl Intermediates by "Hard" Nucleophiles

Mechanism and Stereochemistry

"Hard" nucleophiles, such as organometallic compounds of main group metals, attack directly the electrophilic palladium center, thereby generating via transmetallation an allyl(organyl)palladium(II) complex (see Scheme 2). Subsequent reductive elimination affords a palladium(0)-olefin complex, which readily releases the allylated product and the catalytically active palladium(0) species.

Owing to this inner sphere mechanism, trapping of the π-allyl complex by "hard" nucleophiles occurs with a net retention of configuration. Therefore, starting from an appropriate substrate that allows to follow the stereochemistry such as a *cis* cyclohexenyl acetate derivative, an overall inversion – via inversion followed by retention – is observed, as shown by Negishi in the reaction with an alkenylalane (Scheme 35) [24].

Range of "Hard" Nucleophiles

The palladium-catalyzed allylation of "hard" nucleophiles has been mostly carried out with aryl- and alkenylmetals involving Al [164], B [165, 166], Mg [167], Si [168], Sn [43], Zn [164] and Zr [164, 169]. Additionally, trimethylsilyl cyanide has been successfully used for the cyanation of various allyl carbonates (Scheme 36) [170].

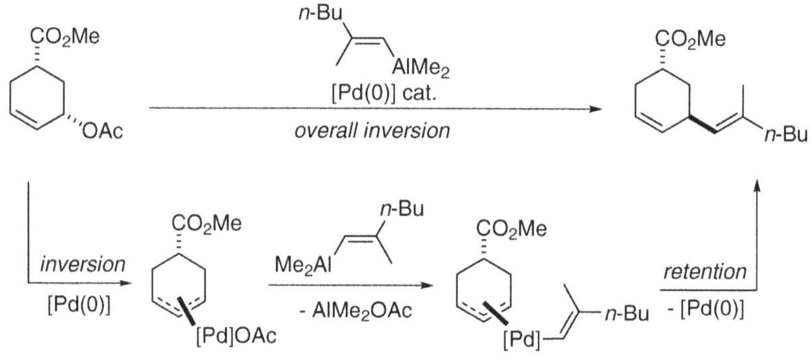

Scheme 35 Proof of single inversion mechanism in the allylic substitution with "hard" nucleophiles

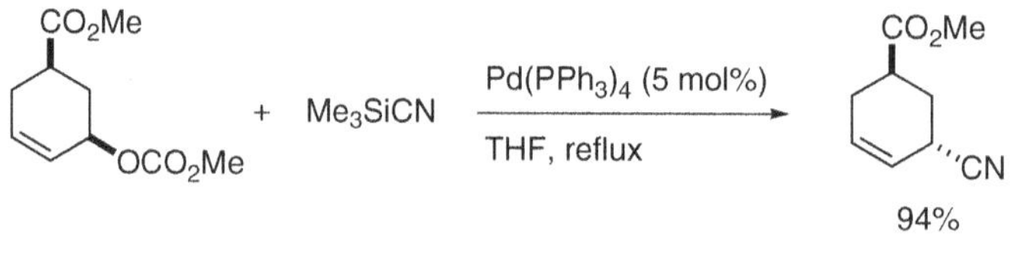

Scheme 36 Example of palladium catalyzed allylic cyanation

Palladium-catalyzed allylic reductions are useful transformations yielding alkenes with overall inversion of configuration. Hydride sources such as $LiAlH_4$ [171], $LiBHEt_3$ [39], Bu_3SnH [172] react as "hard" nucleophiles via transmetallation followed by reductive elimination [173]. In addition, ammonium formates [174, 175] afford a transient palladium-formate complex and the delivery of the hydride occurs via a concerted mechanism [173].

3 Regioselectivity and Memory Effects

When the transiently generated η^3-allyl complex is unsymmetrically substituted the issue of site-selectivity comes into play. In this regard, we can conceptually classify the strategies allowing nucleophile site selection according to intramolecular- and ligand-assistance.

3.1 Intramolecularly Directed Regioselectivity

Efficient approaches to direct the regioselectivity in the palladium-catalyzed allylic substitution reaction have been obtained using substrates capable of coordinating to the metal. Thus, for example Krafft's [176–178], and Yoshida's [179] groups reported that a homoallylically located tertiary amine or thioether and 2-pyridyldimethylsilyl group can efficiently direct the nucleophile to the allylic terminus proximal to the heteroatom. This interesting site selectivity can be interpreted in terms of *trans* influence of the chelated η^3-allyl intermediate, wherein the nucleophile preferentially attacks the longer and, consequently, more reactive Pd–C bond (Scheme 37).

Cook and coworkers studied the allylic substitution of chiral 5-vinyloxazolidinones using phthalimide as the nucleophile. The high and constant regio- and stereoselectivity favoring the 1,2-diamine derivatives of *syn*-configuration can be accounted for on the basis of a hydrogen bond involving the transiently protonated amide moiety and the oxyanion of the deprotonated imide nucleophile (Scheme 38) [180, 181].

3.2 Ligand-Directed Regioselectivity

In the absence of specific directing issues incorporated in the substrate (see above) the regioselectivity of the palladium-catalyzed allylic alkylation depends on the interplay of steric and stereoelectronic factors involving the nucleophile, the ancillary ligand, as well as the allyl fragment [182].

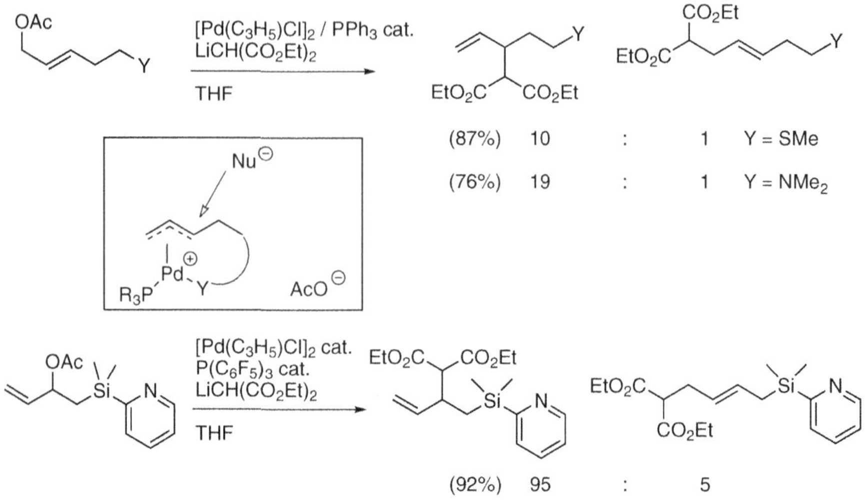

Scheme 37 Intramolecularly-directed palladium catalyzed allylic substitutions

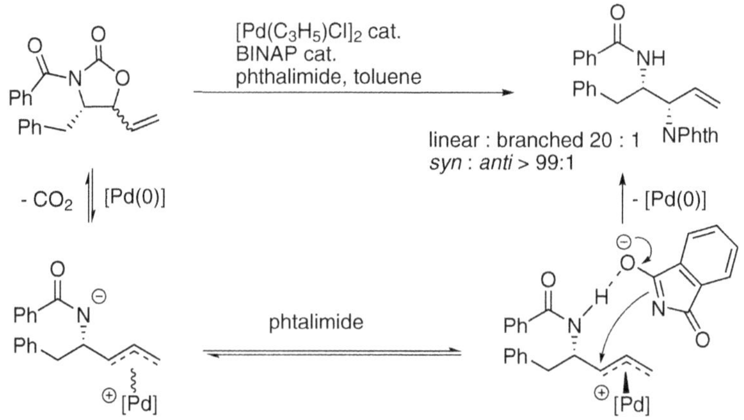

Scheme 38 H-bond directed palladium catalyzed allylic substitution

Two linear (*E* or *Z*) and one branched allylic precursors can oxidatively add to a Pd(0) complex to afford a *syn* or an *anti* configured monosubstituted η^3-allylpalladium intermediate. Following attack of the nucleophile can take place either at the terminal or the internal position of the allyl fragment to give the linear (*E* or *Z*) or the branched product, respectively. This latter reactivity is of particular importance as it lends itself to the development of asymmetric versions. In particular, the linear *E* substrate is expected to generate a *syn* allyl complex, which in turn may generate the linear *E* and/or the branched product. The linear *Z* substrate will initially form an *anti* configured allyl complex, which may react directly with the nucleophile to give the *Z* linear and/or the branched product. If the

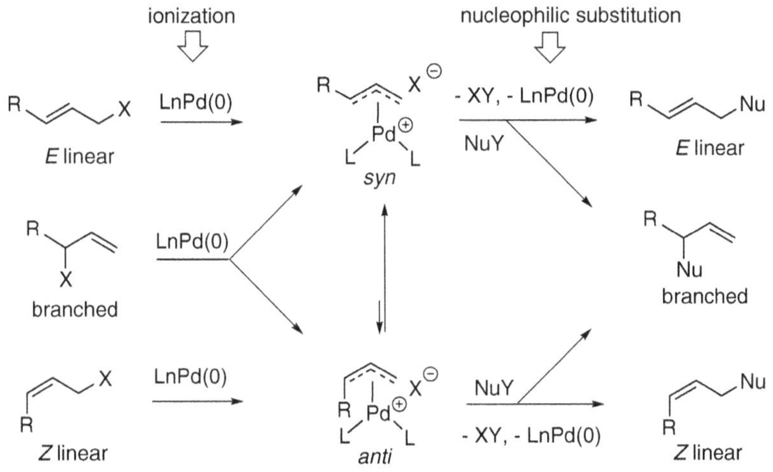

Scheme 39 Possible scenarios in the generation and reactivity of a cationic (chloride-free, see later) monosubstituted η^3-allylpalladium intermediate

syn and the *anti* η^3-allyl intermediates are intercepted without prior equilibration, full stereoretention will be attained. If, on the other hand, the rate of nucleophilic attack is lower than that of *anti-syn* equilibration, the final ratio of products will depend on the thermodynamically averaged intrinsic regioselectivity of each η^3-allyl isomer. Usually, the *syn* π-allyl isomer is more stable than the *anti* one, and the trapping by the malonate anion is slower than equilibration (Scheme 39). As a result, whereas starting from the *E* linear isomer a high degree of stereoretention is normally observed, the *Z* isomer often undergoes (at least partial) *anti-syn* equilibration prior to react. As a corollary, according to this mechanism the branched allylic substrate is expected to give a mixture of final products which will depend on the relative amount and reactivity of the *syn* and *anti* π-allyl intermediates involved.

The scenario is thus rather intricate. Indeed, the regioselectivity may depend on the interplay of several factors such as: (a) the structure and the nature (donor *vs* acceptor) of the allylic fragment and of the ligand, (b) the symmetry and bite angle of the ligand if it is bicoordinating, (c) its stoichiometry of complexation if it is monocoordinating (number of ligands coordinated), and, (d) the position (ionization *vs* nucleophilic substitution) of the rate-determining step.

Under "classical" conditions the structure (and thus the behavior) of a transiently generated allyl complex is independent of the relative position occupied by the unsaturation and the leaving group in the substrate. So, for example, the same *syn* π-allyl complex may derive from the *E* linear as well as the branched product. This latter tends to react with nucleophiles predominantly at the unsubstituted allylic terminus with a linear-to-branched ratio usually directly proportional to the bulkiness of the nucleophile.

A stoichiometric study of allylic alkylation using the sodium anion of methyl diethyl malonate and bidentate ligands of different nature incorporated in a xanthene backbone, which allows large bite angles, revealed that the strong π-acceptor *P,P*-based ligand Xantphos (bite angle 111°) induced mainly the formation of branched products, whereas the use of the weak π-acceptor *N,N*-based ligand yielded mainly the linear product [183].

A way to obtain regioselectivity is by enhancing the nonsymmetry of the allyl moiety [184, 185] or by using unsymmetrical bidentate ligands such as *P,N* ligands.

Pioneering work by Vitagliano and coworkers showed that when a *P,N* ligand was used, the allyl terminus *trans* to the π-accepting phosphorus atom was more electrophilic than the position *trans* to the nitrogen atom [186, 187]. More recent computational studies confirmed the "*trans*-to-P" effect of *P,N* ligands and predict that shift toward earlier transition states should maximize this outcome [188, 189].

The "*trans*-to-P" effect turned out to be of general validity and was later exploited in asymmetric catalysis involving symmetrical π-allyl fragments using enantiopure chiral *P,N* ligands [190–193].

Coming back to the alkylation of unsymmetrical π-allyl fragments, fine tuning of an appropriate *P,N* ligand was shown to direct the nucleophilic attack toward the branched product (Scheme 40).

This was obtained by using an appropriate *P,N* ligand bearing a π-acidic phosphite [194] so as to shift the mechanism of the nucleophilic substitution from a S_N2 type to a more cationic S_N1 type. As an additional bonus, the chiral ligand may match chirality thereby allowing a good enantioselection (Scheme 41) [195–199].

Another way of increasing the branched-to-linear ratio using bidentate *P,N* ligands is to increase their bite angle [200].

Whereas selectivity for the branched isomer, possibly in enantioenriched form, is normally the targeted result, attainment of optimal selectivity for the linear isomer is also challenging. This was obtained using PPh_3 in the presence of a catalytic amount of lithium iodide [201]. This result implies a mechanism featuring substitution on an iodide-ligated allyl complex bearing a *trans* arrangement between the phosphine and the internal allylic terminus, and this independently of starting substrate. Such "*cis*-to-P" attacks are not favored with bulkier monophosphines (see later) (Scheme 42).

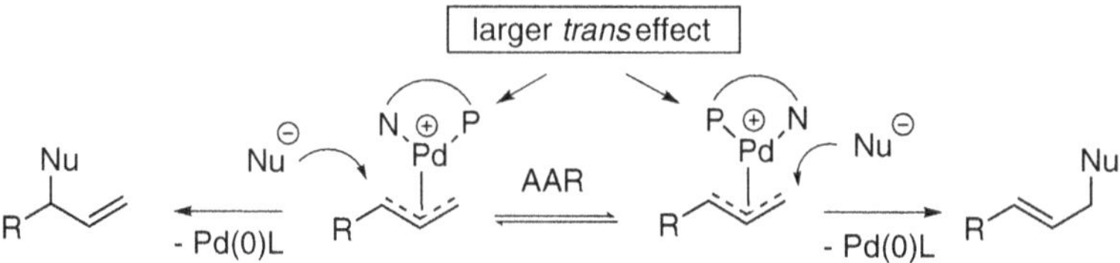

Scheme 40 Role of a *P,N* ligand in the regioselectivity of a palladium catalyzed allylic substitution

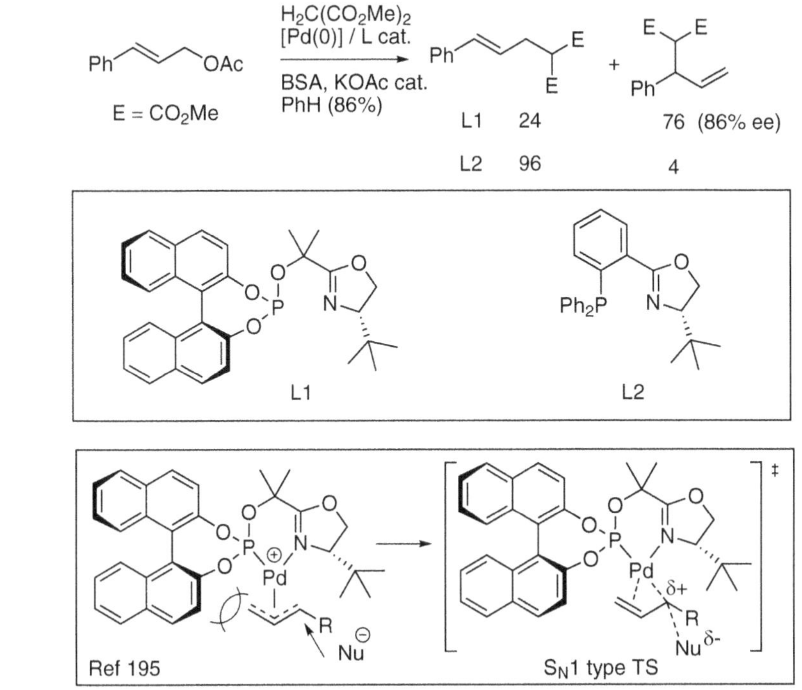

Scheme 41 Branched- and enantio-selective palladium catalyzed allylic substitution via a phosphite-type *P,N* ligand

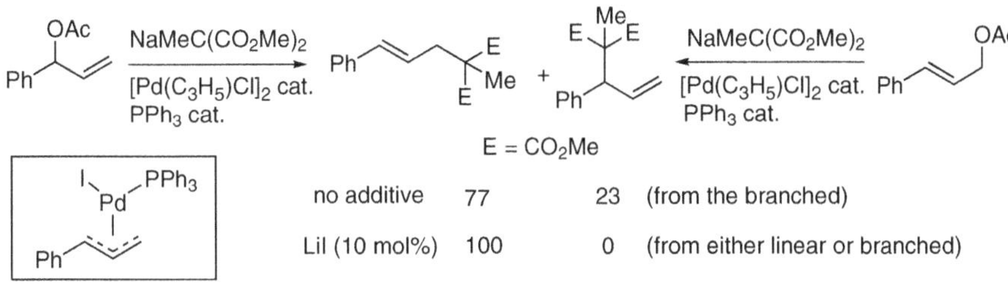

Scheme 42 Iodide effect in the regioselectivity of palladium catalyzed allylic substitution

In monosubstituted allylic substrates the resulting *syn* π-allylpalladium intermediates display a strong preference for terminal attack, whereas the corresponding *anti* isomers, proviso that do not equilibrate to the *syn* one (see above), show an intrinsic preference for internal nucleophilic substitution [202]. This reactivity is highly interesting due to the chirality of the branched products and can be obtained either using special ligands that (thermodynamically) favor the *anti* isomer [203, 204] or generating the *anti* isomer in the presence of very reactive nucleophilic partners capable of intercepting it before it equilibrates to the (usually) more stable

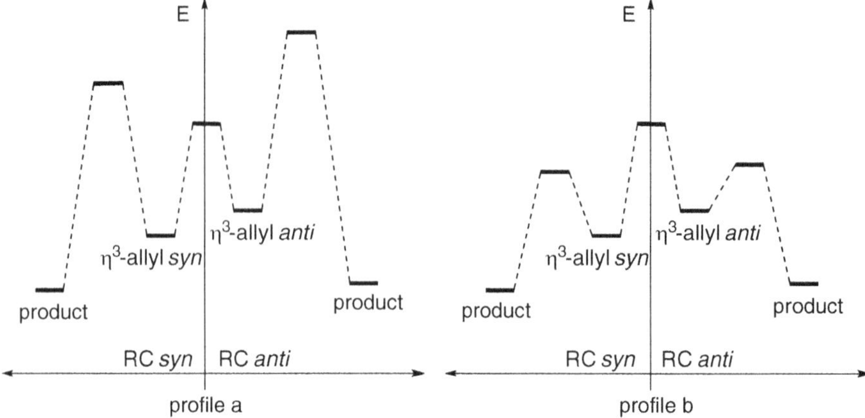

Fig. 7 Qualitative energy profiles of allylic alkylations wherein: (**a**) *anti*-to-*syn* isomerization is fast compared to the nucleophilic attack, and (**b**) the nucleophile can trap the kinetically generated *syn* and *anti* η^3-allyl complexes before they can equilibrate

syn isomer [205, 206].[3] When using malonate anions, *anti*-to-*syn* isomerization is normally fast compared to the nucleophilic attack (Fig. 7, profile a). As a consequence, product distribution results from the equilibrium mixture of the η^3-allyl complexes, in which the *syn* complex is usually enriched. On the other hand, more reactive nucleophiles can trap the η^3-allyl complexes before they can equilibrate (Fig. 7, profile b).

This is the case of the highly reactive zinc amino acid ester enolate **A** [207] which can intercept an *anti* configured η^3-allyl complex before it isomerizes to the more stable *syn* isomer. Noteworthy, this is possible only if the substitution is fast enough (Scheme 43, left). Indeed, if the reaction becomes slower, due for example to steric bulkiness in the η^3-allyl complex, nucleophilic attack is preceded by *anti-to-syn* isomerization (Scheme 43, right).

3.3 Memory Effects

The "canonical" mechanism in Pd-catalyzed allylic alkylations calls for the involvement of a unique bicoordinated cationic η^3-allyl-Pd intermediate, independently of the structure of the allylic precursors, which may differ in the reciprocal location of leaving group and unsaturation (Scheme 44, left). Electronic reasons intrinsically favor trapping by the nucleophile at the more substituted allylic

[3]Metal-catalyzed allylic alkylation to afford branched products can be obtained using transition metals other than palladium. However, this topic is beyond the scope of the present chapter.

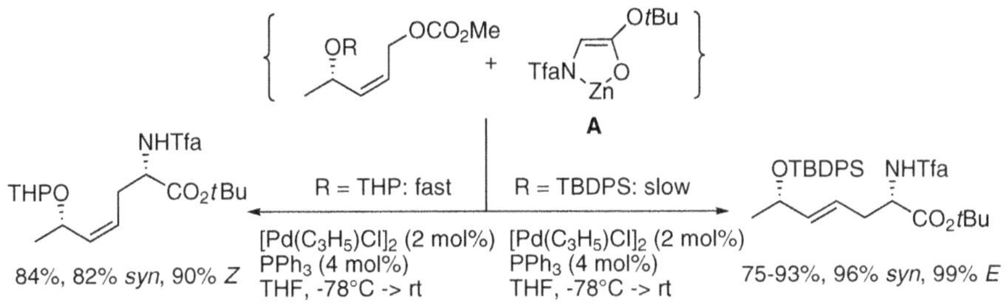

Scheme 43 Generation and interception of an *anti* η^3-allyl palladium complex before (*left*) or after (*right*) its equilibration to the more stable *syn* complex

terminus, whereas increase of steric reasons proportionally increases the preference for the linear product.

However, Fiaud and Malleron [208] recognized in a pioneering work that the above mechanism might not always hold [209]. Such diversion from the classical behavior, named thereafter "memory effect" [210], was later confirmed by several authors and refers to cases wherein the nucleophile reacts at the allylic carbon atom originally occupied by the leaving group [79, 211, 212]. Such effects were observed in the presence of particularly bulky mono-phosphines.

More than one theory has been put forward to explain the reactivity difference between the isomeric allylic substrates, including the involvement of a tight ion pair with the leaving group [210], reaction through η^1-allyl intermediates [213], and unequal *trans* effects arising from unsymmetrical ligation [193, 214].

In close relation with the last point, it was found that catalytic amounts of chloride anions were capable of boosting the oxidative addition of the slow reactive enantiomer of cyclopentenyl pivaloate in the presence of the TSL [215]. Subsequent experimental and theoretical studies allowed to conclude that chloride anions are able to trigger a competitive pathway without involving the classical cationic (see above) η^3-allyl intermediates. The initially formed and highly reactive anionic Pd(0) phosphine complex [216] oxidatively adds to the allylic substrate to generate a neutral chloride-coordinated η^3-allyl complex, which is then attacked by the nucleophile to afford the product (Scheme 44, right) [217].

3.3.1 Regiochemical Memory Effects

If we now come back to consider an allylic alkylation reaction starting from the three possible, linear (*E* or *Z*) or branched, isomers (Scheme 39), the presence of chloride anions unveils a new more complex scenario, wherein the allyl intermediates are duplicated [218]. Thus, taking into account the effect of phosphine, which favors departure of the leaving group *trans* to itself, four allyl complexes can be preferentially generated. Whereas the *E* and the *Z* linear substrates give rise specifically to the *syn*-from-*E* and *anti*-from-*Z* isomers,

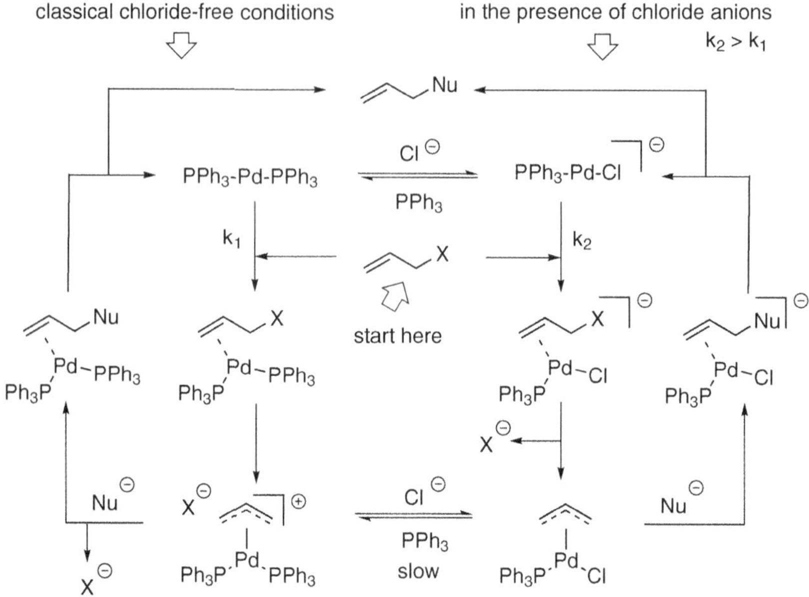

Scheme 44 Different η^3-allylic intermediates involved in the Pd-catalyzed allylic substitution reaction, in the absence (*left*) or presence (*right*) of chloride anions

respectively, the branched isomer may give rise to a mixture of *syn*-from-*branched* and *anti*-from-*branched* complex, whose ratio depends on the conformational population during its formation from the allyl complex. Computations (R = R' = Me) predict a strong bias for *trans*-to-P trapping for the first three complexes, whereas in the *anti*-from-*Z* isomer the intrinsic tendency for internal attack and the *trans* effect mismatch each other. Thus, provided that the transiently generated η^3-allyl complexes are intercepted as soon as they are generated, the presence of chloride anions can, when starting from the linear or branched isomer, bring about a substantial memory effect (Scheme 45). Of course, a slow apparent allyl rotation relative to nucleophilic addition is a *conditio sine qua non* for the existence of this type of memory. Indeed, a fast apparent allyl rotation would wipe out this memory effect channeling the reaction toward the most reactive allyl complex (Curtin–Hammett conditions).

Such a picture has been obtained analyzing the product distribution from some allylic triads and by comparing the results obtained with and without added chloride anions. In the absence of chloride anion, the product distribution from the branched substrate is close to a linear interpolation between the results deriving from the two linear isomers. On the other hand, in the presence of chloride anions, the amount of the branched product deriving from the branched substrate is almost twice as much as from either of the other two linear substrates. This clearly shows that chloride anions boost the selectivity of a branched-to-branched transformation. This result allows to unambiguously distinguish this memory effect from an inherent internal

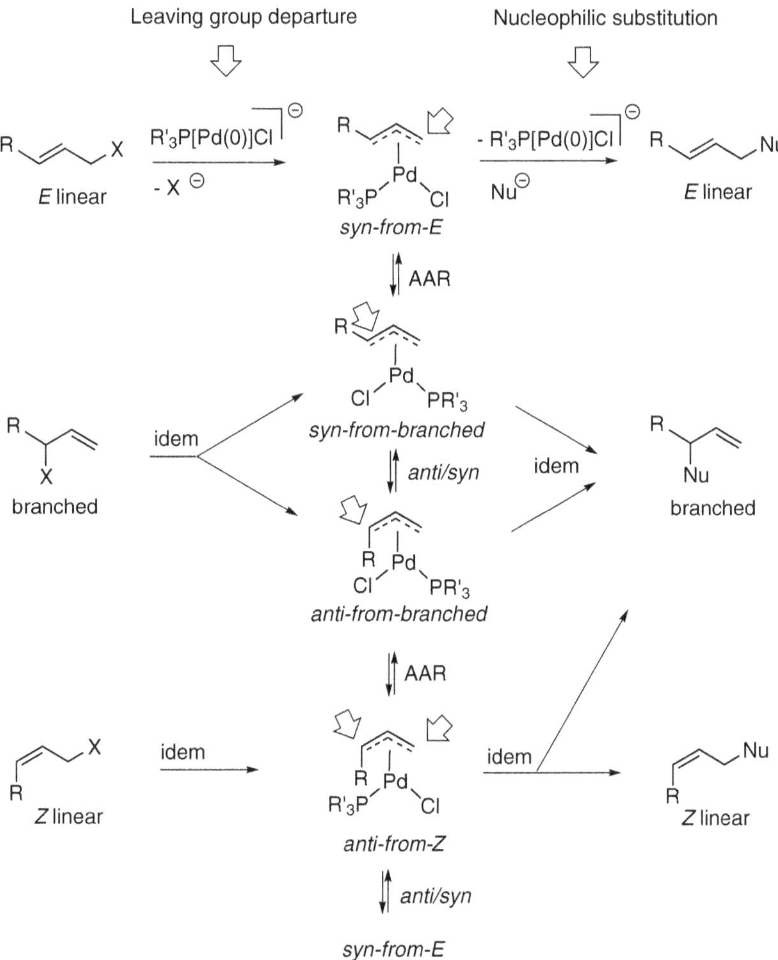

Scheme 45 Possible scenarios in the generation and reactivity of a neutral (chloride containing) monosubstituted η^3-allylpalladium intermediate

attack preference (Scheme 46) and suggests the origin of other branched-to-branched transformations performed in the presence of chloride anions published in the literature. Of course, the action of alternative mechanisms of memory effect such as tight ion pairing with the leaving group remains possible.

A similar situation is observed when using bulky ligands, which, for steric reasons, allow only mono-coordination to palladium and thus generation of unsymmetrical η^3-allyl complexes. In this case, a "memory-type" linear-to-linear and branched-to-branched behavior can be observed. However, in the presence of chloride anions the isomerization rate between the isomeric η^3-allyl complexes is enhanced and reaches Curtin–Hammett conditions. Under these conditions the

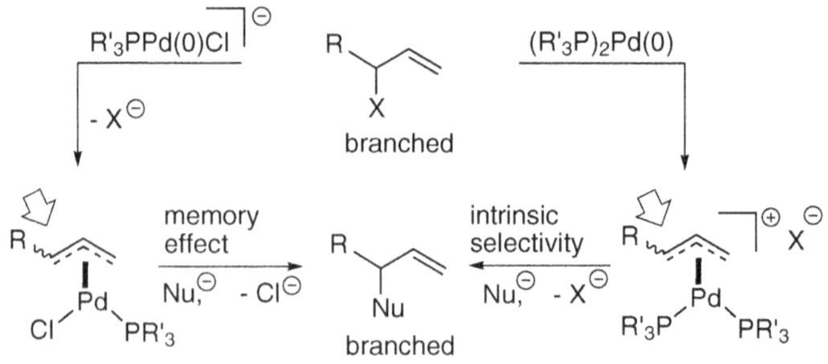

Scheme 46 Intrinsic versus memory-effect-assisted regioselectivity in the Pd-catalyzed allylic substitution reaction

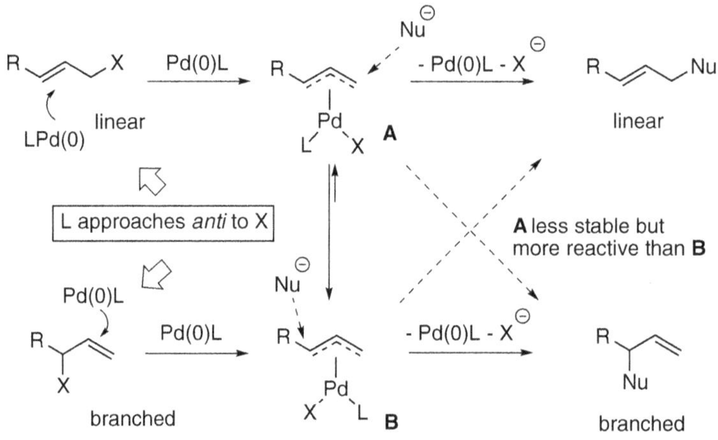

Scheme 47 Mechanistic rationale for the regiochemical memory effect observed in the presence of particular bulky ligands

more reactive, though less stable, *cis* isomer **A** will preferentially react to give the linear isomer as the major product (Scheme 47).

A few representative examples are in accord with the above considerations (Table 3) [80, 219, 220].

This behavior can be understood assuming the formation of a neutral η^3-allyl complex wherein the lone coordinated *P* ligand occupies a position *trans* to the departed group [221]. Such a stereochemistry is in accord with the notion that oxidative addition of allylic substrates to a Pd(0)L complex is the microscopic reverse of the reductive elimination step, which needs in turn a *cis* disposition of the reacting groups.

Privileged entrance of the nucleophile from the same side as originally occupied by the leaving group [221], and prior to a possible apparent allyl rotation, would thus account for the difference between a canonical (Scheme 48, upper part) and a memory retaining mechanism (Scheme 48, lower part).

Table 3 Examples of regiochemical memory effect observed with some monodentate bulky ligands

Ph–CH=CH–CH₂OAc (linear) or Ph–CH(OAc)–CH=CH₂ (branched) $\xrightarrow[\text{[Pd(0)] cat, THF, rt}]{\text{NaCMe(CO}_2\text{Me)}_2}$ linear product + branched product

	Ligand	Substrate	Linear		Branched	Ref
canonical (ionic)	L1	linear	91	:	09	219
	L1	branched	92	:	08	219
memory effect	L2	linear	79	:	21	220
	L2	branched	23	:	77	220
	L3	linear	97	:	03	80
	L3	branched	33	:	67	80
chloride effect	L3 + Cl^-	linear	99	:	01	80
	L3 + Cl^-	branched	84	:	16	80

L1: PPh_3 L2: (*R*)-MeO-MOP L3:

With this type of allyl complexes memory effect can be also accompanied by an efficient enantiodiscrimination and concomitant resolution of the starting racemic substrate, as elegantly shown by Hayashi and coworkers (Scheme 49) [220].

This result can be rationalized assuming that exchange of the complexed allyl face in monosubstituted allyl complexes (via $\eta^3 - \eta^1_{\text{(C–C rot)}} - \eta^3$) is fast compared to apparent allyl rotation (via $\eta^3 - \eta^1_{\text{(C–Pd rot)}} - \eta^3$) and that isomer **A** reacts regio- and stereoselectively, and faster than **B** (Scheme 50).

Bulky trialkylphosphines such as Cy_3P, i-Pr_3P, and t-Bu_3P [222, 223] also allow obtaining high proportions of branched product. Furthermore, a good level of stereoretention has been demonstrated when starting from an enantioenriched

Canonical mechanism
Nu⊖ - Pd(0)L2 -X⊖
Nu⊖ - Pd(0)L2 -X⊖
Pd(0)L2
Pd(0)L2
start here
common intermediate
start here
linear
branched
Memory effect
LPd(0)
L enters trans to leaving group
Pd(0)L
Nu attacks trans to L
fast
- Pd(0)LX
AAR
slow or disfavored
Nu attacks trans to L
fast
- Pd(0)LX

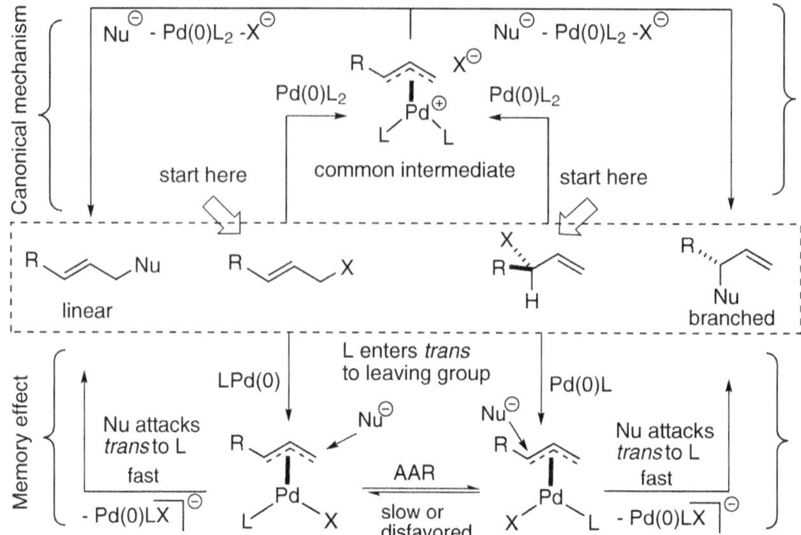

Scheme 48 Regioselectivity in the trapping of monosubstituted allyl complexes. *Upper part*: canonical mechanism; *lower part*: ideal memory effect mechanism

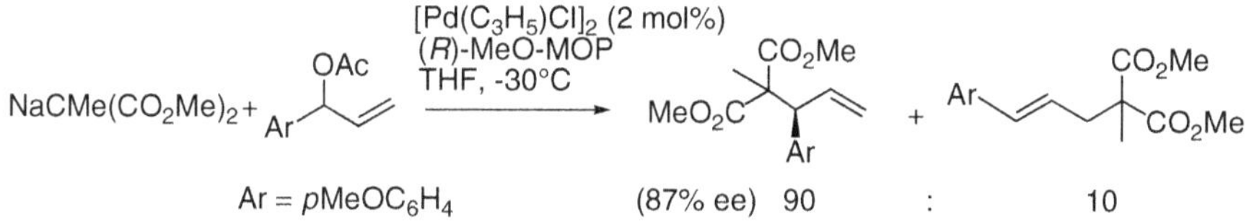

$[Pd(C_3H_5)Cl]_2$ (2 mol%)
(R)-MeO-MOP
THF, -30°C
$NaCMe(CO_2Me)_2$ +
Ar = pMeOC$_6$H$_4$
(87% ee) 90 : 10

Scheme 49 Example of memory effect with concomitant substrate resolution and enantiodiscrimination

start here
Pd(0)L*
Pd(0)L*
$\eta^3-\eta^1(CC)-\eta^3$
fast
- Pd(0)LX
AAR
slow or disfavored
B
A
C

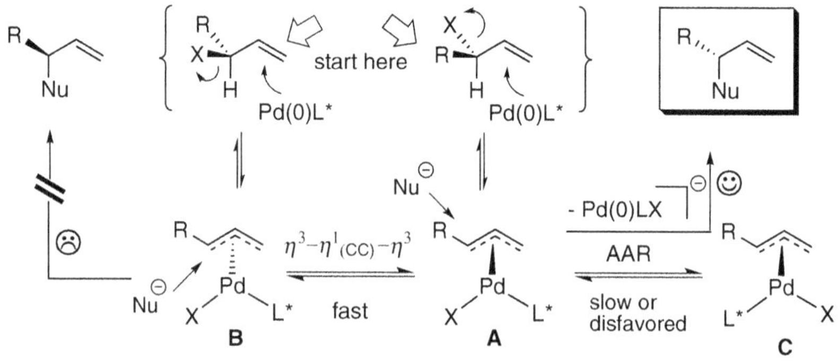

Scheme 50 Mechanistic rationale for the memory effect involving concomitant substrate resolution and enantiodiscrimination

Table 4 Regiochemical memory effect observed with Cy_3P

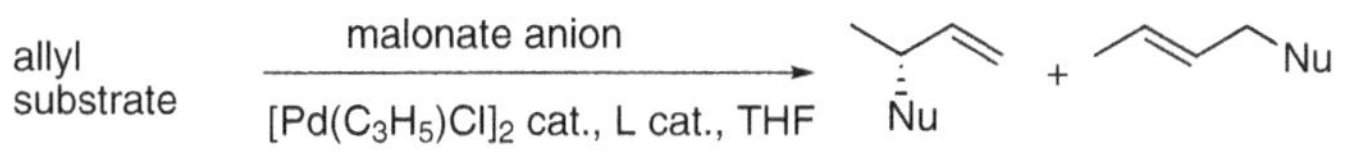

allyl substrate —(malonate anion; $[Pd(C_3H_5)Cl]_2$ cat., L cat., THF)→ branched (Nu) + linear (Nu)

entry	allyl substrate[a]	malonate anion	L	branched	:	linear	Ref.
1	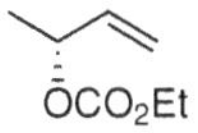 branched, OCO_2Et	$NaHC(CO_2Me)_2$	PCy_3	87 (82% ee)	:	13	222
2	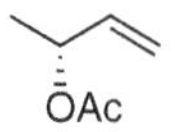 branched, OAc	$NaHC(CO_2Me)_2$	PCy_3	94 (64% ee)	:	6	223
3	branched, OAc	$NaHC(CO_2Me)_2$	PPh_3	50 (15% ee)	:	50	223
4	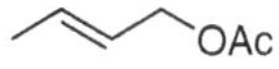 linear, OAc	$NaHC(CO_2Me)_2$	PCy_3	57	:	43	223
5	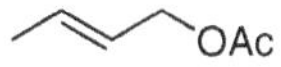 linear, OAc	$NaEtC(CO_2Et)_2$	PCy_3	8	:	92	223

a) allyl carbonate: 91% ee

substrate (Table 4, entry 1) [224]. The behavior of Cy_3P is quite peculiar, as it allows obtaining adequate amounts of branched product even from the linear substrate (Table 4).

The good regio- and stereoselectivity of the branched substrate can be qualitatively accounted for assuming the formation of a substantial amount of the *anti* allyl complex which, thanks to the *trans*-to-P effect and its *anti* substituent, is doubly biased for attack at the internal carbon. On the other hand, the *E* linear substrate is forced to generate only the *syn* η^3-allyl complex, wherein the strong *trans*-to-P effect and the steric factor mismatch one another (Scheme 51). In such a situation, with a rather small nucleophile such as malonate anion the two factors almost compensate each other giving a nearly random mixture of products (Table 4, entry 4). On the other hand, in the presence of a bulkier nucleophile the steric issue becomes the dominant factor and a good linear preference is restored (Table 4, entry 5).

In contrast to what observed with bulky monophosphines, experiments of malonate anion allylation in the presence of the bidentate *P,N* ligand $Ph_2P(CH_2)_2NMe_2$ show no significant memory effects. Indeed, despite the major difference in the *trans* influence between *P* and *N* (*P* > *N*, see above) apparent allyl rotation is fast with respect to nucleophilic substitution [95].

When Enantioselectivity and Regioselectivity Play Together

When unsymmetrical 1-monosubstituted allyl complexes are involved, the rules governing regio- and enantioselectivity interpenetrate, and analysis becomes rather complex. In this case, a chiral (enantiopure) ligand can direct the regioselectivity in a way that an achiral ligand cannot, as it can favor attack of the nucleophile on a specific allyl terminus of one of the two diastereomeric allyl complexes.

This case study is well-exemplified in the palladium-catalyzed substitution of (*E*)-hex-2-enyl methyl carbonate or hex-1-en-3-yl methyl carbonate by 4-methoxyphenol, using the TSL, as studied by Trost and Toste [225]. Herein, the chiral ligand shows a moderate kinetic preference for ionizing and adding the nucleophile from the same side, in line with the fact that the two reactions are the microscopic reverse of the other. However, the use of appropriate reaction conditions (polar solvents, halides as additives) favors the generation of one enantiomer of the branched isomer independently of the structure of the starting substrate (Scheme 52).

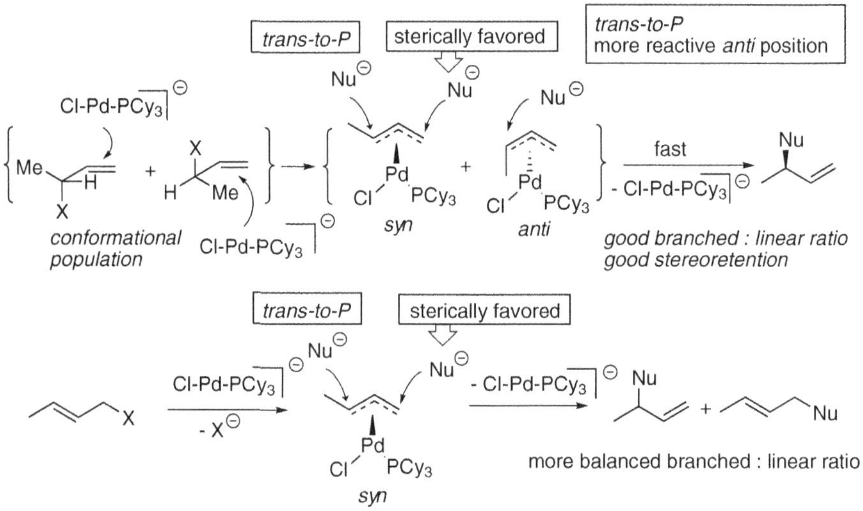

trans-to-P sterically favored *trans-to-P* more reactive *anti* position

Cl-Pd-PCy3 Me H X H Me conformational population Cl-Pd-PCy3 Nu Nu Pd Cl PCy3 *syn* Nu Pd Cl PCy3 *anti* fast - Cl-Pd-PCy3 Nu

good branched : linear ratio good stereoretention

trans-to-P sterically favored

X Cl-Pd-PCy3 - X Nu Nu Pd Cl PCy3 *syn* - Cl-Pd-PCy3 Nu + Nu

more balanced branched : linear ratio

Scheme 51 Mechanistic rationale for the regiochemical memory effect observed in the presence of Cy_3P

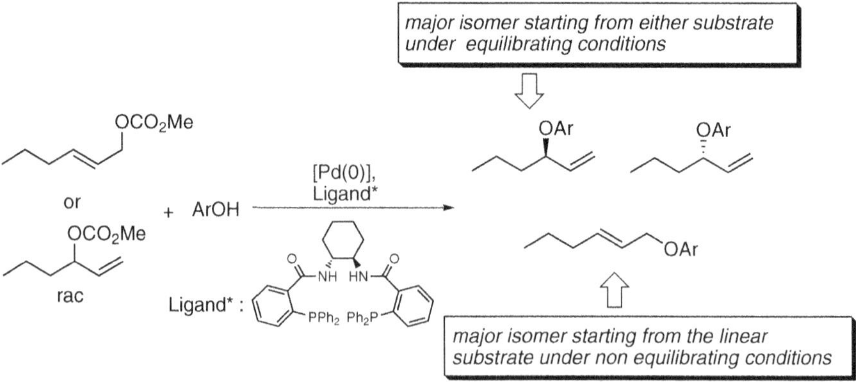

Scheme 52 Influence of a chiral ligand such as TSL on the regioselectivity of the Pd-catalyzed allylic substitution

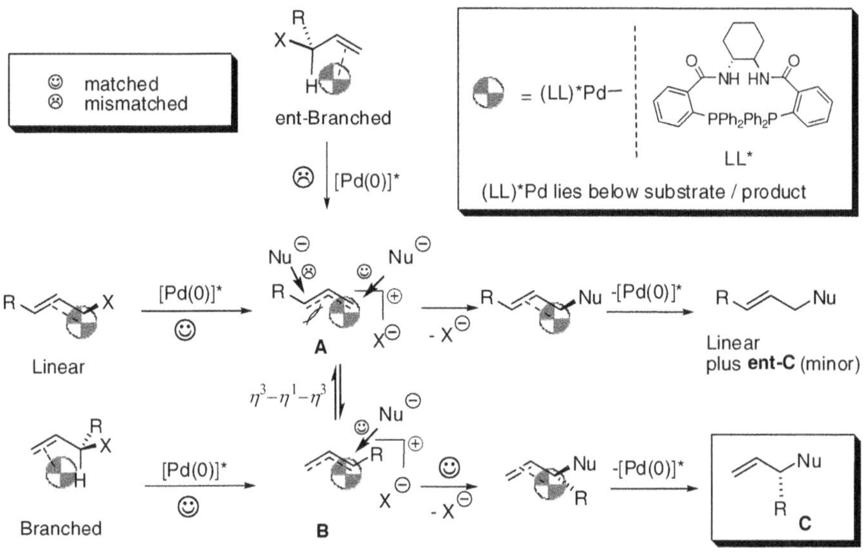

Scheme 53 Rationale accounting for the influence of TSL on the regioselectivity of the Pd-catalyzed allylic substitution

These results can be rationalized assuming that under the former conditions a moderate memory effect is at work. Under the latter conditions, the kinetically generated allyl complex **A** deriving from either the linear isomer or the mismatched branched enantiomer ent-branched, can equilibrate to the more stable and more reactive allyl complex **B**, which will eventually generate the preferred enantiomeric final product **C**. On the other hand, in the matched branched enantiomeric substrate, the leaving group and the entering nucleophile can easily slide out or in, respectively, through the open quadrant of the C_2 symmetric ligand via intermediate **B** (Scheme 53).

3.3.2 Stereochemical Memory Effects

When η^3-allylpalladium complexes bearing 1,3-symmetrically substituted allyl ligands are generated under canonical (no memory) conditions, the same allyl complex will be generated from either enantiomer of the starting allylic substrate. As a result, enantiodiscrimination using a chiral ligand will be exclusively due to regioselection of the nucleophilic attack on the allyl complex (Scheme 54, see Sect. 4.2.1).

However, under particular circumstances, it may happen that the two starting enantiomers give rise to two different η^3-allylpalladium complexes that do not collapse to a unique η^3-allylpalladium complex and whose equilibration between them is forbidden or slow. In this case, preferential attack of the nucleophile on each complex from the side left open by the leaving group will inexorably work against the intrinsic induction of the ligand, affording less enantioenriched products (Scheme 55) [210]. Such a "stereochemical memory effect" is clearly deleterious when asymmetric catalysis is desired, as a fast racemization of the starting substrates is necessary for the success of the process [226]. Like in the case of the above mentioned regiochemical memory, this effect can be the result of tight ion pairing with the leaving group, and/or due to unequal *trans* effects arising from unsymmetrical ligation on neutral π-allyl complexes [227].

Thus, for example, an enantioenriched sample of (*S*)-cyclopentenyl acetate reacts with malonate anion in the presence of the two enantiomeric forms [(*R*,*R*) or (*S*,*S*)] of the TSL, giving different ee's of the allylated product with a preference for attack at the position formerly occupied by the leaving group (Scheme 56) [210].

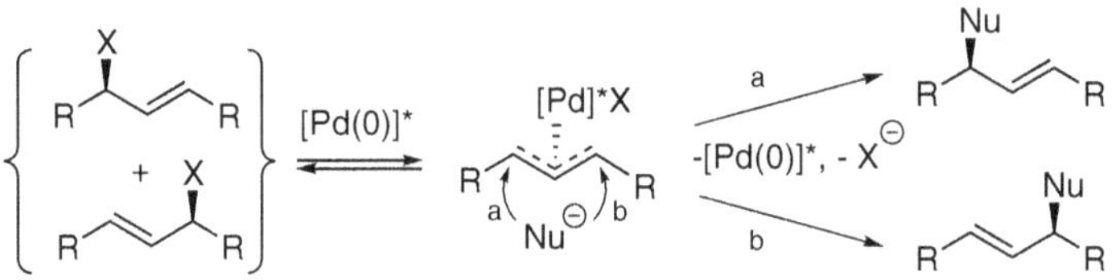

Scheme 54 Enantioselection coincides with regioselection

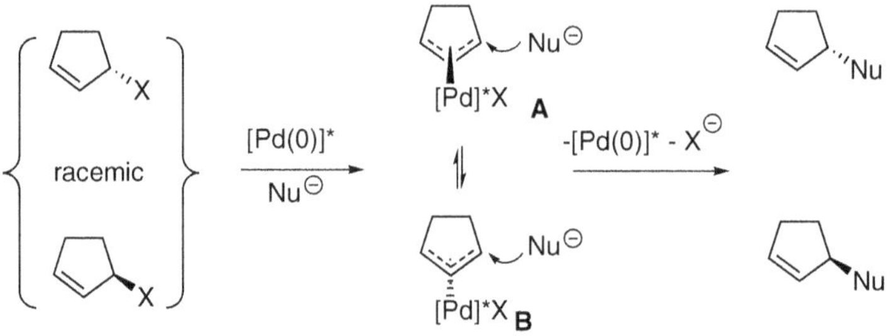

Scheme 55 Slow (A⇆B): Memory effect → low enantioselection. Rapid (A⇆B): no memory effect → enantioselection may be good

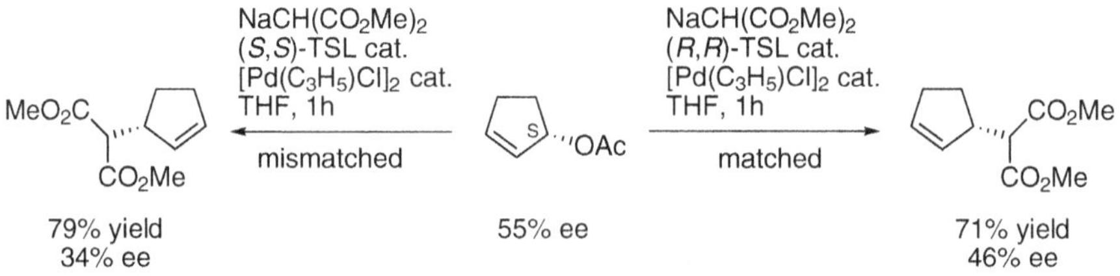

Scheme 56 Example of stereochemical memory effect with TSL

Although inhibition of equilibration between the starting enantiomeric substrates is clearly the reason of the memory effects, the detailed nature of these complexes is more complex and may vary from case to case. The participation of a neutral monophosphine-coordinated allyl complex (see above) has been often invoked. However, a detailed study of the axially chiral MAP and MOP ligands, [79] known to afford strong memory effects, indicated the involvement of cationic complexes featuring a bidentate (*P*,*C*) coordination (Scheme 16). Furthermore, the nature of the solvent affects the rate of interconversion between the diastereomeric complexes.

Study of the structure of the Pd(0)-TSL and of the symmetrically substituted η^3-allyl palladium – STL complexes showed in both cases monomer – oligomer equilibria, featuring *P*,*P*- as well as *P*,*O*-coordination at palladium atom. With this ligand it has been suggested that the "memory effect" may be mainly due to a difference in reactivity between the monomeric and the oligomeric palladium(0) complexes towards the enantiomers of the substrate, in combination with a difference in enantioselectivity between the monomeric and the oligomeric η^3-allyl intermediates when reacting with the nucleophile [228–230]. This implies, *inter alia*, that the ee value of the final product depends on the catalyst concentration. Indeed, lower catalyst loadings, favoring monomer, are expected to lead to higher selectivities [231–233].

Detection of the Stereochemical Memory Effect

Detection of a stereochemical memory effect normally requires preparing one enantiomer of the starting substrate and studying its asymmetric allylic alkylation (AAA) in the presence of one and the other enantiomeric ligands. Alternatively, one may synthesize both enantiomers of the substrate and study their AAA in the presence of one selected enantiomer of the chiral ligand. However, Lloyd–Jones developed a smart technique that allows skipping the use of enantiopure substrates [234, 235]. This method is based on the submission of a racemic (*S*/*R*), but regioselectively D-labeled, substrate to AAA in the presence of the ligand in enantiomerically pure as well as racemic form. Assuming a double inversion

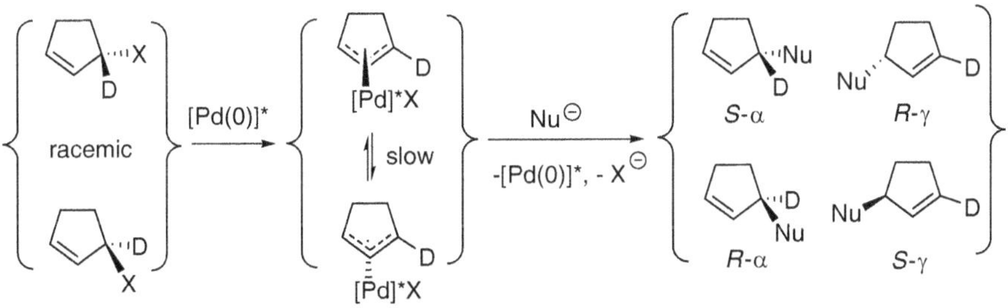

Scheme 57 Detecting the stereochemical memory effect

mechanism (Scheme 26), each enantiomer is expected to afford a couple of α- and γ-products (S-α/R-γ and S-γ/R-α). If no memory effect is at work, and assuming the absence of isotopic effect, the two ratios will be identical and represent the intrinsic enantioselectivity of the catalyst when a *meso* complex is involved. On the other hand, a discrepancy between these two ratios would suggest the presence of a (α or γ) memory effect, associated to one or both manifolds. The recognition of matched and mismatched pairing implies that different α/γ ratios are expected when using one enantiopure ligand or its racemic form. In fact, while in the presence of the enantiopure ligand both the enantiomers of the substrate will experience matched and mismatched pairing, with the racemic one, both the enantiomers of the substrate will have the chance to find their own matched enantiomeric ligand (Scheme 57).

4 Asymmetric Allylic Alkylation

AAA is an exciting field of research and numerous applications have been dedicated to the synthesis of biologically relevant targets [16, 17, 132, 236, 237]. Depending on the nature of the substrate, the nature of the nucleophile, and the reaction conditions, and according to the classical catalytic cycle presented in Scheme 1, enantiodiscrimination in AAA may occur during: (a) olefin-to-metal coordination/oxidative addition (Fig. 8, profile a); (b) nucleophilic attack (Fig. 8, profile b) (see for example [238]).

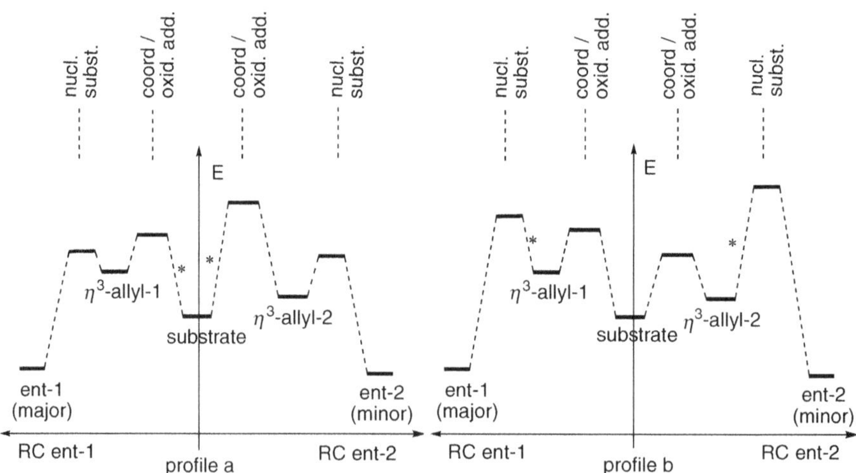

Fig. 8 Qualitative energy profiles for Pd-catalyzed asymmetric allylic substitutions. The starred paths are the rate- and enantiodetermining steps. It is assumed that the η^3-allyl complex is higher in energy than the starting substrate. Possible *anti-syn* isomerizations are neglected

4.1 *Oxidative Addition Is the Enantiodiscriminating Step*

Oxidative addition is the enantiodiscriminating step when there is a structural bias of the substrate for it and in general when the nucleophilic attack is fast compared to the interconversion of diastereomeric π-allyl complexes. In this case, enantioselectivity is likely to be influenced by the nature of the leaving group. Accordingly, a poorer leaving group will induce a more retarded transition state with enhanced steric constraint, expected to lead to greater enantioselectivity, as demonstrated by Fiaud and Legros [239].

4.1.1 Ionization of Enantiotopic Leaving Groups

For *meso* substrates with two enantiotopic leaving groups or achiral substrates with two geminal enantiotopic leaving groups, enantiodiscrimination of the leaving groups during the coordination/oxidative addition step in the presence of chiral ligand generates chiral π-allyl complexes. These complexes were found to react with nucleophile following the classical double inversion process (Scheme 58).

In their total synthesis of Agelastatin, Trost and Dong [240] exploited a palladium-catalyzed AAA starting from a *meso* diBoc-activated cyclopentene (Scheme 59). A high ee and yield were obtained in the presence of the TSL.

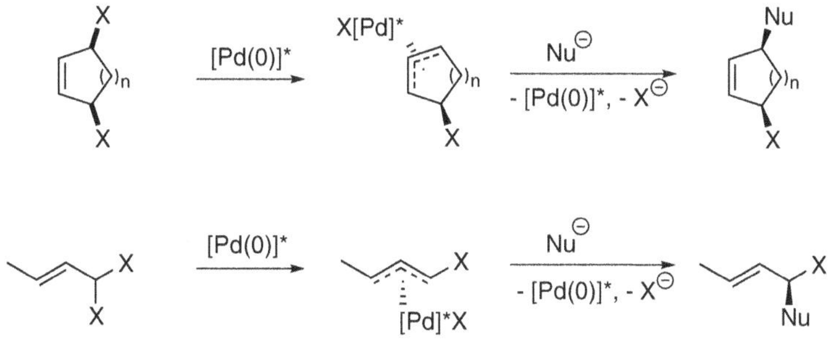

X [Pd(0)]* X[Pd]* Nu⊖ - [Pd(0)]*, - X⊖ Nu X

X [Pd(0)]* X [Pd]*X Nu⊖ - [Pd(0)]*, - X⊖ X Nu

Scheme 58 Ionization of enantiotopic leaving groups

OBoc + Br–(pyrrole, NH)–CO2Me; $[Pd(C_3H_5)Cl]_2$ (1.25 mol%), Ligand (1.9 mol %), Cs_2CO_3, CH_2Cl_2, rt → 83% (92% ee)

Ligand : NH HN, PPh_2 Ph_2P

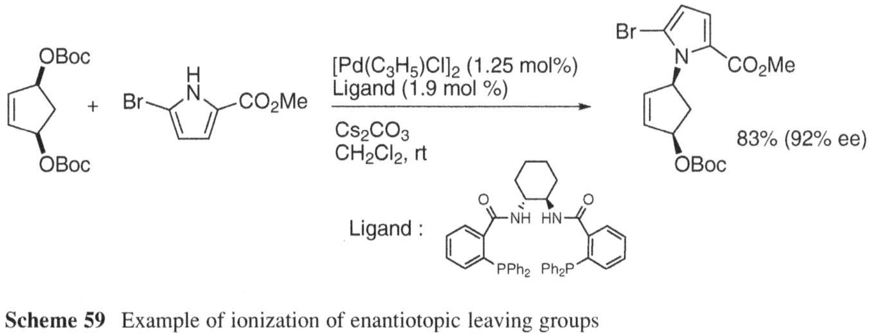

Scheme 59 Example of ionization of enantiotopic leaving groups

4.1.2 Slowly π-σ-π Equilibrating η^3-Allylpalladium Complexes

Monosubstituted η^3-Allylpalladium Complexes

Starting from achiral primary allylic substrates, if π-allyl interconversion is slow compared to nucleophilic attack, then solely the oxidative addition may be the enantiodiscriminating step. This is the case if the chiral catalyst selects preferentially one of the enantiotopic faces of the alkene directing the substitution toward the branched product (Scheme 60).

In the frame of their asymmetric synthesis of Vitamin E, Trost and Asakawa reported an example of intramolecular palladium-catalyzed AAA wherein oxidative addition was the enantiodiscriminating step (Scheme 61). Significatively, all the conducted experiments that accelerate trapping of the π-allyl complex, thereby limiting its racemization, increased the ee of the product. Furthermore, when the rate of racemization was accelerated by the use of additives, or by a higher catalyst concentration, a decrease of the ee was observed [241].

Kinetic Resolution of Secondary Allylic Substrates

Oxidative addition of some substrates leads to diastereomeric π-allyl complexes that do not interconvert or interconvert slowly compared to nucleophilic attack. In this case, due to the double inversion process, a kinetic resolution may be observed if coordination or ionization of one enantiomer is faster than that of the other one (Scheme 62).

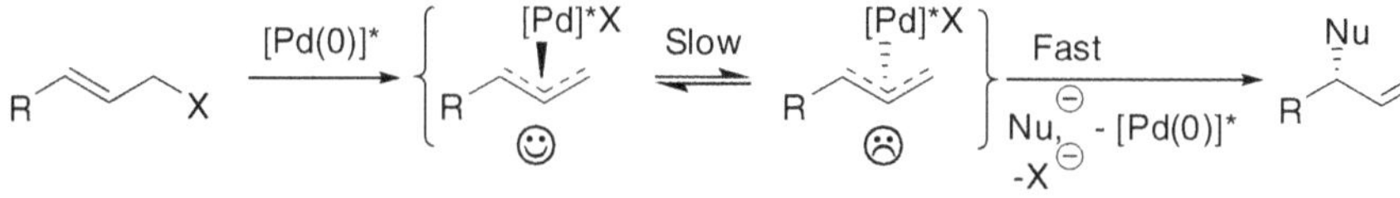

[Pd(0)]* [Pd]*X Slow [Pd]*X Fast Nu, -X - [Pd(0)]* Nu R X R R R

Scheme 60 π-allyl palladium complex interconversion is slow compared to nucleophilic attack

BnO OH OCO_2CH_3 $Pd_2dba_3{\cdot}CHCl_3$ (1 mol%) Ligand (3 mol %) NEt_3 (1.5 equiv.) CH_2Cl_2, rt BnO O 89% (86% ee)

Ligand : NH HN O O PPh_2 Ph_2P

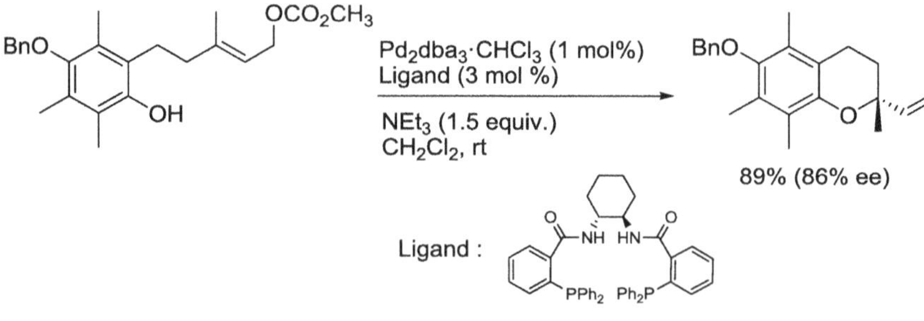

Scheme 61 Example of slow π-allyl palladium complex interconversion compared to nucleophilic attack

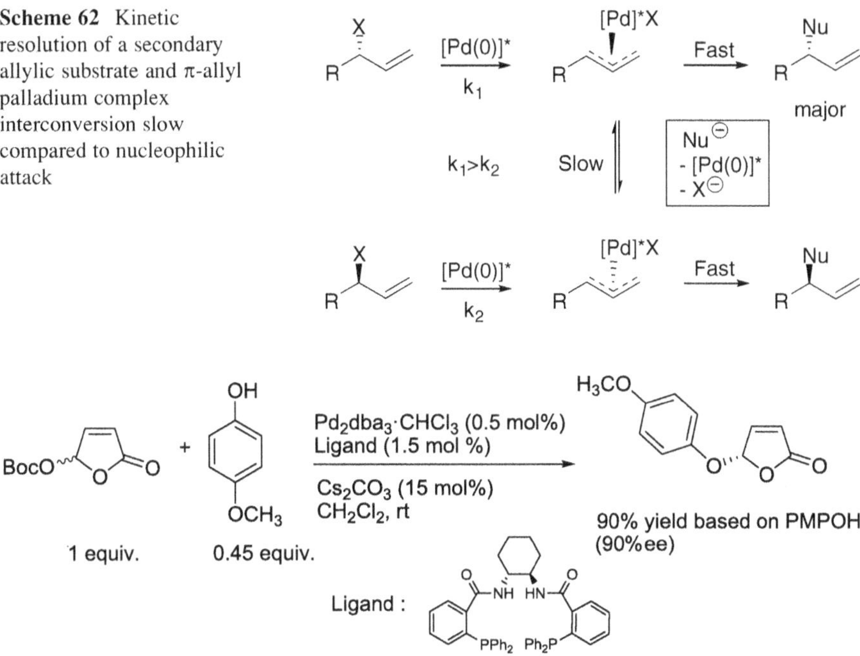

Scheme 62 Kinetic resolution of a secondary allylic substrate and π-allyl palladium complex interconversion slow compared to nucleophilic attack

Scheme 63 Example of Pd catalyzed allylic alkylation involving a kinetic resolution of the substrate and π-allyl palladium complex interconversion slow compared to nucleophilic attack

The above case study has been experimentally put into practice by Trost and Toste [242], who reported an example of kinetic resolution using an oxygen-based nucleophile (Scheme 63).

4.2 Nucleophilic Attack Is the Enantiodiscriminating Step

4.2.1 Desymmetrization of *meso*-π-Allyl Complexes

Either enantiomer of allylic substrates with identical substituents at C1 and C3 afford the same π-allyl complex after oxidative addition onto palladium(0) complex. In the presence of a chiral ligand, the C1 and C3 termini of the *meso* allylic moiety are diastereotopic and thus may exhibit different reactivities toward nucleophiles. Therefore, a suitable chiral ligand can control the regioselectivity of the nucleophilic attack and thus induce the preferential formation of one product enantiomer over the other. In this way, a racemic substrate can be converted into an enantioenriched product (Scheme 64).

AAA's have been intensively studied with 1,3-diphenylpropenyl acetate, which has become the benchmark substrate when exploring new ligands. However, success was also encountered when using various acyclic and cyclic [98] precursors bearing identical substituents at C1 and C3. Indeed, Trost and Oslob reported the intramolecular asymmetric allylic amination of an eight-membered ring substrate. The cyclization proceeded in high yield and ee in the presence of a Trost *P,N*-ligand and afforded an advanced precursor of (−)-Anatoxin-a (Scheme 65) [243].

4.2.2 (Pro)^1-Chiral Nucleophiles

The use of a pronucleophile carrying a prostereogenic reactive carbon atom is another way to accomplish an AAA. Noteworthy, as "soft" nucleophiles add *anti* to the palladium center, they remain segregated by the π-allyl moiety from the chiral environment of the ligand. Nevertheless, if a suitable chiral catalyst is able to selectively activate one face of the prostereogenic nucleophile, one enantiomer of the allylated product may be selectively generated (Scheme 66).

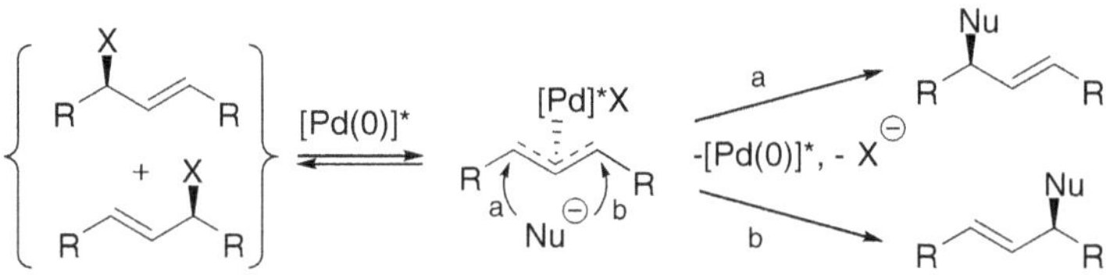

X R R + X R R [Pd(0)]* [Pd]*X R R a Nu b a -[Pd(0)]*, - X⊖ b Nu R R Nu R R

Scheme 64 Enantioselection coincides with regioselection

TsHN CO₂Me OCO₂Me Pd₂(dba)₃·CHCl₃ (2.5 mol%) Ligand (7.5 mol%) CH₂Cl₂, 0°C Ligand: NH HN PPh₂ N TsN MeO₂C 90% (88% ee)

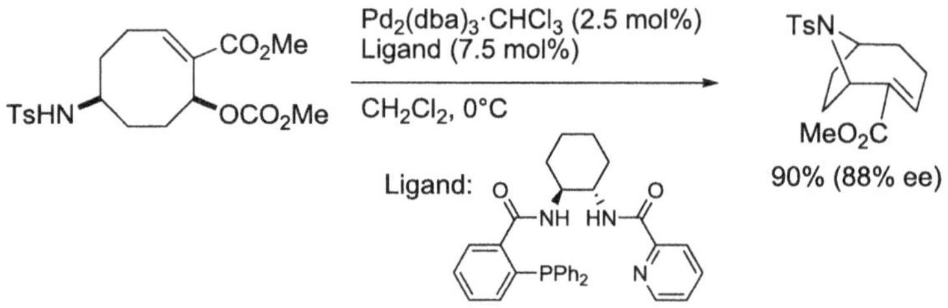

Scheme 65 Example of Pd catalyzed allylic alkylation wherein enantioselection coincides with regioselection

X [Pd(0)]* [Pd]*X R¹ R² - [Pd(0)]*, - X⊖ R¹ R² + R¹ R²

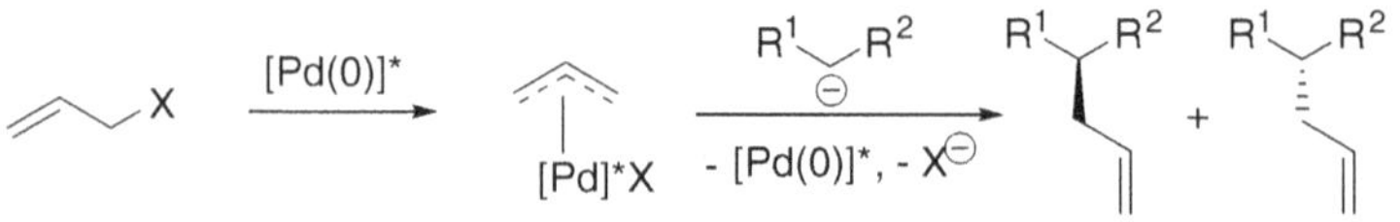

Scheme 66 Generation of a stereogenic center on the reactive carbon atom of the pronucleophile

Bai and coworkers reported the enantioselective palladium-catalyzed bicycloannulation of a β-ketoester, affording the key intermediate in their synthesis of (−)-huperzine A. In the presence of a ferrocenylphosphine ligand, a high yield and ee were achieved (Scheme 67) [244].

Trapping of a π-allyl moiety may lead to the desymmetrization of a *meso* nucleophile if the reaction occurs selectively at one of the two enantiotopic nucleophilic sites. Using a suitable chiral catalyst, it is possible to control the selectivity of the monoallylation and thus to induce the preferential formation of one product enantiomer over the other (Scheme 68).

Taguchi and coworkers reported the desymmetrization of a *meso*-vicinal bissulfonamide, affording the corresponding *N*-monoallylated product in high yield and ee (Scheme 69) [245].

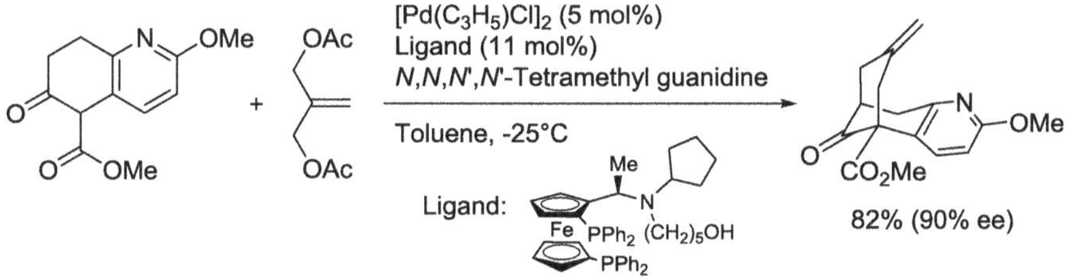

[Pd(C3H5)Cl]2 (5 mol%)
Ligand (11 mol%)
N,N,N',N'-Tetramethyl guanidine
Toluene, -25°C
Ligand:
82% (90% ee)

Scheme 67 Example of Pd catalyzed allylic alkylation involving generation of a stereogenic center on the reactive carbon atom of the pronucleophile

[Pd(0)]*
[Pd]*X
Base
- [Pd(0)]* - X⊖

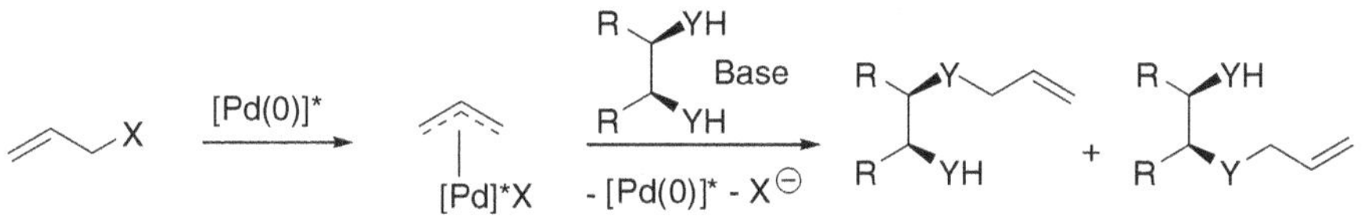

Scheme 68 Desymmetrization of a *meso* pro-nucleophile

1 equiv.
Trs = 2,4,6-(i-Pr)3C6H2SO2
[Pd(C3H5)Cl]2 (3.6 mol%)
Ligand (7.3 mol%)
t-BuOK (1 equiv.)
Toluene-dioxane, -15°C
Ligand:
85% (93% ee)

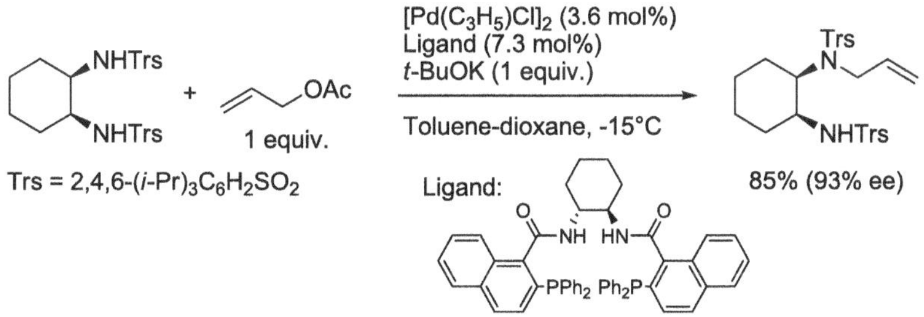

Scheme 69 Example of Pd catalyzed allylic alkylation involving desymmetrization of a *meso* pro-nucleophile

4.2.3 Monosubstituted η^3-Allylpalladium Complexes

Oxidative addition of an achiral primary allylic precursor, or of the corresponding chiral isomeric secondary precursor is expected to provide the same monosubstituted π-allyl derivative (unless in the presence of a memory effect). In such complexes π-σ-π equilibration (Scheme 14) (that switches the complexed enantiotopic allyl faces with invisible *syn-anti* isomerization) is fast compared to nucleophilic substitution and a chiral catalyst may allow preferential attack of the nucleophile to one of the two rapidly equilibrating π-allyl intermediates (Scheme 70, $k_1 \neq k_2$, $k_3 \neq k_4$). As Curtin–Hammett conditions operate, the different thermodynamic stability of the transient π-allyl intermediates is irrelevant for the stereochemical outcome.

Hayashi and coworkers reported the preparation of enantioenriched 4-vinyl oxazolidinones starting from symmetrical 2-butenylene dicarbamates (Scheme 71) [246]. Interestingly, both (*E*)- and (*Z*)-isomers of the substrate afford the cyclized product in essentially the same yields and ee's, indicating that the interconversion of the intermediate π-allylpalladium complexes is fast compared to nucleophilic attack.

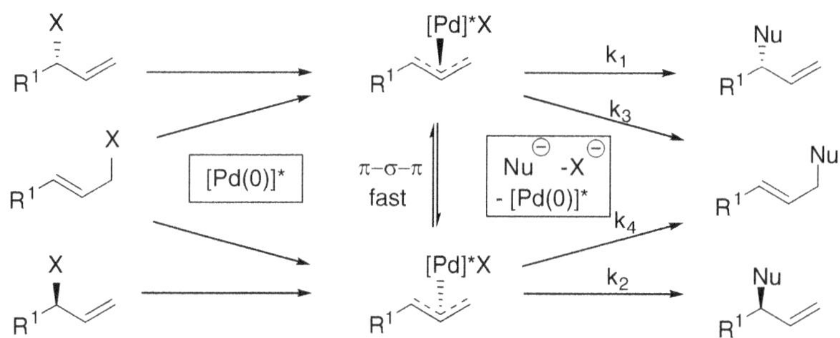

Scheme 70 Fast equilibration of the transient π-allyl palladium complexes (Curtin–Hammett conditions)

(*E*) or (*Z*)

$Pd_2dba_3 \cdot CHCl_3$ (1.5 mol%)
Ligand (3.3 mol%)
THF, reflux, 2h

from (*E*): 80%(73% ee)
from (*Z*): 92%(77% ee)

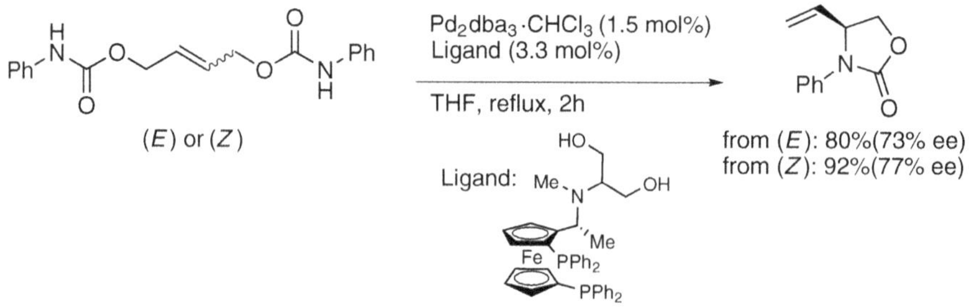

Scheme 71 Example of Pd catalyzed allylic alkylation involving fast equilibration of the transient π-allyl palladium complexes

4.2.4 η^3-Allylpalladium Complexes with Different Substituents at C1 and C3

In the case of allylic substrates bearing different substituents at C1 and C3, oxidative addition of both starting enantiomers leads to diastereomeric π-allylpalladium complexes, displaying opposite configurations at the newly generated planar stereogenic unit. Subsequent nucleophilic trapping of each π-allyl intermediate *anti* to the palladium center affords one enantiomeric set of the two possible regioisomeric products (Scheme 72).

Interconversion of such 1,3-unsymmetrical π-allyl complexes is necessary to achieve an enantioselective reaction. However, these intermediates cannot racemize via a π-σ-π process, as this would induce *syn-anti* isomerization.

As a consequence, (excluding kinetic resolution) an enantioenriched product can be obtained from this type of racemic substrate if allyl enantioface exchange can occur via an alternative mechanism, such as a palladium(0)-catalyzed allyl exchange (see Sect. 2.2.2.4). Enantioselection may then be dictated by the nucleophilic trapping associated with the highest rate ($k_1 \neq k_2$).

During their study toward the total synthesis of (−)-Aflatoxins B_1 and B_{2a}, Trost and Toste were able to turn the kinetic resolution of γ-acyloxybutenolides into a dynamic kinetic resolution (compare Scheme 63 and Scheme 73). Addition of a halide source (*n*-Bu_4NCl 30 mol%) leads, in the presence of 1 equivalent of the phenol, to 84% ee and 74% yield [242].

The halide is believed to promote the equilibration of the diastereomeric π-allyl complexes through the mechanism presented in Scheme 74. It is important to note that this easy addition not only turns a *simple* into a *dynamic* kinetic resolution, but also switches the enantiodiscriminating step from oxidative addition to nucleophilic attack.

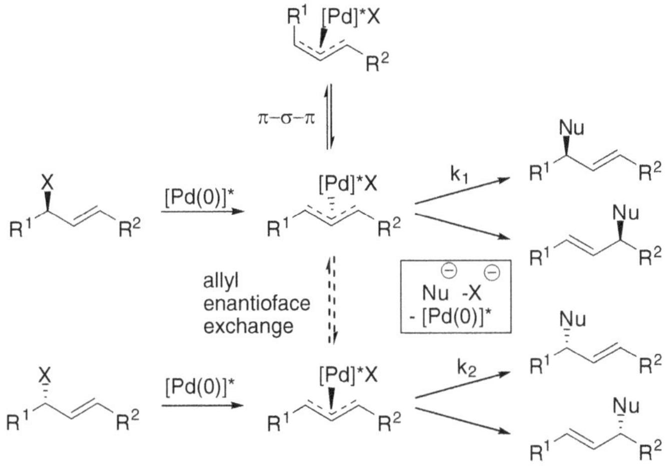

Scheme 72 Generation and reactivity of a non symmetrically 1,3-disubstituted η^3-allylpalladium complex

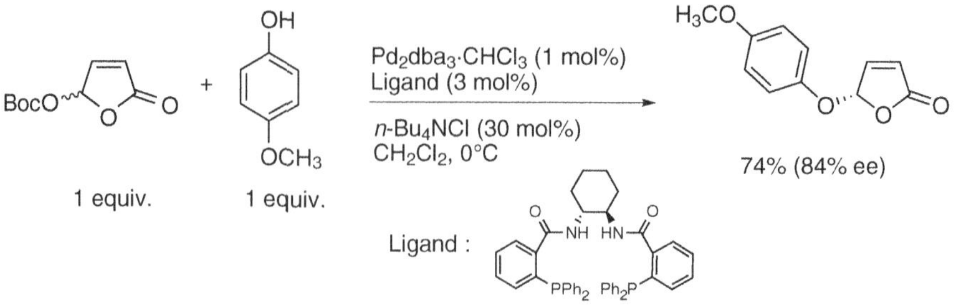

Scheme 73 Achievement of a dynamic kinetic resolution of a non symmetrically 1,3-disubstituted η^3-allylpalladium complex

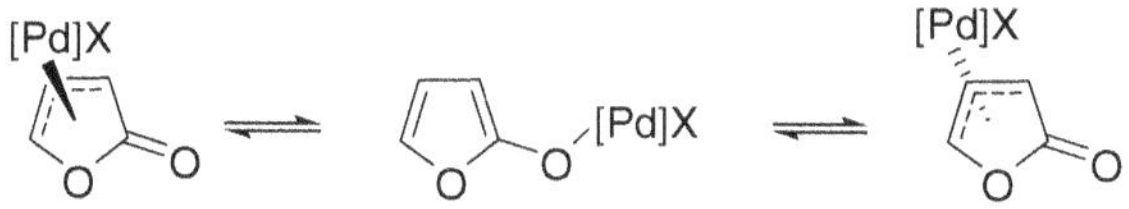

Scheme 74 Mechanistic rationale accounting for an effective dynamic kinetic resolution of a non symmetrically 1,3-disubstituted η^3-allylpalladium complex

5 Conclusion

In conclusion, this chapter aimed at introducing and highlighting the main features of the palladium-catalyzed allylation reaction. This transformation, whose subtle mechanistic details are still to be completely unveiled, turns out to be an extremely useful synthetic tool, allowing nucleophiles of different nature to be allylated in a regio-, stereo- and/or enantioselective way. This chapter clearly shows that a satisfactory understanding of this reaction has to take into account the behavior of the fleeting η^3-allylpalladium complexes as well as the relative kinetic data of the steps composing the overall transformation. Finally, it is worth noting that allylation via palladium catalysis displays features that are peculiar to this metal, and the use of transition metals other than palladium is often associated to alternative, yet not less interesting features.

Acknowledgments The authors warmly thank Professors Francesco Sannicolò and Guy Lloyd-Jones for stimulating discussions.

References

1. Tsuji J et al (1965) Organic syntheses by means of noble metal compounds XVII. Reaction of π-allylpalladium chloride with nucleophiles. Tetrahedron Lett 6:4387–4388
2. Atkins KE et al (1970) Palladium catalyzed transfer of allylic groups. Tetrahedron Lett 11:3821–3824

3. Hata G et al (1970) Palladium-catalyzed exchange of allylic groups of ethers and esters with active-hydrogen compounds. J Chem Soc D, Chem Commun 1392–1393
4. Trost BM, Strege PE (1977) Asymmetric induction in catalytic allylic alkylation. J Am Chem Soc 99:1649–1651
5. Trost BM, Lautens M (1982) Molybdenum catalysts for allylic alkylation. J Am Chem Soc 104:5543–5545
6. Trost BM, Hung MH (1983) Tungsten-catalyzed allylic alkylations. New avenues for selectivity. J Am Chem Soc 105:7757–7759
7. Takeuchi R, Kashio M (1997) Highly selective allylic alkylation with a carbon nucleophile at the more substituted allylic terminus catalyzed by an iridium complex: an efficient method for constructing quaternary carbon centers. Angew Chem Int Ed 36:263–265
8. Tsuji J et al (1984) Allylation of carbonucleophiles with allylic carbonates under neutral conditions catalyzed by rhodium complexes. Tetrahedron Lett 25:5157–5160
9. Zhang S et al (1993) Ruthenium complex-catalyzed allylic alkylation of carbonucleophiles with allylic carbonates. J Organomet Chem 450:197–207
10. Kurosawa H (1979) η-Allylmetal chemistry. Part 7. Allylic alkylation catalysed by platinum complexes. Isolation of rigid (σ-allyl)(pentane-2,4-dionato)platinum(II) complexes. J Chem Soc, Dalton Trans 939–943
11. Cuvigny T, Julia M (1983) Alkylations allyliques catalysees au nickel. J Organomet Chem 250:C21–C24
12. Consiglio G et al (1983) Nickel catalysed asymmetric coupling reaction between allyl phenyl ethers and Grignard reagents. J Chem Soc, Chem Commun 112–115
13. Fouquet G, Schlosser M (1974) Improved carbon-carbon linking by controlled copper catalysis. Angew Chem Int Ed 13:82–83
14. Roustan JL et al (1979) Reactions d'alkylation de chlorures, formate et acetates allyliques catalyses par des complexes du fer et du cobalt. Tetrahedron Lett 20:3721–3724
15. Auburn PR et al (1985) Asymmetric synthesis. Asymmetric catalytic allylation using palladium chiral phosphine complexes. J Am Chem Soc 107:2033–2046
16. Trost BM, Crawley ML (2003) Asymmetric transition-metal-catalyzed allylic alkylations: applications in total synthesis. Chem Rev 103:2921–2944
17. Trost BM (2004) Asymmetric allylic alkylation, an enabling methodology. J Org Chem 69:5813–5837
18. Trost BM (1996) Designing a receptor for molecular recognition in a catalytic synthetic reaction: allylic alkylation. Acc Chem Res 29:355–364
19. Pfaltz A, Lautens M (1999) Allylic substitution reactions. In: Jacobsen EN, Pfaltz A, Yamamoto H (eds) Comprehensive asymmetric catalysis. Springer, Heidelberg
20. Evans LA et al (2008) Counterintuitive kinetics in Tsuji-Trost allylation: ion-pair partitioning and implications for asymmetric catalysis. J Am Chem Soc 130:14471–14473
21. Amatore C et al (2007) Palladium(0)-catalyzed allylic aminations: kinetics and mechanism of the reaction of secondary amines with cationic $[(\eta^3\text{-allyl})PdL_2]^+$ complexes. Organometallics 26:1875–1880
22. Brookhart M et al (1992) $[(3,5\text{-}(CF_3)_2C_6H_3)_4B]^-[H(OEt_2)_2]^+$: a convenient reagent for generation and stabilization of cationic, highly electrophilic organometallic complexes. Organometallics 11:3920–3922
23. Trost BM, Verhoeven TR (1976) Allylic substitutions with retention of stereochemistry. J Org Chem 41:3215–3216
24. Matsushita H, Negishi E (1982) anti-Stereospecificity in the palladium-catalysed reactions of alkenyl- or aryl-metal derivatives with allylic electrophiles. J Chem Soc, Chem Commun 160–161
25. Yamamoto T et al (1981) Oxidative addition of allyl acetate to palladium(0) complexes. J Am Chem Soc 103:5600–5602
26. Amatore C et al (2005) Rate and mechanism of the reaction of (E)-PhCHCH-CH(Ph)-OAc with palladium(0) complexes in allylic substitutions. Organometallics 24:1569–1577

27. Amatore C et al (1999) Evidence of the reversible formation of cationic π-Allylpalladium(II) complexes in the oxidative addition of allylic acetates to palladium(0) complexes. Chem Eur J 5:466–473
28. Vitagliano A et al (1991) Convenient synthesis of cationic (η^3-allyl)palladium complexes. Preparative and stereochemical aspects. Organometallics 10:2592–2599
29. Tsuji J et al (1982) Facile palladium catalyzed decarboxylative allylation of active methylene compounds under neutral conditions using allylic carbonates. Tetrahedron Lett 23:4809–4812
30. Tsuji J et al (1985) Allylic carbonates. Efficient allylating agents of carbonucleophiles in palladium-catalyzed reactions under neutral conditions. J Org Chem 50:1523–1529
31. Ozawa F et al (1992) Preparation and reactions of (π-allyl)palladium and - platinum-carbonate complexes. Organometallics 11:171–176
32. Amatore C et al (2000) Oxidative addition of allylic carbonates to palladium(0) complexes: reversibility and isomerization. Chem Eur J 6:3372–3376
33. Jeffrey PD, McCombie SW (1982) Homogeneous, palladium(0)-catalyzed exchange deprotection of allylic esters, carbonates and carbamates. J Org Chem 47:587–590
34. Tsuji J et al (1981) Regioselective 1,4-addition of nucleophiles to 1,3-diene monoepoxides catalyzed by palladium complex. Tetrahedron Lett 22:2575–2578
35. Trost BM, Molander GA (1981) Neutral alkylations via palladium(0) catalysis. J Am Chem Soc 103:5969–5972
36. Takahashi K et al (1972) Palladium-catalyzed exchange of allylic groups of ethers and esters with active hydrogen compounds. II. Bull Chem Soc Jpn 45:230–236
37. Trost BM, Keinan E (1979) An approach to primary allylic amines via transition-metal-catalyzed reactions. Total Synthesis of (+/−)-Gabaculine. J Org Chem 44:3451–3457
38. Trost BM et al (1980) Allyl sulfones as synthons for 1,1- and 1,3-dipoles via organopalladium chemistry. J Am Chem Soc 102:5979–5981
39. Hutchins RO, Learn K (1982) Regio- and stereoselective reductive replacement of allylic oxygen, sulfur, and selenium functional groups by hydride via catalytic activation by palladium(0) complexes. J Org Chem 47:4380–4382
40. Tanigawa Y et al (1982) Palladium(0)-catalyzed allylic alkylation and amination of allylic phosphates. Tetrahedron Lett 23:5549–5552
41. Tamura R, Hegedus LS (1982) Palladium(0)-catalyzed allylic alkylation and amination of allylnitroalkanes. J Am Chem Soc 104:3727–3729
42. One N et al (1982) Palladium-catalysed allylic alkylations of allylic nitro-compounds. J Chem Soc, Chem Commun 821–822
43. Sheffy FK, Stille JK (1983) Palladium-catalyzed cross-coupling of allyl halides with organotins. J Am Chem Soc 105:7173–7175
44. Paolobelli AB et al (1993) Palladium-catalyzed alkylation of allylic nitrates derived from ceric ammonium nitrate promoted oxidative addition of trimethylsilyloxy-cyclopropanes to 1,3-butadiene. Tetrahedron Lett 34:6333–6336
45. Muzart J (2005) Palladium-catalysed reactions of alcohols. Part B: formation of C-C and C-N bonds from unsaturated alcohols. Tetrahedron 61:4179–4212
46. Muzart J (2007) Procedures for and possible mechanisms of Pd-catalyzed allylations of primary and secondary amines with allylic alcohols. Eur J Org Chem 3077–3089
47. Imada Y et al (2002) Palladium-catalyzed asymmetric alkylation of 2,3-alkadienyl phosphates. Synthesis of optically active 2-(2,3-alkadienyl)malonates. Chem Lett 31:140–141
48. Trost BM et al (2005) Dynamic kinetic asymmetric allylic alkylations of allenes. J Am Chem Soc 127:14186–14187
49. Roberts JS, Klabunde KJ (1977) The direct synthesis of η^3-$ArCH_2PdCl$ compounds by the oxidative addition of $ArCH_2$-chlorine bonds to palladium atoms. J Am Chem Soc 99:2509–2515

50. Legros J, Fiaud JC (1992) Palladium-catalyzed nucleophilic substitution of naphthylmethyl and 1-naphthylethyl esters. Tetrahedron Lett 33:2509–2510
51. Kuwano R et al (2003) Palladium-catalyzed nucleophilic benzylic substitutions of benzylic esters. J Am Chem Soc 125:12104–12105
52. Kuwano R (2009) Catalytic transformations of benzylic carboxylates and carbonates. Synthesis 1049–1061
53. Liegault B et al (2008) Activation and functionalization of benzylic derivatives by palladium catalysts. Chem Soc Rev 37:290–299
54. Stevens RR, Shier GD (1970) π-benzylbis(triethylphosphine) palladium(II) tetrafluoroborate. J Organomet Chem 21:495–499
55. Shimizu I, Tsuji J (1984) Palladium-catalyzed synthesis of 2,3-disubstituted allylamines by regioselective aminophenylation or aminoalkenylation of 1,2-dienes. Chem Lett 233–236
56. Ahmar M et al (1984) Synthese de dienes-1,3 et de styrenes fonctionnalises par carbopalladation catalytique d'allenes. Tetrahedron Lett 25:4505–4508
57. Ma S (2006) Transition-metal-catalyzed reactions of allenes. Pure Appl Chem 78:197–208
58. Patel BA et al (1978) Palladium-catalyzed arylation of conjugated dienes. J Org Chem 43:5018–5020
59. O'Connor JM et al (1983) Some aspects of palladium-catalyzed reactions of aryl and vinylic halides with conjugated dienes in the presence of mild nucleophiles. J Org Chem 48:807–809
60. Robinson SD, Shaw BL (1963) 919. Transition metal-carbon bonds. Part I. π-Allylic palladium complexes from butadiene and its methyl derivatives. J Chem Soc 4806–4814
61. Robinson SD, Shaw BL (1964) 961. Transition metal-carbon bonds. Part II. π -Allylic and related complexes from some cyclic 1,3-dienes. J Chem Soc 5002–5008
62. Bäckvall JE et al (1984) Stereo- and regioselective palladium-catalyzed 1,4-diacetoxylation of 1,3-dienes. J Org Chem 49:4619–4631
63. Bäckvall JE et al (1985) Stereo- and regioselective palladium-catalyzed 1,4-acetoxychlorination of 1,3-dienes. 1-Acetoxy-4-chloro-2-alkenes as versatile synthons in organic transformations. J Am Chem Soc 107:3676–3686
64. Lupin MS et al (1966) Transition metal-carbon bonds. Part VII. The formation of π-allylic-palladium complexes from allenes and palladium halides and the reversed reactions. J Chem Soc A 1687–1691
65. Bäckvall J, Jonasson C (1997) Palladium-catalyzed 1,2-oxidation of allenes. Tetrahedron Lett 38:291–294
66. Parshall GW, Wilkinson G (1962) Mesityl oxide complexes of palladium and platinum. Inorg Chem 1:896–900
67. Trost BM, Fullerton TJ (1973) New synthetic reactions. Allylic alkylation. J Am Chem Soc 95:292–294
68. Chen MS, White MC (2004) A sulfoxide-promoted, catalytic method for the regioselective synthesis of allylic acetates from monosubstituted olefins via C-H oxidation. J Am Chem Soc 126:1346–1347
69. Young AJ, White MC (2008) Catalytic intermolecular allylic C-H alkylation. J Am Chem Soc 130:14090–14091
70. Lin S et al (2008) Intra/intermolecular direct allylic alkylation via Pd(II)-catalyzed allylic C-H activation. J Am Chem Soc 130:12901–12903
71. Jensen T, Fristrup P (2009) Toward efficient palladium-catalyzed allylic C-H alkylation. Chem Eur J 15:9632–9636
72. Jazzar R et al (2010) Functionalization of organic molecules by transition-metal-catalyzed C (sp3)-H activation. Chem Eur J 16:2654–2672
73. Vrieze K (1975) In: Jackman LM, Cotton FA (eds) Dynamic nuclear magnetic resonance spectroscopy. Academic, New York
74. Pregosin PS, Salzmann R (1996) Structure and dynamics of chiral allyl complexes of Pd(II): NMR spectroscopy and enantioselective allylic alkylation. Coord Chem Rev 155:35–68

75. Solin N, Szabó KJ (2001) Mechanism of the $\eta^1 \rightarrow \eta^3 \rightarrow \eta^1$ isomerization in allylpalladium complexes: solvent coordination, ligand and substituent effects. Organometallics 20: 5464–5471
76. Pericàs MA et al (2002) Modular bis(oxazoline) ligands for palladium catalyzed allylic alkylation: unprecedented conformational behaviour of a bis(oxazoline) palladium η^3-1,3-diphenylallyl complex. Chem Eur J 8:4164–4178
77. Ogasawara M et al (2002) Effects of bidentate phosphine ligands on *syn-anti* isomerization in π–allylpalladium complexes. Organometallics 21:4853–4861
78. Faller JW, Wilt JC (2005) Regioselectivity in the palladium/(*S*)-BINAP(S)-catalyzed asymmetric allylic amination: reaction scope, kinetics and stereodynamics. Organometallics 24:5076–5083
79. Lloyd-Jones GC et al (2000) Diastereoisomeric cationic π-allylpalladium-(P, C)-MAP and MOP complexes and their relationship to stereochemical memory effects in allylic alkylation. Chem Eur J 6:4348–4357
80. Boele MDK et al (2004) Bulky monodentate phosphoramidites in palladium-catalyzed allylic alkylation reactions: aspects of regioselectivity and enantioselectivity. Chem Eur J 10:6232–6246
81. Camus J-M et al (2004) Allylpalladium(II) complexes with aminophosphane ligands: solution behaviour and X-ray structure of *cis*-[Pd(η^3-$CH_2CHCHPh$){$Ph_2PCH_2CHPhNH$ (2,6-$C_6H_3iPr_2$)}][PF_6]. Eur J Inorg Chem 1081–1091
82. Kumar PGA et al (2005) Bonding in palladium(II) and platinum(II) allyl MeO- and H-MOP complexes. Subtle differences *via* ^{13}C NMR. Organometallics 24:1306–1314
83. Kawatsura M et al (2000) Palladium-catalyzed asymmetric reduction of racemic allylic esters with formic acid: effects of phosphine ligands on isomerization of π–allylpalladium intermediates and enantioselectivity. Tetrahedron 56:2247–2257
84. Mandal SK et al (2003) Diastereoisomerism in palladium(II) allyl complexes of P, P-, P, S-, and S, S-donor ligands, $Ph_2P(E)N(R)P(E')Ph_2$ [R = $CHMe_2$ or (S)-*CHMePh; E = E4 = lone pair or S]: solution behaviour, X-ray crystal structure and catalytic allylic alkylation reactions. J Organomet Chem 676:22–37
85. Hayashi T et al (1986) Stereo- and regiochemistry in palladium-catalyzed nucleophilic substitution of optically active (E) and (Z)-allylic acetates. J Org Chem 51:723–727
86. Bäckvall JE et al (1991) On the mechanism of palladium(0)-catalyzed reactions of allylic substrates with nucleophiles. Origin for the loss of stereospecificity. Isr J Chem 31:17–24
87. Pedersen TM et al (2001) Enantioconvergent synthesis by sequential asymmetric Horner-Wadsworth-Emmons and palladium-catalyzed allylic substitution reactions. J Am Chem Soc 123:9738–9742
88. Strand D et al (2006) Divergence en route to nonclassical annonaceous acetogenins. Synthesis of pyranicin and pyragonicin. J Org Chem 71:1879–1891
89. Filipuzzi S et al (2008) Structure, bonding and dynamics of several palladium η^3-allyl carbene complexes. Organometallics 27:437–444
90. You S-L et al (2002) Role of planar chirality of S, N- and P, N ferrocene ligands in palladium-catalyzed allylic substitutions. J Org Chem 67:4684–4695
91. Faller JW et al (2001) Rearrangement in allylpalladium complexes with hemilabile chelating ligands. Helv Chim Acta 84:3031–3042
92. Guerrero A et al (2004) Apparent allylic rotation in new allylpalladium(II) complexes with pyrazolyl N-donor ligands. Eur J Inorg Chem 549–556
93. Jalón FA et al (2005) Apparent allyl rotation and Pd-N bond rupture in allylpalladium complexes with N-donor ligands – evidence of an associative mechanism. Eur J Inorg Chem 100–109
94. Montoya V et al (2007) New (η^3-Allyl)palladium complexes with pyridylpyrazole ligands: synthesis, characterization, and study of the influence of N1 substituents on the apparent allyl rotation. Organometallics 26:3183–3190

95. Johansson C et al (2010) Memory and dynamics in Pd-catalyzed allylic alkylation with P, N-ligands. Tetrahedron: Asymmetry 21:1585–1592
96. Kollmar M et al (2001) (Phosphanyloxazoline)palladium complexes, part I: (η^3-1,3-dialkylallyl)(phosphanyloxazoline) palladium complexes: x-ray crystallographic studies, NMR investigations, and quantum-chemical calculations. Chem Eur J 7:4913–4927
97. Cho C-W et al (2006) Studies on the structure and equilibration of (π-allyl)palladium complexes of phosphino(oxazolinyl)ferrocene ligands. Tetrahedron: Asymmetry 17:2240–2246
98. Butts CP et al (2009) Structure-based rationale for selectivity in the asymmetric allylic alkylation of cycloalkenyl esters employing the Trost 'Standard Ligand' (TSL): isolation, analysis and alkylation of the monomeric form of the cationic η^3-cyclohexenyl complex $[(\eta^3\text{-c-}C_6H_9)Pd(TSL)]^+$. J Am Chem Soc 131:9945–9957
99. Granberg KL, Bäckvall J-E (1992) Isomerization of (π-allyl)palladium complexes via nucleophilic displacement by palladium(0). A common mechanism in palladium(0)-catalyzed allylic substitution. J Am Chem Soc 114:6858–6863
100. Takahashi T et al (1984) Chirality transfer from C-O to C-C in the palladium catalyzed ScN' reaction of (*E*)- and (*Z*)-allylic carbonates with carbonucleophiles. Tetrahedron Lett 25:5921–5924
101. Lemaire S et al (2004) Pyrrolizidine alkaloids by intramolecular palladium-catalysed allylic alkylation: synthesis of (±)-isoretronecanol. Eur J Org Chem 2840–2847
102. Fiaud JC, Legros JY (1987) New method for the classification of nucleophiles in the palladium-catalyzed substitution of allylic acetates. J Org Chem 52:1907–1911
103. Trost BM, Verhoeven TR (1980) Allylic alkylation. Palladium-catalyzed substitutions of allylic carboxylates. Stereo- and regiochemistry. J Am Chem Soc 102:4730–4743
104. Trost BM, Verhoeven TR (1976) New synthetic reactions. Catalytic vs. stoichiometric allylic alkylation. Stereocontrolled approach to steroid side chain. J Am Chem Soc 98:630–632
105. Tsuji J et al (1980) Preparation of five- and six-membered cyclic ketones by the palladium-catalyzed cyclization reaction. Application to methyl dihydrojasmonate synthesis. Tetrahedron Lett 21:1475–1478
106. Giambastiani G, Poli G (1998) Palladium catalyzed alkylation with allylic acetates under neutral conditions. J Org Chem 63:9608–9609
107. Poli G et al (1999) Palladium-catalyzed allylic alkylations via titanated nucleophiles: a new early-late heterobimetallic system. J Org Chem 64:2962–2965
108. Trost BM, Murphy DJ (1985) A model for metal-templated catalytic asymmetric induction via π-allyl fragments. Organometallics 4:1143–1145
109. Trost BM, Matthew L, Crawley ML (2011) Enantioselective allylic substitutions in natural product synthesis. Top Organomet Chem doi:10.1007/3418_2011_13
110. Michelet V et al (2002) Synthesis of natural products and biologically active compounds via allylpalladium and related derivatives. In: Negishi E (ed) Handbook of organopalladium chemistry for organic synthesis, vol 2. Wiley, New York
111. Trost BM, Verhoeven TR (1977) Cyclizations via organopalladium intermediates. Macrolide formation. J Am Chem Soc 99:3867–3868
112. Bui The Thuong M et al (2007) New access to kainic acid via intramolecular palladium-catalyzed allylic alkylation. Synlett 1521–1524
113. Wade PA et al (1982) Palladium catalysis as a means for promoting the allylic C-alkylation of nitro compounds. J Org Chem 47:365–367
114. Braun M, Meier T (2006) New developments in stereoselective palladium-catalyzed allylic alkylations of preformed enolates. Synlett 661–676
115. Åkermark B, Jutand A (1981) Addition of ketone enolates to π-allylpalladium compounds. Stereochemistry and scope of the reaction. J Organomet Chem 217:C41–C43
116. Fiaud JC, Malleron JL (1981) Ketone enolates as nucleophiles in palladium-catalysed allylic alkylation. J Chem Soc, Chem Commun 1159–1160

117. Elliot MR et al (1998) Regio- and stereoselective palladium(0)-catalyzed alkylation of vinyloxiranes with non-stabilized lithium esters enolates nucleophiles. A direct access to highly functionalized allylic alcohols. Tetrahedron Lett 39:8849–8852
118. Braun M et al (2000) Diastereoselective and enantioselective palladium-catalyzed allylic substitution with nonstabilized ketone enolates. Angew Chem Int Ed 39:3494–3497
119. Negishi E et al (1982) Highly regio- and stereospecific palladium-catalyzed allylation of enolates derived from ketones. J Org Chem 47:3188–3190
120. Negishi E, John RA (1983) Countercation effects on the palladium-catalyzed allylation of enolates. J Org Chem 48:4098–4102
121. Trost BM, Keinan E (1980) Enolstannanes as electrofugal groups in allylic alkylation. Tetrahedron Lett 21:2591–2594
122. Trost BM, Self CR (1984) On the palladium-catalyzed alkylation of silyl-substituted allyl acetates with enolates. J Org Chem 49:468–473
123. Tsuji J et al (1983) Palladium-catalyzed allylation of ketones and aldehydes with allylic carbonates via silyl enol ethers under neutral conditions. Chem Lett 1325–1326
124. Kazmaier U, Zumpe FL (1999) Chelated enolates of amino acid esters - efficient nucleophiles in palladium-catalyzed allylic substitutions. Angew Chem Int Ed 38:1468–1470
125. Shimizu I et al (1980) Palladium-catalyzed rearrangement of allylic esters of acetoacetic acid to give γ, δ-unsaturated methyl ketones. Tetrahedron Lett 21:3199–3202
126. Tsuda T et al (1980) Facile generation of a reactive palladium(II) enolate intermediate by the decarboxylation of palladium(II) β-ketocarboxylate and its utilization in allylic acylation. J Am Chem Soc 102:6381–6384
127. Tsuji J et al (1983) Palladium-catalyzed allylation of ketones and aldehydes via allyl enol carbonates. Tetrahedron Lett 24:1793–1796
128. Tunge JA, Burger EC (2005) Transition metal catalyzed decarboxylative additions of enolates. Eur J Org Chem 1715–1726
129. You SL, Dai LX (2006) Enantioselective palladium-catalyzed decarboxylative allylic alkylations. Angew Chem Int Ed 45:5246–5248
130. Fiaud JC, Aribi-Zouioueche L (1982) Stereochemistry in the palladium-catalyzed rearrangement of some cyclohex-2-enyl acetoacetates. Tetrahedron Lett 23:5279–5282
131. Trost BM et al (2009) Palladium-catalyzed decarboxylative asymmetric allylic alkylation of enol carbonates. J Am Chem Soc 131:18343–18357
132. Trost BM et al (2010) Catalytic asymmetric allylic alkylation employing heteroatom nucleophiles: a powerful method for C-X bond formation. Chem Sci 1:427–440
133. Nagano T, Kobayashi S (2009) Palladium-catalyzed allylic amination using aqueous ammonia for the synthesis of primary amines. J Am Chem Soc 131:4200–4201
134. Murahashi SI et al (1988) Palladium(0)-catalyzed hydroxylamination of allyl esters. Synthesis of N-allylhydroxylamines and secondary allylamines. Tetrahedron Lett 29:2973–2976
135. Murahashi SI et al (1986) Palladium(0) catalyzed azidation and amination of allyl acetates. Selective synthesis of allyl azides and primary allyl amines. Tetrahedron Lett 27:227–230
136. Inoue Y et al (1985) Direct N-allylation of amides with 2-allylisourea catalyzed by palladium (0). Bull Chem Soc Jpn 58:2721–2722
137. Cerezo S et al (1998) Palladium(0)-catalyzed allylation of highly acidic and non-nucleophilic arenesulfonamides, sulfamide and cyanamide. I. Tetrahedron 54:14869–14884
138. Byström SE et al (1985) Synthesis of protected allylamines via palladium-catalyzed amide addition to allylic substrates. Tetrahedron Lett 26:1749–1752
139. Connell RD et al (1988) An efficient, palladium-catalyzed route to protected allylic amines. J Org Chem 53:3845–3849
140. Inoue Y et al (1984) N-Allylation of imides catalyzed by palladium(0). Bull Chem Soc Jpn 57:3021–3022
141. Trost BM, Sudhakar AR (1987) A cis hydroxyamination equivalent: application to the synthesis of (−)-acosamine. J Am Chem Soc 109:3792–3794

142. Trost BM et al (1988) A transition-metal-controlled synthesis of (±)-aristeromycin and (±)-2',3'-diepi-aristeromycin. An unusual directive effect in hydroxylations. J Am Chem Soc 110:621–622
143. Deardorff DR et al (1985) A palladium-catalyzed route to mono- and diprotected cis-2-cyclopentene-1,4-diols. Tetrahedron Lett 26:5615–5618
144. Trost BM, Organ MG (1994) Deracemization of cyclic allyl esters. J Am Chem Soc 116:10320–10321
145. Trost BM et al (1993) Triphenylsilanol as a water surrogate for regioselective Pd catalyzed allylations. Tetrahedron Lett 34:1421–1424
146. Trost BM, McEachern EJ (1999) Inorganic carbonates as nucleophiles for the asymmetric synthesis of vinylglycidols. J Am Chem Soc 121:8649–8650
147. Trost BM, Scanlan TS (1986) Synthesis of allyl sulfides via a palladium mediated allylation. Tetrahedron Lett 27:4141–4147
148. Auburn PR et al (1986) Homogeneous catalysis. Production of allyl alkyl sulphides by palladium mediated allylation. J Chem Soc, Chem Commun 146–147
149. Lu X, Ni Z (1987) Palladium-catalyzed synthesis of allylic and benzylic sulfides from the corresponding dithiocarbonates. Synthesis 66–68
150. Goux C et al (1992) Synthesis of allyl aryl sulphides by palladium(0)-mediated alkylation of thiols. Tetrahedron Lett 33:8099–8102
151. Maitro G et al (2006) Preparation of allyl sulfoxides by palladium-catalyzed allylic alkylation of sulfenate anions. J Org Chem 71:7449–7454
152. Inomata K et al (1981) Regio- and stereocontrolled synthesis of allylic p-tolyl sulfones catalyzed by palladium(0) complex. Chem Lett 1357–1360
153. Divekar S et al (1999) Palladium-catalyzed synthesis of allylic thioacetates. A convenient access to allylic thiols. Tetrahedron 55:4369–4376
154. Lüssem BJ, Gais HJ (2004) Palladium-catalyzed enantioselective allylic alkylation of thiocarboxylate ions: asymmetric synthesis of allylic thioesters and memory effect/dynamic kinetic resolution of allylic esters. J Org Chem 69:4041–4052
155. Moreno-Mañas M et al (1993) Palladium-catalyzed allylation of pyrimidine-2,4-diones (uracils) and of 6-membered heterocyclic ambident sulfur nucleophiles. Tetrahedron 49:1457–1464
156. Arredondo Y et al (1993) Palladium-catalyzed allylation of 5-membered heterocyclic ambident sulfur nucleophiles. Tetrahedron 49:1465–1470
157. Yamada Y et al (1979) Palladium-catalyzed thiono-thiolo allylic rearrangement of O-allyl phosphoro- and phosphonothionates. Tetrahedron Lett 20:5015–5018
158. Hiroi K et al (1984) Palladium-catalysed allylic sulphinate-sulphone rearrangements. Asymmetric induction in the palladium-catalysed transfer of chiral sulphinates to sulphones. J Chem Soc, Chem Commun 303–305
159. Tamaru Y et al (1990) Palladium-catalyzed [2,3] rearrangement of alkyl allyl sulfites to alkyl allylsulfonates. J Org Chem 55:1823–1829
160. Fiaud JC (1983) The palladium-catalysed reaction of lithium diphenylthiophosphides with allylic carboxylates. A stereoselective synthesis of phosphine sulfides. J Chem Soc, Chem Commun 1055–1056
161. Butti P et al (2008) Palladium-catalyzed enantioselective allylic phosphination. Angew Chem Int Ed 47:4878–4881
162. Moreno-Mañas M, Trius A (1981) Double bond formation by one pot palladium induced reactions between aldehydes, allylic alcohols and triphenylphosphine. Tetrahedron Lett 22:3109–3112
163. Malet R et al (1992) Palladium-catalyzed preparation of dialkyl allylphosphonates. A new preparation of diethyl 2-oxoethylphosphonate. Synth Commun 22:2219–2228
164. Matsushita H, Negishi E (1981) Palladium-catalyzed stereo- and regiospecific coupling of allylic derivatives with alkenyl- and arylmetals. A highly selective synthesis of 1,4-dienes. J Am Chem Soc 103:2882–2884

165. Miyaura N et al (1980) The palladium-catalyzed cross-coupling reaction of 1-alkenylboranes with allylic or benzylic bromides. Convenient syntheses of 1,4-alkadienes and allylbenzenes from alkynes via hydroboration. Tetrahedron Lett 21:2865–2868
166. Yatagai H (1980) The reaction of alkenylboranes with palladium acetate. Stereoselective synthesis of olefinic derivatives. Bull Chem Soc Jpn 53:1670–1676
167. Hayashi T et al (1981) Regioselective allylation of a Grignard reagent catalysed by phosphine-nickel and -palladium complexes. J Chem Soc, Chem Commun 313–314
168. Yoshida J et al (1978) Stereoselective preparation of 1,4-dienes by palladium catalyzed allylation of (*E*)-alkenylpentafluorosilicates. Application to total synthesis of (±)-recifeiolide. Tetrahedron Lett 19:2161–2164
169. Hayasi Y et al (1981) Ligand control in palladium-catalyzed coupling reactions between organozirconium compounds and allylic species. Tetrahedron Lett 22:2629–2632
170. Tsuji Y et al (1998) Palladium-complex-catalyzed cyanation of allylic carbonates and acetates using trimethylsilyl cyanide. Organometallics 17:4835–4841
171. Jones DN, Knox SD (1975) Stereochemistry of formation and reduction of π-allyl palladium chloride complexes from steroidal olefins. J Chem Soc, Chem Commun 165–166
172. Keinan E, Greenspoon N (1982) Organo tin nucleophiles III. Palladium catalyzed reductive cleavage of allylic heterosubstituents with tin hydride. Tetrahedron Lett 23:241–244
173. Mandai T et al (1994) A novel method for stereospecific generation of natural C-17 stereochemistry and either C-20 epimer in steroid side chains by palladium-catalyzed hydrogenolysis of C-17 and C-20 allylic carbonates. Tetrahedron 50:485–486
174. Hey H, Arpe HJ (1973) Removal of allyl groups by formic acid catalyzed by (triphenylphosphane)palladium. Angew Chem Int Ed Engl 12:928–929
175. Tsuji J, Yamakawa T (1979) A convenient method for the preparation of 1-olefins by the palladium catalyzed hydrogenolysis of allylic acetates and allylic phenyl ethers with ammonium formate. Tetrahedron Lett 20:613–616
176. Krafft ME et al (1998) Heteroatom-directed, palladium-catalyzed, regioselective allylation: substitution with inversion. J Org Chem 63:1748–1749
177. Krafft ME et al (1998) Regioselective additions to π-allyl metal complexes. Pure Appl Chem 70:1083–1090
178. Krafft ME, Lucas MC (2003) Palladium-catalyzed, heteroatom assisted, regioselective cyclizations. Chem Commun 1232–1233
179. Itami K et al (2001) Regioselective catalytic allylic alkylation directed by removable 2-PyMe$_2$Si group. J Am Chem Soc 12:6957–6958
180. Cook GR et al (2003) Hydrogen bond directed highly regioselective palladium-catalyzed allylic substitution. J Am Chem Soc 125:5115–5120
181. Poli G, Madec D (2004) Unusual regioselectivities in palladium-catalyzed allylic substitution. Chemtracts - Org Chem 17:104–114
182. Poli G, Scolastico C (1999) New modes of regiocontrol in palladium-catalyzed allylic alkylations. Chemtracts - Org Chem 12:822–836
183. van Haaren RJ et al (2002) The effect of ligand donor atoms on the regioselectivity in the palladium catalyzed allylic alkylation. Inorganica Chimica Acta 327:108–115
184. Oslob JD et al (1997) Steric influences on the selectivity in palladium-catalyzed allylation. Organometallics 16:3015–3021
185. Branchadell V et al (1999) Density functional study on the regioselectivity of nucleophilic attack in 1,3-disubstituted (diphosphino)(η^3-allyl)palladium cations. Organometallics 18:4934–4941
186. Åkermark B et al (1984) Alkylation of (π-allyl)palladium systems. Mechanism and regiocontrol. Organometallics 3:679–682
187. Åkermark B et al (1987) Ligand effects and nucleophilic addition to (η^3-allyl)palladium Complexes. A carbon-13 Nuclear Magnetic Resonance Study. Organometallics 6:620–628
188. Goldfuss B (2006) Electronic differentiations in palladium-catalyzed allylic substitutions. J Organomet Chem 691:4508–4513

189. Lange DA, Goldfuss B (2007) Electronic differentiation competes with transition state sensitivity in palladium-catalyzed allylic substitutions. Beilstein J Org Chem. doi:10.1186/1860-5397-3-36
190. Dawson GJ et al (1993) Asymmetric palladium catalysed allylic substitution using phosphorus containing oxazoline ligands. Tetrahedron Lett 34:3149–3150
191. Sprinz J, Helmchen G (1993) Phosphinoaryl- and phosphinoalkyloxazolines as new chiral ligands for enantioselective catalysis: very high enantioselectivity in palladium catalyzed allylic substitutions. Tetrahedron Lett 34:1769–1772
192. von Matt P, Pfaltz A (1993) Chiral phosphinoaryldihydrooxazoles as ligands in asymmetric catalysis: Pd-catalyzed allylic substitution. Angew Chem Int Ed 32:566–568
193. Brown JM et al (1994) Mechanistic and synthetic studies in catalytic allylic alkylation with palladium complexes of 1-(2-diphenylphosphino-1-naphthyl)isoquinoline. Tetrahedron 50:4493–4506
194. Ansell J, Wills M (2002) Enantioselective catalysis using phosphorus-donor ligands containing two or three P–N or P–O bonds. Chem Soc Rev 31:259–268
195. Prétôt R, Pfaltz A (1998) New ligands for regio- and enantiocontrol in Pd-catalyzed allylic alkylations. Angew Chem Int Ed 37:323–325
196. Hilgraf R, Pfaltz A (1999) Chiral bis(N-tosylamino)phosphine- and TADDOL-phosphite-oxazolines as ligands in asymmetric catalysis. Synlett 1814–1816
197. Helmchen G, Pfaltz A (2000) Phosphinooxazolines – a new class of versatile, modular P, N-ligands for asymmetric catalysis. Acc Chem Res 33:336–345
198. Pamies O et al (2005) New phosphite-oxazoline ligands for efficient Pd-catalyzed substitution reactions. J Am Chem Soc 127:3646–3647
199. You SL et al (2001) Highly regio- and enantioselective Pd-catalyzed allylic alkylation and amination of monosubstituted allylic acetates with novel ferrocene P, N-ligands. J Am Chem Soc 123:7471–7472
200. van Haaren RJ et al (2000) Bite angle effect of bidentate P–N ligands in palladium catalysed allylic alkylation. J Chem Soc, Dalton Trans 1549–1554
201. Kawatsura M et al (1998) Regiocontrol in palladium-catalysed allylic alkylation by addition of lithium iodide. J Chem Soc, Chem Commun 217–218
202. Sjögren M et al (1994) Stereo- and regiocontrol in palladium-catalyzed allylic alkylation using 1,10-phenanthrolines as ligands. Organometallics 13:1963–1971
203. Åkermark B et al (1990) Ligand-induced selective stabilization of the anti isomer in (η^3-allyl)palladium complexes: an attempt to control the E-Z stereochemistry in palladium-promoted allylic substitutions. J Am Chem Soc 112:4587–4588
204. Sjögren MPT et al (1992) Selective stabilization of the *anti*-isomer of η3-allyl palladium and platinum complexes. Organometallics 11:3954–3964
205. Kazmaier U, Zumpe F (2000) Palladium-catalyzed allylic alkylations without isomerization-dream or reality? Angew Chem Int Ed 39:802–804
206. Kazmaier U et al (2008) Influences on the regioselectivity of palladium-catalyzed allylic alkylations. Chem Eur J 14:1322–1329
207. Krämer K et al (2006) Isomerization-free allylic alkylations of terminal π-allyl palladium complexes. J Org Chem 71:8950–8953
208. Fiaud JC, Malleron JL (1981) To what extent is a π-allylic intermediate involved in some palladium-catalyzed alkylations ? Tetrahedron Lett 22:1399–1402
209. Trost BM, Schmuff NR (1981) On the mechanism of allylic alkylations catalyzed by palladium. Tetrahedron Lett 22:2999–3000
210. Trost BM, Bunt RC (1996) On the question of the symmetry of formally symmetrical π-(allyl)palladium cationic intermediates in allylic alkylations. J Am Chem Soc 118:235–236
211. Lloyd-Jones GC, Stephen SC (1998) Memory effects in Pd-catalyzed allylic alkylation: stereochemical labeling through isotopic desymmetrization. Chem Eur J 4:2539–2549
212. Poli G, Scolastico C (1999) Memory effects in Pd-catalyzed allylic alkylations. Chemtracts - Org Chem 12:837–845

213. Jutand A (2003) The use of conductivity measurements for the characterization of cationic Palladium complexes and for the determination of kinetic and thermodynamic data in Palladium-catalyzed reactions. Eur J Inorg Chem 2017–2040
214. Sprinz J et al (1994) Catalysis of allylic substitutions by Pd complexes of oxazolines containing an additional P, S, or Se Center. X-ray crystal structures and solution structures of chiral π-allyl palladium complexes of phosphinoaryloxazolines. Tetrahedron Lett 35:1523–1526
215. Lloyd-Jones GC, Stephen SC (1998) Chloride ion effects on kinetic resolution in Pd-catalysed allylic alkylation. Chem Commun 2321–2322
216. Kozuch S et al (2005) What makes for a good catalytic cycle? A theoretical study of the role of an anionic palladium(0) complex in the cross coupling of an aryl halide with an anionic nucleophile. Organometallics 24:2319–2330
217. Fristrup P et al (2008) On the nature of the intermediates and the role of chloride ions in Pd-catalyzed allylic alkylations: added insight from density functional theory. J Phys Chem A 112:12862–12867
218. Fristrup P et al (2006) Deconvoluting the memory effect in Pd-catalyzed allylic alkylation: effect of leaving group and added chloride. Chem Eur J 12:5352–5360
219. Hayashi T et al (1997) Regio- and enantio-selective allylic alkylation catalysed by achiral monophosphine–palladium complex. Chem Commun 561–562
220. Hayashi T et al (1998) Retention of regiochemistry of allylic esters in palladium-catalyzed allylic alkylation in the presence of a MOP ligand. J Am Chem Soc 120:1681–1687
221. Goldfuss B, Kazmaier U (2000) Electronic differentiations in palladium alkene complexes: trans-phosphine preference of allylic leaving groups. Tetrahedron 56:6493–6496
222. Blacker AJ (1999) Use of tricyclohexylphosphine to control regiochemistry in palladium-catalyzed allylic alkylation. Org Lett 1:1969–1971
223. Acemoglu L, Williams JMJ (2001) Remarkable ligand effects in regioselective palladium-catalysed allylic substitution reactions. Adv Synth Catal 343:75–77
224. Faller JW, Sarantopoulos N (2004) Retention of configuration and regiochemistry in allylic alkylations via the memory effect. Organometallics 23:2179–2185
225. Trost BM, Toste FD (1999) Regio and enantioselective allylic alkylation of an unsymmetrical substrate: a working model. J Am Chem Soc 121:4545–4554
226. Wang Y et al (2003) Backbone effect of MAP ligands on their coordination patterns with palladium(II). Organometallics 22:1856–1862
227. Svensen N et al (2007) Memory effects in palladium-catalyzed allylic alkylations of 2-cyclohexen-1-yl acetate. Adv Synth Catal 349:2631–2640
228. Butts CP et al (1999) Robust and catalytically active mono- and bis-Pd-complexes of the 'Trost modular ligand'. Chem Commun 1707–1708
229. Fairlamb IJS et al (2002) Analysis of stereochemical convergence in asymmetric Pd-catalysed allylic alkylation reactions complicated by halide and memory effects. Chem Eur J 8:4443–4453
230. Lloyd-Jones GC et al (2004) Coordination of the Trost modular ligand to palladium allyl fragments: oligomers, monomers, and memory effects in catalysis. Pure Appl Chem 76:589–601
231. Fairlamb IJS, Lloyd-Jones GC (2000) On the effect of catalyst loading in Pd-catalysed allylic alkylation. Chem Commun 2447–2448
232. Trost BM, Surivet JP (2000) Diastereo- and enantioselective allylation of substituted nitroalkanes. J Am Chem Soc 122:6291–6292
233. Gais HJ et al (2003) Highly selective palladium catalyzed kinetic resolution and enantioselective substitution of racemic allylic carbonates with sulfur nucleophiles: asymmetric synthesis of allylic sulfides, allylic sulfones, and allylic alcohols. Chem Eur J 9:4202–4221
234. Lloyd-Jones GC (2001) Isotopic desymmetrisation as a stereochemical probe. Synlett 161–183

235. Fairlamb IJS et al (2004) Isotopic desymmetrization in the study of homogeneous catalysis. Phos Sulf Sil 179:907–910
236. Trost BM, Van Vranken DL (1996) Asymmetric transition metal-catalyzed allylic alkylations. Chem Rev 96:395–422
237. Lu Z, Ma S (2008) Metal-catalyzed enantioselective allylation in asymmetric synthesis. Angew Chem Int Ed 47:258–297
238. Bantreil X et al (2011) γ- and δ-Lactams via palladium-catalyzed intramolecular allylic alkylation: enantioselective synthesis, NMR investigations, and DFT rationalization. Chem Eur J 17:2885–2896
239. Fiaud JC, Legros JY (1990) Substrate leaving group control of the enantioselectivity in the palladium-catalyzed asymmetric allylic substitution of 4-alkyl-1-vinylcyclohexyl derivatives. J Org Chem 55:4840–4846
240. Trost BM, Dong G (2006) New class of nucleophiles for palladium-catalyzed asymmetric allylic alkylation. Total synthesis of agelastatin A. J Am Chem Soc 128:6054–6055
241. Trost BM, Asakawa N (1999) An asymmetric synthesis of the vitamin E core by Pd catalyzed discrimination of enantiotopic alkene faces. Synthesis 1491–1494
242. Trost BM, Toste FD (2003) Palladium catalyzed kinetic and dynamic kinetic asymmetric transformations of γ-acyloxybutenolides. Enantioselective total synthesis of (+)-aflatoxin B_1 and B_{2a}. J Am Chem Soc 125:3090–3100
243. Trost BM, Oslob JD (1999) Asymmetric synthesis of (−)-Anatoxin-a via an asymmetric cyclization using a new ligand for Pd-catalyzed alkylations. J Am Chem Soc 121:3057–3064
244. He XC et al (2001) Studies on the asymmetric synthesis of huperzine A. Part 2: highly enantioselective palladium-catalyzed bicycloannulation of the β-keto-ester using new chiral ferrocenylphosphine ligands. Tetrahedron: Asymmetry 12:3213–3216
245. Kitagawa O et al (2006) Catalytic asymmetric desymmetrization of *meso*-diamide derivatives through enantioselective *N*-allylation with a chiral π-allyl Pd catalyst: improvement and reversal of the enantioselectivity. J Org Chem 71:2524–2527
246. Hayashi T et al (1988) Asymmetric cyclization of 2-butenylene dicarbamates catalyzed by chiral ferrocenylphosphine-palladium complexes: catalytic asymmetric synthesis of optically active 2-amino-3-butenols. Tetrahedron Lett 29:99–102

Top Organomet Chem (2012) 38: 65–94
DOI: 10.1007/3418_2011_8

Published online: 21 June 2011

Computational Insights into Palladium-Mediated Allylic Substitution Reactions

Jonatan Kleimark and Per-Ola Norrby

Abstract Allyl palladium complexes have a rich chemistry. Many aspects of their structure and reactivity have been studied computationally. This chapter gives an overview of the history in this field, from structural studies and the effect of ligands and substituents, to the rich reactivity of the title complexes. The latter includes complex formation, reactions with nucleophiles and electrophiles, and dynamic equilibria. An important focus area has been the Tsuji–Trost reaction, in particular asymmetric versions thereof. A brief overview of computational methods, aimed at modeling novices, can be found in the introduction.

Keywords Density functional theory · Molecular mechanics · Palladium-mediated allylation · Quantum mechanics · Reaction mechanisms

Contents

J. Kleimark and P.-O. Norrby (✉)
Department of Chemistry, University of Gothenburg, Kemigården 4, 412 96 Göteborg, Sweden
e-mail: pon@chem.gu.se

1 Introduction

Historically, allyl metal complexes (e.g., allyl Grignard reagents) have been considered strong nucleophiles, but already in the early 70s Tsuji demonstrated that (η^3-allyl)palladium complexes reacted as allylic electrophiles with stabilized carbanions such as malonates [1, 2]. This could be considered an umpolung, since allyl palladium complexes can be obtained from reaction of nucleophilic allyls, such as Grignards, with Pd^{II} salts. The nucleophilic attack on the (η^3-allyl)palladium moiety produces Pd^0, and by introduction of electrophilic allylic substrates capable of reacting with Pd^0, such as acetates, a catalytic process can be realized. Palladium will bind to two auxiliary ligands throughout the process (Scheme 1), allowing for efficient reactivity control by ligand design. The allylation process allows formation of a wide range of bonds (e.g., C–C, C–N, and C–O), but the reaction is subject to regio- and stereo-chemical selectivity issues. Therefore, there has been a great interest in subjecting the reaction to detailed theoretical studies to elucidate the underlying factors controlling the selectivity.

2 Quantum Mechanics

Computational studies have mostly focused on the Pd-allyl complexes, the species that reacts with the incoming nucleophiles (Scheme 1). The earliest studies utilized the Extended Hückel Molecular Orbital (EHMO) method. This qualitative quantum mechanical (QM) method does not allow accurate energy calculations, and therefore cannot reliably be used to determine geometries by energy minimization.

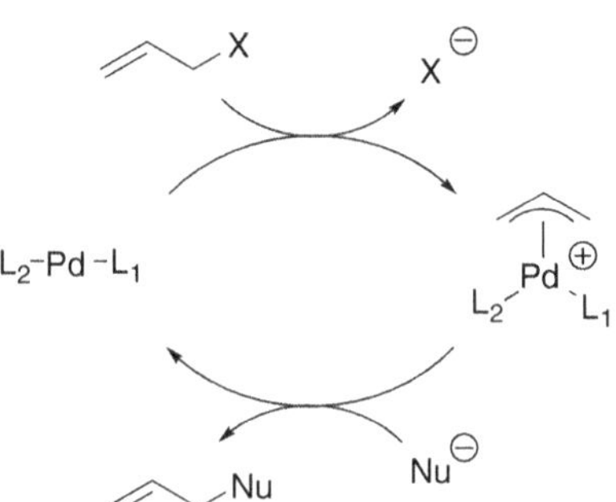

Scheme 1 Palladium-catalyzed allylation

However, when based on accurate structures, for example from neutron diffraction or, if hydrogen positions are corrected, from X-ray crystallography, the EHMO method can be used to determine shapes and relative energy levels of frontier orbitals, allowing for a correlation with chemical reactivity. Other semiempirical approaches were developed where strongly simplified QM schemes were augmented with empirical parameters, but in general the reliability for transition metal complexes was too low to allow anything except qualitative studies.

2.1 Hartree–Fock Calculations

The Hartree–Fock (HF) method enabled a breakthrough in organic chemistry. Using the HF method, the Schrödinger equation can be solved iteratively, with the approximation that each electron only interacts with the average density of all other electrons. In simple terms, the HF method ignores the fairly important fact that as an electron moves to any specific point in space, all other electrons must move away from that point, so-called *electron correlation*. Despite this strong approximation, HF recovers a large part of the total energy of the molecule, and the remaining error is fairly constant, so that the relative energy of different atom configurations is reasonably accurate. In chemical terms, this means that molecular geometries and reaction barriers can frequently be reasonably well described by the HF method. However, many cases are known where the variation in electron correlation is large enough to give significant errors when the HF results are compared to experiments. Specifically, the high density of electrons in heavier elements such as transition metals tends to make the HF method too unreliable. For closed shell species (such as most Pd complexes), it is still possible to obtain qualitative results, but quantitative predictions of molecular geometries or relative reactivities generally require more accurate methods.

2.2 Correlated Calculations

Even though the HF method gives an approximate wavefunction, it is still possible to evaluate it using the Schrödinger equation. More specifically, it is possible to introduce small perturbations of the wavefunction to obtain an improved solution. One way to introduce well-controlled perturbations is to mix the HF ground state with other low-energy states. The most easily implemented way of doing this uses the HF excited states. Compared to HF calculations, such correlated calculations are costly, but can lead to significantly improved results. These methods are generally termed Many-Body Perturbation Theory (MBPT), or Møller–Plesset (MP) theory. In the simplest implementation, the HF ground state is mixed with all single and double electron excitations, MP2. Even better results, but at vastly increased cost, are obtained if 3, 4, or 5 electrons are excited simultaneously, the MP3, MP4, and MP5 methods. Many studies of metal complexes have been performed at the MP2 level, frequently with validation of selected results at the MP4 level, or truncated versions thereof. It is possible to extrapolate perturbation theory, or to optimize the excited states variationally to arrive at even better wavefunctions, but such methods invariably carry a very high cost, and are in practice limited to very small model systems. Several such methods, such as CCSD(T) and MCPF, have been applied to Pd complexes [3], but detailing such methods goes beyond the scope of this text. The interested reader is referred to several illuminating texts on this subject [4, 5].

2.3 Density Functional Theory

An alternative to wavefunction theory that has gained in popularity in later years is based on Density Functional Theory (DFT). In short, DFT does not solve the Schrödinger equation for a wavefunction, but rather an equivalent equation for the electron density. For a purist, DFT is deficient in that we do not know the exact form of the underlying equation, even though we know many of the properties it must have, and we cannot systematically improve it in the same way we can with wavefunction theory. However, several breakthroughs in the understanding of DFT have now led to a method that is as fast as HF, but with an accuracy on a par with MP2.[1] Introduction of nonlocal terms (i.e., making the energy dependent also on the gradient of the density) produced a method that could at least equal HF in accuracy. In the 90s, Becke solved some of the known deficiencies in DFT by combining it with HF theory, producing a hybrid method that has been the standard ever since. Currently, a majority of all published computational studies of palladium complexes are based on Becke's hybridization scheme, in many combinations, of which B3LYP [6–8] seems to be the most popular. More recent improvements include accounting for van der Waals dispersion forces [9–12], a weakly attractive interaction that was unimportant for the small models used in early studies, but with a strong impact on relative energies in medium to large systems [13].

[1]This point can be discussed, the methods certainly are not equal. In some cases, MP2 will give better energies, in some cases a hybrid method such as B3LYP is more accurate, and simpler pure DFT methods such as BP86 generally give good geometries. All the methods can give large and very different errors in pathological cases. However, it is probably safe to say that if three different methods such as MP2, a GGA functional such as BP86, and a hybrid method such as B3LYP all agree, the result will be reliable.

2.4 Basis Sets

All quantum chemical approaches require mathematical descriptions of the spatial distribution of electrons, that is, orbitals. The most common approach is to construct the molecular orbitals from idealized atomic orbitals, the Linear Combination of Atomic Orbitals (LCAO) approach. Other methods have been used, such as plane wave and finite element approaches, but only LCAO will be discussed here. In the simplest cases (e.g., EHMO), rigid atomic orbitals are used directly as the basis set. However, accurate calculations require that the orbitals can change size and shape to accommodate shifts of electrons in a molecular environment. The most important modification is to include atomic orbitals of different sizes, allowing the self-consistent field (SCF) procedure to optimize the contribution of each. As a simple example, a calculation might include two different sizes of 1s-orbital on hydrogen. When the hydrogen is directly bound to Pd, it becomes more hydridic, and the program will accommodate the added electron density by including more of the large 1s-orbital, whereas a hydrogen bound to oxygen would use more of the smaller orbital (and correspondingly larger orbitals on oxygen). It is quite common to use one fixed orbital of each type for core electrons, and two different sizes for each orbital in the valence shell, a so-called split valence basis set. If each orbital exists in two differently sized versions, the basis set is termed double-ζ (acronym: DZ), if three copies are employed, it is triple-ζ (TZ), and so on. Sometimes a very large orbital, a so-called diffuse orbital, is added to accommodate anions. This is usually indicated by adding a "+" in the basis set name, or preceding the name by "aug-."

The shape of orbitals can be modified by mixing in a small amount of an orbital with a higher quantum number, similar to forming sp-hybrids. For example, adding a few percent of an appropriately sized p-orbital to an s-orbital will produce a new, s-like orbital that has been shifted away from the atomic nucleus, and therefore is better at forming bonds. Similarly, adding a small amount of d-orbital character to a p-orbital produces a slightly bent, p-like orbital with better ability to form π-bonds. This is called polarization, and is a requirement for good descriptions of chemical bonds. Almost all calculations today add at least a set of d-orbitals to all elements heavier than helium, and if hydrogen bonding is important, p-orbital polarization on hydrogen is required.

The many core electrons on transition metals cause a huge increase in computational time without adding much to the accuracy since they are fairly constant under most chemical changes. It is therefore common practice to replace the core electrons by a so-called Effective Core Potential (ECP).

Calculations at the HF or DFT level generally require at least a split valence basis set with polarization. Frequently, larger basis sets are tested without further geometry optimization (so-called single point calculations), to check whether the conclusions are stable with respect to improvements in the method. Correlated methods such as MP2 require better basis sets. Since this is very expensive (MP2 scales with N^5, where N is the number of orbitals employed), correlated calculations have sometimes been reported with smaller basis sets, but a healthy dose of skepticism is recommended when evaluating such results.

2.5 Solvent

The QM methods described so far are basically valid only for molecules or complexes isolated from the surrounding, in vacuo or, more popular, in "gas phase". However, homogeneous catalysis occurs in solvent. Most early studies ignored the solvent, or introduced one or a few solvent molecules in specific orientations. For a few metal complexes with a low and constant polarity, this treatment can be valid, but for ionic reactions, errors can be huge when the solvent is ignored. Structures, even of ionic intermediates, are usually good as long as they have no more than a single charge, but barriers for combination of two ionic species with opposite charge simply cannot be calculated in the gas phase; the combination is usually barrierless in the absence of solvent. Inclusion of several explicit solvent molecules rapidly makes the calculation intractable, in particular since all possible orientations of the solvent molecules must be sampled. An approximation that has gained in popularity is to employ continuum solvation models. Many brands of these models are available, but one of the most popular, the polarizable continuum model (PCM) [14, 15] encloses the entire molecule by a cavity dotted by point charges that have been parameterized to give a fair representation of the *average* influence of the solvent, alleviating the need for extensive sampling. Application of these models can be tricky, in particular for reactions where cavities are at the point of merging, and they certainly do not capture all aspects of solvation, but in general they improve results and allow location of structures that would undergo electrostatic collapse in the absence of solvent. In palladium allyl chemistry, solvation models were not employed until about a decade ago [16], but are now an essential component in describing these reactions.

2.6 Molecular Mechanics

Quantum mechanics is a great aid in understanding chemistry at the molecular level, but it has one serious drawback: the computational cost. Even today, accurate methods cannot be reliably applied to systems with more than a couple of hundred atoms, and in the 80s when QM was first applied to allyl-Pd complexes, the limit was more like 10 atoms. An alternative, introduced already in the 50s, is to ignore the movement of the electrons, assume that they are locked in chemical bonds with known properties, and look just at the movement of the atoms in the force field provided by the bonds. This method, molecular mechanics (MM), has been a standard tool in organic chemistry and biochemistry since the 70s. Today, MM allows studies of systems with several thousand atoms, and location of millions of conformers, but application in organometallic and inorganic chemistry has been more limited [17]. MM is strongly reliant on availability of accurate parameters describing the chemical bonds making up the structure. Bonds in organic structures are fairly consistent, meaning that parameters (e.g., the ideal length of a bond between two sp^3-hybridized carbons) can be transferred between different structures with reasonable reliability. Bonds to metals are much more varied, and usually parameters have to be derived for each system of interest, severely limiting the applicability of MM. However, (η^3-allyl)palladium complexes are consistent enough to allow derivation of widely applicable parameters, *vide infra*.

3 Structure

Computationally, structures are obtained by changing the geometry until the energy is minimized. To do this efficiently, the program must be able to calculate the derivatives of the energy with respect to atomic coordinates. Today, modeling programs do this automatically for all but the most complex correlation methods. Structure prediction only depends on being able to calculate relative energies of very similar structures, and therefore can be reliable already at fairly low levels of theory, with the caveat that strong charge separation requires the use of accurate solvation models. The best performance/cost ratio for this purpose today is obtained from DFT methods, but for closed shell palladium complexes, even the HF structures used in some earlier studies were qualitatively correct. Before these methods became generally available, computational studies generally employed crystal structures, focusing only on the shape and energy of frontier orbitals and charge distribution (*vide infra*).

3.1 η^3-Allyl Complexes

An early effort to use computational aid to gain insight into structures of palladium complexes was made in beginning of the 90s, when Norrby et al. developed an MM force field for the (η^3-allyl)palladium moiety (Fig. 1). The parameterization of the force field was based on X-ray structures, MCPF calculations for small model systems, and equilibrium data from NMR [18]. The resulting force field was able to reproduce the structures of known (η^3-allyl)palladium complexes, and used to predict *syn-anti* equilibria for substituted (η^3-crotyl)palladium complexes with

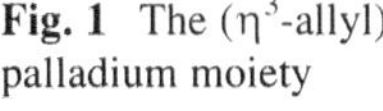
Fig. 1 The (η^3-allyl) palladium moiety

2,9-disubstituted phenantrolines as ligands. The results were in agreement with experimental results. The methodology has also been applied to the related (η^3-cyclohexenyl)palladium system, and proved to be able to solve the intricate equilibrium between the different boat- and chair-like conformations [19]. The force field has been further developed and several publications dealing with different ligands have followed [20, 21].

Using semiempirical methods, Pregosin et al. calculated the structure of a simplified (η^3-allyl)palladium, where the palladium atom was represented with a positive point charge. This rather crude method could reproduce the effect of substituents on the bonds in the allyl moiety. The effect was described as an electronic effect [22].

With the development of computational methods and resources, new methods could be employed. Szabó used MP2 and MP4 to calculate the structure of (η^3-allyl)palladium without any ligands on Pd [23]. This structure was compared with results from the more advanced MCPF level [18]. The bond distances and angles from the two methods agreed well. For example, the Pd-C(allyl) distances differ by no more than 0.014 Å.

Several different studies have been carried out where DFT calculations has been used to examine (η^3-allyl)palladium systems. Many diverse ligand combinations have been employed, such as P,P [24, 25], P,N [26, 27] and P,S [28]. In most DFT studies, the focus has been more on reactivity than on structure, and they will therefore be discussed in more detail later.

Through the work of Sakaki and Kurosawa, the nature of the dinuclear palladium (I) π-allyl (Fig. 2) has been investigated. When comparing to the monometallic species, the biggest difference is the much shorter Pd-C(allyl) bonds. Using MP2 calculations, this structural characteristic, along with other structural properties could be reproduced [29, 30].

Some effort has been put into sorting out one special case of the (η^3-allyl) palladium family, the (η^3-allyl)$_2$Pd structure (Fig. 3). The *cis:trans* ratio was the focus of a study by Casarin et al. The DFT calculations gave only a small energy difference of 0.8 kJ/mol between the geometries, with the *trans* orientation as the most favored [31]. Similar complexes with Pt, favoring *trans*, and Ni, favoring *cis*, were also investigated in the same study. The result for (η^3-allyl)$_2$Pd has been

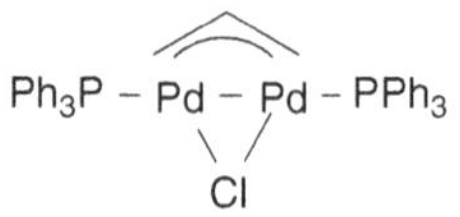

Ph_3P – Pd – Pd – PPh_3
Cl

Fig. 2 A dinuclear palladium complex

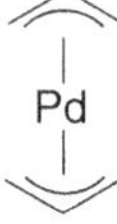

Fig. 3 (η^3-allyl)$_2$Pd

confirmed in a study by Szabó, although a slightly larger energy difference of 3 kJ/mol was reported [32]. For further studies on the bisallyl-palladium complexes, see Sect. 5.2 of this chapter concerning *Nucleophilic allylic substitution.*

3.2 η^1-Allyl Complexes

The allylic moiety can coordinate to palladium in two different ways, in a η^3-fashion, where the coordination is a combination of a σ-bond (alkyl-like) and a π-bond (alkene-like), or through a single σ-bond, giving a η^1-complex. Interestingly enough, even though η^3-allyl complexes are electrophilic, η^1-allyl seem to react as nucleophiles (*vide infra*).

Szabó has investigated the structure of the nucleophilic allylpalladium complexes computationally [32]. The bisallyl-palladium complex was scrutinized and, as mentioned earlier, an energy difference of ca 3 kJ/mol is found between the *cis* and *trans* forms. The nucleophilic reactivity of these complexes is observed in the presence of phosphines, modeled by Szabó as the PH_3 ligand. Addition of the ligand forces the bisallyl-palladium complex to go from the (η^3,η^3)-form to the (η^3,η^1)-form (Scheme 2). This process is found to be an exothermic process.

When expanding the calculations to the $(\eta^3,\eta^1$-octadienyl)palladium complex, the same process is also exothermic. The η^1 bond is found to be between the terminal carbon and the Pd, favored by 27 kJ/mol compared to the internal carbon.

The ligand effect is studied by exchanging the PH_3 (π-acceptor) for a NH_3 (σ-donor). The complexation energy for the σ-donor ligand is approximately 50 kJ/mol lower, indicating that the (η^3,η^1)-complexes are more easily formed using π-acceptor ligands. No significant structural differences could be found between the complexes with different ligands.

3.3 Calculated Structures vs. X-Ray and NMR

To verify the accuracy of the calculated structures, comparisons with X-ray structures and NMR results are useful tools. NMR is helpful when trying to elucidate unknown structures. It has the advantage over X-ray that it gives information about the solution structure, which in most cases is more relevant to reactivity than the solid structure. The downside is that the information is more difficult to analyze, is

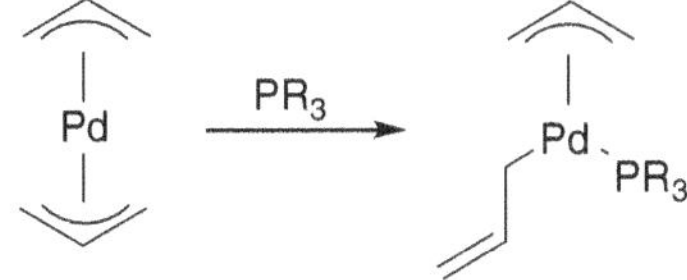

Scheme 2 The exothermic $(\eta^3,\eta^3)-(\eta^3,\eta^1)$ transformation

less accurate than X-ray analysis, and never shows the entire picture. Nonetheless, several studies have been carried out within the field.

Pregosin and co-workers used NOE results together with MOLCAD to derive a solution structure of a [(η^3-pinene)palladium BINAP] complex [33]. In a further study, good structural agreement between the experimental and calculated (MM) results could be shown for a similar structure. This time, the predictions from the calculations could be verified from NOE studies [34].

X-ray crystallography yields structures that are easy to compare to computational models. Distances, angles, and torsions can be evaluated, even if one has to pay attention to the potential pitfalls of solid state properties, such as packing effects.

In a study by Pregosin et al., a β-pinene allyl coordinated to a Pd-(BINAP) complex was studied. MM2 calculations were employed, and the results were compared with the existing X-ray structure. The calculated structure differs only slightly from the experimental structure, the main difference being the rotations of the phenyls in the BINAP-part. These results were further extended in an attempt to calculate the solvent structure of a similar complex carrying a (S,S)-Chiraphos ligand instead of the BINAP ligand (Fig. 4). The chiral array of the Chiraphos ligand proved to be different from the BINAP, as expected from experimental results [22].

Jonasson et al. employed DFT calculations when studying *cis*- and *trans*-4-acetoxy-[η^3-(1,2,3)-cyclohexenyl]palladium chloride dimers. The results were compared with X-ray structures and showed very small deviations from the measured distances in the crystal. Except for the Pd-Cl bond in the *cis*-structure, the deviation for all calculated parameters is within 0.03 Å. Several important features, such as the asymmetric allyl-palladium bonding were reproduced in the calculations [35].

In another example, Helmchen and co-workers investigated different allyl-phosphanyloxazoline)palladium complexes with both ab initio and DFT calculations and weighted these findings against the experimental results from NMR investigations and X-ray structures. Ab initio calculations were used when optimizing the different isomers, and the structural agreement was good, even if some bonds were slightly too long. To analyze the order of the energies of the different complexes, single point calculations using DFT were used. The ratios between the different isomers, obtained from NMR studies, were well reproduced [26, 27].

At the MP2 level of calculation, Sakaki et al. performed a study where they compared calculated structures of $PdCl(\eta^3\text{-}C_3H_4R)(PH_3)$ (R=H, Me, CN) and $[Pd(\eta^3\text{-}C_3H_5)(PH_3)_2]^+$ with X-ray structures. Good agreement was achieved, but it was important to include d-polarization to reach this result [30].

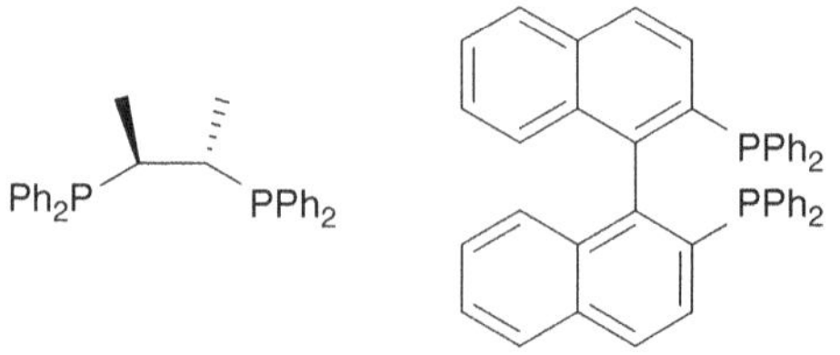

Ph2P PPh2

PPh2
PPh2

Fig. 4 Chiraphos and BINAP

3.4 Ligand Effects on Structure

Ligands can effect the (η^3-allyl)palladium complex in different ways, mainly by electronic *trans* influence and through steric effects. The *trans* influence of different ligands affects the bond length of the *trans* carbon-palladium bond in the (η^3-allyl) palladium complex. For example, in complexes of ligands with one phosphine and one other heteroatom (e.g., PN-ligands), the Pd–C bond *trans* to phosphorus is generally longer than the other terminal Pd–C bond. This effect is reproduced by calculations, for example in a study employing MP2/MP4 by Szabó, where a simple (η^3-allyl)palladium is modeled with a series of small ligands: Cl, F, NH_3, $H_2C{=}CH_2$, PH_3. The pure σ-donors, F^-, Cl^- and NH_3 induces a shortening of the *trans* Pd–C bond, with a difference of at least 0.07 Å between the strongest donor, F^-, and NH_3. The π-acceptor ligands $H_2C{=}CH_2$ and PH_3, on the other hand, elongates the *trans* Pd–C bonds compared to the unligated complex [23].

In another study, Blöchl et al. investigated a (η^3-allyl)palladium with two different ligands, PH_3 and a *N*-coordinated pyrazole. The asymmetry in the palladium carbon bond lengths was reproduced using the projector augmented wave method [36].

Ligands coordinated to the (η^3-allyl)palladium complex give rise to steric effects, which can affect the *syn-anti* configuration of a substituted allyl. The *syn* configuration is normally the most stable [37], but the ratio between the configurations differs when a bulky ligand is used. In a study by Norrby et al., the effect of substituted 1,10-phenantroline ligands was investigated. Employing a modified MM2 force field with a special parameter set for the (η^3-allyl)palladium moiety [18], it was found that substituents which provided bulk *in* the coordination plane would enhance the *anti* preference. This effect was also verified experimentally, yielding as much as 81% *anti* configuration with a simple 1-butenyl group. On the other hand, bulk outside the coordination plane gave a much lower preference for *anti* complexes [37]. Further studies by the same group revealed that a careful parameterization is needed to achieve good predications of different (η^3-allyl) palladium complexes. On the other hand, when a well-parameterized force field is in hand, the accuracy and versatility is very good [21].

4 Dynamics

Dynamics in (η^3-allyl)palladium complexes have mainly been studied by NMR techniques. The timescale of the relevant dynamic processes is suitable for investigation by coalescence or saturation transfer techniques. Some slower processes can also be seen by purification of one isomer followed by direct monitoring of the equilibration over minutes or hours. These were in fact some of the earliest exchange processes to be studied by NMR, and were reviewed already in 1975 by Vrieze [38]. The fast dynamic processes of (η^3-allyl)Pd complexes can roughly be

Fig. 5 Observable exchange processes in (η^3-allyl)Pd complexes

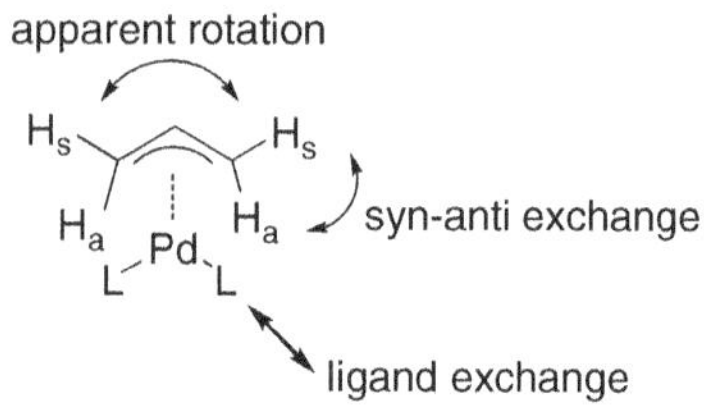

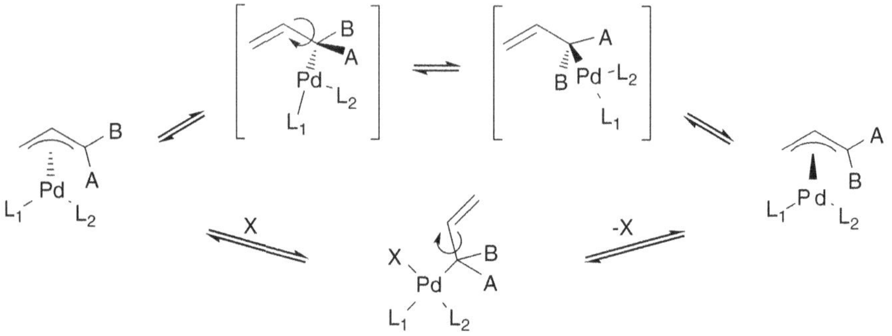

Scheme 3 η^3-η^1-η^3 isomerization, with and without assistance by added ligand X

divided into three classes: *syn-anti* exchange, apparent rotation, and ligand exchange (Fig. 5). There is also a slow exchange of allyl groups between Pd centers (*vide infra*).

4.1 Syn-anti *exchange*

As opposed to most other unsaturated systems, the terminal substituents on the (η^3-allyl)Pd moiety readily exchange on the NMR timescale. The process occurs through a bond rotation in a (η^1-allyl)Pd complex, and is therefore also known as η^3-η^1-η^3 or, in older nomenclature, π-σ-π isomerization, Scheme 3. The rate of the reaction is dramatically increased by added ligand, such as halide ions, since these coordinate to Pd and stabilize the η^1 complex [38]. As already discussed, when the ligands are strong enough to out-compete η^3 coordination, the square planar η^1 complexes can be observed [39–41]. It is important to point out that the η^3-η^1-η^3 isomerization does not exchange the positions of the ligands in relation to the allyl termini, even in the absence of added ligand. The Pd center stays square planar throughout the process, even in the unobservable η^1 complexes without added ligands where the geometry is more properly described as T shaped [42]. This is illustrated in Scheme 3, where it can be seen that ligand L_1 stays *trans* to the exchanging terminus. We also note that the chirality of the exchanging terminus stays constant, whereas the two remaining allyl carbons are inverted; the process can also be described as an exchange of coordinating π-faces of the alkene moiety in the (η^1-allyl)Pd complex.

Several computational studies have included an evaluation of the η^3-η^1 equilibration, mostly for cases where an added ligand coordinates to the empty coordination site in the (η^1-allyl)Pd complex [39–41]. When no additional ligand was present in the calculations, the (η^1-allyl)Pd complex was found to be a transition state for bond rotation, not a stable energy minimum [42], and thus cannot partake in additional processes, such as site rearrangements of the ligands. In the complexes studied, the rotation from η^3 to η^1 coordination mode will occur in the direction that will let the central hydrogen on the allyl interact agostically with the empty coordination site on Pd.

4.2 Apparent Rotation

The apparent rotation was first studied by ^{1}H NMR, and was initially termed "*syn-syn, anti-anti* exchange," from the observable behavior of the protons [38]. However, the overall process can also be seen as an exchange between the two auxiliary ligands on the (η^3-allyl)Pd complex. The process is distinct from the *syn-anti* exchange, and can easily be faster, in particular for cationic (η^3-allyl)Pd complexes in the presence of halide anions [43]. The exchange in complexes with chiral ligands clearly shows that the apparent rotation does not involve breaking of any palladium–carbon bonds [43, 44]. In some cases, an apparent rotation can be obtained by rapid ligand exchange (*vide infra*), but it has been shown that bidentate ligands stay coordinated throughout the process [45]. The mechanism has been suggested to be Berry pseudo-rotation in a penta-coordinated complex obtained by addition of a halide anion to Pd (the allyl is considered a bidentate ligand here, Scheme 4) [38]. This mechanism has been supported by X-ray of a postulated intermediate for chloride-assisted apparent rotation in an (η^3-allyl)Pd complex with

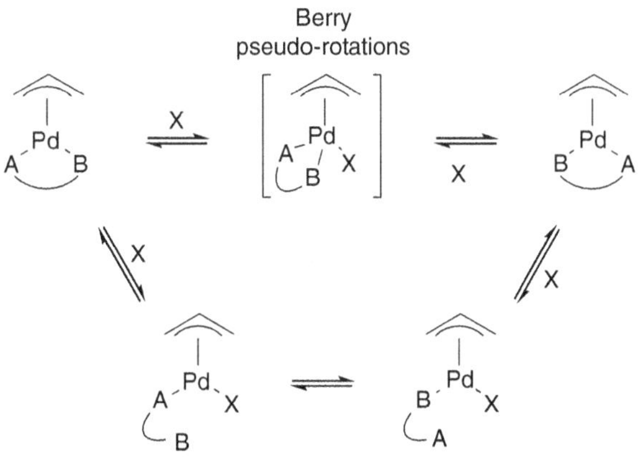

Berry
pseudo-rotations

Scheme 4 Proposed mechanisms for apparent rotation

a rigid, bidentate ligand [43]. An alternative suggestion from Pregosin and coworkers involve breaking the bond to one arm of the bidentate ligand, followed by ligand rearrangement and reformation of the bond to Pd [44]. The bond breaking to one ligand arm was conclusively shown by NMR exchange (Scheme 4) [46]. A recent computational study of the dynamics in (η^3-allyl)Pd complexes at the B3LYP level identified transition states that seemed to correspond to Berry pseudo-rotation processes, but QRC calculations [47] clearly showed breaking of palladium–ligand bonds [42], indicating that the apparent rotation is indeed composed of three successive ligand substitution processes. Reevaluation of the X-ray structure from Hansson et al. [43]. indicates that one bond to a dmphen nitrogen has indeed been broken, but is kept close to the palladium by the rigid framework of the dmphen ligand.

4.3 Ligand Exchange

In the presence of excess ligands, most (η^3-allyl)Pd complexes will undergo relatively rapid ligand exchange. This also includes reversible breaking of halide- or carboxylate-bridged dimers by added ligand. Ligand exchange can be associative (Scheme 5) or dissociative (Scheme 6). Both types of processes have recently been studied by DFT methods [42]. Associative processes were found to be favored, at least for reaction of cationic (η^3-allyl)Pd complexes with halide or neutral complexes with a neutral ligand. Successive ligand displacements can give the appearance of allyl rotation, even in the case of bidentate ligands (Scheme 4). There is a common misconception in the literature that associative ligand exchange at square planar complexes is initiated by ligand donation into an empty orbital (d_z2 or p_z), which is perpendicular to the coordination plane, an erroneous corollary of the 18-electron "rule." Electronic structure calculations clearly show that this picture is incorrect; there are *no* low-lying empty orbitals in d^8 complexes, they are coordinatively saturated with two ω-bonds [48]. Associative processes in square planar complexes go through a trigonal bipyramidal transition state, without any pre-complexation of the approaching ligand. For trigonal bipyramidal (η^3-allyl)Pd transition states, the allyl spans one axial and one equatorial position, the spectator ligand occupies the remaining axial position, and the incoming and leaving ligands each lie in an equatorial position [42].

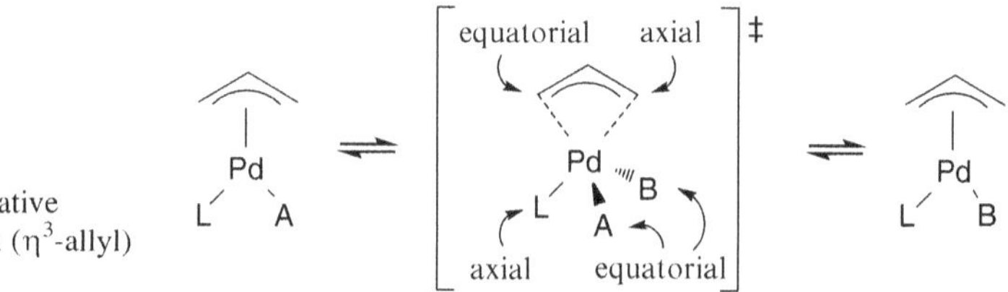

Scheme 5 Associative ligand exchange in (η^3-allyl) Pd complexes

Scheme 6 Dissociative ligand exchange in (η^3-allyl) Pd complexes, also showing plausible but slow mechanisms for apparent rotation

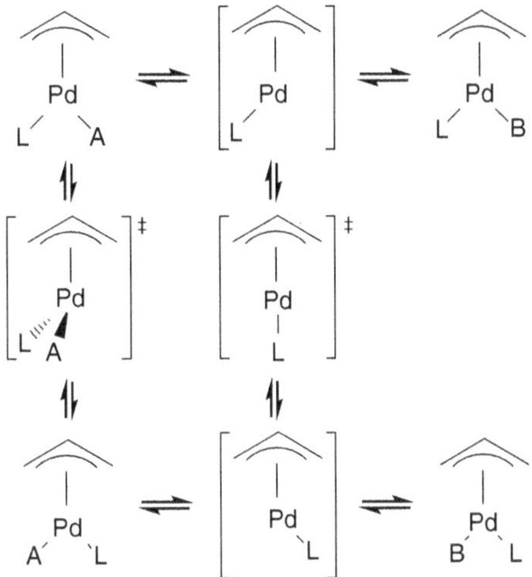

Dissociative processes were found to have a higher barrier [42], despite the known tendency of the DFT method to favor dissociation of dative ligands. The dissociation barrier was not insurmountable, only high compared to associative processes. In the complete absence of competing ligands, exchange would still be possible through dissociative processes. However, in such a situation, apparent rotation would still not occur through site exchange of the remaining ligand (Scheme 6, middle); the direct rotation without ligand dissociation has a slightly lower barrier (Scheme 6, left). We also note that the strong favoring of associative paths, by >80 kJ/mol, makes it likely that an associative exchange with a weak ligand such as a solvent molecule would be preferred compared to the purely dissociative path.

4.4 Allyl Exchange

None of the dynamic processes discussed so far allows for loss of stereochemical information in η^3-cycloalkenyl complexes. Still, such complexes have been seen to lose stereochemical integrity over time [49–52]. The process is presumably slow under catalytic conditions, but it has been invoked to rationalize loss of stereospecificity [50, 51]. The postulated mechanism is nucleophilic attack of a Pd^0 complex on an (η^3-allyl)Pd (Scheme 7). To our knowledge, there are no published computational studies of this process, but preliminary calculations using a small model system[2] indicates that the binuclear complex is a favored and symmetric ground

[2]B3LYP/LACVP* using PH_3 as a ligand model, gas phase calculations on the system depicted in Scheme 7.

Scheme 7 Allyl exchange through nucleophilic attack by Pd^0

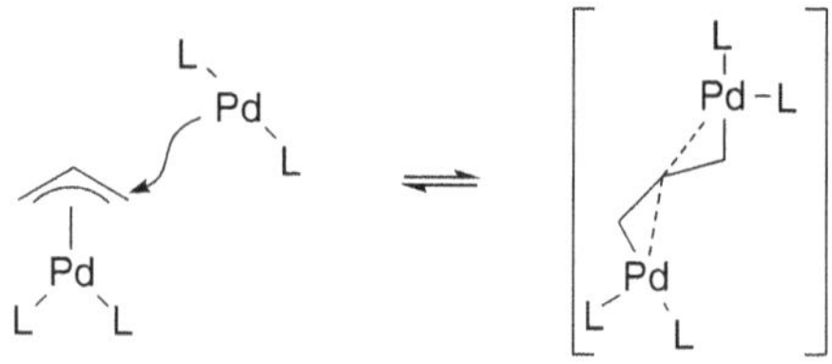

state, formed virtually without a barrier. We note that the binuclear complex in Scheme 7 could be seen as a *trans* analog of the known complex depicted in Fig. 2. The surprisingly facile allyl exchange process raises the question why this effect is not seen in more Pd-catalyzed allylic alkylations, where it should lower stereospecificity drastically. We can speculate that the low concentration of Pd^0 under catalytic conditions lowers the rate of this stereochemical scrambling, but since the concentration is still high enough to drive the regular catalytic process, we must further postulate that Pd^0 has a strong preference for reaction with the allylic substrate, a reaction which is obviously fast enough to out-compete the scrambling depicted in Scheme 7. A further clue to the mechanism comes from the observation that the scrambling that *has* been observed can be suppressed by addition of chloride anions in the form of lithium chloride [50, 51].

5 Reactivity

Allylation reactions can be both electrophilic and nucleophilic. Both types have been studied computationally. Here, we will first review palladium-assisted electrophilic allylation (e.g., the Tsuji–Trost reaction), which has been the subject of numerous studies since the 80s. We will then go through studies of nucleophilic allyl-palladium complexes.

5.1 Electrophilic Allyl

In general, Pd-assisted allylations are ionic. In a "standard" Tsuji–Trost reaction (Scheme 8), a neutral Pd^0 complex coordinates a neutral allyl carboxylate. S_N2-like ionization produces a cationic (η^3-allyl)Pd complex and a carboxylate anion. The incoming nucleophile is frequently an anionic malonate. Reaction of the nucleophile with the cationic (η^3-allyl)Pd complex produces a neutral alkene complex that dissociates the product and enters a new catalytic cycle. A proper description of this seemingly simple reaction requires accounting for the surrounding solvent [16], since the combination of the two oppositely charged species occurs without a barrier in the gas phase. Only in more recent studies have the necessary solvation

Scheme 8 Catalytic cycle of palladium-assisted allylic substitution

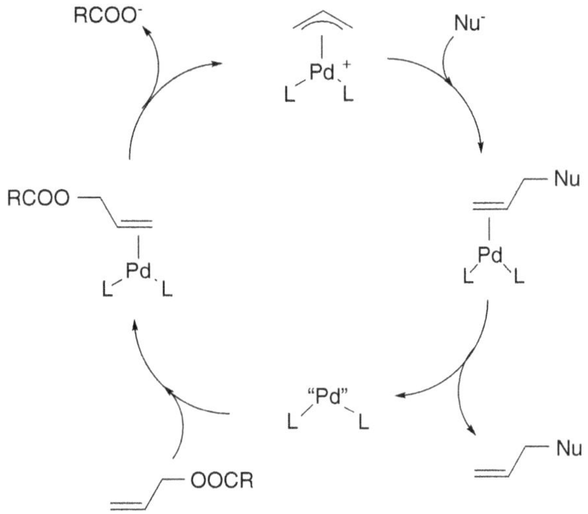

models been available, and their utilization is not without problems. Earlier studies have circumvented these problems by limiting the investigation to the (η^3-allyl)Pd complex itself, attempting to correlate reactivity with frontier orbitals, charge distribution, or ligand-imposed structural distortions. Another approach has been to ensure minimal charge change by utilizing neutral model nucleophiles (e.g., ammonia). A study by Hagelin et al. showed that transition states could be found even for anionic nucleophiles, and also indicated that the transition state in solvent is significantly different from the gas phase transition state even for neutral nucleophiles [16].

5.1.1 Regioselectivity

Nucleophilic attack on (η^3-allyl)metal species can occur at any of the three carbons. Attack at the central carbon is generally followed by reductive elimination, forming cyclopropanes. However, with (η^3-allyl)palladium complexes, reaction with nucleophiles generally occur at the terminal positions. Already in the beginning of the 1980s, Eisenstein and Curtis could explain this observation using EHMO calculations. The study concluded that the observed regioselectivity was under frontier orbital control, and not charge control [53].

When using certain σ-donor ligands with less stabilized carbon nucleophiles, cyclopropanes can be produced by nucleophilic attack at the central carbon followed by reductive elimination [54]. The reason for this selectivity is examined by Bäckvall and co-workers and was found to be under frontier orbital control. MP2 calculations show that π-allyl complexes with σ-donor ligands, primarily N-ligands, have a low-lying empty symmetrical orbital with a large coefficient on the central carbon [55, 56].

An orbital description of the (η^3-allyl)palladium moiety was given by Szabó, explaining the preference for the terminal attack. The activating effect of ligands was also rationalized, showing how reactivity increased when exchanging a σ-donor with a π-acceptor ligand [23].

One of the most studied problems in (η^3-allyl)palladium concerns *which* allyl terminus is attacked in unsymmetric complexes. EHMO calculations have also been applied to this problem, showing increased reactivity of the carbon with the longest Pd–C bond [57].

The regioselectivity is sensitive to a variety of steric effects. Norrby and co-workers correlated MM-calculated structural features with experimental selectivities in the first application of QSAR methods to asymmetric catalysis [58]. The most important effects were found to be the previously mentioned Pd–C bond length, enforced rotation of the allyl toward a product-like geometry, and direct interactions with the incoming nucleophile.

5.1.2 Different Nucleophiles

Experimentally, many different nucleophiles have been employed in the allylic substitution reaction. The most common are carbon nucleophiles, usually malonates, but amines, phosphines, sulfinates, alcohols, and other heteroatom nucleophiles have been used to yield new carbon–heteroatom bonds.

Computationally, only a few types of nucleophiles have been studied. The most common is the carbon nucleophile, mostly represented by a malonate [28, 59–65]. In addition to this, not only amine nucleophiles have been the focus of some investigations [36, 63, 66], but also fluoride [16], cyanide [16], and acetate [42] have been investigated. We note that when the “nucleophile” is also a leaving group, this step is in fact the reverse of the ionization step (Scheme 8), and the calculated barriers can be directly applied to selectivities in the ionization.

Intramolecular allylation has been used successfully to create new rings, but only a few such cases have been studied computationally. Madec et al. studied reactions with tethered carbanion nucleophiles in systems where both 5- and 7-membered ring products had been observed experimentally. The high selectivity for 5-membered rings observed with an amide linker compared to the low selectivity with a methylene linker could be fully rationalized by rigidity observed in the computational model [67]. The modeling also predicted a strong counterion effect, and allowed design of a much more reactive system by removal of the chelating sodium counterion, either by crown ether or a two-phase reaction where sodium stayed in the water phase [67].

5.1.3 Substituent Effects

Substituted allyls are synthetically important in palladium-catalyzed allylic substitution reactions. The substitution pattern of the allyls gives implications for both the steric and electronic interactions. When the substituents on the two termini differ, both regio- and stereoselectivity is influenced by the substituent itself, and by the interaction between the substituent and ligands on palladium. Substituents can occupy *syn* and *anti* positions (Fig. 5), with important differences in selectivity [37, 43, 65], but in many cases only *syn* complexes have been considered in computational studies.

The β-substituted (η^3-allyl)palladium has been extensively studied by Szabó and co-workers. A new type of electronic interaction, occurring between the allylic β-substituents and palladium is described and characterized. The structure and stability, as well as the regiochemistry of a nucleophilic attack are influenced by this interaction [68–72].

5.1.4 Ligand Effects

The ligand effect on reactivity was discussed already in the beginning of the 1980s after the emergence of the synthetic use of the allylic substitution reaction. Sakaki et al. studied the nucleophilic attack on a (η^3-allyl) palladium complex. The increased reactivity of the π-acceptor ligands was examined using semiempirical calculations, and explained with their ability to stabilize the important reacting orbitals and influencing the Pd–C bond strength [73].

Ligands with several heteroatoms, such as the bisthiazoline depicted in Fig. 6, can give different chelation options. In this situation, where a competition between the chelations was present, calculations have been helpful in sorting out the mechanistic possibilities. For the ligand in Fig. 6, DFT was used to deduce that the *N,N*-chelation was the lowest in energy, in agreement with experimental observations [60].

Each ligand will mostly affect structure and reactivity of the allyl terminus *trans* to itself. Thus, when employing two different ligands, or one heterodentate ligand, the relative reactivity of the two termini of the (η^3-allyl)palladium moiety will be affected. The effect can also be observed in the Pd^0 complex, where Goldfuss and Kazmaier showed that the leaving group prefers a position *trans* to the phosphine in a complex with both Cl^- and PH_3 ligands [74]. In a more detailed study, Svensen et al. investigated the equilibria in systems with halides and mono-dentate phosphines, and found strong *trans* effects leading to significant memory effects [75].

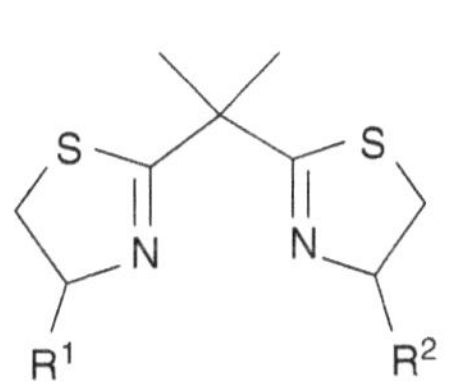

S S N N R¹ R²

Fig. 6 A bisthiazoline ligand

5.1.5 Enantioselectivity

There can be several mechanisms by which enantioselectivity can be obtained in the Tsuji–Trost reaction [2]. However, the most common technique is to utilize a chiral ligand in combination with a substrate that yields a symmetrically substituted (η^3-allyl)palladium intermediate (e.g., 1,3-diphenyl, or cycloalkenyl). In this case, the enantioselectivity is determined by the regioselectivity in the nucleophilic attack. For other substrates, enantioselectivity can arise from desymmetrization of starting materials with enantiotopic leaving groups, or from selective reaction of one intermediate only combined with rapid equilibration between isomeric intermediates. In each case, modeling can aid understanding the underlying selectivity-determining interactions. Computational studies can be divided into those trying to understand the selectivity from observable intermediates and those calculating the selectivity-determining transition states explicitly.

Pregosin et al. showed in the beginning of the 90s, by means of MM2 calculation, that the two chiral ligands BINAP and Chiraphos gave, for steric reasons, rise to different chiral arrays when coordinating to a (η^3-allyl)palladium complex [22].

Trost and coworkers used modeling in the program CAChe to show that the enantiomeric excess originating from a *meso*-like diphosphino ligand (Fig. 7) was due to the possibility of two different propeller forms [76].

In an MM2* study, Barbaro et al. investigated chiral diphosphino ligands, such as BINAP, extensively. The various chiral pockets of different ligands were visualized, and shown to be well defined and stable, but the correlation between the experimental ee's and ground state populations was not good, probably due to differences in reactivity between the complexes [77].

Peña-Cabrera et al. have demonstrated the ability of MM to assist in the rational design of asymmetric catalysis, using a modified MM2 force field [65]. A series of chiral phenantroline ligands were studied and the enantioselectivities were predicted from relative energies and enforced rotation in the (η^3-allyl)palladium intermediate. The experimental results confirmed the predictions [65]. In a subsequent study by Oslob et al., stereoselectivity could be quantitatively correlated with calculated structural features [58].

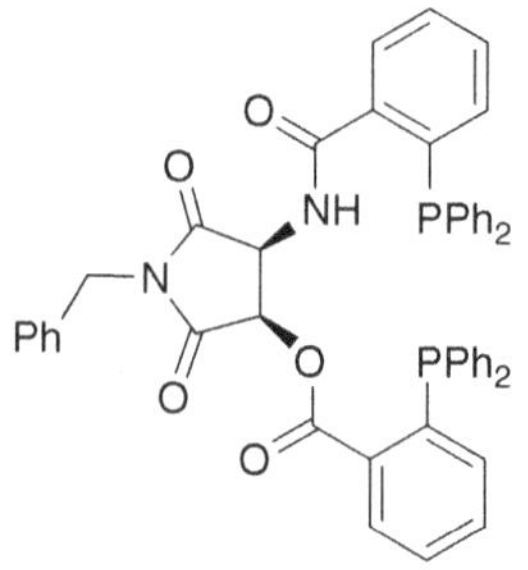

Fig. 7 *Meso*-like compound giving stereoselectivity through two propeller forms

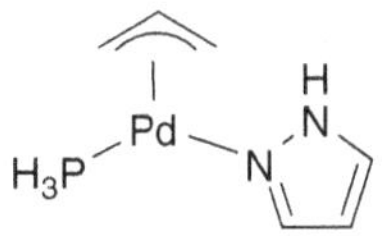

Fig. 8 The "unconstrained" (η^3-allyl)palladium complex used to investigate the electronic effect difference in PN-ligands

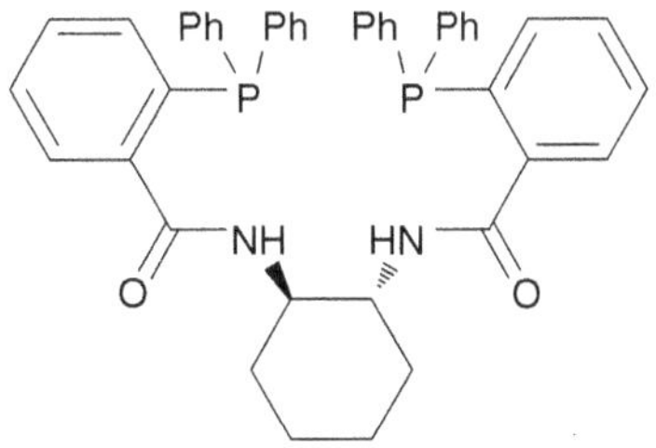

Fig. 9 A Trost modular ligand

Moberg et al. showed that the enantioselectivity could be dependent on the conformation of the ligand in a study of chiral pyridinoxazolines. Two different conformational minima could be found computationally and they gave rise to different steric environments leading to a stereoselective allylic alkylation [64].

Betz et al. have studied dithiazoline ligands in asymmetric allylic substitution (Fig. 6). The structure of the (η^3-allyl)palladium complex was calculated, and the enantioselectivity as well as (N,N) vs. (N,S) competition is rationalized [60].

In a DFT study by Blöchl concerning bidentate phosphine-pyrazole ligands, the steric effects proved to be the reason for the high stereoselectivity. The bulky ligands force the allyl out of the "square planar" plane opening up for attack on one of the terminal carbons [36]. The authors also studied the importance of the *trans* effects in a small model system, and concluded that nucleophilic attack *trans* to phosphorus is favored by 8 kJ/mol, corresponding to a 20:1 preference for this attack at room temperature (Fig. 8).

The Trost modular ligand (Fig. 9) is one of the most frequently used ligands in asymmetric allylic alkylation. The origin of the powerful stereoselective ability of this ligand has been the focus of a study by Butts et al. The identification of a hydrogen-bond interaction, which influences both the ionization and nucleophilic attack, proves to be the source of the observed selectivity. This conclusion opens up the possibility for further development of this reaction including design of new ligands [61].

In their quest for new P,N-ligands, Andersson and co-workers performed DFT calculations to verify the observed enantiomeric excess. The resulting enthalpies were in agreement with the experimental observations, and the calculated structure was verified with NOE contacts [63].

The previously mentioned dinuclear palladium catalysts have been used in enantioselective palladium-catalyzed allylic substitution reactions, giving a moderate ee of 57%. The stereoselectivity is due to steric reasons, the strained arrangement assumed by the catalytic complex giving different environment for the terminal allylic carbons [62].

5.1.6 Reductive Elimination

In the allylic substitution, the nucleophilic attack takes place on the allyl and the role for the metal catalyst is to provide coordination sites for the allyl to form the (η^3-allyl)palladium complex. For very basic nucleophiles, the reaction is believed to follow another pathway. Both the allyl and nucleophile coordinate to the metal catalyst, and a reductive elimination yields the final product.

Some effort has been put in to calculations regarding the feasibility of this reaction. Sakaki et al. performed MP2/MP4 calculations on a $MH(\eta^1\text{-}C_3H_5)(PH_3)$ system (M=Pd, Pt). The η^1-complex was found to be less stable, by at least 9 kcal/mol, than the corresponding η^3-complex, but the reductive elimination from the η^1-Pd complex had a small barrier of 5.4 kcal/mol, and was exothermic by almost 30 kcal/mol, making it a feasible reaction. The Pt complex, on the other hand, had a barrier for the reductive elimination of more than 20 kcal/mol [78].

In a further study, $Pd(XH_3)(\eta^3\text{-}C_3H_5)(PH_3)$ complexes (X=C, Si, Ge, Sn) were studied by the same group. The ability to do a reductive elimination was calculated at MP2/MP4 or CCSD(T) level. It was found that carbon gave the highest E_A with 23.3 kcal/mol; the other three had smaller E_A, 11–13 kcal/mol. The high E_A for C is explained with a large distortion energy required to increase the Pd–C bond length. The other three groups (Si, Ge, Sn) has much smaller energy costs when distorting the Pd–X bond, probably due to the hypervalency of these elements [79].

Mendez et al. studied an intramolecular diallyl reaction for the formation of five-membered rings through an improved variant of the Oppolzer cyclization (Scheme 9). A computational study was carried out to investigate the nature of the reductive elimination between the two allylic pieces. DFT calculations suggest that, in the presence of excess PH_3, two η^1-allyls react through the terminal carbons, in a somewhat unorthodox reductive elimination. These findings were supported by experimental evidence [80].

5.2 *Nucleophilic Allyl*

The umpolung of the allylpalladium reactivity, where the allyl shows nucleophilic abilities, has been known for several decades [81]. In the middle of the 90s, Yamamoto and co-workers could identify bisallyl-Pd structures through NMR

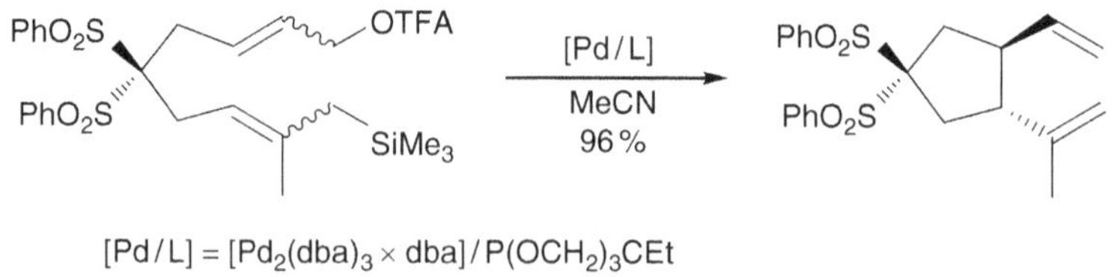

PhO2S
PhO2S
OTFA
SiMe3
[Pd/L]
MeCN
96 %
PhO2S
PhO2S

[Pd/L] = [Pd2(dba)3 × dba]/P(OCH2)3CEt

Scheme 9 Modified Oppolzer cyclization investigated by Mendez et al.

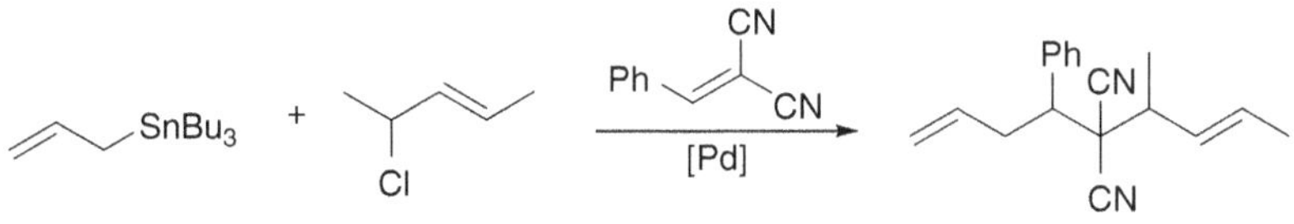

Scheme 10 Substituent effects on allylation

studies, and these reacted with aldehydes in a nucleophilic fashion [82]. This opened the field for a whole new type of reactivity. Further studies showed that the bisallyl-palladium complexes could undergo an initial *electrophilic* attack on one of the allyl moieties, followed by a *nucleophilic* attack on the other one. In other words, bisallyl-palladium complexes can be described as *amphiphilic* reagents, that is both nucleophilic and electrophilic species [32].

5.2.1 Substituent Effects

In a study by Solin et al., the effects of substitution on the allyls are examined. The reaction studied was between the above-mentioned bisallylpalladium complex, which reacts with benzylidenemalonitrile, yielding substituted 1,7-octadienes with very high regioselectivity (Scheme 10) [83].

With the aid of DFT calculations, the nature of the (η^3,η^1)-bisallylcomplex was revealed. It was found that the unsubstituted allyl is the one forming the η^1-bond. This form is favored by at least 12 kJ/mol over the competing coordination pattern. The energy barrier for the electrophilic attack was also found to be 12 kJ/mol lower for the former complex. The high regioselectivity was ascribed to the electronic effects of the alkyl substituents, the terminal alkyls have two important effects: destabilization of the η^3,η^1-bis-allylpalladium intermediate when the η^1-moiety is substituted and an increase in the activation barrier of the electrophilic attack caused by the alkyl substitution.

It has been speculated that a (η^1-allyl)palladium complex could also be attacked by nucleophiles, and that this could help explain some observed memory effects [39, 40]. However, two separate computational studies have shown that approaching a nucleophile to an η^1-allyl forces it into η^3-allyl form, if necessary expelling another ligand on palladium. In experimental terms, this indicates the η^1-allyl group cannot undergo reaction with nucleophiles, but must equilibrate to the η^3-allyl form before reaction can take place [84, 85].

5.2.2 Different Electrophiles

Experimentally, several different electrophiles have been employed in the electrophilic allylic substitution [86]. The examples involving calculations are, on the

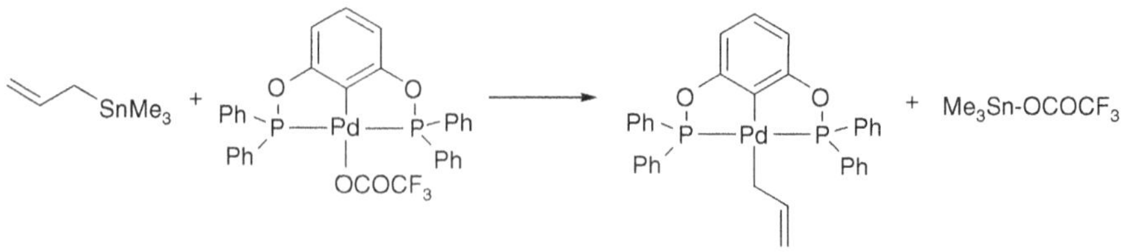

Scheme 11 Synthesis of a (η^1-allyl)palladium with a pincer ligand

other hand, rather few. Szabó has studied acetic acid and formaldehyde as electrophiles. Both species were found to have small energy barriers for the electrophilic attack [32]. Low activation energy was also found for the attack of an NCO-electrophile in a study by Solin et al. [87].

5.2.3 Palladium Pincer Complexes

In development of electrophilic allylic substitution, Szabó searched for complexes where a mono-allyl palladium species would be forced into the η^1 binding mode. This can be achieved using ligands that require three of the available coordination sites on the Pd (Scheme 11), the so-called "pincer" ligands. This leaves one site open for coordination, which causes the allyl to coordinate in a η^1-fashion. The allyl-pincer complex is synthesized *via* a transmetallation from an allylstannane (Scheme 11), and can sequentially be reacted with an electrophile to achieve an allylation. Both imines and aldehydes can be used as electrophiles.

Calculations on these systems have identified the η^1allyl-pincer complex as the active catalyst and concluded that the electrophilic attack proceeds with a low activation barrier on the γ-position of the η^1-allyl [88].

6 Summary

Molecular modeling has been shown to be a strong support in understanding all aspects of palladium-assisted allylation reactions. Structures of intermediates can be calculated with good accuracy at moderate levels of theory. Substantially more problematic is quantitative calculation of reaction selectivities, in particular when the reagents are ionic and oppositely charged. However, advances in DFT and use of continuum solvation now allow also these types of reactions to be understood on a molecular level. Among the most recent additions to the computational arsenal is the correction for dispersive van der Waals interactions, allowing a more accurate treatment of large complexes and bimolecular reactions.

References

1. Hartwig J (2010) Organotransition metal chemistry. University Science Books, Sausalito, CA
2. Trost BM, Crawley ML (2003) Asymmetric transition-metal-catalyzed allylic alkylations: applications in total synthesis. Chem Rev 103:2921
3. Dedieu A (2000) Theoretical studies in palladium and platinum molecular chemistry. Chem Rev 100:543
4. Cramer CJ (2004) Essentials of computational chemistry: theories and models, 2nd edn. Wiley-VCH, Weinheim
5. Jensen F (2006) Introduction to computational chemistry, 2nd edn. Wiley-VCH, Weinheim
6. Becke AD (1993) Density-functional thermochemistry. 3. The role of exact exchange. J Chem Phys 98:5648
7. Lee CT, Yang WT, Parr RG (1988) Development of the Colle-Salvetti correlation-energy formula into a functional of the electron-density. Phys Rev B 37:785
8. Stephens PJ, Devlin FJ, Chabalowski CF, Frisch MJ (1994) Ab-initio calculation of vibrational absorption and circular-dichroism spectra using density-functional force-fields. J Phys Chem 98:11623
9. Grimme S (2004) Accurate description of van der Waals complexes by density functional theory including empirical corrections. J Comput Chem 25:1463
10. Zhao Y, Truhlar DG (2008) The M06 suite of density functionals for main group thermochemistry, thermochemical kinetics, noncovalent interactions, excited states, and transition elements: two new functionals and systematic testing of four M06-class functionals and 12 other functionals. Theor Chem Acc 120:215
11. Averkiev BB, Zhao Y, Truhlar DG (2010) Binding energy of d(10) transition metals to alkenes by wave function theory and density functional theory. J Mol Catal A-Chem 324:80
12. Grimme S, Antony J, Ehrlich S, Krieg H (2010) A consistent and accurate ab initio parametrization of density functional dispersion correction (DFT-D) for the 94 elements H-Pu. J Chem Phys 132:154104
13. Nilsson Lill SO (2009) Application of dispersion-corrected density functional theory. J Phys Chem A 113:10321
14. Miertus S, Scrocco E, Tomasi J (1981) Electrostatic interaction of a solute with a continuum – a direct utilization of abinitio molecular potentials for the prevision of solvent effects. Chem Phys 55:117
15. Cammi R, Tomasi J (1995) Remarks on the use of the apparent surface-charges (ASC) methods in solvation problems – iterative versus matrix-inversion procedures and the renormalization of the apparent charges. J Comput Chem 16:1449
16. Hagelin H, Akermark B, Norrby PO (1999) A solvated transition state for the nucleophilic attack on cationic η^3-allylpalladium complexes. Chem-Eur J 5:902
17. Comba PH, Hambley TW (2009) Molecular modeling of inorganic compounds, 3rd edn. Wiley-VCH, Weinheim
18. Norrby PO, Akermark B, Haeffner F, Hansson S, Blomberg M (1993) Molecular mechanics (MM2) parameters for the (η^3-Allyl)palladium moiety. J Am Chem Soc 115:4859
19. Akermark B, Oslob JD, Norrby PO (1995) Conformations of (η^3-Cyclohexenyl)palladium systems – a molecular mechanics (MM2) study. Organometallics 14:1688
20. Hagelin H, Svensson M, Akermark B, Norrby PO (1999) Molecular mechanics (MM3*) force field parameters for calculations on palladium olefin complexes with phosphorus ligands. Organometallics 18:4574
21. Hagelin H, Akermark B, Norrby PO (1999) New molecular mechanics (MM3*) force field parameters for calculations on (η^3-allyl)palladium complexes with nitrogen and phosphorus ligands. Organometallics 18:2884
22. Pregosin PS, Ruegger H, Salzmann R, Albinati A, Lianza F, Kunz RW (1994) X-ray-diffraction, multidimensional NMR-spectroscopy, and MM2* calculations on chiral allyl complexes of palladium(II). Organometallics 13:83

23. Szabo KJ (1996) Effects of the ancillary ligands on palladium-carbon bonding in (η^3-allyl) palladium complexes. Implications for nucleophilic attack at the allylic carbons. Organometallics 15:1128
24. Branchadell V, Moreno-Manas M, Pajuelo F, Pleixats R (1999) Density functional study on the regioselectivity of nucleophilic attack in 1,3-disubstituted (diphosphino) (η^3-allyl)palladium cations. Organometallics 18:4934
25. Delbecq F, Lapouge C (2000) Regioselectivity of the nucleophilic addition to (η^3-allyl) palladium complexes. A theoretical study. Organometallics 19:2716
26. Kollmar M, Goldfuss B, Reggelin M, Rominger F, Helmchen G (2001) (Phosphanyloxazoline)palladium complexes, part I: (η^3-1,3-dialkylallyl)(phosphanyloxazoline)palladium complexes: X-ray crystallographic studies, NMR investigations, and quantum-chemical calculations. Chem-Eur J 7:4913
27. Kollmar M, Steinhagen H, Janssen JP, Goldfuss B, Malinovskaya SA, Vazquez J, Rominger F, Helmchen G (2002) (Phosphanyloxazoline)palladium complexes, part II – (η^3-Phenylallyl) (phosphanyloxazoline)palladium complexes: X-ray crystallographic studies, NMR investigations, and ab initio/DFT calculations. Chem-Eur J 8:3103
28. Kleimark J, Johansson C, Olsson S, Håkansson M, Hansson S, Åkermark B, Norrby PO (2011) Sterically governed selectivity in palladium-assisted allylic alkylation. Organometallics 30:230
29. Kurosawa H, Hirako K, Natsume S, Ogoshi S, Kanehisa N, Kai Y, Sakaki S, Takeuchi K (1996) New insights into structures, stability, and bonding of μ-allyl ligands coordinated with Pd-Pd and Pd-Pt fragments. Organometallics 15:2089
30. Sakaki S, Takeuchi K, Sugimoto M, Kurosawa H (1997) Geometries, bonding nature, and relative stabilities of dinuclear palladium(I) pi-allyl and mononuclear palladium(II) pi-allyl complexes. A theoretical study. Organometallics 16:2995
31. Casarin M, Pandolfo L, Vittadini A (2001) UV-photoelectron spectra of [M(η^3-C3H5)(2)] (M = Ni, Pd, Pt) revisited: a quasi-relativistic density functional study. Organometallics 20:754
32. Szabo KJ (2000) Umpolung of the allylpalladium reactivity: mechanism and regioselectivity of the electrophilic attack on bis-allylpalladium complexes formed in palladium-catalyzed transformations. Chem-Eur J 6:4413
33. Ruegger H, Kunz RW, Ammann CJ, Pregosin PS (1991) 2D-NOESY studies on [Pd- η^3-C10h15)((R)(+)-Binap)]+ defining the relative spatial orientation of the phenyl and π-allyl moieties. Magn Reson Chem 29:197
34. Ammann CJ, Pregosin PS, Ruegger H, Albinati A, Lianza F, Kunz RW (1992) Chemistry of Pdii complexes containing both Binap (2'2'-bis(diphenylphosphino)binaphthyl) and a η^3-pinene or η^3-allyl ligand - NMR-studies on [Pd(η^3-C10H15)(S(-)Binap)] (CF3SO3) – the crystal-structure of [Pd(η^3-C10H15)(4,4'-dimethylbipyridine)](CF3SO3). J Organomet Chem 423:415
35. Jonasson C, Kritikos M, Backvall JE, Szabo KJ (2000) Asymmetric allyl-metal bonding in substituted (η^3-allyl)palladium complexes: X-ray structures of cis- and trans-4-acetoxy-[η^3-(1,2,3)-cyclohexenyl]palladium chloride dimers. Chem-Eur J 6:432
36. Blochl PE, Togni A (1996) First-principles investigation of enantioselective catalysis: asymmetric allylic amination with Pd complexes bearing P,N-ligands. Organometallics 15:4125
37. Sjogren M, Hansson S, Norrby PO, Akermark B, Cucciolito ME, Vitagliano A (1992) Selective stabilization of the antiisomer of (η^3-Allyl)palladium and (η^3-Allyl)platinum complexes. Organometallics 11:3954
38. Vrieze K (1975) Dynamic nuclear magnetic resonance spectroscopy. Academic Press, New York
39. Jutand A, Mosleh A (1995) Rate and mechanism of oxidative addition of aryl triflates to zerovalent palladium complexes – evidence for the formation of cationic (sigma-aryl)palladium complexes. Organometallics 14:1810

40. Roy AH, Hartwig JF (2004) Oxidative addition of aryl sulfonates to Palladium(0) complexes of mono- and bidentate phosphines. Mild addition of aryl tosylates and the effects of anions on rate and mechanism. Organometallics 23:194
41. Szabo KJ (2004) Palladium-catalyzed electrophilic allylation reactions via bis(allyl)palladium complexes and related intermediates. Chem-Eur J 10:5269
42. Johansson C, Lloyd-Jones GC, Norrby PO (2010) Memory and dynamics in Pd-catalyzed allylic alkylation with P,N-ligands. Tetrahedron-Asymmetry 21:1585
43. Hansson S, Norrby PO, Sjogren MPT, Akermark B, Cucciolito ME, Giordano F, Vitagliano A (1993) Effects of phenanthroline type ligands on the dynamic processes of (η^3-allyl)palladium complexes – molecular-structure of (2,9-dimethyl-1,10-phenanthroline)[(1,2,3-η)-3-methyl-2-butenyl]-chloropalladium. Organometallics 12:4940
44. Albinati A, Kunz RW, Ammann CJ, Pregosin PS (1991) 2D NOESY of palladium π-allyl complexes – reporter ligands, complex dynamics, and the X-ray structure of [Pd(η^3-C4H7)(Bpy)](CF3SO3) (Bpy = Bipyridine). Organometallics 10:1800
45. Crociani B, Dibianca F, Giovenco A, Boschi T (1987) Reactions of pyridine-2-carbaldimines with chloro-bridged palladium(II) and platinum(II) 2-methylallyl dimers – solution behavior of the cationic complexes [M(η^3-2-MeC3H4)(Py-2-CH=NR)]+. Inorg Chim Acta 127:169
46. Gogoll A, Ornebro J, Grennberg H, Backvall JE (1994) Mechanism of apparent π-allyl rotation in (π-allyl)palladium complexes with bidentate nitrogen ligands. J Am Chem Soc 116:3631
47. Goodman JM, Silva MA (2003) QRC: a rapid method for connecting transition structures to reactants in the computational analysis of organic reactivity. Tetrahedron Lett 44:8233
48. Weinhold FL, Landis F (2005) Valency and bonding. Cambridge University Press, Cambridge
49. Amatore C, Gamez S, Jutand A, Meyer G, Moreno-Manas M, Morral L, Pleixats R (2000) Oxidative addition of allylic carbonates to palladium(0) complexes: reversibility and isomerization. Chem-Eur J 6:3372
50. Cantat T, Agenet N, Jutand A, Pleixats R, Moreno-Manas M (2005) The effect of chloride ions on the mechanism of the oxidative addition of cyclic allylic carbonates to Pd-0 complexes by formation of neutral [(η^1-allyl)PdClL2] complexes. Eur J Org Chem 2005:4277
51. Granberg KL, Backvall JE (1992) Isomerization of (π-allyl)palladium complexes via nucleophilic displacement by palladium(0) – a common mechanism in palladium(0)-catalyzed allylic substitution. J Am Chem Soc 114:6858
52. Martin JT, Oslob JD, Akermark B, Norrby PO (1995) A detailed study of (η^3-cyclohexenyl) palladium systems. Acta Chem Scand 49:888
53. Curtis MD, Eisenstein O (1984) A molecular-orbital analysis of the regioselectivity of nucleophilic-addition to η^3-allyl complexes and the conformation of the η^3-allyl ligand in L3(CO)2(η^3-C3H5)MoII complexes. Organometallics 3:887
54. Hegedus LS, Darlington WH, Russell CE (1980) Cyclopropanation of ester enolates by π-allylpalladium chloride complexes. J Org Chem 45:5193
55. Aranyos A, Szabo KJ, Castano AM, Backvall JE (1997) Central versus terminal attack in nucleophilic addition to (π-allyl)palladium complexes. Ligand effects and mechanism. Organometallics 16:1058
56. Szabo KJ (1998) Effects of polar gamma-substituents on the structure and stability of palladacyclobutane complexes. J Mol Struc-Theochem 455:205
57. Ward TR (1996) Regioselectivity of nucleophilic attack on [Pd(allyl)(phosphine)(imine)] complexes: a theoretical study. Organometallics 15:2836
58. Oslob JD, Akermark B, Helquist P, Norrby PO (1997) Steric influences on the selectivity in palladium-catalyzed allylation. Organometallics 16:3015
59. Bantreil X, Prestat G, Moreno A, Madec D, Fristrup P, Norrby PO, Pregosin PS, Poli G (2011) γ- and δ-lactams through palladium-catalyzed intramolecular alkylation: enantioselective synthesis, NMR investigation, and DFT rationalization. Chem-Eur J 17:2885
60. Betz A, Yu L, Reiher M, Gaumont AC, Jaffres PA, Gulea M (2008) (N, N) vs. (N, S) chelation of palladium in asymmetric allylic substitution using bis(thiazoline) ligands: a theoretical and experimental study. J Organomet Chem 693:2499

61. Butts CP, Filali E, Lloyd-Jones GC, Norrby PO, Sale DA, Schramm Y (2009) Structure-based rationale for selectivity in the asymmetric allylic alkylation of cycloalkenyl esters employing the Trost 'Standard Ligand' (TSL): isolation, analysis and alkylation of the monomeric form of the cationic η^3-cyclohexenyl complex [(η^3-c-C6H9)Pd(TSL)](+). J Am Chem Soc 131:9945
62. Calabro G, Drommi D, Bruno G, Faraone F (2004) Effect of chelating vs. bridging coordination of chiral short-bite P-X-P (X = C, N, O) ligands in enantioselective palladium-catalysed allylic substitution reactions. Dalton Trans 2004:81
63. Mazuela J, Paptchikhine A, Tolstoy P, Pamies O, Dieguez M, Andersson PG (2010) A new class of modular P, N-ligand library for asymmetric Pd-catalyzed allylic substitution reactions: a study of the key Pd-π-allyl intermediates. Chem-Eur J 16:620
64. Moberg C, Bremberg U, Hallman K, Svensson M, Norrby PO, Hallberg A, Larhed M, Csoregh I (1999) Selectivity and reactivity in asymmetric allylic alkylation. Pure Appl Chem 71:1477
65. PenaCabrera E, Norrby PO, Sjogren M, Vitagliano A, DeFelice V, Oslob J, Ishii S, ONeill D, Akermark B, Helquist P (1996) Molecular mechanics predictions and experimental testing of asymmetric palladium-catalyzed allylation reactions using new chiral phenanthroline ligands. J Am Chem Soc 118:4299
66. Piechaczyk O, Thoumazet C, Jean Y, le Floch P (2006) DFT study on the palladium-catalyzed allylation of primary amines by allylic alcohol. J Am Chem Soc 128:14306
67. Madec D, Prestat G, Martini E, Fristrup P, Poli G, Norrby PO (2005) Surprisingly mild "enolate-counterion-free" Pd(0)-catalyzed intramolecular allylic alkylations. Org Lett 7:995
68. Macsari I, Szabo KJ (1999) Nature of the interactions between the β-silyl substituent and allyl moiety in (η^3-allyl)palladium complexes. a combined experimental and theoretical study. Organometallics 18:701
69. Szabo KJ (1996) Effects of β-substituents and ancillary ligands on the structure and stability of (η^3-allyl)palladium complexes. implications for the regioselectivity in nucleophilic addition reactions. J Am Chem Soc 118:7818
70. Szabo KJ (1997) Nature of the interactions between polar β-substituents and palladium in η^3-allylpalladium complexes – a combined experimental and theoretical study. Chem-Eur J 3:592
71. Szabo KJ (2001) Nature of the interaction between β-substituents and the allyl moiety in (η^3-allyl)palladium complexes. Chem Soc Rev 30:136
72. Szabo KJ, Hupe E, Larsson ALE (1997) Stereoelectronic control on the kinetic stability of β-acetoxy-substituted (η^3-allyl)palladium com plexes in a mild acidic medium. Organometallics 16:3779
73. Sakaki S, Nishikawa M, Ohyoshi A (1980) A palladium-catalyzed reaction of a π-allyl ligand with a nucleophile – an MO study about a feature of the reaction and a ligand effect on the reactivity. J Am Chem Soc 102:4062
74. Goldfuss B, Kazmaier U (2000) Electronic differentiations in palladium alkene complexes: *trans*-phosphine preference of allylic leaving groups. Tetrahedron 56:6493
75. Svensen N, Fristrup P, Tanner D, Norrby PO (2007) Memory effects in palladium-catalyzed allylic Alkylations of 2-cyclohexen-1-yl acetate. Adv Synth Catal 349:2631
76. Trost BM, Breit B, Organ MG (1994) On the nature of the asymmetric induction in a palladium-catalyzed allylic alkylation. Tetrahedron Lett 35:5817
77. Barbaro P, Pregosin PS, Salzmann R, Albinati A, Kunz RW (1995) 1,3-diphenylallyl complexes of palladium(II) – NMR, X-ray, and catalytic studies. Organometallics 14:5160
78. Sakaki S, Satoh H, Shono H, Ujino Y (1996) Ab initio MO study of the geometry, $\eta^3 \Leftrightarrow \eta^1$ conversion, and reductive elimination of a palladium(II) η^3-allyl hydride complex and its platinum(II) analogue. Organometallics 15:1713
79. Biswas B, Sugimoto M, Sakaki S (1999) Theoretical study of the structure, bonding nature, and reductive elimination reaction of Pd(XH3)(η^3-C3H5) (PH3) (X = C, Si, Ge, Sn). Hypervalent behavior of group 14 elements. Organometallics 18:4015

80. Mendez M, Cuerva JM, Gomez-Bengoa E, Cardenas DJ, Echavarren AM (2002) Intramolecular coupling of allyl carboxylates with allyl stannanes and allyl silanes: a new type of reductive elimination reaction? Chem-Eur J 8:3620
81. Hara M, Ohno K, Tsuji J (1971) Palladium-catalysed hydrosilation of olefins and polyenes. J Chem Soc D - Chem Comm 1971:247
82. Nakamura H, Asao N, Yamamoto Y (1995) Palladium-catalyzed and platinum-catalyzed addition of aldehydes with allylstannanes. J Chem Soc Chem Commun 1995:1273
83. Solin N, Narayan S, Szabo KJ (2001) Control of the regioselectivity in catalytic transformations involving amphiphilic bis-allylpalladium intermediates: mechanism and synthetic applications. J Org Chem 66:1686
84. Fristrup P, Ahlquist M, Tanner D, Norrby PO (2008) On the nature of the intermediates and the role of chloride ions in Pd-catalyzed allylic alkylations: added insight from density functional theory. J Phys Chem A 112:12862
85. Garcia-Iglesias M, Bunuel E, Cardenas DJ (2006) Cationic (η^1-allyl)-palladium complexes as feasible intermediates in catalyzed reactions. Organometallics 25:3611
86. Aydin J, Kumar KS, Sayah MJ, Wallner OA, Szabo KJ (2007) Synthesis and catalytic application of chiral 1,1'-bi-2-naphthol- and biphenanthrol-based pincer complexes: selective allylation of sulfonimines with allyl stannane and allyl trifluoroborate. J Org Chem 72:4689
87. Solin N, Narayan S, Szabo KJ (2001) Palladium-catalyzed tandem bis-allylation of isocyanates. Org Lett 3:909
88. Solin N, Kjellgren J, Szabo KJ (2004) Pincer complex-catalyzed allylation of aldehyde and imine substrates via nucleophilic η^1-allyl palladium intermediates. J Am Chem Soc 126:7026

Top Organomet Chem (2012) 38: 95–154
DOI: 10.1007/3418_2011_9

Published online: 14 June 2011

Palladium-Catalyzed Enantioselective Allylic Substitution

Ludovic Milhau and Patrick J. Guiry

Abstract Palladium-catalyzed allylic substitution is one of the main reactions for testing new chiral ligands. The most relevant examples from the work published in the period 2007 to mid-2010 are reviewed. The vast majority of the work published within this timeframe relies upon the application of chiral ligands for asymmetric induction. The recent advances in the development and applications of new chiral P–P, P–N, P–O, P–S, N–N, N–S, S–S, and NHC ligands are covered and are the main focus of this chapter. Other aspects of enantioselective palladium allylic alkylations are discussed in the subsequent sections, for example, heterogeneous catalysis, the use of chiral salt additives, and recent applications in kinetic resolution.

Keywords Asymmetric allylic substitution · Chiral ligands · Enantioselective coupling reaction · Palladium catalysis · Tsuji–Trost reaction

Contents

L. Milhau and P.J. Guiry (✉)
Centre for Synthesis and Chemical Biology, School of Chemistry and Chemical Biology, University College Dublin, Dublin, Ireland
e-mail: patrick.guiry@ucd.ie

1 Introduction

Palladium-catalyzed allylic substitution (Scheme 1) is one of the most popular catalyzed reactions in organic chemistry and one of the main reactions for testing new chiral ligands [1–3]. Although a few applications exist in protecting group chemistry (i.e., deprotection of allyl protected alcohols [4] or decarboxylative formate reduction [5, 6]), the main use of this reaction is the introduction of a new substituent in the allyl position. A number of allylic substrates can be reacted with a variety of carbon and heteroatom nucleophiles. Since the first stoichiometric examples reported by Tsuji [7, 8] and Trost [9, 10] in 1965 and 1973 respectively, the reaction has been widely studied in terms of chemo-, regio-, and enantioselectivity.

This chapter intends to highlight the recent advances in enantioselective palladium-catalyzed allylic substitutions. Because of the sheer number of publications on a

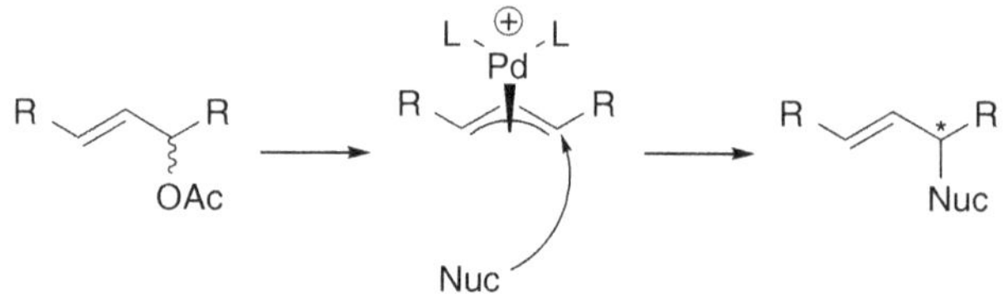

Scheme 1 Generic asymmetric allylic alkylation

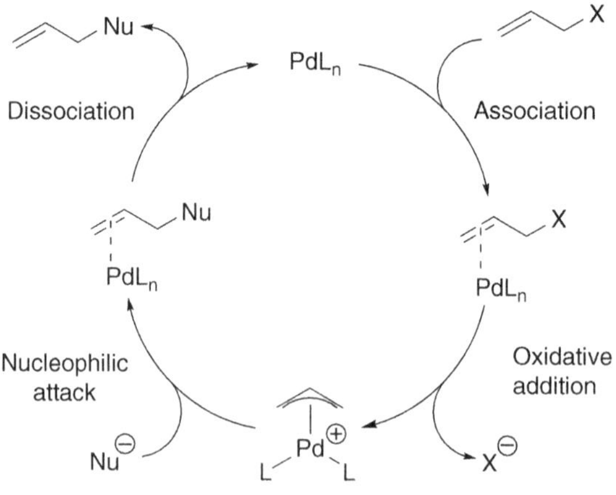

Scheme 2 General mechanism of an asymmetric allylic alkylation

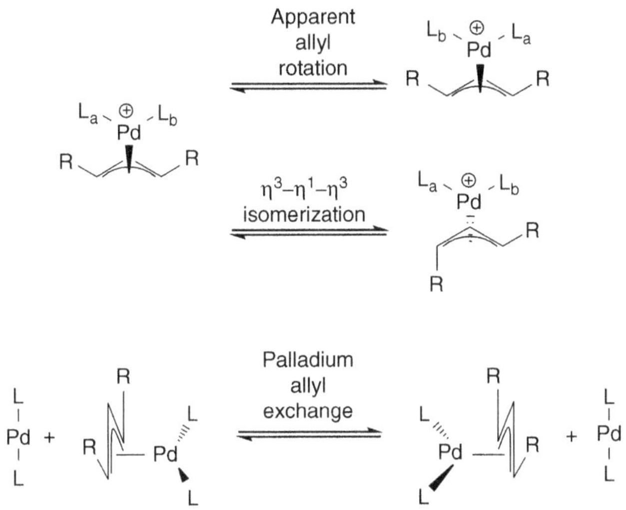

Scheme 3 Possible interconversions of palladium allyl complexes

topic which has been abundantly reviewed [11–14], we have restricted ourselves to work published in the period 2007 to mid-2010, and selected what appeared to us to be the most relevant examples.

The mechanism of the reaction may vary according to the conditions used, but the general steps are the complexation of the olefin to palladium, followed by oxidative addition, nucleophilic attack on the π-allyl and dissociation of the product (Scheme 2) [15].

The relative speed of each of these steps also affects the outcome of the reaction. After the two first reversible steps, the palladium allyl complex can undergo conformational interconversions [3]. Among the best known processes, the $\eta^3-\eta^1-\eta^3$ isomerization leads to the *syn/anti* interconversion and the apparent "allyl rotation" leads to an *endo/exo* isomerization (Scheme 3). The palladium-allyl exchange, although usually slow compared to the other mechanisms, results in an inversion of configuration on all three allyl carbons. If nucleophilic attack is slow, the more stable product will then be obtained.

Soft nucleophiles tend to attack one of the terminal carbons of the allylpalladium complex, whereas hard nucleophiles directly attack the palladium. As a consequence, soft nucleophiles tend to lead to net retention of configuration, and hard nucleophiles to inversion. This is nicely illustrated by a recent example by Spilling [16], showing a perfect retention of enantiomeric excess through a palladium-catalyzed allylic alkylation (Scheme 4).

The general mechanism discussed does not account for the so-called memory effect. In the case of two different allyl substrates (**1** or **2**) supposed to give the same π-allylpalladium complex, the reaction gives two different mixtures of product (**3** or **4**) (Table 1) [17]. This has been rationalized by suggesting that a chloride ion

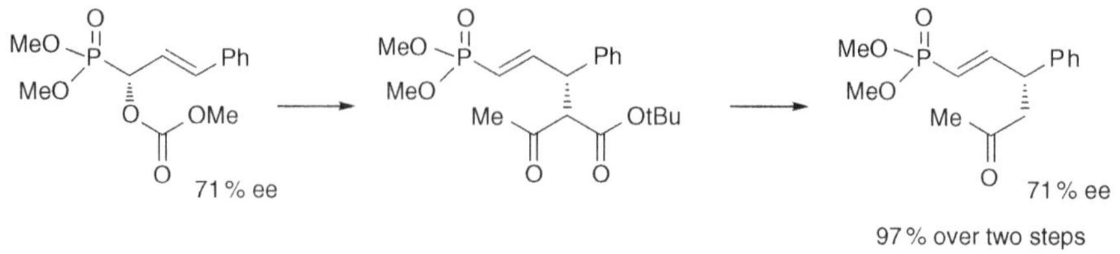

Scheme 4 Enantiomeric excess retention through allylic alkylation

Table 1 Illustration of the memory effect

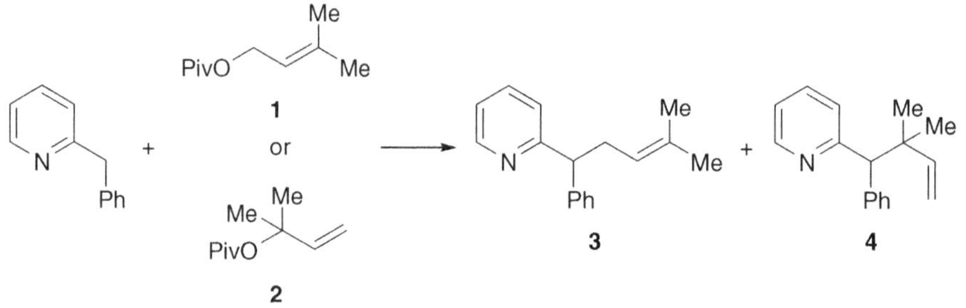

Entry	Substrate	Yield %	Linear **3**/branched **4**
1	**1**	74	>19:1
2	**2**	71	1:1.6

replaces one of the two phosphine ligands L bound to palladium forming a contact ion pair and leading to an unsymmetrical π-allylpalladium complex [18].

Different methods and approaches can be used to induce enantioselectivity. Next to the use of chiral ligands or additives, the use of an already chiral substrate or nucleophile to induce the chirality of a new center has been reported. Chiral substrates are mainly used for specific purposes, that is the synthesis of a class of compound. Kazmaier and co-workers have recently used palladium-catalyzed allylic substitution as the key step in the synthesis of glycoamino acids and glycopeptides starting from chiral substrates (Scheme 5) [19]. However, most of the applications of asymmetric allylic substitution are in total synthesis, but these will not be detailed here and will be dealt elsewhere in this book [20].

Very few occurrences of chiral nucleophiles have been reported. The most recent example to our knowledge is the amination of tosylate **5** (Scheme 6), but the selectivity was low [21].

Aside from these few results, the vast majority of the work published within the timeframe of this chapter relies upon the application of chiral ligands for asymmetric induction in allylic alkylations. The recent advances in the development and applications of new chiral ligands will be covered in Sect. 2, and will be the main focus of this chapter. Other aspects of enantioselective palladium-catalyzed allylic substitution will be discussed in the subsequent sections, for example, heterogeneous

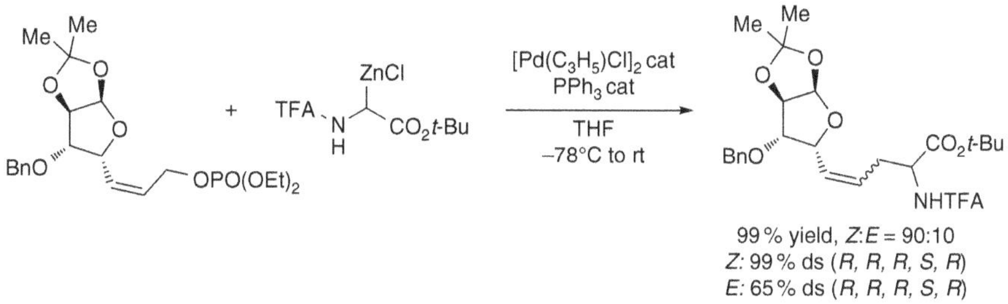

Scheme 5 Asymmetric allylic alkylation employing a chiral substrate

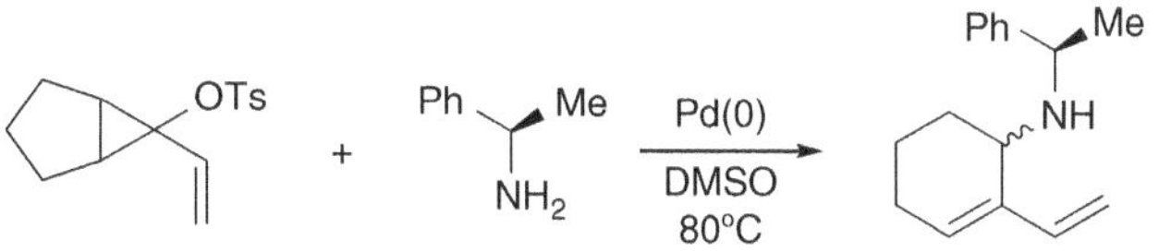

Scheme 6 Allylic alkylation with a chiral nucleophile

catalysis (Sect. 3), the use of chiral salt additives (Sect. 4) and recent applications in kinetic resolution (Sect. 5).

2 Chiral Ligands

Chiral ligands will be classified according to the donor atoms interacting with palladium. A mechanistic distinction must be made between homotopic and heterotopic ligands. When there are two identical donor atoms, with identical geometrical surroundings, only steric interactions govern the selectivity of allylic substitution. When the two atoms are different or have a different electronic environment, electronic effects come into play. In the case of the widely developed phosphorus–nitrogen bidentate ligands, a "trans to P" effect is often observed with the nucleophilic attack occurring preferentially on one side of the π-allyl complex (Fig. 1) [22, 23]. The stronger the Pd-allyl interaction is, the more selective the reaction will be. When there is a close Pd-allyl distance, the transition structure is cation-like and provides an optimal electronic differentiation, whereas more distant Pd-allyl units give more ene-like transition structures. This effect can be amplified by the presence of electron withdrawing groups on the ligand [22, 24].

To compare the different ligands, we will report for each of them the best result for the standard catalyzed reaction of 1,3-diphenylpropenyl acetate **6** with the conjugate base of dimethyl malonate **7** (Scheme 7), which we will abbreviate to Standard AAA. For clarity, we do not specify the experimental conditions for each of the examples, these being available in the original papers referenced throughout

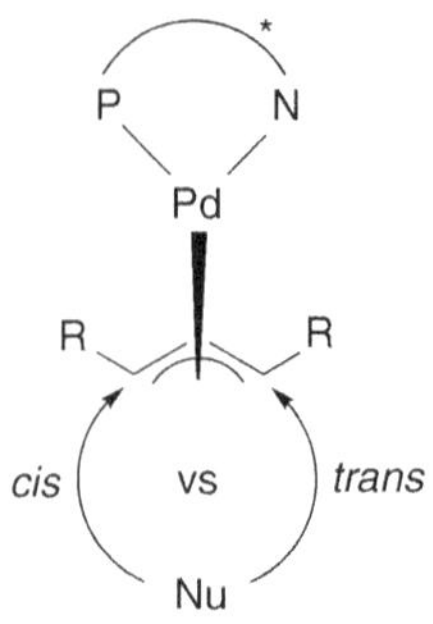

Fig. 1 *Trans* or *cis* nucleophilic attack

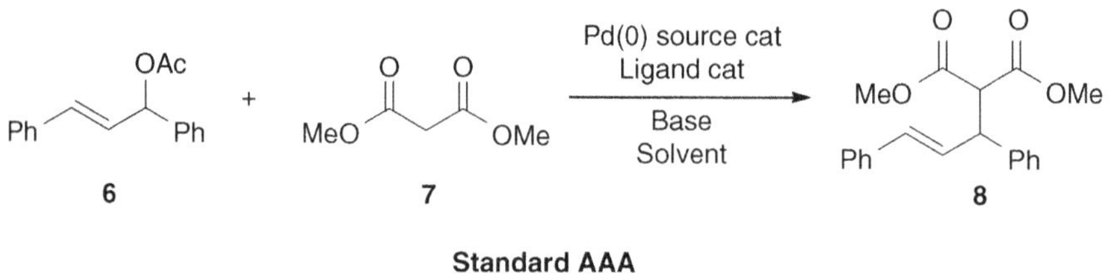

Scheme 7 *De facto* standard asymmetric allylic alkylation

this chapter. If the results for this standard reaction are not available, the reaction showing the best results for the palladium-ligand complex will be given. Furthermore, if a ligand displays spectacular results in a reaction other than our reference one, it will also be discussed.

2.1 P–P Ligands

P–P ligands are one of the largest classes of ligands used in asymmetric catalysis. Aside from several well-established and proven scaffolds, some brand new designs have emerged in the past few years. However, the majority of recent publications is devoted to the slight modification of known ligand structures to fine tune the catalyst reactivity.

An instance of structural variation is the modification of DIOP **L1**. First synthesized by Berens in 1995 [25], the DIOP analogs **L2** have been recently applied to allylic alkylation reactions. Its backbone is more rigid than in DIOP, due to a double anomeric effect, and higher enantioselectivities were therefore observed (Table 2, entries 1 and 2) [26].

Efforts have also focused on the development of new reaction conditions, as exemplified by the use of organic carbonate as solvent with PhanePhos **L3** [27], but with limited success (Table 2, compare entries 3 and 4).

Table 2 Unusual carbonate solvent used in AAA

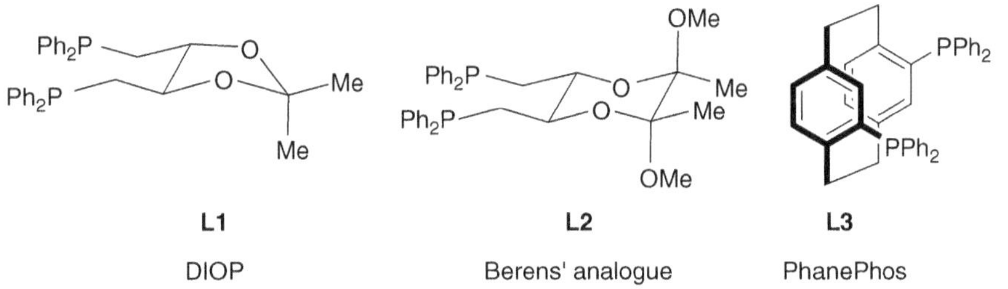

Standard AAA				
Entry	Ligand	Solvent	Conversion %	ee %
1	**L1**	CH_2Cl_2	71	6
2	**L2**	CH_2Cl_2	29	60
3	**L3**	CH_2Cl_2	80	44
4	**L3**	$C(O)(OEt)_2$	98	35

2.1.1 Trost's Ligands

Trost's ligands are undoubtedly the most successful in asymmetric allylic substitutions. Most of the total synthesis applications using a chiral ligand for this class of reaction use Trost's standard ligand **L4**. Computational methods and X-ray structures have shown that the palladium atom is bonded through the phosphorus atoms, while the nitrogen and oxygen atoms are not involved [28, 29]. It uses the phenyl groups on the phosphorus atoms to transfer the chiral information to the substrate (Fig. 2).

This initial model to explain the sense of asymmetric induction has been improved as a result of the mechanistic studies of Lloyd-Jones and Norrby. Two phenomena add to the steric interactions due to the phenyl groups: pro-*S* delivery of the nucleophile can be facilitated by hydrogen bonding with one amide N–H, and pro-*R* delivery can be facilitated by an escort ion M^+ binding to one amide carbonyl (Fig. 3) [28].

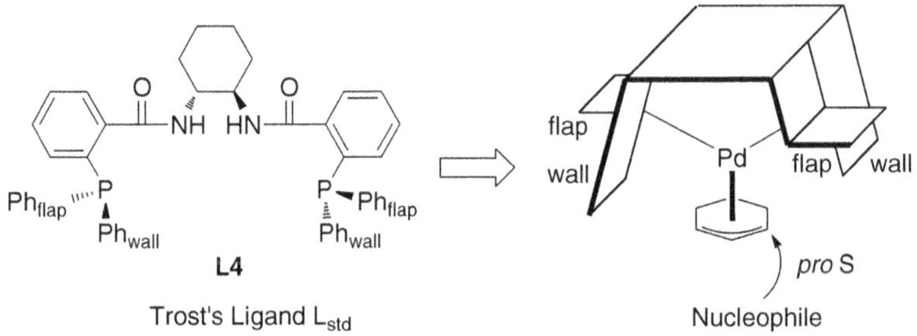

Fig. 2 Proposed steric environment of palladium complexed by Trost's ligand **L4**

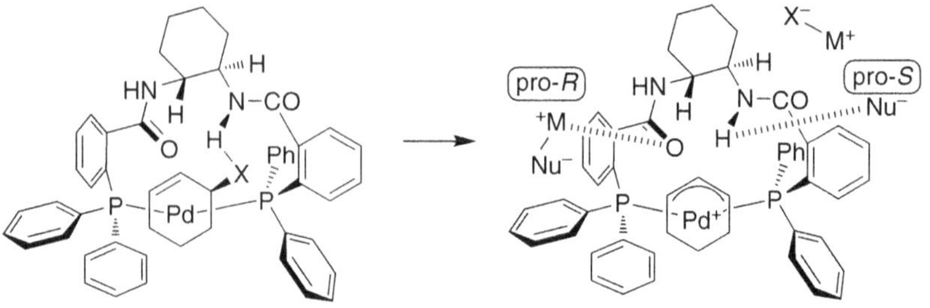

Fig. 3 Pro-*S* and pro-*R* nucleophile delivery with Trost's ligand **L4**

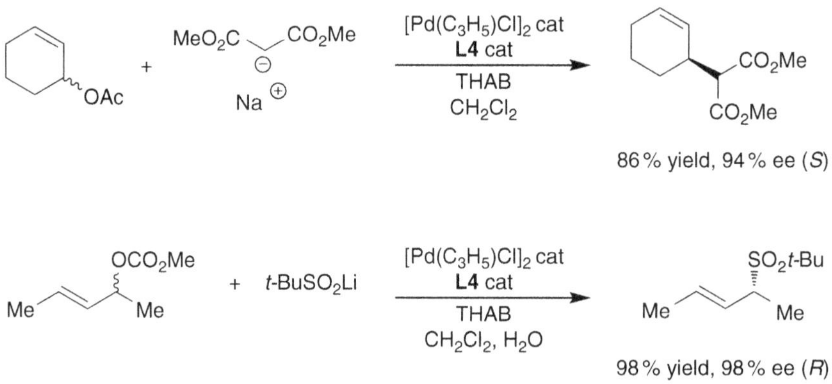

86 % yield, 94 % ee (*S*)

98 % yield, 98 % ee (*R*)

Scheme 8 Uses of Trost's ligand **L4** in asymmetric allylic substitutions

Ligand **L4** has been successfully used in alkylations [30], as well as in sulfonations [31] (Scheme 8). Tetraheptylammonium bromide (THAB) was used as an additive to increase the solubility of the nucleophilic salt in the solvent and high yields and enantioselectivities were obtained.

This sulfonation can also be used as an intramolecular rearrangement [31]. Allylic sulfinates **9** generate a π-allylpalladium complex and the leaving group becomes the nucleophile. The size of the substrate ring can be expanded and the *t*-Bu group can be exchanged by other hindering groups, while keeping high yields and selectivities (Scheme 9). This reaction is very versatile as the starting material can be obtained from the corresponding racemic alcohol and racemic alkyl sulfinyl chloride [32].

The transformation of O-allyl thiocarbamates **10** into S-allyl thiocarbamates **11** is another variation (Scheme 9). As for the previous example, the starting substrates are readily available from the racemic alcohol and now an isothiocyanate.

A noteworthy discovery was made during a study by Trost [33]: the base used in the generation of the nucleophile was found to influence the overall allylic alkylation (Table 3). The originally proposed mechanism suggests that in the case of an enolate obtained with Cs_2CO_3 as base, the *Si* face attack generates less steric

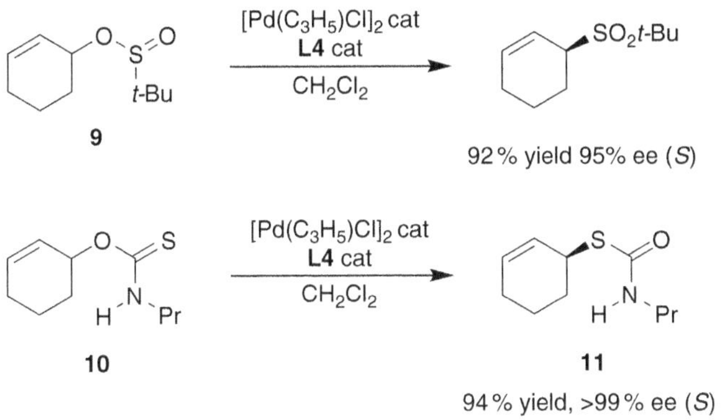

Scheme 9 In situ-generated sulfur nucleophile with Trost's ligand **L4**

Table 3 Counter-ion directed AAA

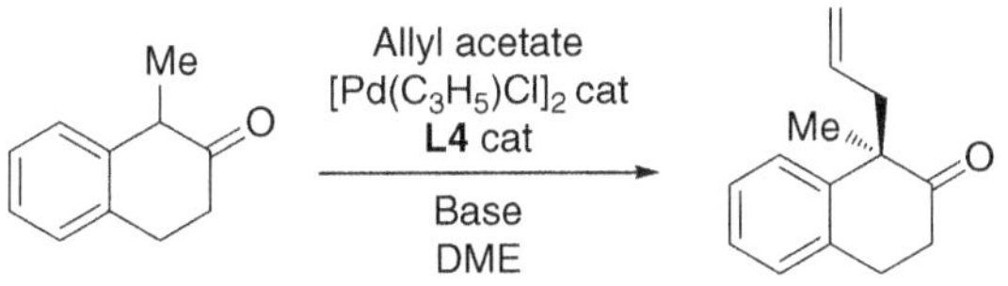

Entry	Base	Yield %	ee % (config.)
1	LDA	90	33 (*S*)
2	Cs_2CO_3	91	45 (*R*)

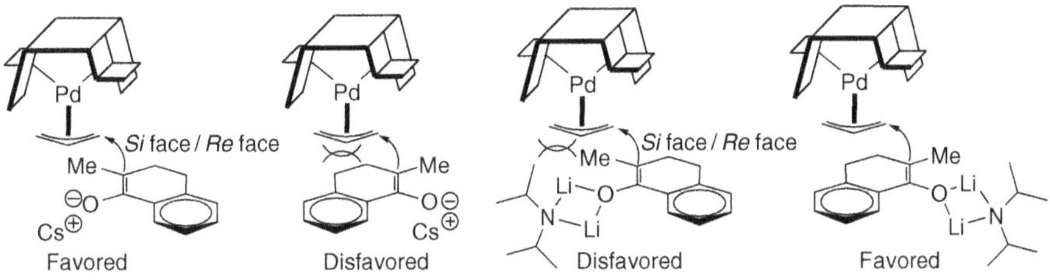

hindrance than the *Re* face attack. However, the LDA-generated nucleophile carries a bulky counterion, which causes the attack via the *Re* face to be favored. This explanation is now challenged by the work of Lloyd-Jones and Norrby [28].

The parent Trost ligand has been modified in an effort to increase activity/selectivity. For example, **L5** is a very powerful ligand for the palladium-catalyzed decarboxylative asymmetric allylic substitution (DAAS) of enol carbonates **12** (Scheme 10). It afforded the desired product **13** without the formation of the corresponding unsubstituted ketone **14** by simple decarboxylation [33].

Ligand **L5** has also shown very good results in the reaction of unstabilized nucleophiles [34]. As an example, a hard nucleophile was obtained from

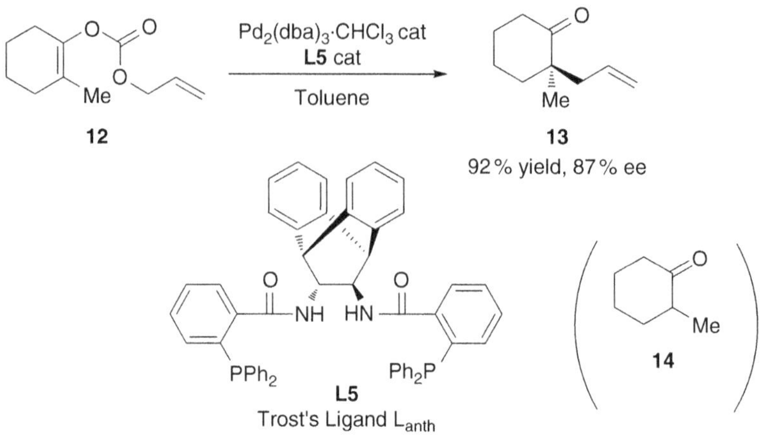

Scheme 10 Trost's **L5** used in DAAS

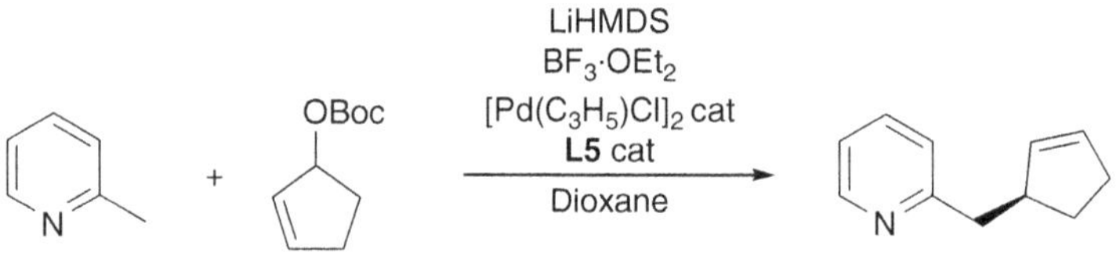

Scheme 11 Unstabilized nucleophile used with ligand **L5**

a 2-methylpyridine – BF$_3$ complex in the presence of LiHMDS which reacted with allylic substrate **15** to give the product **16** with very high enantiomeric excess (Scheme 11).

The Trost ligand analog **L6** is another good ligand for decarboxylative allylic substitution. It has been used in the synthesis of α-hydroxyaldehydes [35], versatile building blocks in natural product synthesis [36]. It afforded product **17** with high selectivity, limiting the formation of the isomeric **18**. The rates k_1 and k_2 are affected by the ligand used, causing the formation of different mixtures with **L5** and **L6** (Table 4).

L6 has also been used on bisallylic carbonates, such as Morita–Baylis–Hillman adducts **19**. These adducts are useful building blocks and the focus of much research [37]. Using phenol as a nucleophile, the desired product **20** was obtained in high regio- and enantioselectivities (Scheme 12). This is presumably helped by the presence of 5% tetrabutylammonium triphenydifluorosilicate (TBAT), as it is known that fluoride additives attack the palladium atom of the π-allyl complex, resulting in greater $\eta^3-\eta^1-\eta^3$ equilibration [38].

Table 4 Regioselectivity in DAAS with Trost's ligands

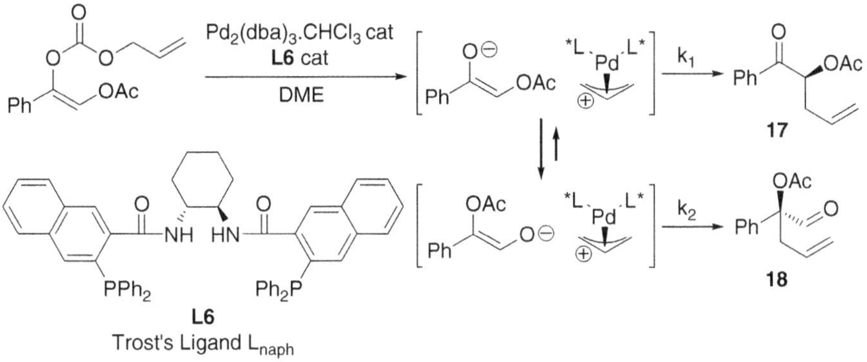

Entry	Ligand	Yield %	**17/18**	**17** ee %
1	**L5**	27	3/2	25 (*R*)
2	**L6**	99	49/2	90 (*R*)

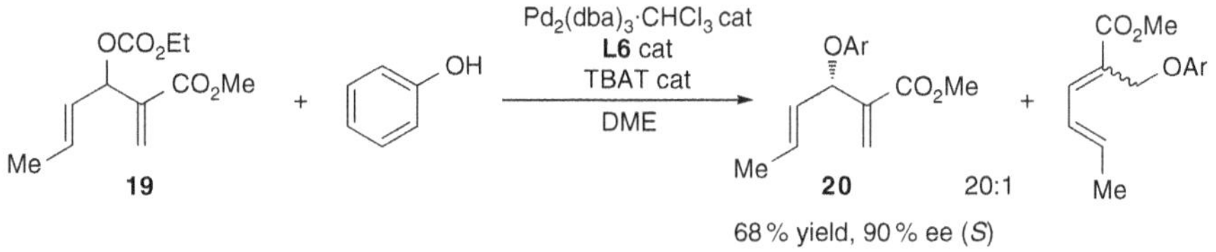

Scheme 12 DAAS using Morita–Baylis–Hillman adducts as substrates

2.1.2 Metallocenes

Metallocenes form a second class of ligands, where elements of chirality are brought by the substitution pattern around the cyclopentadienyl rings. Josiphos-type ligands **L7** have the two phosphorus atoms linked with the same ring, thus displaying planar and central chirality. Variations of the phosphorus substituents allowed for the tuning of the allylic phosphination reaction affording good to high levels of enantioselectivity (Table 5) [39].

In contrast, Zhang's group has been developing metallocene ligands **L8**, where the two phosphorus atoms are on both cyclopentadienyl rings. By changing the metal involved (Fe or Ru), the P–P distance can be modified, affecting the selectivity of the alkylation and amination, as depicted in Table 6 [40].

Interestingly in the presence of Pd(II), the conformer **C1** is exclusively formed (Fig. 4). The ester substituents on each ring are close to the reaction site and can thus influence the selectivity of the allylic substitution. However, $-CH_2OH$, $-CH_2OAc$ and $-CH_2OMe$ substituents all showed decreased performances in catalysis [41].

Table 5 Josiphos-type ligands **L7** used in asymmetric allylic substitution

Entry	R	R′	Yield %	ee %
1	Ph	Cy	68	87
2	3,5-$(CF_3)_2C_6H_3$	Cy	92	82
3	Ph	*t*-Bu	78	69
4	Cy	Ph	40	80
5	Ph	1,5-bicyclooctane	65	82

Table 6 Zhang's metallocene ligands **L8** used in asymmetric allylic substitution

Entry	Nucleophile	Base	Metal	Solvent	Yield %	ee %
1	$CH_2(CO_2Me)_2$	BSA	Ru	CH_2Cl_2	99	96
2	$CH_2(CO_2Me)_2$	BSA	Fe	CH_2Cl_2	>90	88
3	H_2NBn	–	Ru	Toluene	73	94
4	H_2NBn	–	Fe	Toluene	99	99

The well-known diphosphine ligand BPPFA **L9** developed by Kumada and Hayashi [42] for cross coupling and also applied to allylic substitutions has been modified by the introduction of different R groups and hemilabile donor atoms

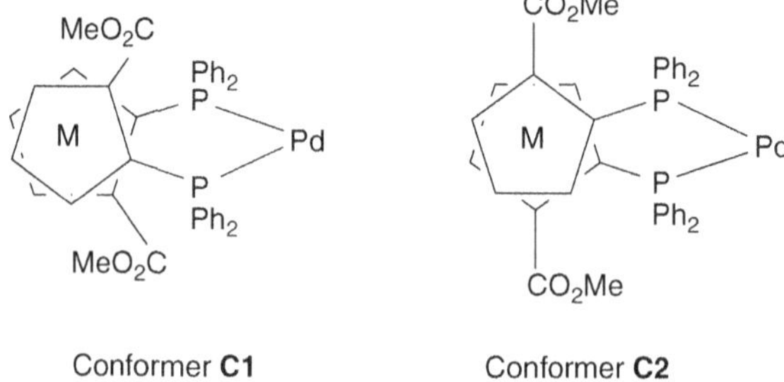

Fig. 4 Possible conformers of Zhang's ligands with palladium

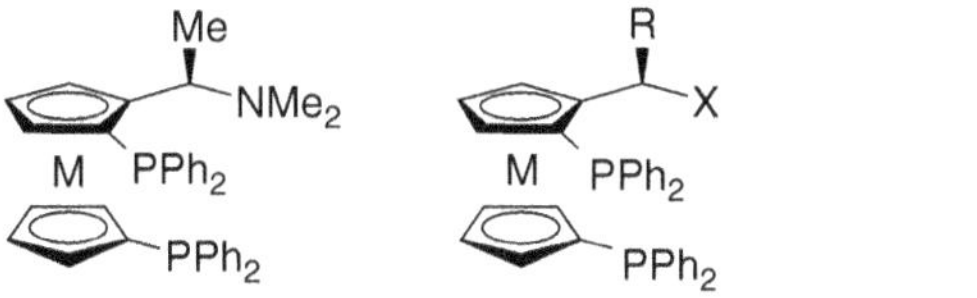

L9
BPPFA

L10 R = Ph X = NMe_2
L11 R = Ph X = OAc
L12 R = Ph X = OMe
L13 R = Me X = OH

Fig. 5 BPPFA **L9** and derivatives **L10–13**

Table 7 BPPFA **L9** and derivatives **L10–13** results in AAA

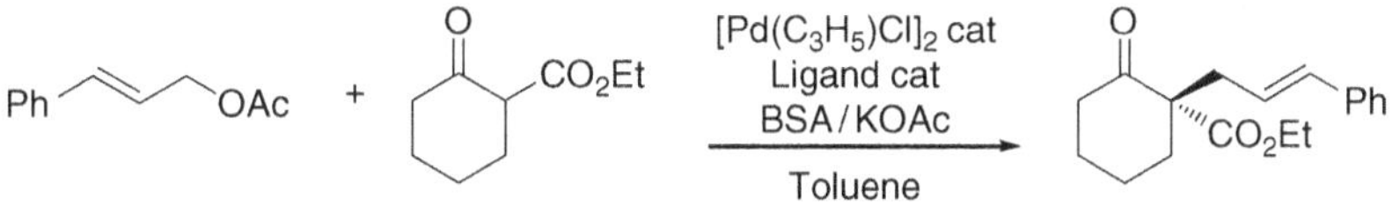

Entry	Ligand	Conversion %	ee % (config.)
1	**L9**	100	8 (*S*)
2	**L10**	100	53 (*S*)
3	**L11**	100	23 (*S*)
4	**L12**	100	29 (*S*)
5	**L13**	100	85 (*R*)

L10–L13 (Fig. 5). The optimal ligand was **L13** (85% ee) with reversed enantioselectivity from **L9** to **L11**, presumably due to a strong interaction with the incoming nucleophile [43].

These ligands have been used with chiral nucleophiles, creating a new chiral center on the attacking carbon [43]. At room temperature, perfect regioselectivity was observed and very short reaction times were needed to achieve full conversion (Table 7).

Sebesta has introduced an imidazolium unit onto the N-substituent of the BPPFA parent ligand to increase the solubility in ionic liquids of the resultant ligand **L14** and subsequent catalytic intermediates [44]. By using a mixture of ionic liquid and dichloromethane, aminations were performed with up to 92% ee, whereas the use of pure dichloromethane afforded an ee of 70% and lower yields (Table 8).

2.1.3 Axially Chiral Ligands

The reference in P–P axially chiral ligands is BINAP **L15**, even if more efficient ligands exist nowadays for specific purposes. It shows great results in terms of selectivity and yields in allylic alkylations [45] and fine tuning of the conditions adapts the reactivity (Table 9, entries 1 to 2). The related axially chiral ligands **L16–L19** were applied to this transformation and (*S*)-MeO-BIPHEP (**L17**) gave the

Table 8 Sebesta's BPPFA derivative **L14** used in asymmetric allylic substitution

Entry	Solvent	Yield %	ee %
1	CH_2Cl_2	19	70 (*R*)
2	[bdmim]PF_6 and CH_2Cl_2	70	92 (*R*)

Table 9 C_2-symmetric axially chiral P–P ligands used in AAA

Entry	Ligand	R	LG	Yield %	dr	ee % (config.)
1	(*S*)-**L15**	Ph	Ac	97	99/1	99 (*R*, *S*)
2	(*S*)-**L15**	H	CO_2Me	73	–	97 (*R*)
3	(*S*)-**L16**	H	CO_2Me	68	–	94 (*R*)
4	(*S*)-**L17**	H	CO_2Me	76	–	98 (*S*)
5	(*R*)-**L18**	H	CO_2Me	65	–	91 (*R*)
6	(*R*)-**L19**	H	CO_2Me	40	–	92 (*R*)

highest ee of 98% (entry 4). The more electron deficient DIFLUORPHOS **L19** showed a significantly reduced yield (entry 6).

A selective intramolecular alkylation has also been reported with these ligands [46]. Although intermolecular asymmetric allylic alkylations are a well-established

Table 10 C_2-symmetric axially chiral P–P ligands used in intramolecular AAA

$[Pd(C_3H_5)Cl]_2$cat, Ligand cat, TBABr, KOH aq./CH_2Cl_2: **21** → **22**

Entry	Ligand	Yield %	ee %
1	**L15**	88	72 (*S*, *R*)
2	**L17**	80	72 (*S*, *R*)
3	**L19**	0	–

Table 11 Electronically differentiated axially chiral P–P ligands used in AAA

$[Pd(C_3H_5)Cl]_2$cat, Ligand cat, BSA/KOAc, DMF

L20 X = PPh_2
L21 X = $P(O)Ph_2$

Entry	Ligand	Yield %	ee %
1	**L20**	46	29
2	**L21**	88	53

synthetic method, intramolecular versions have been scarcely reported, only Genet [47] being able to control two stereogenic centers in this step. Moderate ee's of 72% have been obtained with BINAP **L15** and MeOBIPHEP **L17**, while no reaction was observed with DIFLUORPHOS **L19** in the conversion of allylic acetate **21** to lactam **22** (Table 10).

In these C_2-symmetric ligands, stereochemical information is transferred by steric interactions alone. By differentiating the two binding sites, one could expect to increase the selectivity of the reaction. The related ligand SEGPHOS **L20** has been transformed into the mono-phosphine oxide ligand **L21** [48], affording increased yield and ee in allylic alkylation (Table 11). However, no structural information was given on the binding mode of the ligand, whether through the two phosphorus atoms or through one phosphorus and one oxygen.

The phosphorus atoms do not need to be directly linked with the chiral biaryl unit for the chiral information to be transferred. In ligand **L22**, two possible conformers exist [49]. The equatorial conformer **L22eq** is kinetically favored

Table 12 Conformationally fixed ligands **L22** used in AAA

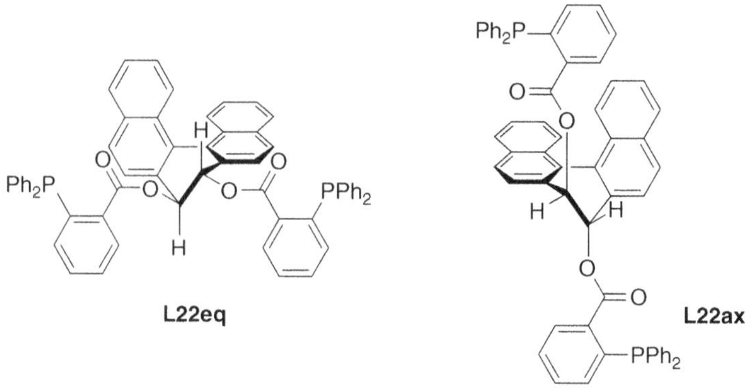

Standard AAA					
Entry	Nucleophile	Base	Ligand	Yield %	ee %
1	$CH_2(CO_2Me)_2$	BSA/KOAc	**L22ax**	93	82 (*S*)
2	$CH_2(CO_2Me)_2$	BSA/KOAc	**L22eq**	95	36 (*S*)
3	H_2NBn	None	**L22ax**	91	70 (*R*)
4	H_2NBn	None	**L22eq**	91	90 (*R*)

during its synthesis, and interconversion does not occur at room temperature. However, the axial conformer **L22ax** can be obtained through an alternate synthesis involving very bulky silyl protecting groups. They have both been tested for their enantiodifferentiating ability in allylic alkylation and amination (Table 12). Curiously, **L22ax** was best for allylic alkylations (ee 82%), whereas **L22eq** was optimal for amination (ee 90%) using diphenylallyl acetate **15** as a standard substrate (**Standard AAA**, see Scheme 7).

2.1.4 Phosphinites, Phosphites, and Phosphoramidites

Phosphorus-based ligands are not limited to phosphines and other functional groups involving heteroatoms can be used. Diphosphinites are easily prepared from the corresponding diol and chlorodiphenylphosphine and such derivatives of ascorbic acid (**L23**) have been synthesized and used in allylic substitutions with interesting results [50]. Enantiomeric excesses of up to 91% and good yields have been obtained in reactions of malonates, making these ligands a cheap and promising new class (Table 13).

C_2-symmetric chiral diphosphites **L24** derived from carbohydrates have also been used, with turnover frequencies above 22,000 h^{-1} and ee's of 98% (Table 14) [51]. Two phosphites are formed from the chiral carbohydrate scaffold by reaction with a pro-atropisomeric unit, which freely rotates at room temperature.

Table 13 Ascorbic acid-derived diphosphite ligand **L23** used in AAA

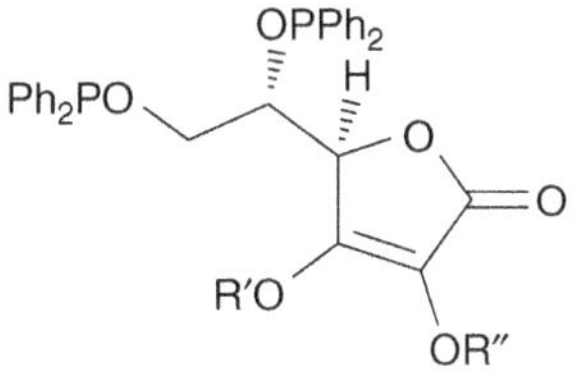

L23

Standard AAA					
Entry	Ligand	R′	R″	Yield %	ee %
1	**L23-1**	H	H	75	2 (*R*)
2	**L23-2**	H	Bn	84	91 (*R*)
3	**L23-3**	Bn	Bn	85	69 (*R*)
4	**L23-4**	Bn	H	90	20 (*R*)

Table 14 C_2-symmetric diphosphite **L24** used in AAA

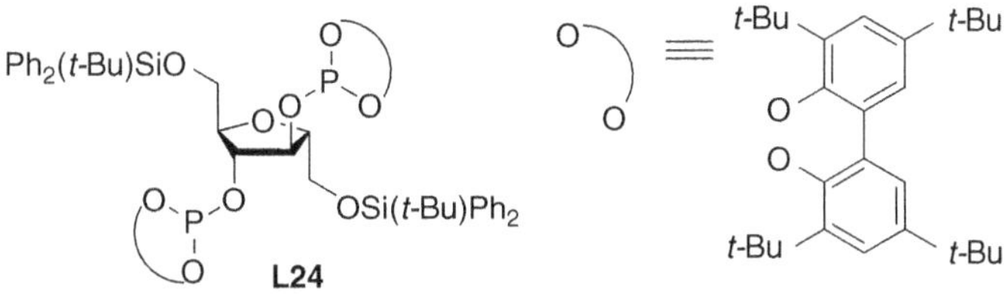

L24

Standard AAA			
Entry	Nucleophile	Yield %	ee %
1	$CH_2(CO_2Me)_2$	100	98 (*S*)
2	H_2NBn	100	98 (*R*)

Other central scaffolds can be used, such as a diol derived from dihydroanthracene [52]. In allylic alkylations (Fig. 6), both versions of **L25** with X = C(O) or X = CH_2 gave the same results, probably due to the distance between the binding sites and the site of variation.

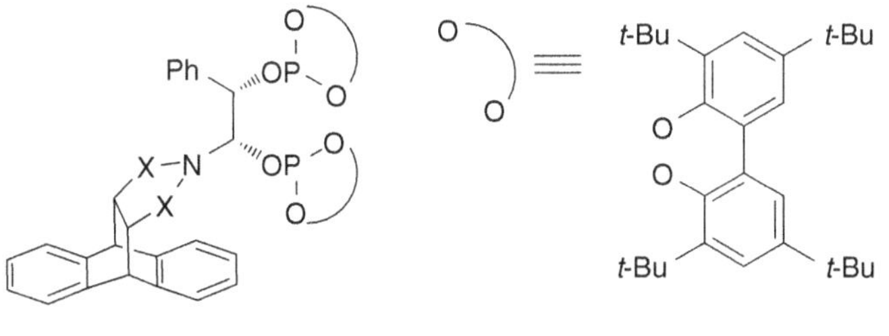

L25 **Standard AAA:** 93 % yield, 90 % ee (*R*)

Fig. 6 Dihydroanthracene-derived diphosphite ligand **L25** used in AAA

Table 15 Influence of steric hindrance on **L26** and **L27** used in AAA

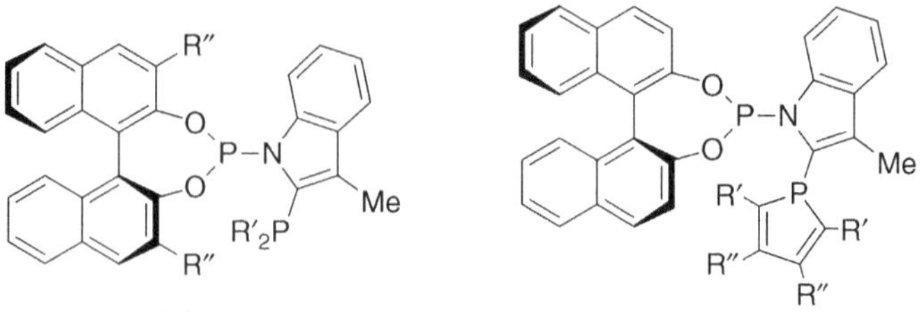

Standard AAA					
Entry	Ligand	R′	R″	Conversion %	ee % (config.)
1	**L26-1**	Ph	H	100	43 (*S*)
2	**L26-2**	*o*-Tol	H	100	66 (*S*)
3	**L26-3**	Ph	TMS	15	90 (*S*)
4	**L27-1**	Ph	H	99	32 (*S*)
5	**L27-2**	H	Me	100	56 (*S*)

In the case of IndolPHOS **L26** [53], the enantioselectivity increased with increasing steric hindrance of both the phosphine and the phosphoramidite units. With IndolPhospholes **L27,** the two reported examples did not allow for trends in the influence of the bulkiness of the phosphole unit to be identified (Table 15).

The two phosphorus atoms can also belong to phosphoramidites. Sugar-based ligands **L28** have been developed by Dieguez and Pamies [54]. Hybrids of phosphite and phosphoramidites of the same class have also been explored [55–57].

Changing one phosphoramidite into a phosphite (Table 16, entries 1 and 2) gave an opposite selectivity (*S* instead of *R*), hinting for two different mechanisms or for a strong electronic differentiation between the two phosphorus atoms. The different chiral elements of the ligands all have an influence on the selectivity of the reaction. One must note that the apparent change in selectivity between alkylation and amination is due to the priority of the substituents in the Cahn–Ingold–Prelog system, the actual sense of asymmetric induction is the same (entries 5 and 6). Other sugar-derived ligands such as **L29** showed good enantioselectivity (entry 7) [56].

Pamies and Dieguez have also developed noncyclic scaffolds. They reasoned that bulky substituents are required in α–position to the phosphite (R′) and on the 3- and 3′-positions of the biaryl units [58, 59]. Excellent enantioselectivities were obtained for alkylations (Table 17, entries 1-4) for alkylations (entires 1–4) as well as aminations (entry 5), while the selectivity dropped dramatically in case of the unsubstituted biphenyl ligand (entry 6).

Table 16 Dieguez and Pamies sugar-based P–P ligands **L28** and **L29** used in AAA

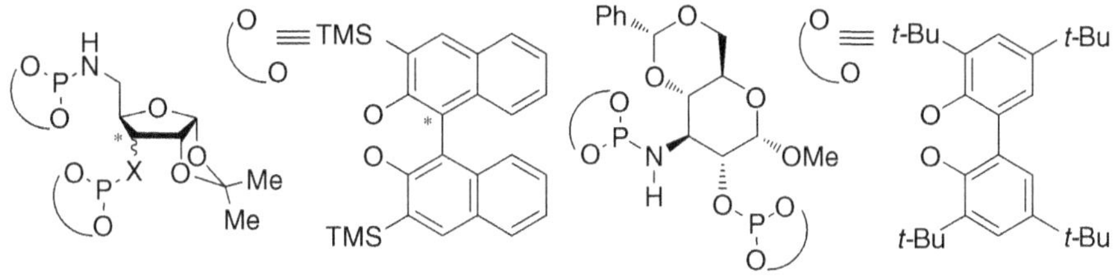

Standard AAA						
Entry	Ligand	X	C*	Ax*	Yield %	ee % (config.)
1	**L28-1**	NH	*S*	*R*	78	75 (*R*)
2	**L28-2**	O	*S*	*R*	12	80 (*S*)
3	**L28-3**	O	*S*	*S*	12	12 (*S*)
4	**L28-4**	O	*R*	*R*	71	6 (*S*)
5	**L28-5**	O	*R*	*S*	96	98 (*S*)
6[a]	**L28-5**	O	*R*	*S*	98	97 (*R*)
7	**L29**	–			100	85 (*R*)

[a]Amination with H_2NBn as nucleophile

Table 17 Dieguez and Pamies acyclic P–P ligands **L30** used in AAA

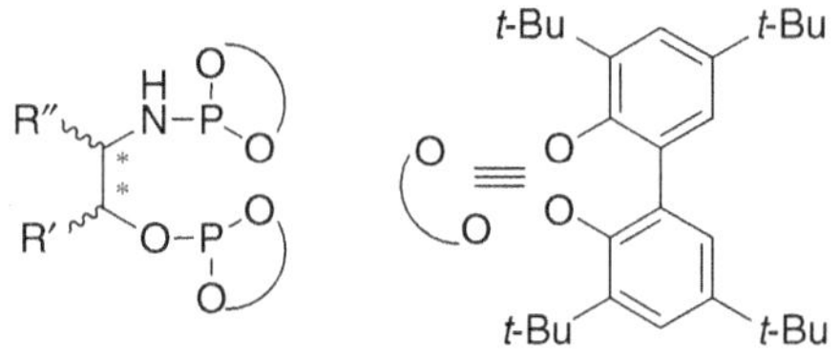

Standard AAA				
Entry	R′	R″	Yield %	ee % (config.)
1	(*S*)-Ph	(*R*)-Ph	100	94 (*R*)
2	(*R*)-Ph	(*S*)-Ph	100	96 (*S*)
3	H	(*S*)-Ph	100	68 (*S*)
4	(*R*)-Ph	H	100	86 (*S*)
5[a]	(*S*)-Ph	(*R*)-Ph	100	96 (*S*)
6[b]	(*S*)-Ph	(*R*)-Ph	41	6 (*R*)

[a]Amination with H_2NBn as nucleophile
[b]Diol : biphenyl-2,2′-diol

2.1.5 P-Chiral Ligands

In all the previous cases, there was some measure of symmetry around the phosphorus atoms, with at least two identical substituents. P-chiral ligands have been less frequently applied in catalysis. Recently, asymmetric bis-diamidophosphites **L31** and **L32** have been prepared [60, 61], with promising results (Fig. 7).

Good enantioselectivities have also been achieved with less complex structures, for example, *t*-Bu-QUINOXP* **L33** (Fig. 8) [62].

2.1.6 Supramolecular Ligands

Although previous examples have shown that the influence of the steric environment on the selectivity of the reaction decreases when the distance to the binding site increase, a few supramolecular ligands exist. Porphyrins can be used to complex a planar chiral diphosphite through its amine functionality as in **L34** (Fig. 9) [63]. However, only an enantiomeric excess of 18% toward the *S* isomer was achieved at full conversion in the typical reaction with dimethyl malonate **16**. A porphyrin dimer **L35** has been investigated and it afforded a better ee of 45%, which is postulated to be due to the phosphite axial chirality as the opposite atropisomer gave the other hand of alkylated product in preference [64].

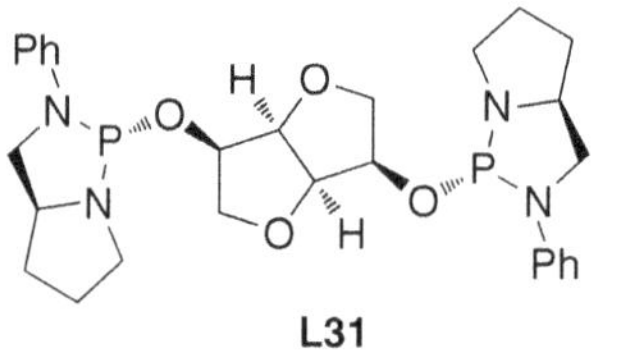

L31

Standard AAA: 72% yield, 98% ee (*S*)

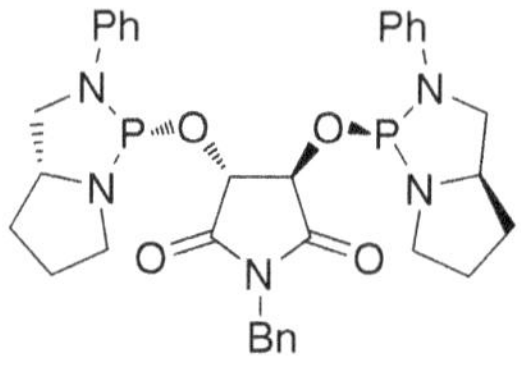

L32

98% yield, 96% ee (*R*)

Fig. 7 P-Chiral ligands **L31** and **L32** used in AAA

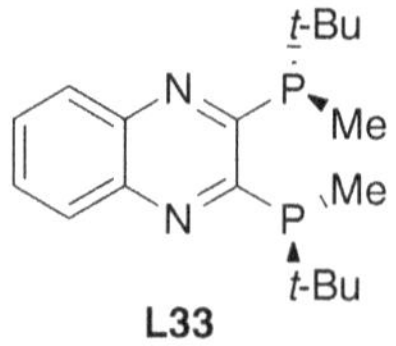

L33

Standard AAA: 85% yield, 92% ee (*S*)

Fig. 8 P-Chiral *t*-Bu-QUINOXP* ligand **L33** used in AAA

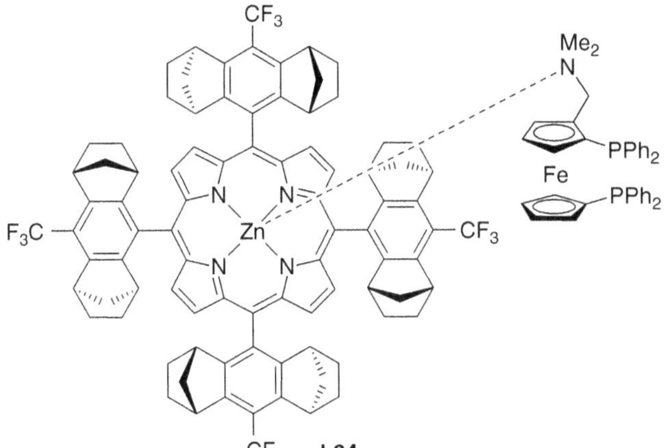

CF3 L34

Standard AAA: 99% yield, 18% ee (*S*)

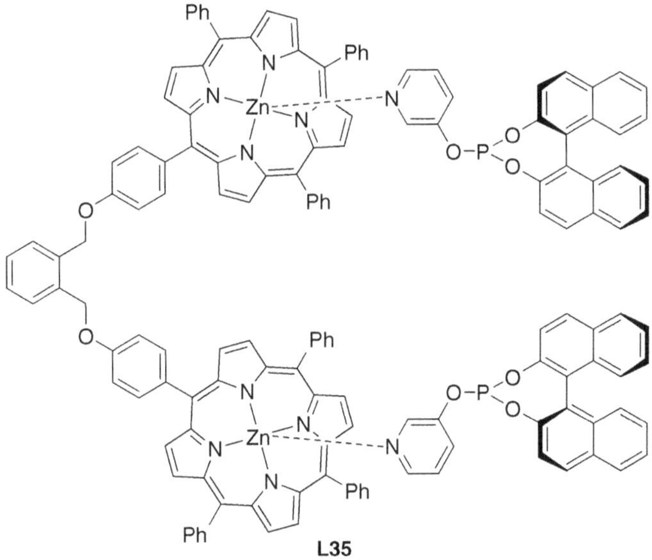

L35

Standard AAA: 99% yield, 45% ee (*S*)

Fig. 9 Porphyrin-based ligands **L34** and **L35** used in AAA

A protein-based approach to enantioselective allylic substitution has recently been studied [65]. The diphosphine ligand **L36** contains a biotin unit, a cofactor to avidin proteins, forming an interesting environment for the reaction site (Fig. 10). The best recorded result was 95% yield, with 90% ee.

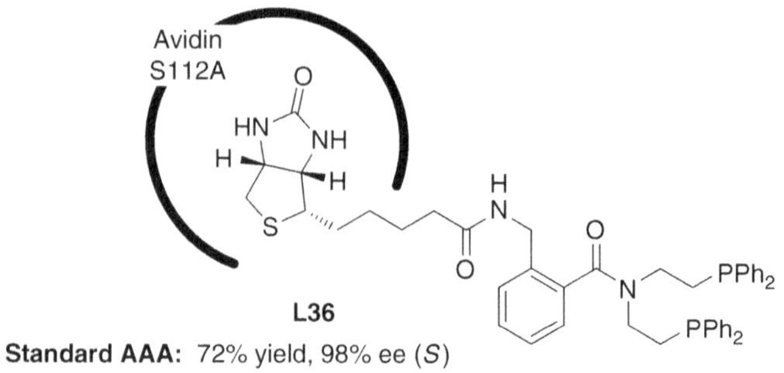

Fig. 10 Protein-cofactor complex used as a ligand in AAA

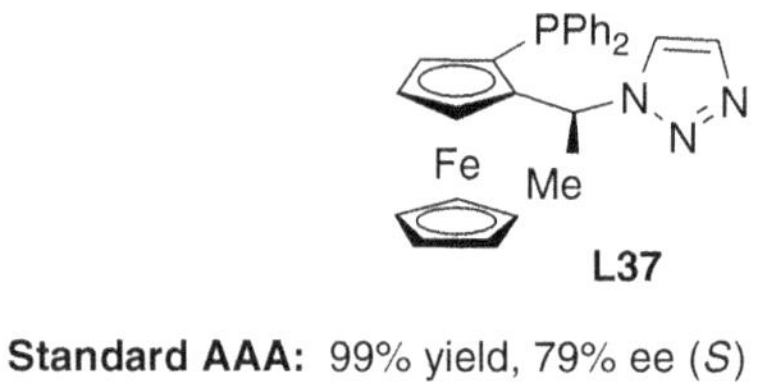

Fig. 11 P–N variation of the Josiphos ligand used in AAA

2.2 *P–N Ligands*

P–N compounds form a very rich class of ligands in asymmetric catalysis [23]. To give a clear overview of this area, the ligands will be first classified according to the characteristic scaffold features and miscellaneous remaining ligands by the nitrogen group present.

2.2.1 Metallocenes and Planar Chirality

Similar to the concept of Josiphos **L7** discussed in a preceding Sect. 2.1.2, P–N ligands can be derived from metallocenes with phosphorus and nitrogen atoms substituents on the same ring, with an additional chiral center on a side chain. ClickFerrophos **L37** displayed moderate selectivity in asymmetric allylic alkylation (Fig. 11) [66].

Oxazolines are a recurring feature in P–N ligands. **L38**, with an iron or ruthenium central atom, gave very good enantiomeric excesses, although the *R,S*/*S,S* selectivity was poor (Table 18) [67].

Symmetrical versions such as **L39** exist, but they have been less efficient in the reaction of **6** with **23** [67]. It has been shown that in solution, each ligand molecule can complex two palladium atoms (**C39-2**) and that none of the P–Pd–P compound **C39-1** can be found [68]. In the standard asymmetric allylic alkylation reaction of

Table 18 Metallocene- and oxazoline-derived ligand **L38** used in AAA

Entry	Metal	Yield %	*R,S* **24**/*S,S* **25**	ee % (*R,S*/*S,S*)
1	Fe	90	54/46	94/95
2	Ru	90	52/48	94/95

Table 19 Symmetric metallocene- and oxazoline-derived ligand **L39** used in AAA and possible palladium complexation

Standard AAA					
Entry	Ligand	Metal	R	Yield %	ee % (config.)
1	**L39-1**	Ru	*i*-Pr	95	92 (*S*)
2	**L39-2**	Ru	*t*-Bu	95	98 (*S*)

6, these ligands **L39** have shown good results [68]. It must be noted that a larger R group on the oxazoline induced better enantioselectivity (Table 19).

Alternatively, phosphorus and nitrogen donor atoms can be on different cyclopentadienyl rings of the metallocene. **L40** has a chiral center on the oxazoline moiety, another on the phosphorus atom and axial chirality on the binaphthol unit. This ligand has been used in reactions with enolates generated from **26** and **27** acting as nucleophiles [69, 70]. Excellent enantioselectivities were achieved (Scheme 13).

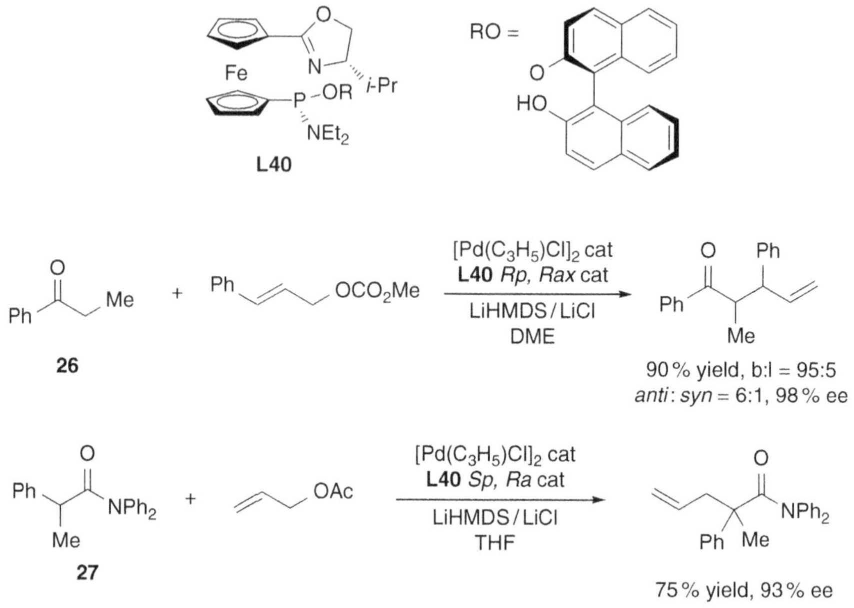

Scheme 13 P-Chiral oxazoline-based ligand **L40** used in AAA

L41 **L42**

Standard AAA: 99% yield, 37% ee (*S*) 92% yield, 96% ee (*Rv*)

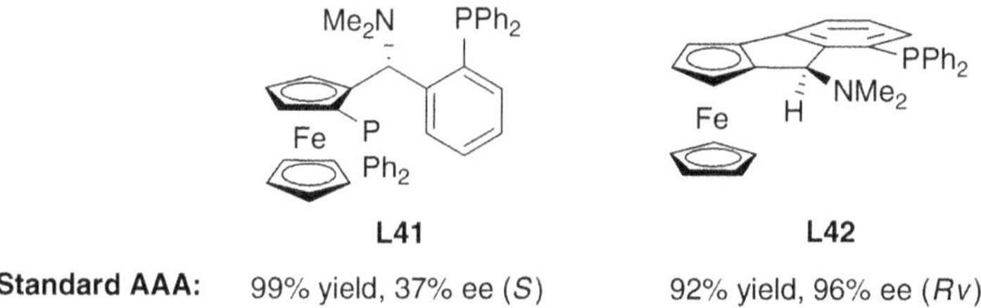

Fig. 12 Taniaphos **L41** and derivative **L42** used in AAA

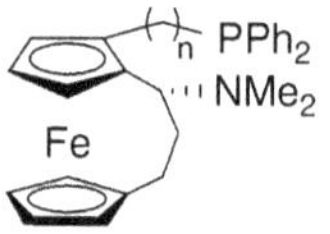

Standard AAA:
L43 n = 0 15 % yield, 7 % ee (*S*)
L44 n = 1 94 % yield, 84 % ee (*S*)

Fig. 13 P–N ligands **L43** and **L44** with a bridged ferrocene unit used in AAA

Taniaphos ligands **L41** and derivates such as **L42** also belong to the P–N ligand category, although Taniaphos itself is a P–N–P ligand. **L42** afforded the desired product with 96% ee, whereas Taniaphos **L41** only gave 37% ee (Fig. 12) [71].

Another variation on the ferrocene motif involves linking of the two cyclopentadienyl rings together, through a propylene chain. The phosphine can be either directly bound to the ring (**L43**) or through a methylene unit (**L44**), the latter being much more efficient in asymmetric allylic alkylation (Fig. 13) [72].

L45 L46

Standard AAA: 84 % yield, 99 % ee (*S*) 60 % yield, 58 % ee (*S*)

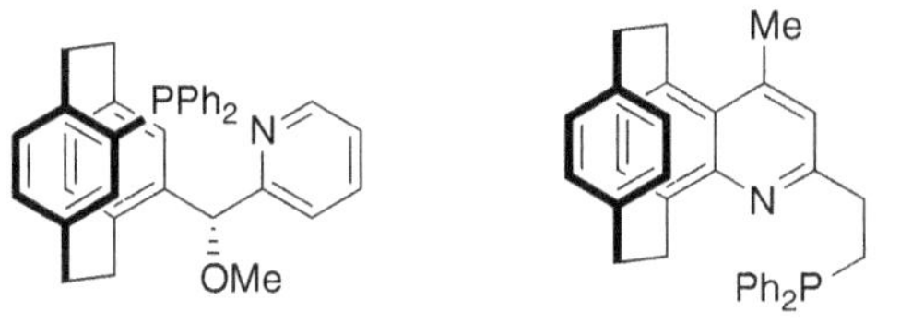

Fig. 14 Other P–N ligands **L45** and **L46** with planar chirality used in AAA

Table 20 Quinazolinap **L47** used in AAA

L47

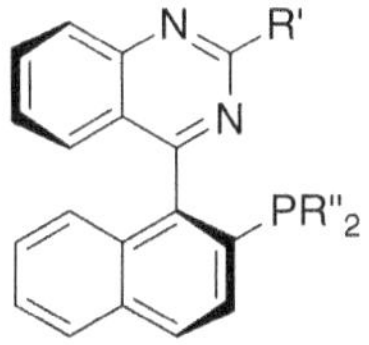

Standard AAA					
Entry	Ligand	R′	R″	Yield %	ee % (config.)
1	**L47-1**	Adm	Ph	88	44 (*R*)
2	**L47-2**	*c*-Bu	Ph	95	89 (*R*)
3	**L47-3**	*i*-Pr	3,5-Xylyl	100	92 (*R*)

Ligands, other than metallocenes, possessing planar chirality have also been investigated. As an example, [2,2]-paracyclophane derived ligands **L45** and **L46** have been very effective with enantioselectivities up to 99% (Fig. 14) [73, 74].

2.2.2 Ligands with Axial Chirality

Quinazolinap ligands of type **L47** have been extensively investigated in asymmetric catalysis by Guiry and co-workers [75–80]. The 3-position of the quinazoline and the substituents of the phosphine have been found to be very important in tuning the reactivity in catalysis. In asymmetric allylic alkylation, the enantioselectivity decreased in the presence of a bulky group such as adamantyl (adm) in the 3-position (Table 20) [81, 82]

In another approach, both of the binding atoms were surrounded by a binaphthyl unit, as in **L48** [83]. Efficiency seemed maximal when the conformation of both binaphthyl units was the same (both *S* or both *R*) (Fig. 15). Mismatching the binaphthyl units reduced the enantioselectivity of the reaction, and replacing a binaphthyl by a biphenyl decreased the yield of the alkylation.

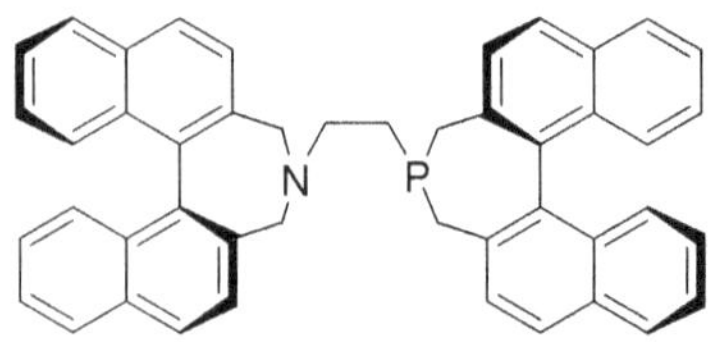

L48

Standard AAA: 100 % yield, 98 % ee (*S*)

Fig. 15 Bis(binaphthyl) P–N ligand **L48** used in AAA

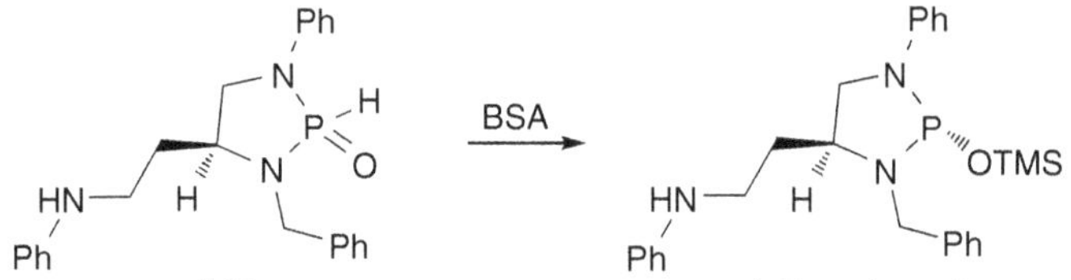

L49 —BSA→ L49 activated

Standard AAA: 94 % yield, 99 % ee (*S*)

Scheme 14 Diaphox **L49** used in AAA

2.2.3 P-Chiral Ligands

The Diaphox ligand **L49** is peculiar as the actual active species binding to the palladium is generated in situ [84]. The surrounding of the phosphorus atom is unsymmetrical, making the P(III) a chiral center. A carbon atom on the main ring is also chiral. Excellent results were obtained in asymmetric allylic alkylation (Scheme 14).

2.2.4 Other Oxazoline-Based Ligands

The Phox ligands of type **L50** are a very versatile class, easily available from a variety of routes [85]. They are very efficient in asymmetric allylic alkylation and have been used with a variety of substrates and nucleophiles (Scheme 15) [26, 86]. The good results observed with nitroalkanes such as **28** are particularly impressive as these nucleophiles are known to be very unreactive in asymmetric allylic alkylations and to generally afford very poor enantioselectivities [86].

Decarboxylative allylic substitutions have also been successfully developed by Stoltz and co-workers with this ligand class. A few unusual properties have been reported with several substrates. In the reaction of substrate **29**, it has been shown by a combination of ^{31}P NMR spectroscopy and X-ray crystallography that the catalytic cycle goes through the formation of a η^1-σ-allyl-palladium species, instead of the usual η^3-π-allyl-palladium (Scheme 16) [87].

In the case of substrate **30** with a fluorine atom on the prochiral carbon, it has been shown that a ratio of ligand to palladium of 1 to 4 afforded the best results (Table 21) [88], although this has not been rationalized to date.

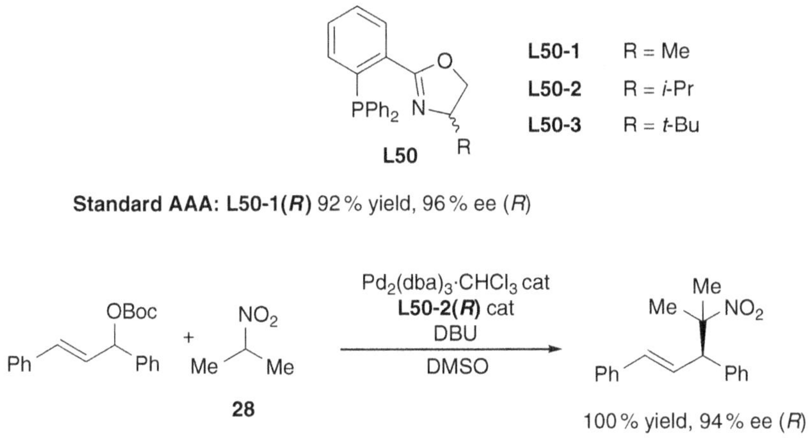

Scheme 15 Phox ligands **L50** used in AAA

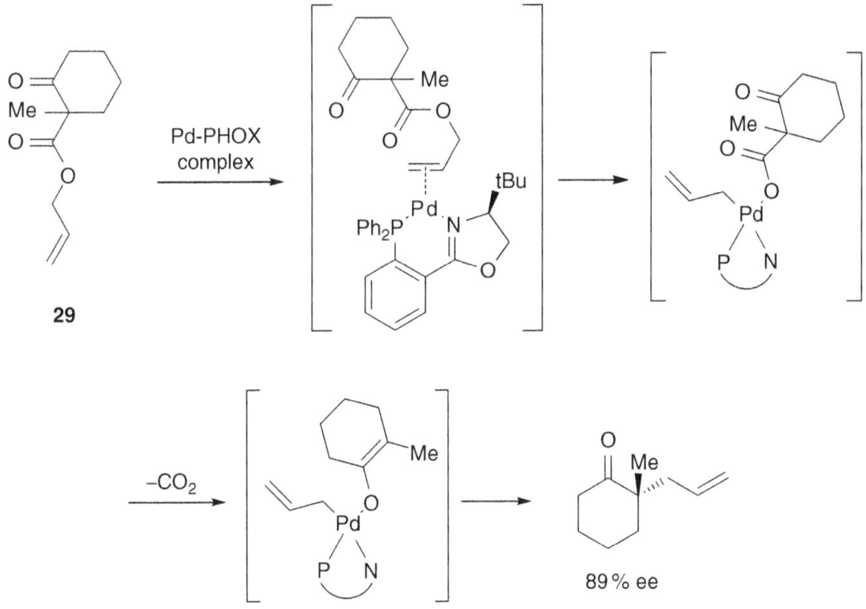

Scheme 16 Phox ligands **L50** used in DAAS

Ring expanding allylations have been reported using allylic alkylation and **L50-3** (*S*) [89]. The postulated mechanism (Scheme 17) shows the ring expanding step being favored in terms of enthalpy with the opening of a strained four-membered ring, and in terms of entropy with the release of carbon dioxide.

The introduction of an ether side chain on the parent ligand **L50** gave rise to the related analog **L51**, which displayed an excellent enantioselectivity (Fig. 16) [90].

Table 21 Unusual ligand to palladium ratio with Phox **L50** used in AAA

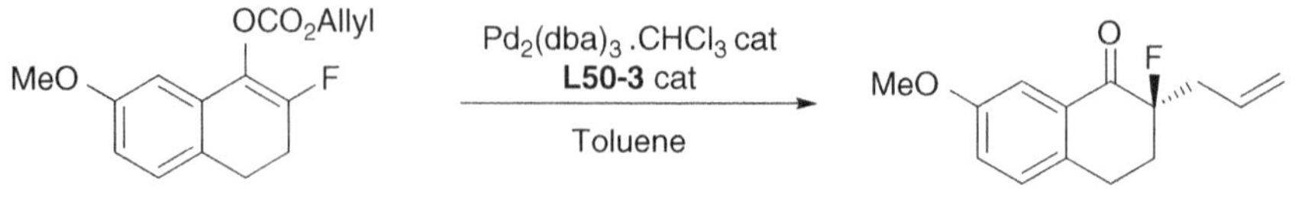

30

Entry	L\Pd ratio	Yield %	ee % (config.)
1	1:4	92	94 (*R*)
2	1.25:1	86	54 (*R*)

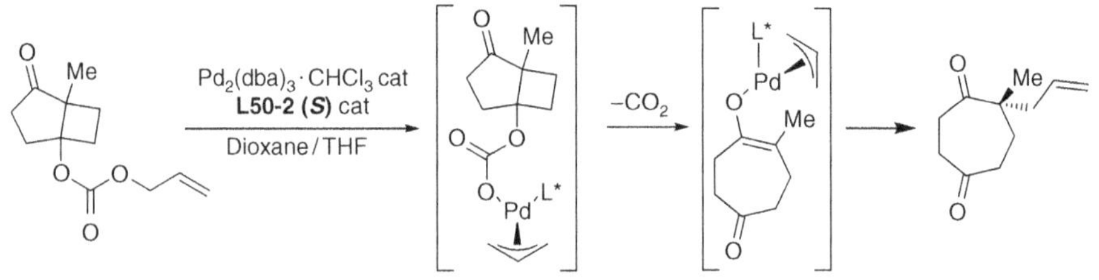

Scheme 17 Postulated mechanism for ring expanding allylation using palladium-Phox **L50**

It has been shown by X-ray crystallography that the ethereal oxygen is too far away from the binding site and thus does not have a hemilabile role.

The basic design of **L50** was also altered by replacing the aromatic ring by a biphenyl. There is free conversion between the two diastereomeric conformers of

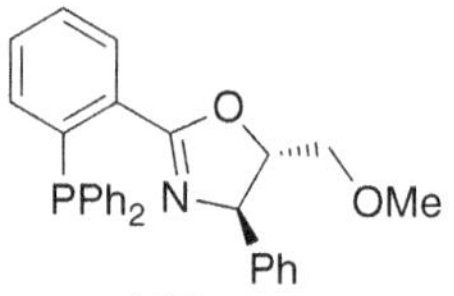

L51

Standard AAA: 99% yield, 97% ee (*R*)

Fig. 16 Phox derivative **L51** used in AAA

L52 in solution, but only one of them complexes to the palladium catalyst (**C52**) (Scheme 18) [91].

Instead of oxazoline-phosphine ligands, oxazoline-phosphites have also been investigated recently. Dieguez and Pamies have developed a library of ligands such as **L53** with ee's up to 92% (Fig. 17) [92]. They have also adapted their work on P–P ligands **L29**, based on previous work by Uemura [93, 94]. Near perfect enantioselectivity was achieved with **L54** [95].

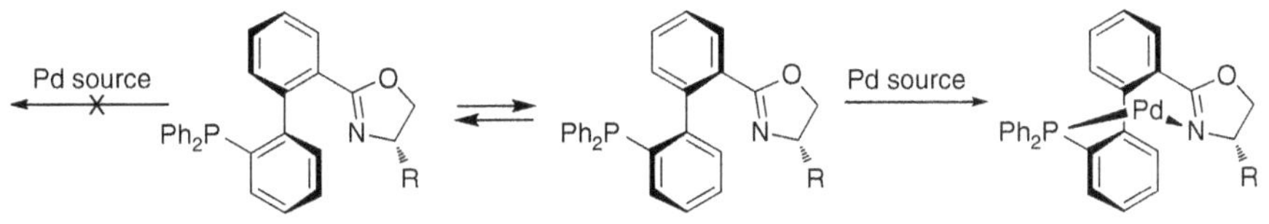

L52 C52

Standard AAA: 99 % yield, 92 % ee (*S*)

Scheme 18 Complex selectivity of Phox derivative **L52**

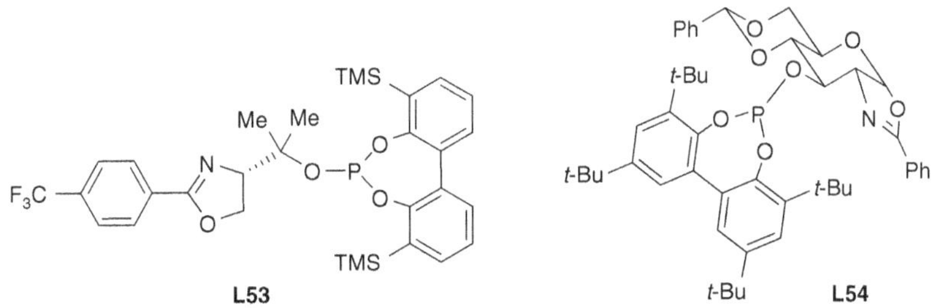

L53 L54

Standard AAA: 100% yield, 92% ee (*S*) 100% yield, 99% ee (*S*)

Fig. 17 Dieguez and Pamies oxazoline-based ligands **L53** and **L54** used in AAA

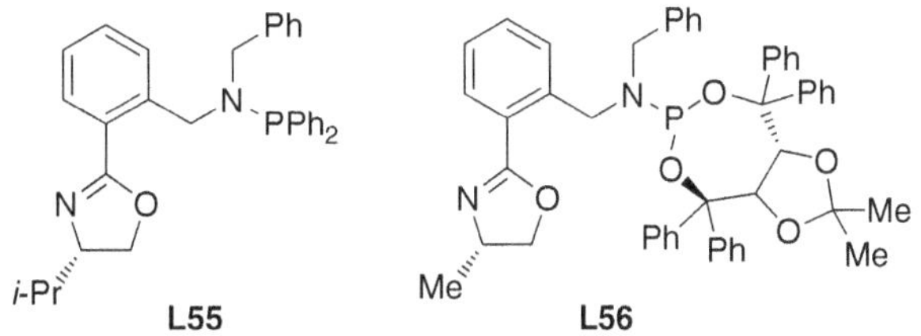

L55 L56

Standard AAA: 64 % yield, 38 % ee (*R*) 71 % yield, 87 % ee (*S*)

Fig. 18 Guiry's oxazoline derived P–N ligands **L55** and **L56** used in AAA

Examples of oxazoline-N-diphenylphosphinoamine **L55** and oxazoline-phosphoramidite **L56** have been reported by Guiry and co-workers with moderate-to-good enantioselectivities (Fig. 18) [96].

Finally, although they are not oxazolines, ligands **L57** have a related structure, and they also gave excellent enantioselectivity (Fig. 19) [97].

2.2.5 Other Pyridine-Based Ligands

The pyridine unit is widespread in many ligand classes and has also been used in combination with a range of phosphorus-based functional groups. For instance, the

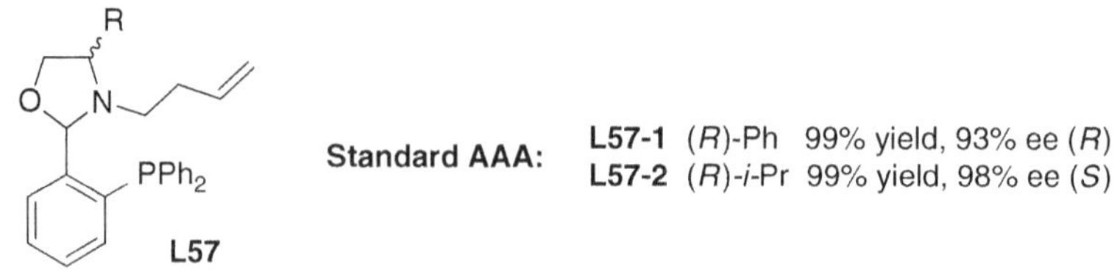
R
O
N
PPh2
L57
Standard AAA:
L57-1 (R)-Ph 99% yield, 93% ee (R)
L57-2 (R)-i-Pr 99% yield, 98% ee (S)

Fig. 19 Dihydrooxazoline-derived P–N ligands using AAA

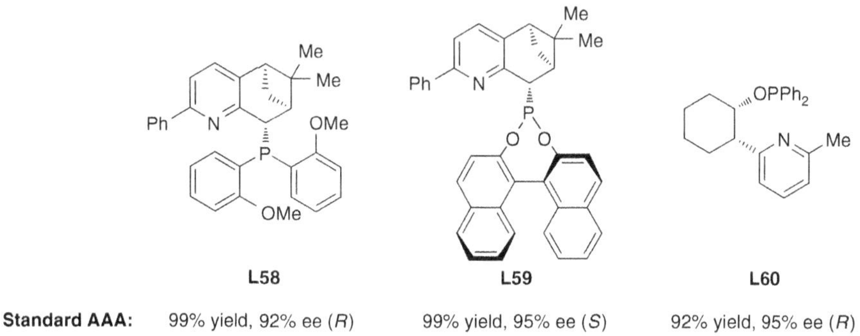
Me
Me
Ph
N
OMe
P
OMe
Ph
N
P
O O
OPPh2
N
Me
L58
L59
L60
Standard AAA:
99% yield, 92% ee (R)
99% yield, 95% ee (S)
92% yield, 95% ee (R)

Fig. 20 Pyridine-based ligands **L58–60** used in AAA

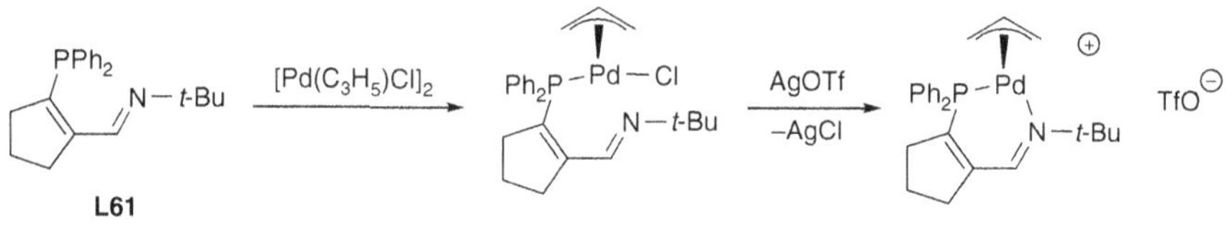
PPh2
N–t-Bu
L61
[Pd(C3H5)Cl]2
Ph2P–Pd–Cl
N–t-Bu
AgOTf
–AgCl
Ph2P–Pd
N–t-Bu
TfO

Scheme 19 Synthesis of Schaffer and Schmidt's P–N ligand **L61**-palladium complex

same scaffold is shared by ligands **L58** and **L59** (Fig. 20) and good selectivity have been reported for both [98, 99]. Remarkably, when using **L59**, a stronger base is needed to achieve high ee. Pyridine-phosphinites, as in **L60,** have also been used with excellent results [100].

2.2.6 Other Imine- and Sulfoximine-Based Ligands

Imines can be valuable components of P–N ligands. Recently, mechanistic insights have been brought by Shaffer and Schmidt who have shown that some 3-iminophosphines **L61** have a hemilabile behavior (Scheme 19) [101]. Thus, the actual mechanism of the reaction may vary from one ligand to the other.

Some success in enantioselectivity has been obtained with similar kinds of ligands (**L62**), but variations of the structure have shown that better results were obtained with **L63** compared to the isomeric **L62** (Fig. 21) [102].

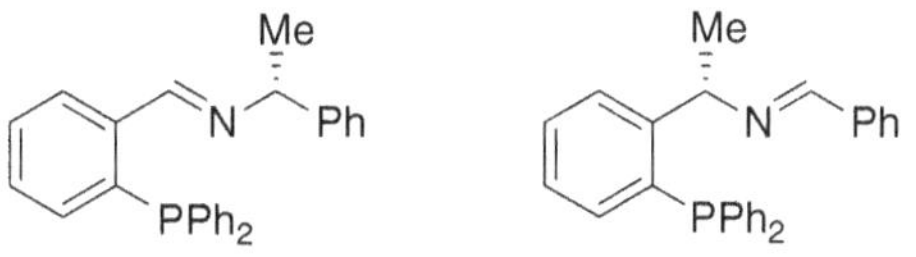

	L62	L63
Standard AAA:	85 % yield, 14 % ee (*R*)	99 % yield, 94 % ee (*S*)

Fig. 21 Isomeric imine-based P–N ligands **L62** and **L63** used in AAA

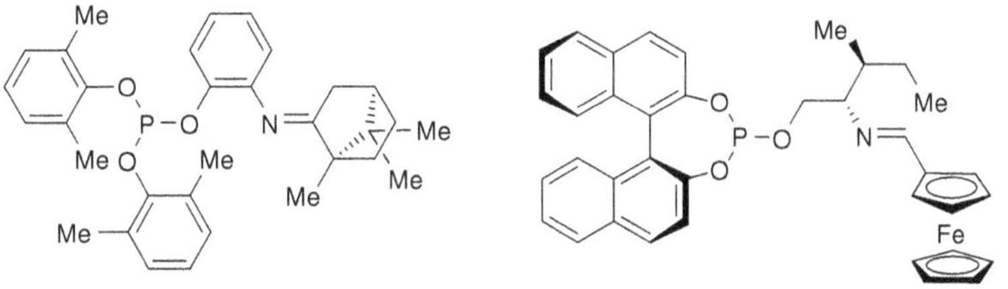

	L64	L65
Standard AAA:	99% yield, 94% ee (*R*)	45% yield, 50% ee (*S*)

Fig. 22 Hindered imine-based P–N ligands **L64** and **L65** used in AAA

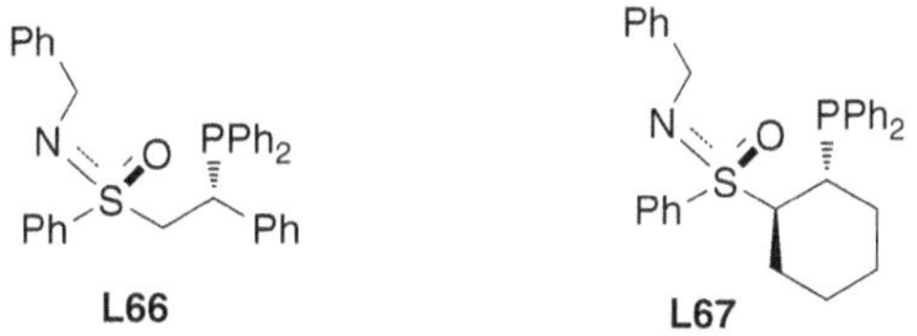

	L66	L67
Standard AAA:	98 % yield, 82 % ee (*R*)	98 % yield, 97 % ee (*R*)

Fig. 23 Sulfoximine-based P–N ligands **L66** and **L67** used in AAA

Hindered imine-containing units have also been combined with phosphites to give ligands which afforded good enantioselectivities [103, 104]. Chirality was either present on the imine side (**L64**), or on the linker between the two binding atoms and on the phosphite substituents (**L65**), with the former ligand class giving higher levels of enantioselectivity (Fig. 22).

Phosphino-sulfoximines **L66** and **L67** have been found to afford moderate-to-good enantioselectivities in the standard allylic alkylation reaction (Fig. 23) [105].

2.2.7 Other Amine-Based Ligands

Tertiary amines have been combined with phosphines (**L68**) [106] or phosphites (**L69**) as bidentate P–N ligands [107]. A few examples with good ee have been

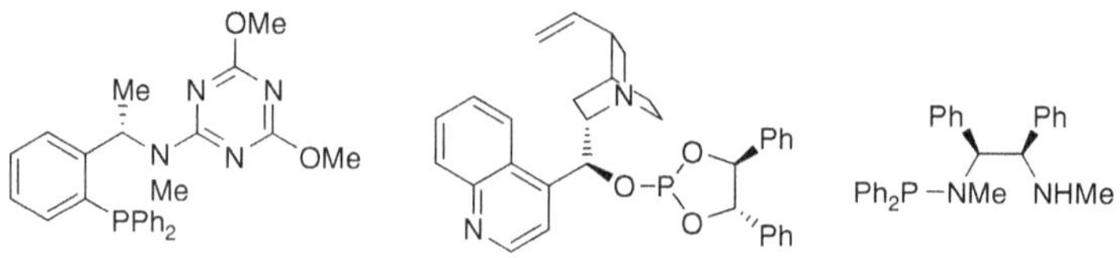

Standard AAA: L68 99% yield, 90% ee (*R*); L69 72% yield, 94% ee (*R*); L70 98% yield, 93% ee (*R*)

Fig. 24 Tertiary and secondary amine-based P–N ligands **L68–70** used in AAA

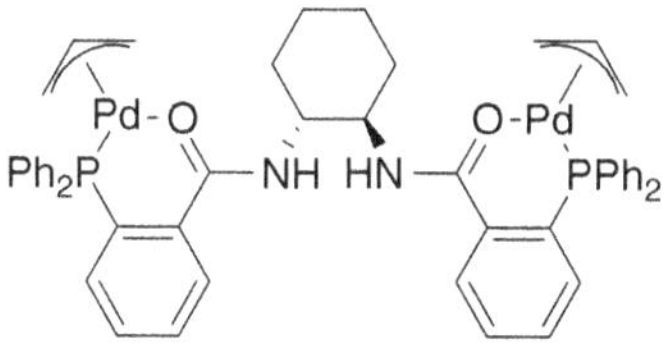

C4

Fig. 25 P–O coordinated Trost's ligand **L4**-palladium complex

reported (Fig. 24). Bujoli and Petit have developed a secondary amine ligand combined with an N-diphenylphosphine moiety **L70**, which exhibited a high enantioselectivity [108].

2.3 *Other P-Based Ligands*

The majority of phosphorus-based ligands belongs to the P–P or the P–N classes. However, several examples of monodentate ligands or polydentate ligands involving heteroatoms or olefins serving as coordination sites exist. These will be discussed below, with the exception of NHC-involving ligands, which will be dealt with in Sect. 2.6.

2.3.1 P-Amide Ligands

Lloyd-Jones and co-workers demonstrated that under certain circumstances an alternate phosphorus-oxygen mode of coordination was possible with Trost's standard ligand **L4**, resulting in a bis-Pd-complex **C4** (Fig. 25) [109]. Recently, a series of phosphine-amide ligands involving this mode of coordination has been prepared.

Burke reported ligand **L71** which possesses a free OH possibly acting as a third binding site, with a moderate ee of 62% in the standard AAA (Fig. 26) [110]. Sugar-based example **L72** has been developed by Framery and co-workers with better asymmetric induction, with an ee of 83% recorded [111]. Framery simplified the

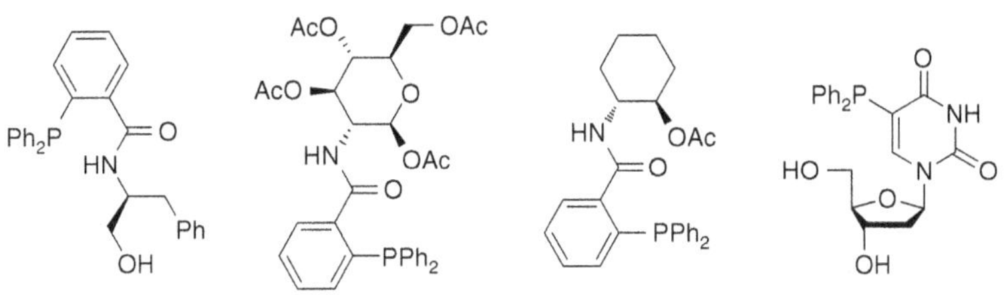

	L71	L72	L73	L74
Standard AAA:	63 % yield, 62 % ee (*S*)	98 % yield, 83 % ee (*R*)	97 % yield, 80 % ee (*S*)	99 % yield, 15 % ee (*S*)
Amination:				99 % yield, 82 % ee (*S*)

Fig. 26 Hemilabile oxygen in diphenylphosphine-containing ligands **L71–74**

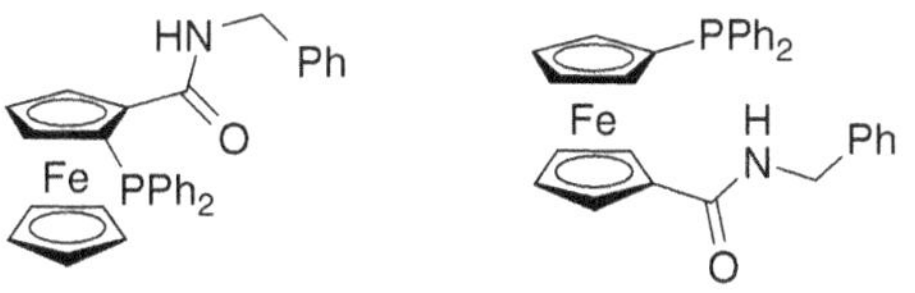

	L75	L76
Standard AAA:	85 % yield, 90 % ee (*R*)	92 % yield, 0 % ee

Fig. 27 Ferocene-based P-amide ligands **L75** and **L76** used in AAA

structure of his sugar-based ligand **L72** and kept only two chiral centers in ligand **L73**. Standard AAA gave an ee of 80%, comparable to the results of **L72** [112]. Interestingly, the sense of enantioselectivity was inversed.

A uracil-based ligand **L74** has also been investigated and is thought to involve a similar mode of coordination. Results were poor in the standard allylic alkylation with an ee of 15%, while amination gave an ee of 82% [113].

Phosphine-amide ferrocene-containing ligands have also been investigated. Stepnicka and co-workers have synthesized ferrocenes with the two functional groups on the same cyclopentadienyl ring (**L75**) or on opposite rings (**L76**). The latter gave racemic mixtures of products when used in standard AAA, but ee's of 90% were obtained with **L75** (Fig. 27) [114].

2.3.2 Other P–O Ligands

Stepnicka has also placed a carboxylic acid and a phosphine in **L77**, based on a ferrocene scaffold related to Josiphos **L7** (Fig. 28). However, a very poor ee of 10% was observed in standard AAA [115]. The N,O-phenylene prolinol derivative **L78** gave a good ee of 90% in the standard AAA [116]. High ee's were also obtained with the phosphine-sulfoxide **L79**, although no information was given about the

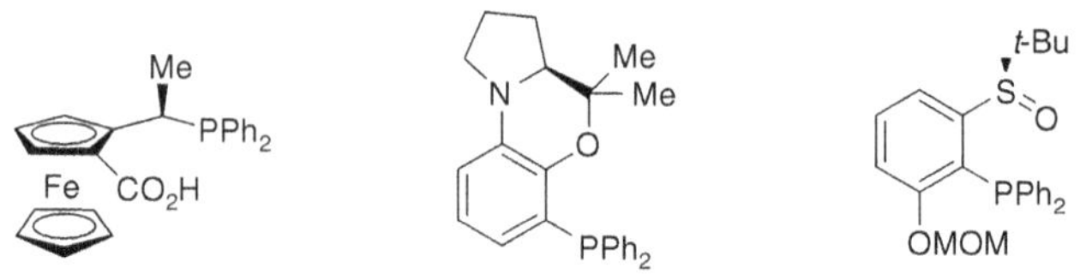

	L77	L78	L79
Standard AAA:	94% yield, 10% ee (*S*)	93% yield, 90% ee (*S*)	68% yield, 93% ee (main *enantiomer not reported)*

Fig. 28 P–O ligands **L77–79** used in AAA

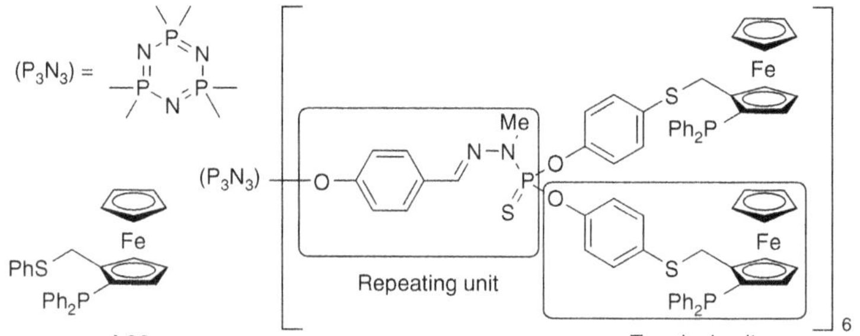

	L80		Terminal unit
Standard AAA:	96% yield, 93% ee (*R*)	1st generation dendrimer	89% yield, 93% ee (*R*)

Fig. 29 Dendrimeric P–S ligands **L80** used in AAA

coordination mode of the ligand (P–O or P–S). Lower ee's up to 76% were obtained in amination [117].

2.3.3 P–S ligands

Efforts have been devoted to the development of ferrocene-derived phosphorus–sulfur ligands. With ligand **L80**, an ee of 93% was obtained (Fig. 29). When a scaffold similar to **L80** was used as the terminal unit of dendrimers of 1st, 2nd, 3rd, and 4th generation, the ee's in the standard AAA stayed consistently above 90% [118].

Alternatively, the thioether can be directly linked with the ring and the phosphorus atom on a side chain. Chan and co-workers developed a series of such ligands, **L81–L83** (Scheme 20). **L82** gave the best result in the standard AAA with an ee of 95%. Similar results were obtained in AAA with indoles as nucleophile, irrespective of the steric or electronic nature of the indole [119].

Fukuzamo and co-workers developed a similar ferrocene P–S ligand **L84** (Scheme 21). A maximal ee of 90% was recorded in standard AAA reactions [120]. Chan and co-workers also altered their design of **L81–L83** and came up with ligand **L85**, where the phosphorus and sulfur atoms were separated by four

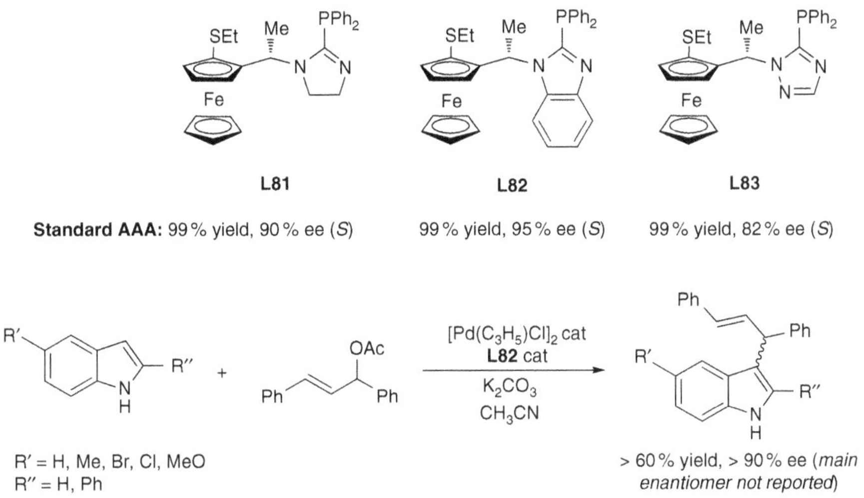

Scheme 20 Large bite-angle ferrocene-based P–S ligands **L81–83** used in AAA

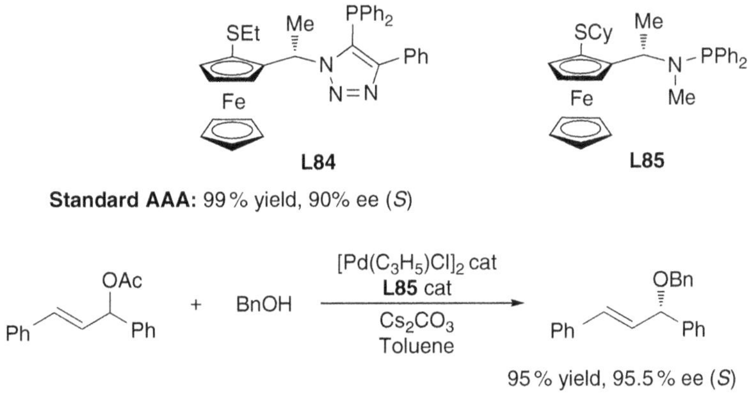

Scheme 21 Altered ferrocene-based P–S ligands **L84** and **L85** used in AAA

atoms instead of five. Very good ee's up to 95.5% were obtained in asymmetric allylic etherifications [121].

A sugar-based P–S ligand **L86** has been developed by Khiar and co-workers (Fig. 30). However, a poor ee of 30% was recorded [122].

2.3.4 Monodentate P Ligands

Although monodentate ligands are usually less effective at stereocontrol than polydentate ligands, several examples have been recently reported. Benzoferrocene **L87** gave rise to ee's up to 51% in asymmetric allylic alkylation (Fig. 31) [123].

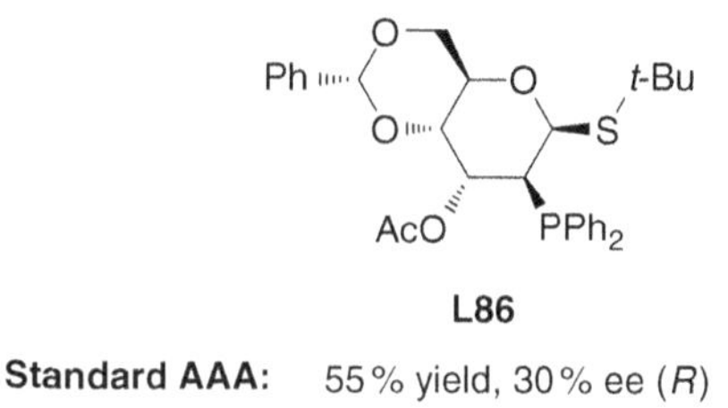

Standard AAA: 55 % yield, 30 % ee (*R*)

Fig. 30 Khiar's sugar-based P–S ligand **L86** used in AAA

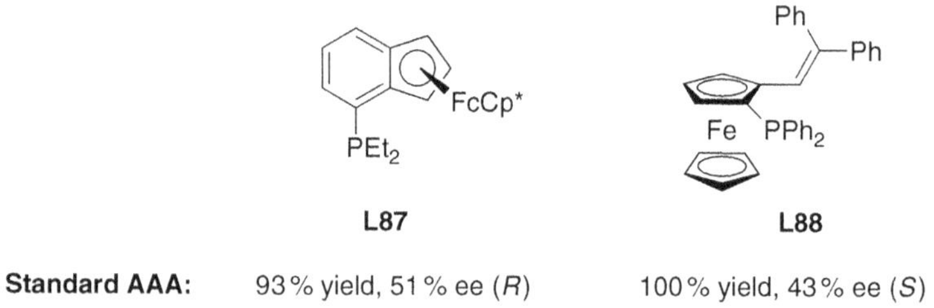

Standard AAA: 93 % yield, 51 % ee (*R*) 100 % yield, 43 % ee (*S*)

Fig. 31 Monodentate ferrocene-based P ligands **L87** and **L88** used in AAA

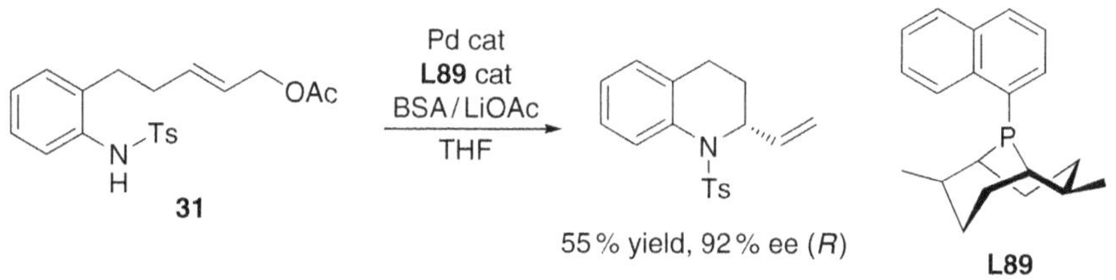

55 % yield, 92 % ee (*R*)

Scheme 22 Hameda's monodentate P ligand **L89** used in intramolecular allylic substitution

Potentially π-coordinating substituents, as used in **L88**, did not increase the enantioselectivity as the maximal recorded ee in AAA was 43% [124].

Hameda's and co-workers have developed an excellent ligand for intramolecular allylic aminations. In the cyclization of **31** ligand (*S*)-9-NapBN **L89** induced ee's up to 92% (Scheme 22), where in similar conditions, BINAP **L15** and Trost's standard ligand **L4** gave ee's of 0% and 17%, respectively [125].

Aside the aforementioned phosphines, other phosphorus functional groups have been used. Phosphite and phosphoramidites were obtained from phosphorus chlorides derived from chiral diols. These phosphorus chlorides were then reacted with alcohols or amines, chiral or not, to give phosphites and phosphoramidites, respectively. Standard AAA results were only recorded for **L90** in the past few years, with an ee of 72% [126]. **L91** was used in asymmetric allylation with an allylic alcohol [127]. In this reaction, triethyl borane promoted the leaving of the allylic oxygen and an enantiomeric excess of 71% was reported (Scheme 23).

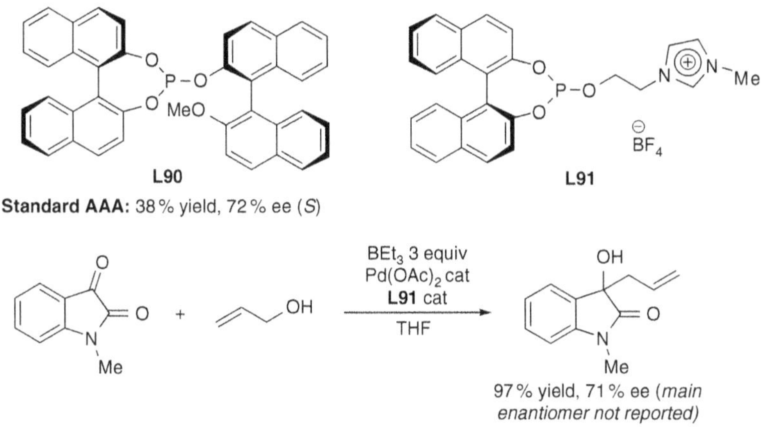

Scheme 23 Axially chiral monodentate P ligands **L90** and **L91** used in AAA

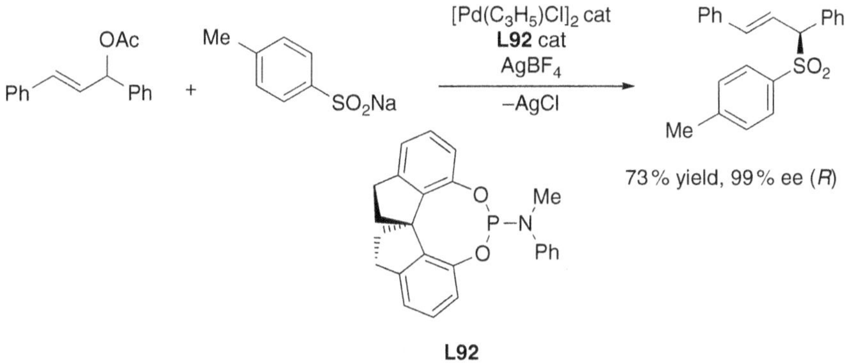

Scheme 24 Phosphoramidite-based monodentate P ligand **L92** used in asymmetric allylic substitution

L92 was used in allylic sulfonylation with *p*-TolSO_2Na [128]. In the presence of BF_4^- as counterion, an exceptional ee of 99% was obtained (Scheme 24).

P-chiral phosphorodiamidites have also been successfully used in asymmetric allylic substitutions, using various side chains (**L93–L96** in Fig. 32). Gavrilov developed ligands **L93–L95** [126, 128, 129], while Lyubimov worked on **L96** [130]. When the oxygen side chain was not chiral, formation of the *S* product was favored in AAA. The chiral side chain of **L94**, with the presence of a methoxy oxygen possibly acting as an hemilabile coordination site, reversed the sense of the enantioselectivity, and the *R* product was obtained. These four ligands gave ee's above 90%.

Mixtures of ligands have also been used, where a 1 to 1 ratio of ligands **L97** and **L98** gave better results in AAA than either of the two ligands used alone [131]. These heteroleptic mixtures gave an ee of 69%, with only one of the two ligands seeming to control the stereochemical outcome of the reaction (Table 22).

	L93	L94	L95	L96
Standard AAA:	89% yield 93% ee (*S*)	71% yield 99% ee (*S*)	79% yield 99% ee (*S*)	100% yield 91% ee (*S*)

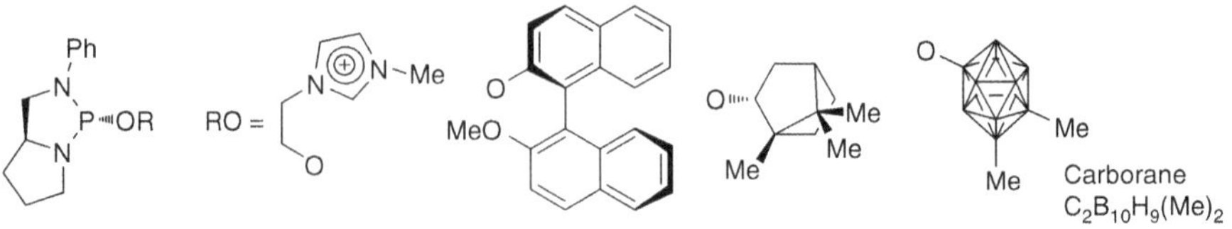

Fig. 32 Phosphorodiamidite-based monodentate ligands **L93–96** used in AAA

Table 22 Monodentate P ligands mixtures used in AAA

L97 **L97**

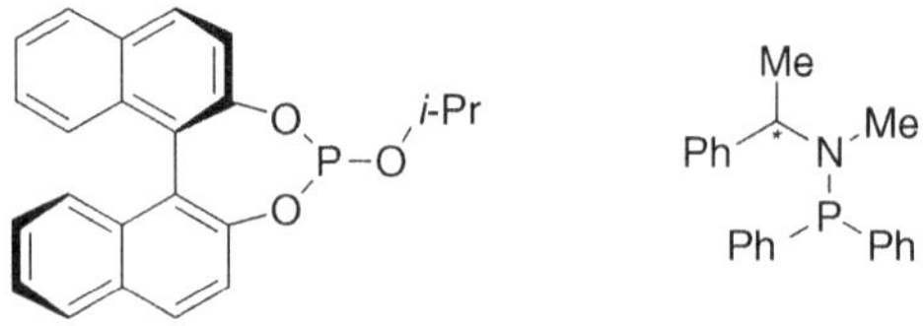

Standard AAA			
Entry	Ligand	Conversion %	ee % (config.)
1	(*R*)-**L97**	100	58 (*S*)
2	(*R*)-**L98**	100	68 (*R*)
3	(*S*)-**L98**	100	68 (*S*)
4	(*R*)-**L97**/(*R*)-**L98**	100	69 (*S*)
5	(*R*)-**L97**/(*S*)-**L98**	100	69 (*S*)

Self-assembling ligands have been recently applied with success by Breit and Börner to Rh-catalyzed asymmetric hydrogenation [132]. These monodentate ligands form complexes in pairs, becoming bidentate supramolecular units (Fig. 33). They were tested in AAA, and while large ligands such as **L99** and **L100** seemed to benefit from the ability to coordinate in pairs, smaller ligands **L101** and **L102** showed better results as monodentate units [133].

2.4 N–N Ligands

Second after phosphorus in the number of ligands used to coordinate to the metal, nitrogen has been the focus of a significant amount of research. With a handful of available functional groups, excellent ligands for asymmetric allylic substitutions have been developed.

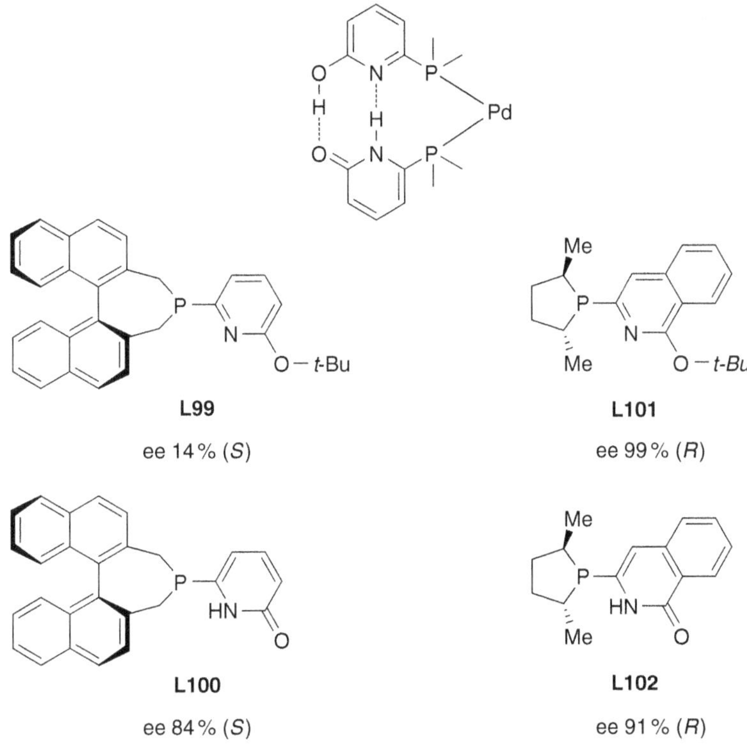

Fig. 33 Self-assembling P ligands

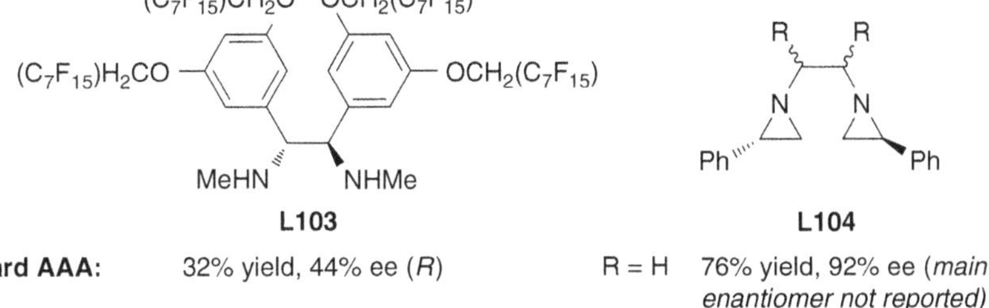

Fig. 34 Diamine and diaziridine-based ligands **L103** and **L104** used in AAA

2.4.1 Amine- and Imine-Based Ligands

An easily recyclable ligand **L103** has been developed by Sinou and co-workers, with fluorinated side chains to allow its recovery by liquid extraction with fluorous solvents. Although the ee's were moderate when used in standard AAA (44%, Fig. 34), the ligand kept its efficiency after recycling [134].

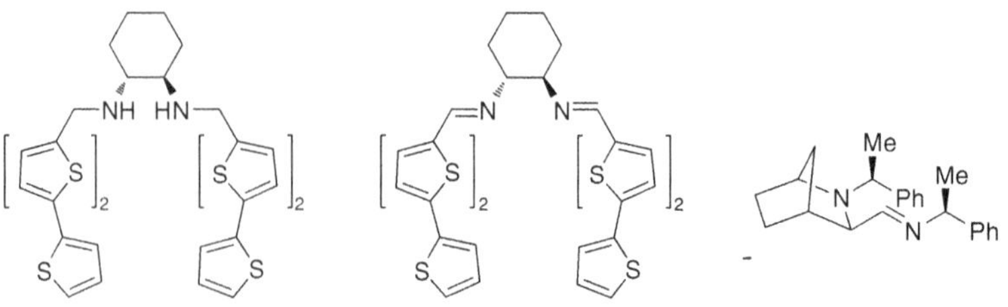

	L105	L106	L107
Standard AAA:	99 % yield, 99 % ee (*S*)	< 10 % yield	95 % yield, 90 % ee (*S*)

Fig. 35 Amine- and imine-containing N–N ligands **L105–107** used in AAA

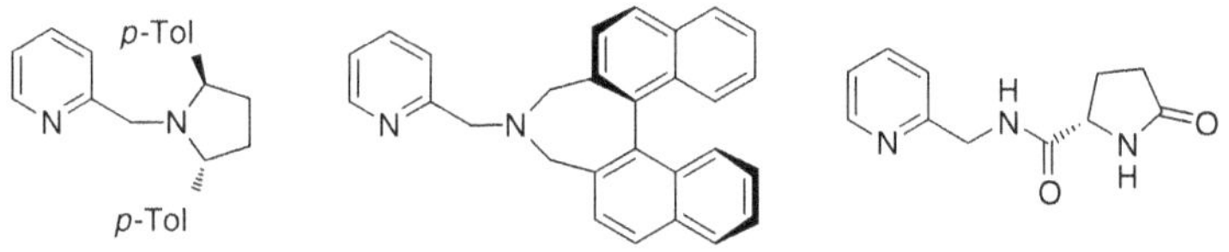

	L108	L109	L110
Standard AAA:	100 % yield, 70 % ee (*R*)	100 % yield, 82 % ee (*S*)	72 % yield, 93 % ee (*S*)

Fig. 36 Pyridine-based N–N ligands **L108–110** used in AAA

Diaziridines have been investigated by Savoia and co-workers, with better results. **L104** (R = H) gave an ee of 92% in AAA (Fig. 34), where all other variations of the R substituents gave comparable yields and lower ee's [135].

The diamine ligand **L105** has shown a near perfect ee of 99% in AAA (Fig. 35) [136]. Interestingly, the related imine version **L106** gave only traces amount of product, with no ee recorded. In both cases, it has been shown that the sulfur atoms do not bind to the palladium. The amine-imine ligand **L107** based on a 2-azanorbornene scaffold gave satisfactory results with an ee of 90% (Fig. 35) [137].

2.4.2 Pyridine-Based Ligands

For chelation purposes, the development of bidentate N–N ligands possessing amine-containing chiral side-chains at the 2-position, such as **L108** and **L109**, has been investigated (Fig. 36). When used in standard AAA, they gave ee's of 70% and 82%, respectively [137, 138].

A ligand with a side chain containing an amide and a lactam (**L110**) has been developed by O'Leary [139]. Very good ee's up to 93% were recorded when used in standard AAA, but information is lacking about the mode of coordination as such compounds may bind through an oxygen instead of the nitrogen of the amide.

Ligands such as **L111** possessing an imidazoline side chain have been developed by Claver, but these gave only moderate yields but excellent ee's in AAA (Fig. 37) [140]. A new version **L112** of the well-known Pybox ligand was prepared and tested by Shibatomi and Iuriosa and gave much better results with an ee of 99% [141].

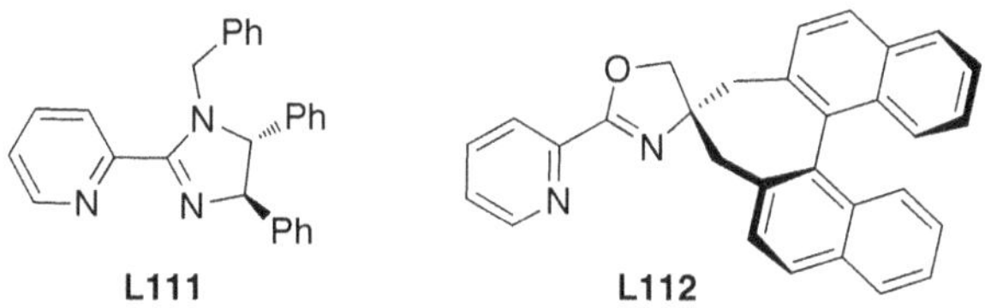

L111 L112

Standard AAA: 46% yield, 99% ee (*S*) 90% yield, 99% ee (*R*)

Fig. 37 Pyridine-imidazole and pyridine-oxazoline N–N ligands **L111** and **L112** used in AAA

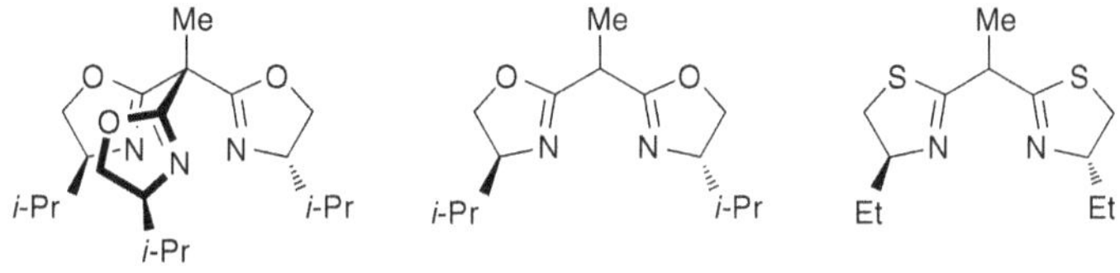

	L113	L114	L115
Standard AAA:	90 % yield, 95 % ee (*main enantiomer not reported*)	89 % yield, 89 % ee	95 % yield, 88 % ee

Fig. 38 Oxazoline- and thiazoline-containing N–N ligands **L113–115** used in AAA

2.4.3 Other Oxazoline-Based Ligands

Gade and Bellemin-Laponnez have investigated a trisoxazoline ligand **L113** [142]. Even though it acts as a bidentate ligand by coordinating through two nitrogen atoms, it induced better ee's than the related bisoxazoline ligand **L114** (Fig. 38). A similar bisthiazoline ligand **L115** has been developed by Gulea and Reiher. This ligand exhibited a good ee of 88% in AAA. However, the mode of coordination is a competition between nitrogen and sulfur [143].

2.5 S, Se, and Te Ligands

Several sulfur-containing ligands have been discussed in the previous sections. Nevertheless, a few examples not belonging to any of the previous classes have also been developed. Thioether containing ligands used in asymmetric allylic substitutions have been reviewed by Dieguez and Martin in 2007 [144], and this section will cover important advances since then.

2.5.1 Nitrogen-Based Ligands

Related to ligand **L44** discussed in Sect. 2.2, thioether **L116** showed a lower ee in AAA, and a reverse sense of selectivity (Fig. 39) [72]. Several benzylamides have

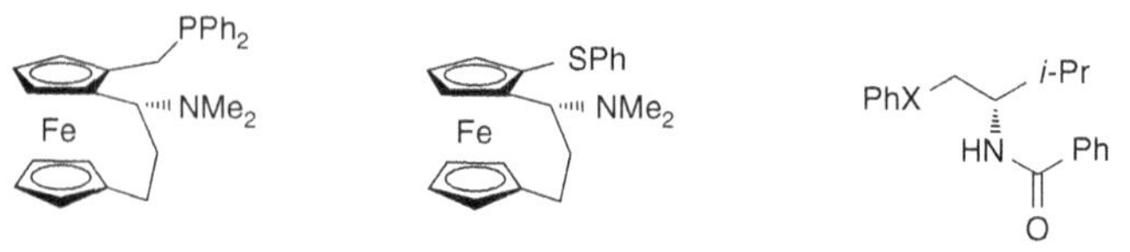

Standard AAA: L44: 94% yield, 84% ee (*S*); L116: 41% yield, 68% ee (*R*); L117: X = Se 97% yield, 98% ee; X = S 94% yield, 86% ee; X = Te 39% yield, 59% ee

Fig. 39 N-Chalcogen ligands **L116** and **L117** used in AAA

Table 23 Comparison of S and Se as donor atoms in N-containing ligands **L118–121**

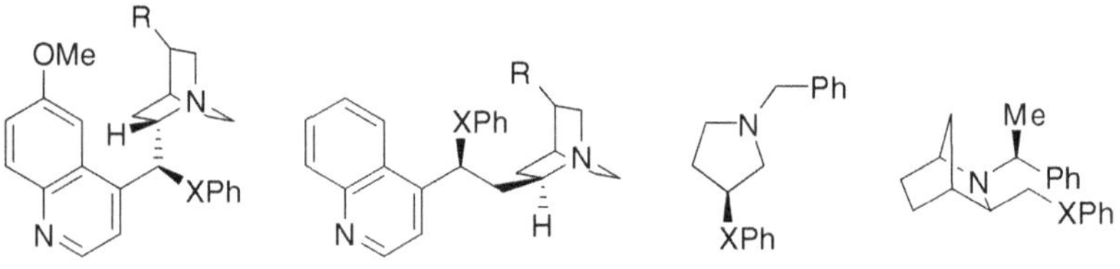

L118 L119 L120 L121

Standard AAA					
Entry	Ligand	X	R	Yield %	ee % (config.)
1	**L118**	Se	Et	17	60 (*R*)
2	**L118**	S	Et	67	46 (*R*)
3	**L118**	Se	$CH{=}CH_2$	23	66 (*R*)
4	**L118**	S	$CH{=}CH_2$	46	56 (*R*)
5	**L119**	Se	Et	16	48 (*S*)
6	**L119**	S	Et	22	62 (*S*)
7	**L119**	Se	$CH{=}CH_2$	20	62 (*S*)
8	**L119**	S	$CH{=}CH_2$	22	76 (*S*)
9	**L120**	Se	–	93	98 (*R*)
10	**L120**	S	–	73	90 (*R*)
11	**L121**	Se	–	55	76 (*S*)
12	**L121**	S	–	62	65 (*S*)

been investigated and the most favorable structure found by Vargas and Broye was ligand **L117** [145, 146]. The yield and enantioselectivity in AAA decreased as they replaced selenium by the related elements, sulfur and tellurium.

Skarzewski and co-workers have investigated several ligand architectures such as *Cinchona* alkaloid derivatives **L118** and **L119**, pyrrolidine derivatives **L120**, and 2-azanorbornene derivatives **L121** (Table 23) [137, 147].

A dithiamine **L122** derived from 2-azanorbornene gave the best result for this class of compound with an ee of 95% (Fig. 40) [137]. Song and Wang have developed an aziridine-thioether ligand **L123** [138] and an enantiomeric excess of 91% was reported.

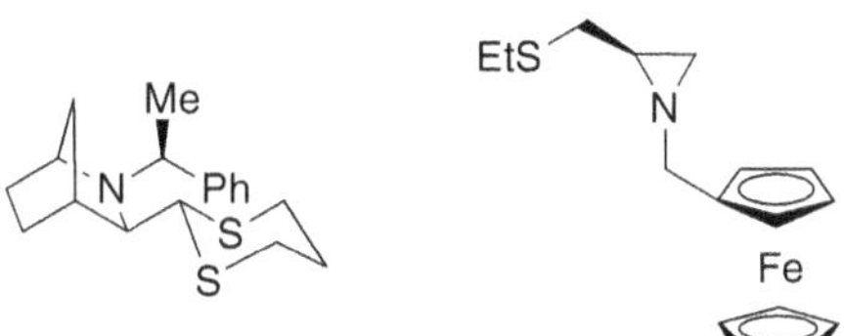

L122 L123

Standard AAA: 94 % yield, 95 % ee (*S*) 95 % yield, 91 % ee (*S*)

Fig. 40 Tertiary amine-based N–S ligands **L122** and **L123** used in AAA

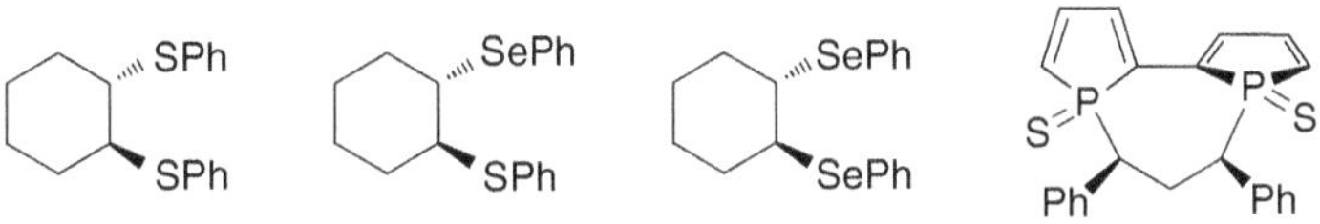

	L124	L125	L126	L127
Standard AAA:	92% yield 30% ee (*R*)	90% yield 50% ee (*R*)	90% yield 34% ee (*R*)	100% yield 66% ee (*R*)

Fig. 41 S and Se-only bidentate ligands **L124–127** used in AAA

2.5.2 S and Se Only Ligands

Moderate ee's were obtained with the ligands series **L124–L126** developed by Skarzewski (Fig. 41) [148]. The S–S ligand **L127** investigated by Gouygou was used in AAA with an ee of 66% [149].

2.6 NHC Ligands

A few studies have been recently published trying to gain insights into the behavior of π-allylpalladium complexes in the presence of NHC ligands with hemilabile nitrogen or phosphorus on a side chain [150–153].

Two applications to asymmetric allylic substitutions have been reported. Williams and co-workers have studied a carbene with an imine side chain acting as hemilabile ligand **L128** (Fig. 42) [154]. A moderate ee of 53% was recorded in standard AAA. Roland and co-workers have worked with the very similar ligand **L129**, where the imine was replaced by a secondary amine and better ee's up to 74% have been observed [155].

3 Heterogeneous Catalysis

Although far less papers are published in the domain of heterogeneous catalysis than homogeneous catalysis, this topic is of importance as these processes are often preferred for large-scale productions. Examples of supported ligands and supported

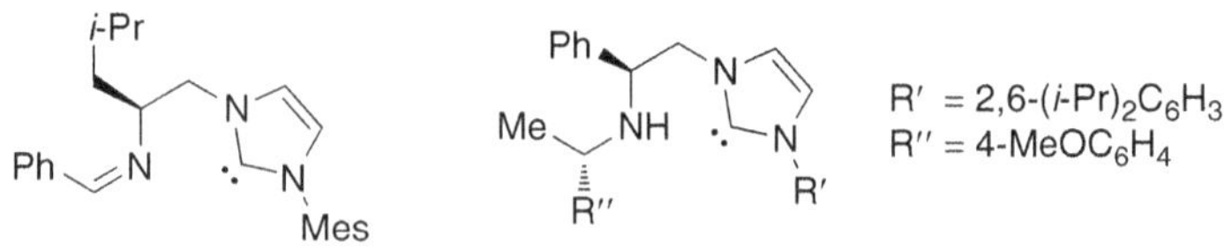

L128 **L129**

Standard AAA: 99% yield, 53% ee (*main enantiomer not reported)* 77% yield, 74% ee (*R*)

Fig. 42 NHC-containing ligands **L128** and **129** with a hemilabile nitrogen on a side chain

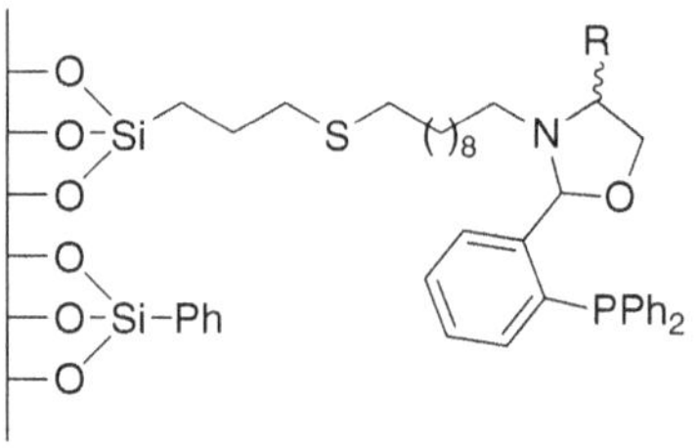

L130

Standard AAA:	**L130-1**	R = (*R*)-Ph	53% yield, 80.5% ee (*R*)
	L130-2	R = (*S*)-i-Pr	82% yield, 86.5% ee (*S*)
	L57-1	R = (*R*)-Ph	99% yield, 93% ee (*R*)
	L57-2	R = (*S*)-i-Pr	99% yield, 98% ee (*S*)

Fig. 43 Silicate-supported PN ligand **L130** used in AAA

catalysts for enantioselective allylic substitutions have been reported, allowing for the easy recovery of the ligand and of the metal, respectively.

3.1 Polymer Supported Ligands

P–N ligand **L57** also exists in supported version **L130** in which the ligand is linked with a silicate MCM-41 polymer (Fig. 43) [97]. Performances of **L130** are lower than those of the original free ligand **L57**.

Other supported bidentate ligands have been studied. Polyethyleneglycol acrylate co-polymer (PEGA) supported **L131** developed by Medal and co-workers afforded ee's of 60% in asymmetric allylic alkylations (Scheme 25) [156]. Uozomi and co-workers investigated **L132**, a polystyrene-polyethyleneglycol (PS-PEG) supported chiral P–N ligand [157, 158]. An outstanding ee of 98% was observed in the asymmetric allylic amination of **96**.

Uozumi also worked with supported monodentate ligands. **L133** afforded the AAA product in 90% enantioselectivity, in a reaction performed in water (Fig. 44) [159].

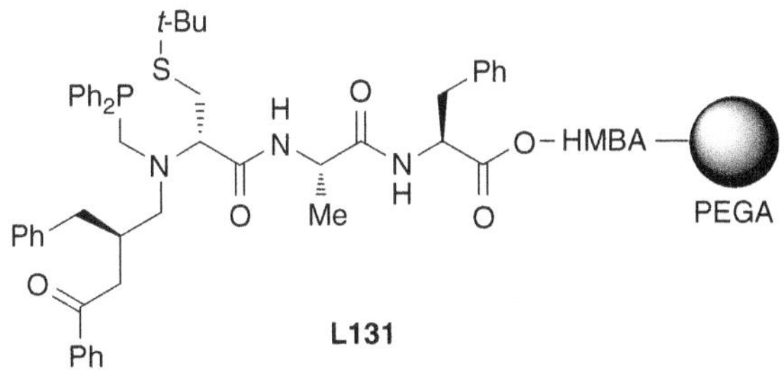

L131

Standard AAA: 94 % yield, 60 % ee (*R*)

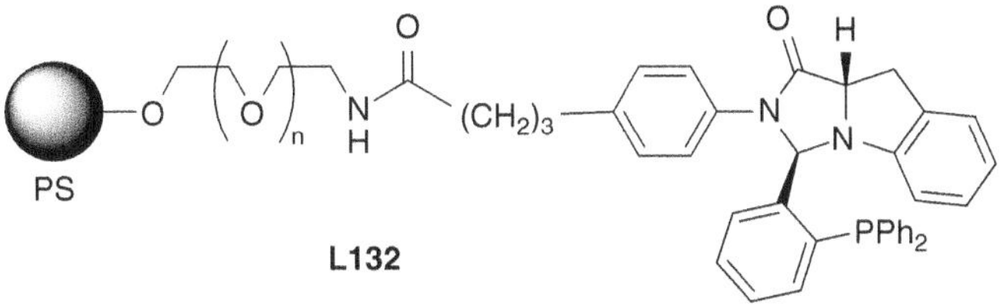

L132

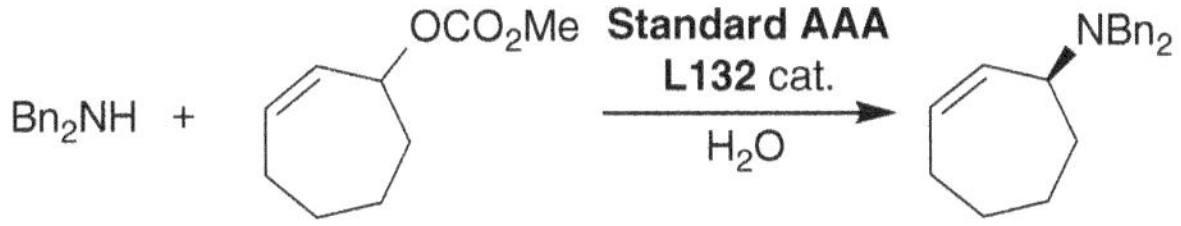

95 **96** **97**

91 % yield, 98 % ee (*S*)

Scheme 25 Polymer-supported bidentate ligands **L131** and **L132** used in asymmetric allylic substitutions

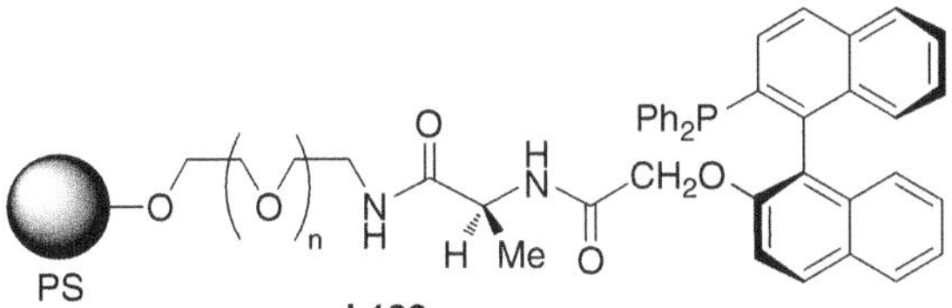

L133

Standard AAA in water: 96% yield, 90% ee (*R*)

Fig. 44 Supported monodentate P ligand **L133** used in AAA

Supported phosphoramidites have also been reported. Kamer and co-workers obtained their best result with **L134**, with an ee of 58% in AAA (Fig. 45) [160]. Zhi-Dong and co-workers had a better ee of 64% in AAA with their ligand **L135**, using N-methyaminomethyl polystyrene resin [161].

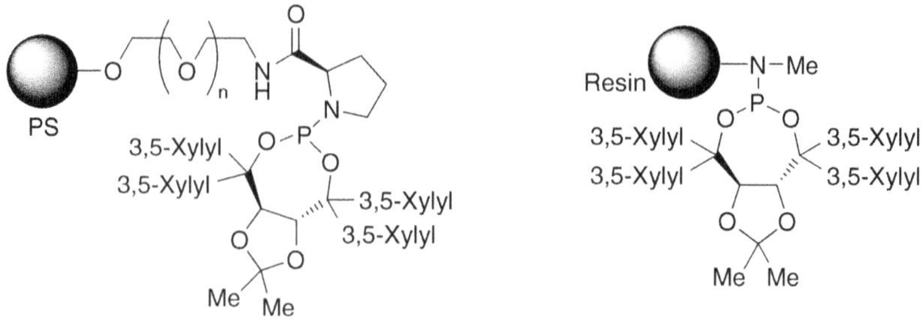

L134 **L135**

Standard AAA: 84 % yield, 58 % ee (*S*) 88 % yield, 64 % ee (*S*)

Fig. 45 Supported phosphoramidite ligands **L134** and **L135** used in AAA

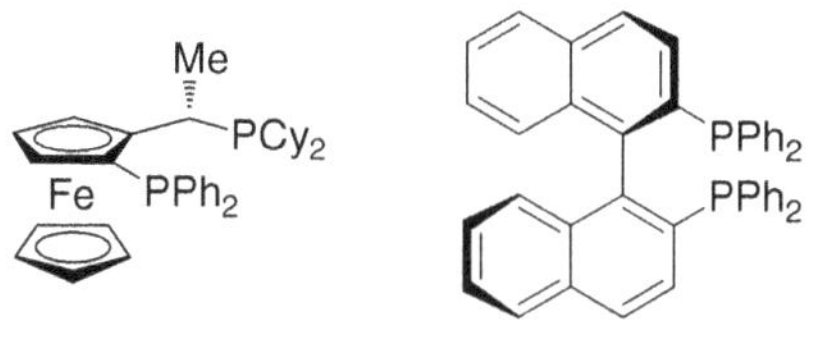

L7 **L15**

Standard AAA with Pd/Al$_2$O$_3$: 13% yield, 88% ee (*S*) 73% yield, 60% ee (*S*)

Fig. 46 Common ligands **L7** and **L15** used in AAA with Al_2O_3-supported palladium

3.2 Solid Phase Palladium

Baiker and co-workers have performed an interesting study on the use of palladium over Al_2O_3 with a variety of ligands [162, 163]. Josiphos **L7** gave an ee of 88%, but with a yield of only 13%. BINAP **L15** was used in AAA with an ee of 60% (Fig. 46).

Another interesting approach was to use colloidal solutions of nanoparticles of palladium. Dieguez, Pamies, Gomez, and Leeuwen have used **L54** with these nanoparticles in AAA and have observed ee's of 80% [164]. Claver, Claudret, Philippot and Gomez again have used a similar system with the diphosphite **L136** and obtained ee's of 89% (Fig. 47) [165]. In these cases, the catalyst loading is superior to standard homogeneous AAA as only surface palladium atoms are involved in catalysis.

4 Chiral Salts Additives

The use of chiral additives as a means of inducing or improving the levels of enantioselectivity in AAA is a topic that has recently been investigated.

Based on the concept of asymmetric counterion directed catalysis (ACDC), (*R*)-3,3′-bis(2,4,6-triisopropylphenyl)-1,1′-binaphthyl-2,2′-diyl hydrogenphosphate

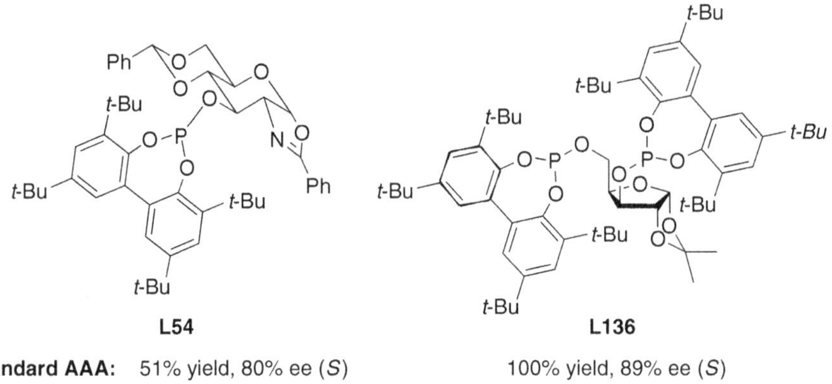

Fig. 47 Ligands **L54** and **L136** used in AAA with colloidal palladium particles

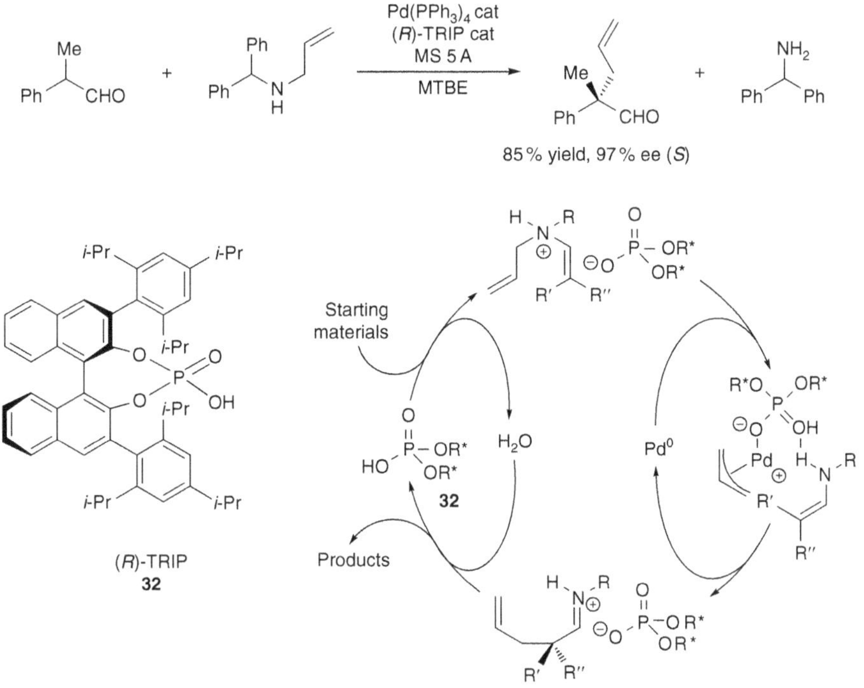

Scheme 26 Mechanism of an ACDC with (*R*)-TRIP **32**

((*R*)-TRIP) **32** was involved in the mechanism of the reaction (Scheme 26) [166]. It promotes the approach of the π-allylpalladium complex by the nucleophile. An ee of 97% was obtained in the α-allylation of aldehydes.

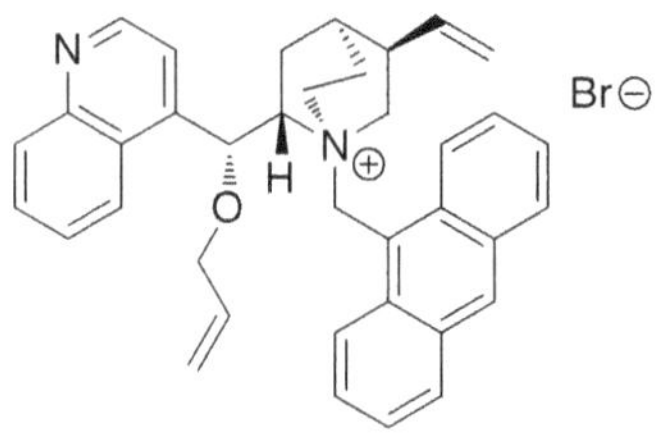

33

Standard AAA: 65 % yield, 73 % ee (*S*)

Fig. 48 Cinchonidinium salt **33** as phase transfer catalysts in AAA

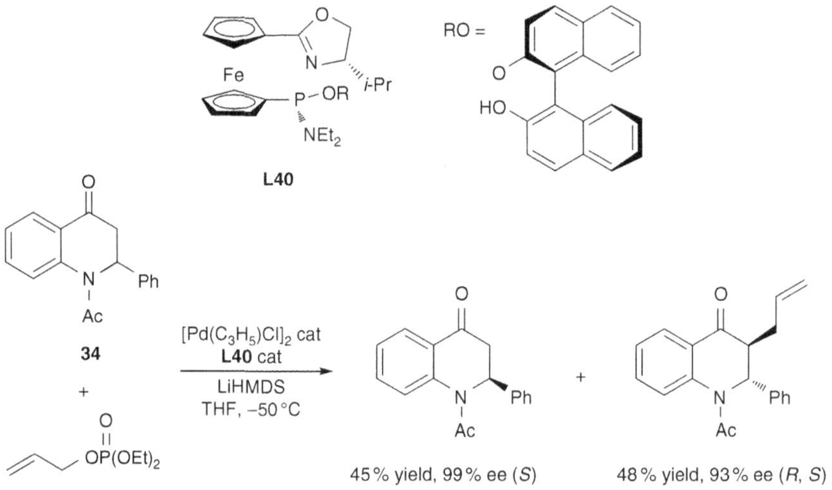

Scheme 27 Kinetic resolution with ferrocenyl oxazoline-based P-chiral P–N ligand **L40**

Cinchonidinium salts have also been used, this time as a phase transfer catalyst [167]. In standard AAA, with salt **33** being the only source of chirality, an ee of 73% was observed (Fig. 48).

5 Kinetic Resolutions

A kinetic resolution describes a reaction in which the enantiomers of a racemic substrate are converted to the chiral product at different rates. In ideal conditions, 50% of the starting material is recovered while a yield of 50% of product is obtained, both compounds in 100% enantiopurity.

5.1 Recent Examples

A few examples have been recently published. In a study by Hou and co-workers, nucleophile **34** was resolved, using **L40** as ligand [168]. The product was isolated

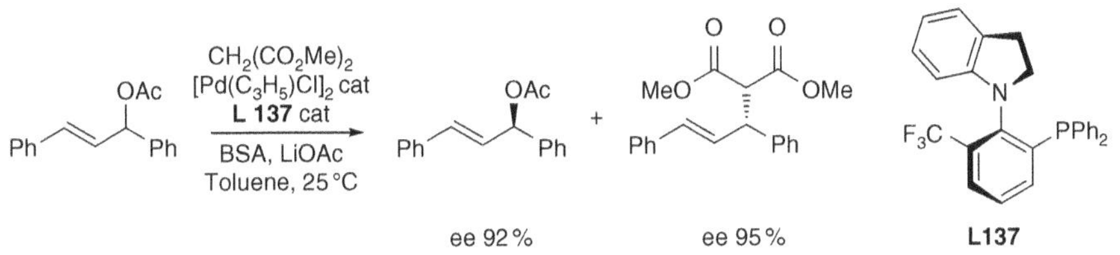

Scheme 28 Kinetic resolution with P–N ligand **L137**

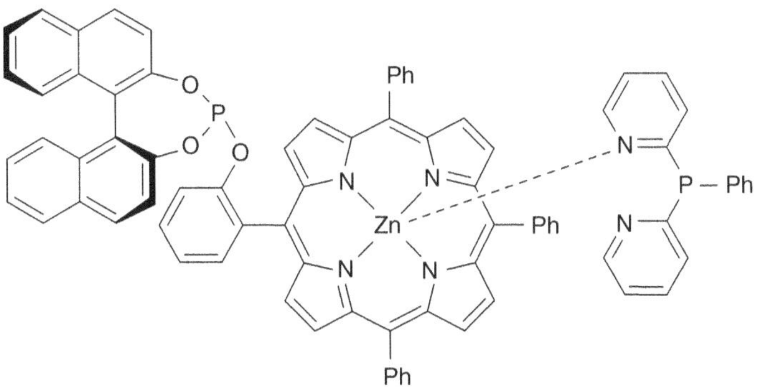

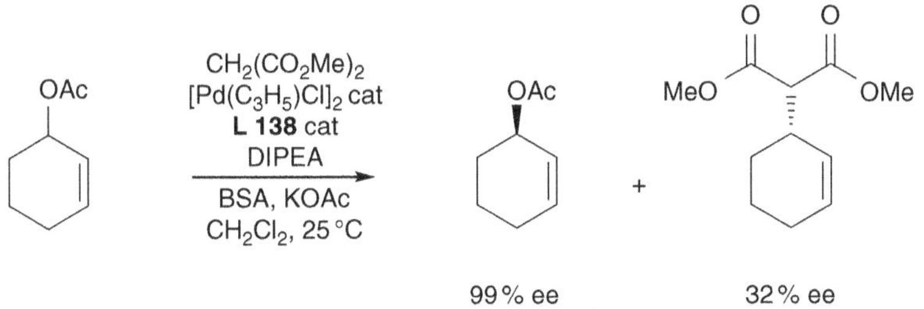

Scheme 29 Kinetic resolution with supramolecular ligand **L138**

in 93% ee, while the remaining nucleophile isomer was 99% enantiopure (Scheme 27).

Most of the recent publications on this topic aim at resolving the allylic substrate. Mino and co-workers have used P–N ligand **L137** in the standard AAA reaction with ee's of 92% and 95% for the starting substrate and the product, respectively (Scheme 28) [169].

Supramolecular ligands have been used for kinetic resolution with moderate success. Reek and co-workers have used porphyrin-based ligand **L138** with racemic cyclohexenyl acetate [170]. Unreacted starting material was recovered in 99% enantiopurity. However, the product was obtained with an ee of only 32% (Scheme 29).

Colloidal solutions of palladium with ligand **L136** have also shown impressive results in kinetic resolution. In this study, the product was obtained with 97% ee, while the starting material was recovered in 89% enantiopurity (Scheme 30) [165].

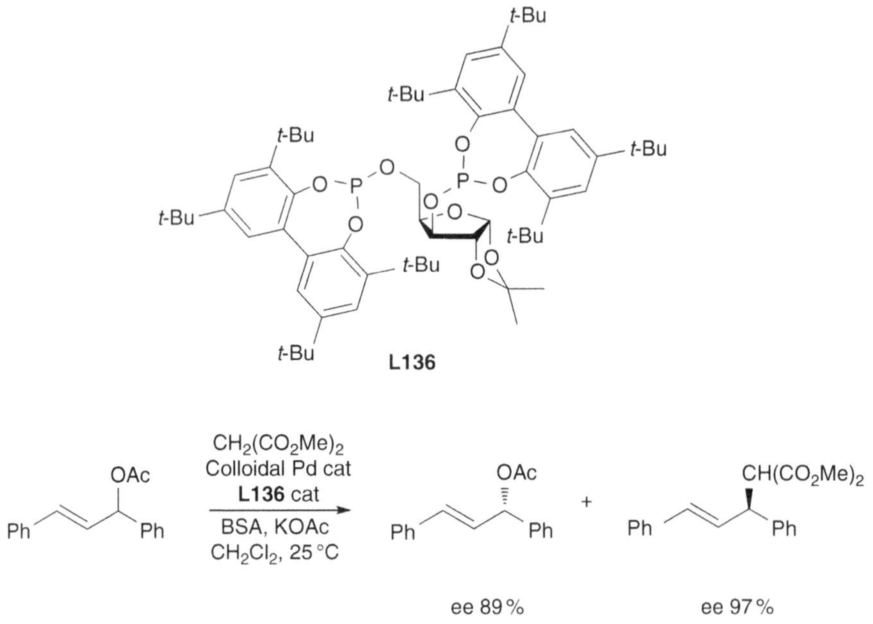

Scheme 30 Kinetic resolution with colloidal palladium and P–P ligand **L136**

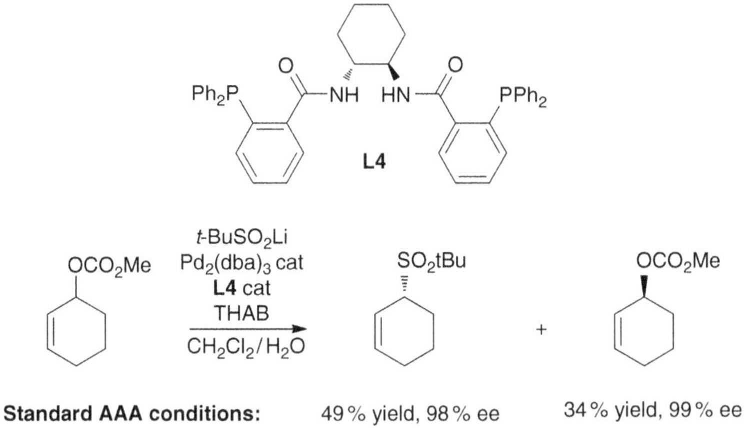

Scheme 31 Trost ligand **L4** used in kinetic resolution

The Trost standard ligand **L4** has been used in various kinetic resolutions with sulfur-containing compounds [31]. For example, the sulfonated product was obtained with an enantiopurity of 98% and the starting allylic carbonate was recovered with an ee of 99% (Scheme 31).

5.2 *Dynamic Kinetic Asymmetric Allylic Alkylations*

In the case of dynamic kinetic resolution, yields above 50% are possible. The reaction involves the racemization of the starting material or of an intermediate, or proceeds through a meso or prochiral intermediate. Trost and Fandrick have reviewed this topic up to the end of 2006 with numerous examples [171]. There have not been any significant advances in this field during the period covered by this chapter.

6 Conclusion

This chapter reports on important developments in enantioselective allylic substitutions covering the period 2007 to mid-2010. During that time, there have been isolated reports on the use of chiral substrates, nucleophiles, and additives, whereas the vast majority of the literature focuses on enantioselective examples where the asymmetric induction originates from a chiral palladium catalyst in a homogeneous medium. Such catalysts rely upon the design, synthesis, and application of chiral ligands, and we have classified the literature based on the donor atoms involved.

The broad utility of the various ligands used, including some supramolecular examples, is demonstrated by the high levels of enantiocontrol induced in a wide range of allylic substitutions, many of which are key steps in total synthesis. However, the majority of the reports study the standard substrate, 1,3-diphenylpropenyl acetate, and its alkylation by dimethyl malonate or amination by benzylamine. Use of this test reaction, which is experimentally and analytically facile to perform, allows researchers to directly compare their ligands in terms of reactivity and enantioselectivity to previously reported ligands. The use of new reaction media for allylic substitution and studies on heterogeneous catalytic approaches have also been investigated.

Although this chapter shows extensive research in over the past three and a half years, few truly significant synthetic breakthroughs have been achieved and the number of publications on this area has decreased somewhat compared to the activity obvious in the 1990s. However, this area is far from being exhausted in terms of interest in both academia and industry, and it is hoped that this chapter will stimulate both the development of new ligand architectures and their application in newly developed palladium-catalyzed enantioselective allylic substitutions.

References

1. (2002) The Tsuji–Trost reaction and related carbon–carbon bond formation reactions. In: Negishi E (ed) Handbook of organopalladium chemistry for organic synthesis, vol 2. Wiley, New York
2. Tsuji J (1995) Reactions of allylic compounds via π-allylpalladium complexes catalyzed by Pd(0). In: Tsuji J (ed) Palladium reagents and catalysts: innovations in organic synthesis. Wiley, Chichester

3. Trost BM, Lee C (2000) Asymmetric allylic alkylation reactions. In: Ojima I (ed) Catalytic asymmetric synthesis, 2nd edn. Wiley, New York
4. Krouzelka J et al (2008) Synthesis of 3-(2-hydroxy-1-phenylethyl)- and 3-(2-hydroxy-2-phenylethyl)adenine, DNA adducts derived from styrene. J Heterocycl Chem 45:789–795
5. Cheng H-Y et al (2007) Regioselective palladium-catalyzed formate reduction of N-heterocyclic allylic acetates. J Org Chem 72:2674–2677
6. Kim JM et al (2009) An expedient aralkylation of Baylis-Hillman adduct via the Pd-catalyzed decarboxylative protonation strategy. Tetrahedron Lett 50:1734–1737
7. Tsuji J et al (1965) Organic syntheses by means of noble metal and compounds. XVII. Reaction of π-allylpalladium chloride with nucleophiles. Tetrahedron Lett 6:4387–4388
8. Tsuji J (1969) Carbon-carbon bond formation via palladium complexes. Acc Chem Res 2:144–152
9. Trost BM, Fullerton TJ (1973) New synthetic reactions. Allylic alkylation. J Am Chem Soc 95:292–294
10. Trost BM (1977) Organopalladium intermediates in organic synthesis. Tetrahedron 33:2615–2649
11. Frost CG et al (1992) Selectivity in palladium catalyzed allylic substitution. Tetrahedron Asymmetry 3:1089–1122
12. Trost BM, Crawley ML (2003) Asymmetric transition-metal-catalyzed allylic alkylations: applications in total synthesis. Chem Rev 103:2921–2943
13. Trost BM (2004) Asymmetric allylic alkylation, an enabling methodology. J Org Chem 69:5813–5837
14. Nicolaou KC et al (2005) Palladium-catalyzed cross-coupling reactions in total synthesis. Angew Chem Int Ed 44:4442–4489
15. Svensen N et al (2007) Memory effects in palladium-catalyzed allylic alkylations of 2-cyclohexen-1-yl acetate. Adv Synth Catal 349:2631–2640
16. Yan B, Spilling CD (2008) Synthesis of cyclopentenones via intramolecular HWE and the palladium-catalyzed reactions of allylic hydroxy phosphonate derivatives. J Org Chem 73:5385–5396
17. Trost BM, Thaisrivongs DA (2009) Palladium-catalyzed regio-, diastereo-, and benzylic allylation of 2-substituted pyridines. J Am Chem Soc 131:12056–12057
18. Fristrup P et al (2008) On the nature of the intermediates and the role of chloride ions in Pd-catalyzed allylic alkylations: added insight from density functional theory. J Phys Chem A 112:12862–12867
19. Kraemer K et al (2009) A straightforward approach towards glycoamino acids and glycopeptides via Pd-catalyzed allylic alkylation. Org Biomol Chem 7:103–110
20. Trost BM, Crawley ML (2011) Enantioselective allylic substitutions in natural product synthesis. Top Organomet Chem. doi: 10.1007/3418_2011_13
21. Jacquet O et al (2009) Enantiotopic discrimination in palladium-mediated nucleophilic substitutions on achiral substrates: chiral ligand versus chiral nucleophile. Synthesis 3047–3050
22. Lange DA, Goldfuss B (2007) Electronic differentiation competes with transition state sensitivity in palladium-catalyzed allylic substitutions. Beilstein J Org Chem 3(36):2007–2086
23. Guiry PJ, Saunders CP (2004) The development of bidentate P, N ligands for asymmetric catalysis. Adv Synth Catal 346:497–537
24. Flanagan SP, Guiry PJ (2006) Substituent electronic effects in chiral ligands for asmmetric catalysis. J Organomet Chem 691:2125–2154
25. Berens U et al (1996) Transacetalization of diethyl tartrate with acetals of α-dicarbonyl compounds: Aa simple access to a new class of C_2-symmetric auxiliaries and ligands. J Org Chem 60:8204–8208
26. Marques CS, Burke AJ (2007) Palladium catalyzed enantioselective asymmetric allylic alkylations using the Berens' DIOP analogue. Tetrahedron Asymmetry 18:1804–1808

27. Schaeffner B et al (2008) Organic carbonates as alternative solvents for palladium-catalyzed substitution reactions. Chem Sus Chem 1:249–253
28. Butts CP et al (2009) Structure-based rationale for selectivity in the asymmetric allylic alkylation of cycloalkenyl esters employing the Trost 'Standard Ligand' (TSL): isolation, analysis and Aalkylation of the monomeric form of the cationic η^3-cyclohexenyl complex $[(\eta^3\text{-c-}C_6H_9)Pd(TSL)]^+$. J Am Chem Soc 131:9945–9957
29. Amatore C et al (2007) On the formation of Pd(II) complexes of Trost modular ligand involving N-H activation or P, O coordination in Pd-catalyzed allylic alkylations. J Organomet Chem 692:1457–1464
30. Fuchs S et al (2007) A highly stereoselective divergent synthesis of bicyclic models of photoreactive sesquiterpene lactones. Eur J Org Chem 7:1145–1152
31. Gais HJ (2007) Palladium-catalyzed allylic alkylation of sulfur and oxygen nucleophiles – asymmetric synthesis, kinetic resolution and dynamic kinetic resolution. In: Enders D, Jaeger KE (eds) Asymmetric synthesis with chemical and biological methods. Wiley-VCH, Weinheim
32. Gontcharov AV et al (1999) tert-Butylsulfonamide. A new nitrogen source for catalytic aminohydroxylation and aziridination of olefins. Org Lett 1:783–786
33. Trost BM et al (2009) Palladium-catalyzed decarboxylative asymmetric allylic alkylation of enol carbonates. J Am Chem Soc 131:18343–18357
34. Trost BM, Thaisrivongs DA (2008) Strategy for employing unstabilized nucleophiles in palladium-catalyzed asymmetric allylic alkylations. J Am Chem Soc 130:14092–14093
35. Trost BM et al (2008) Ligand controlled highly regio- and enantioselective synthesis of α-acyloxyketones by palladium-catalyzed allylic alkylation of 1,2-enediol carbonates. J Am Chem Soc 130:11852–11853
36. Coppola GM, Schuster HF (1997) α-hydroxy acids in enantioselective syntheses. VCH, Weinheim
37. Iwabushi Y et al (1999) Chiral amine-catalyzed asymmetric Baylis–Hillman reaction: a reliable route to highly enantiomerically enriched (α-methylene-β-hydroxy)esters. J Am Chem Soc 121:10219–10220
38. Trost BM, Toste FD (1999) Regio- and enantioselective allylic alkylation of an unsymmetrical substrate: a working model. J Am Chem Soc 121:4545–4554
39. Butti P et al (2008) Palladium-catalyzed enantioselective allylic phosphination. Angew Chem Int Ed 47:4878–4881
40. Liu D et al (2007) Novel C_2-symmetric planar chiral diphosphine ligands and their application in Pd-catalyzed asymmetric allylic substitutions. J Org Chem 72:6992–6997
41. Xie F et al (2008) Reversal in enantioselectivity for the palladium-catalyzed asymmetric allylic substitution with novel metallocene-based planar chiral diphosphine ligands. Tetrahedron Lett 49:1012–1015
42. Hayashi T et al (1974) Asymmetric catalytic hydrosilylation of ketones: preparation of chiral ferrocenylphosphines as chiral ligands. Tetrahedron Lett 15:4405–4408
43. Fukuda Y et al (2007) Asymmetric construction of quaternary carbon stereocenter by Pd-hemilabile ligand-catalyzed allylic substitution. Tetrahedron Lett 48:3389–3391
44. Sebesta R, Bilcik F (2009) Imidazolium-tagged ferrocenyl diphosphanes in allylic substitution with heteroatom nucleophiles. Tetrahedron Asymmetry 20:1892–1896
45. Braun M et al (2008) Palladium-catalyzed diastereoselective and enantioselective allylic alkylations of ketone enolates. Adv Synth Catal 350:303–314
46. Bantreil X et al (2009) Enantioselective γ-lactam synthesis via palladium-catalyzed intramolecular asymmetric allylic alkylation. Synlett 9:1441–1444
47. Giambastiani G et al (1998) A new palladium-catalyzed intramolecular allylation to pyrrolidin-2-ones. J Org Chem 63:804–807
48. Fukuda Y et al (2007) Development of novel hemilabile segphos P-P=O ligands. Chem Pharm Bull 55:955–956

49. Ohmori K et al (2007) Two isolable conformers of dihydropentahelicenediol derivatives: stereochemical property and its utility for asymmetric reactions. Chem Lett 36:328–329
50. Sharma RK et al (2008) Asymmetric allylic alkylation by palladium-bisphosphinites. Tetrahedron Asymmetry 19:655–663
51. Balanta Castillo A et al (2008) An outstanding palladium system containing a C_2-symmetrical phosphite ligand for enantioselective allylic substitution processes. Chem Commun 46: 6197–6199
52. Sanhes D et al (2009) New chiral diphosphites derived from substituted 9,10-dihydroanthracene. Applications in asymmetric catalytic processes. Tetrahedron Asymmetry 20:1009–1014
53. Wassenaar J et al (2009) INDOLPhosphole and INDOLPhos palladium-allyl complexes in asymmetric allylic alkylations. Organomet 28:2724–2734
54. Raluy E et al (2007) Sugar-based diphosphoroamidite as a promising new class of ligands in Pd-catalyzed asymmetric allylic alkylation reactions. J Org Chem 72:2842–2850
55. Raluy E et al (2007) First chiral phosphoroamidite-phosphite ligands for highly enantioselective and versatile Pd-catalyzed asymmetric allylic substitution reactions. Org Lett 9:49–52
56. Mata Y et al (2007) Pyranoside phosphite-phosphoramidite ligands for Pd-catalyzed asymmetric allylic alkylation reactions. Tetrahedron Asymmetry 17:3282–3287
57. Raluy E et al (2009) Modular furanoside phosphite-phosphoroamidites, a readily available ligand library for asymmetric palladium-catalyzed allylic substitution reactions. Origin of enantioselectivity. Adv Synth Catal 351:1648–1670
58. Pamies O et al (2007) New highly effective phosphite-phosphoramidite ligands for palladium-catalyzed asymmetric allylic alkylation reactions. Adv Synth Catal 349:836–840
59. Pamies O, Dieguez M (2008) Screening of a phosphite-phosphoramidite ligand library for palladium-catalysed asymmetric allylic substitution reactions: the origin of enantioselectivity. Chem Eur J 14:944–960
60. Gavrilov KN et al (2008) A P*-chiral bisdiamidophosphite ligand with a 1,4:3,6-dianhydro-D-mannite backbone and its application in asymmetric catalysis. Tetrahedron Lett 49:3120–3123
61. Gavrilov KN et al (2009) P*, P*-Bidentate diastereoisomeric bisdiamidophosphites based on N-benzyltartarimide and their applications in asymmetric catalytic processes. Tetrahedron Asymmetry 20:2490–2496
62. Imamoto T et al (2007) t-Bu-QuinoxP* ligand: applications in asymmetric Pd-catalyzed allylic substitution and Ru-catalyzed hydrogenation. J Org Chem 72:7413–7416
63. Slagt VF et al (2007) Fine-tuning ligands for catalysis using supramolecular strategies. Eur J Inorg Chem 4653–4662
64. Slagt VF et al (2007) Supramolecular bidentate phosphorus ligands based on bis-zinc(II) and bis-tin(IV) porphyrin building blocks. Dalton Trans 2302–2310
65. Pierron J et al (2008) Artificial metalloenzymes for asymmetric allylic alkylation on the basis of the biotin-avidin technology. Angew Chem Int Ed 47:701–705
66. Fukuzawa S-I et al (2007) ClickFerrophos: new chiral ferrocenyl phosphine ligands synthesized by Click chemistry and the use of their metal complexes as catalysts for asymmetric hydrogenation and allylic substitution. Org Lett 9:5557–5560
67. Zhao X et al (2009) Enamines: efficient nucleophiles for the palladium-catalyzed asymmetric allylic alkylation. Tetrahedron 65:512–517
68. Liu D et al (2007) The synthesis of novel C2-symmetric P, N-chelation ruthenocene ligands and their application in palladium-catalyzed asymmetric allylic substitution. Tetrahedron Lett 48:585–588
69. Zhang K et al (2008) Highly enantioselective palladium-catalyzed alkylation of acyclic amides. Angew Chem Int Ed 47:1741–1744
70. Zheng W-H et al (2007) Highly regio-, diastereo-, and enantioselective Pd-catalyzed allylic alkylation of acyclic ketone enolates with monosubstituted allyl substrates. J Am Chem Soc 129:7718–7719

71. Fukuzawa S-I et al (2007) Preparation of chiral homoannularly bridged N, P-ferrocenyl ligands by intramolecular coupling of 1,5-dilithioferrocenes and their application in asymmetric allylic substitution reactions. Eur J Org Chem 33:5540–5545
72. Sebesta R et al (2008) [3]Ferrocenophane ligands with an inserted methylene group. Eur J Org Chem 31:5157–5161
73. Jiang B et al (2008) [2.2]Paracyclophane-derived chiral P, N-ligands: design, synthesis, and application in Palladium-catalyzed asymmetric allylic alkylation. J Org Chem 73:7833–7836
74. Ruzziconi R et al (2007) Quinolinophane-derived alkyldiphenylphosphines: two homologous P, N-planar chiral ligands for palladium-catalysed allylic alkylation. Tetrahedron Asymmetry 18:1742–1749
75. Fleming WJ et al (2009) Axially chiral P-N ligands for the copper catalyzed.beta.-borylation of α, β-unsaturated esters. Org Biomol Chem 7:2520–2524
76. Flanagan SP et al (2005) The preparation and resolution of 2-(2-pyridyl)- and 2-(2-pyrazinyl)-quinazolinap and their application in palladium-catalyzed allylic substitution. Tetrahedron 61:9808–9821
77. Connolly DJ et al (2004) Preparation and resolution of a modular class of axially chiral quinazoline-containing ligands and their application in asymmetric Rhodium-catalyzed olefin hydroboration. J Org Chem 69:6572–6589
78. McCarthy M, Guiry PJ (2000) A new quinazoline-containing axially chiral ligand for asymmetric catalysis. Polyhedron 19:541–543
79. Lacey PM et al (2000) Synthesis and resolution of 2-methyl-Quinazolinap, a new atropisomeric phosphinamine ligand for asymmetric catalysis. Tetrahedron Lett 41:2475–2478
80. McCarthy M et al (1999) The preparation and resolution of 2-phenyl-quinazolinap, a new atropisomeric phosphinamine ligand for asymmetric catalysis. Tetrahedron Asymmetry 10:2797–2807
81. Fekner T et al (2008) Synthesis, resolution, and application of cyclobutyl- and adamantyl-quinazolinap ligands. Eur J Org Chem 30:5055–5066
82. Maxwell AC et al (2008) Electronically varied quinazolinaps for asymmetric catalysis. Org Biomol Chem 6:3848–3853
83. Zalubovskis R et al (2008) Self-adaptable catalysts: substrate-dependent ligand configuration. J Am Chem Soc 130:1845–1855
84. Nemoto T, Hamada Y (2007) Pd-catalyzed asymmetric allylic substitution reactions using P-chirogenic diaminophosphine oxides: DIAPHOXs. Chem Rec 7:150–158
85. Cardillo G et al (2003) Aziridines and oxazolines: valuable intermediates in the synthesis of unusual amino acids. Aldrichimica Acta 36:39–50
86. Maki K et al (2007) Pd-Catalyzed allylic alkylation of secondary nitroalkanes. Tetrahedron 63:4250–4257
87. Lamac M et al (2009) Preparation, coordination and catalytic use of planar-chiral monocarboxylated dppf analogues. New J Chem 33:1549–1562
88. Belanger E et al (2008) Unexpected effect of the fluorine atom on the optimal ligand-to-palladium ratio in the enantioselective Pd-catalyzed allylation reaction of fluorinated enol carbonates. Chem Commun 28:3251–3253
89. Schulz SR, Blechert S (2007) Palladium-catalyzed synthesis of substituted cycloheptane-1,4-diones by an asymmetric ring-expanding allylation (AREA). Angew Chem Int Ed 46:3966–3970
90. Popa D et al (2007) Phosphinooxazolines derived from 3-amino-1,2-diols: highly efficient modular P-N ligands. Adv Synth Catal 349:2265–2278
91. Tian F et al (2009) Phosphine-oxazoline ligands with an axial-unfixed biphenyl backbone: the effects of the substituent at oxazoline ring and P phenyl ring on Pd-catalyzed asymmetric allylic alkylation. Tetrahedron 65:9609–9615
92. Dieguez M, Pamies O (2008) Modular phosphite-oxazoline/oxazine ligand library for asymmetric Pd-catalyzed allylic substitution reactions: scope and limitations-origin of enantioselectivity. Chem Eur J 14:3653–3669

93. Yonehara K et al (1999) Palladium-catalyzed asymmetric allylic substitution reactions using new chiral phosphinite-oxazoline ligands derived from D-glucosamine. J Org Chem 64:9374–9380
94. Yonehara K et al (1999) Palladium-catalysed asymmetric allylic alkylation using new chiral phosphinite–nitrogen ligands derived from D-glucosamine. Chem Commun 5:415–416
95. Mata Y et al (2009) Pyranoside phosphite-oxazoline ligand library: highly efficient modular P,N ligands for palladium-catalyzed allylic substitution reactions. A study of the key palladium allyl intermediates. Adv Synth Catal 351:3217–3234
96. Bronger RPJ, Guiry PJ (2007) Aminophosphine-oxazoline and phosphoramidite-oxazoline ligands and their application in asymmetric catalysis. Tetrahedron Asymmetry 18:1094–1102
97. Guo X-F, Kim G-J (2008) Highly enantioselective allylic alkylation catalyzed by new P, N-chelate ligands from L-valinol. React Kinet Catal Lett 93:325–332
98. Meng X et al (2009) Novel pyridine-phosphite ligands for Pd-catalyzed asymmetric allylic substitution reaction. Catal Commun 10:950–954
99. Meng X et al (2009) Asymmetric hydrogenation and allylic substitution reaction with novel chiral pinene-derived N, P-ligands. Tetrahedron Asymmetry 20:1402–1406
100. Liu Q-B, Zhou Y-G (2007) Synthesis of chiral cyclohexane-backbone P, N-ligands derived from pyridine and their applications in asymmetric catalysis. Tetrahedron Lett 48:2101–2104
101. Shaffer AR, JaR S (2009) Reactivity of (3-iminophosphine)palladium(II) complexes: evidence of hemilability. Organomet 28:2494–2504
102. Huang J-D et al (2007) Readily available phosphine-imine ligands from alpha-phenylethylamine for highly efficient Pd-catalyzed asymmetric allylic alkylation. J Mol Catal A Chem 270:127–131
103. Gavrilov K et al (2007) P,N-bidentate phosphites with a chiral ketimine fragment, their application in enantioselective allylic substitution and comparison with phosphine analogues. Synthesis 1717–1723
104. Gavrilov KN et al (2007) Ferrocenyliminophosphites as easy-to-modify ligands for asymmetric catalysis. Eur J Org Chem 29:4940–4947
105. Lemasson F et al (2007) Synthesis of 1,5-P, N-phosphino-sulfoximines through phospha-Michael reaction of alkenyl sulfoximines and their evaluation as ligands in palladium-catalyzed allylic alkylation. Tetrahedron Lett 48:8752–8756
106. Huang JD et al (2008) Synthesis of novel chiral phosphine-triazine ligand derived from α-phenylethylamine for Pd-catalyzed asymmetric allylic alkylation. Chinese Chem Lett 19:261–263
107. Wang Q-F et al (2008) Facile one-pot synthesis of cinchona alkaloid-based P, N ligands and their application to Pd-catalyzed asymmetric allylic alkylation. Tetrahedron Asymmetry 19:2447–2450
108. Schnitzler V et al (2008) Nitrogen-based chirality effects in novel mixed phosphorus/nitrogen ligands applied to palladium-catalyzed allylic substitutions. Organomet 27: 5997–6004
109. Butts CP et al (1999) Robust and catalytically active mono- and bis-Pd-complexes of the Trost modular ligand. Chem Commun 17:1707–1708
110. Marinho VR et al (2008) Novel chiral P, O-ligands for homogeneous Pd(0) catalyzed asymmetric allylic alkylation reactions. Tetrahedron Asymmetry 19:454–458
111. Glegola K et al (2007) Influence on the enantioselectivity in allylic alkylation of the anomeric position of the phosphine-amide ligands derived from D-glucosamine. Tetrahedron 63:7133–7141
112. Glegola K et al (2009) Palladium-catalyzed asymmetric allylic alkylation using phosphine-amide derived from chiral trans-2-aminocyclohexanol. Phosphorus Sulfur Silicon Relat Elem 184:1065–1075
113. Ropartz L et al (2007) Phosphine containing oligonucleotides for the development of metallodeoxyribozymes. Chem Commun 15:1556–1558

114. Lamac M et al (2007) Preparation of chiral phosphinoferrocene carboxamide ligands and their application to palladium-catalyzed asymmetric allylic alkylation. Organomet 26: 5042–5049
115. Lamac M et al (2007) Synthesis, coordination and catalytic utility of novel phosphanyl-ferrocenecarboxylic ligands combining planar and central chirality. Eur J Inorg Chem 16:2274–2287
116. Jiang B, Huang Z-G (2007) Chiral P, O-ligands derived from N, O-phenylene-prolinols for palladium-catalyzed asymmetric allylic alkylation. Tetrahedron Lett 48:1703–1706
117. Chen J et al (2009) Palladium-catalyzed asymmetric allylic nucleophilic substitution reactions using chiral tert-butanesulfinylphosphine ligands. Tetrahedron Asymmetry 20:1953–1956
118. Routaboul L et al (2007) New phosphorus dendrimers with chiral ferrocenyl phosphine-thioether ligands on the periphery for asymmetric catalysis. J Organomet Chem 692: 1064–1073
119. Cheung HY et al (2007) Enantioselective Pd-catalyzed allylic alkylation of indoles by a new class of chiral ferrocenyl P/S ligands. Org Lett 9:4295–4298
120. Kato M et al (2009) Synthesis of novel ferrocenyl-based P, S ligands (ThioClickFerrophos) and their use in Pd-catalyzed asymmetric allylic substitutions. Eur J Org Chem 30: 5232–5238
121. Lam FL et al (2008) Palladium-(S, pR)-ferroNPS-catalyzed asymmetric allylic etherification: electronic effect of nonconjugated substituents on benzylic alcohols on enantioselectivity. Angew Chem Int Ed 47:1280–1283
122. Khiar N et al (2008) New sulfur-phosphine ligands derived from sugars: synthesis and application in palladium-catalyzed allylic alkylation and in rhodium asymmetric hydrogenation. ARKIVOC 8:211–224
123. Thimmaiah M et al (2007) Novel benzoferrocenyl chiral ligands: Synthesis and evaluation of their suitability for asymmetric catalysis. J Organomet Chem 692:1956–1962
124. Stepnicka P et al (2008) Planar chiral alkenylferrocene phosphanes: Preparation, structural characterization and catalytic use in asymmetric allylic alkylation. J Organomet Chem 693:446–456
125. Hara O et al (2007) Synthesis of 2,6-dimethyl-9-aryl-9-phosphabicyclo[3.3.1]nonanes: their application to asymmetric synthesis of chiral tetrahydroquinolines and relatives. Tetrahedron 63:6170–6181
126. Gavrilov KN et al (2007) MOP-type binaphthyl phosphite and diamidophosphite ligands and their application in catalytic asymmetric transformations. Adv Synth Catal 349: 1085–1094
127. Qiao X-C et al (2009) From allylic alcohols to chiral tertiary homoallylic alcohol: palladium-catalyzed asymmetric allylation of isatins. Tetrahedron Asymmetry 20:1254–1261
128. Gavrilov KN et al (2007) Chiral ionic phosphites and diamidophosphites: a novel group of efficient ligands for asymmetric catalysis. Adv Synth Catal 349:609–616
129. Gavrilov KN et al (2007) Diastereomeric P*-chiral diamidophosphites with terpene fragments in asymmetric catalysis. Tetrahedron Asymmetry 18:2557–2564
130. Lyubimov SE et al (2009) The use of a new carboranylamidophosphite ligand in the asymmetric Pd-catalysed allylic alkylation in organic solvents and supercritical carbon dioxide. J Organomet Chem 694:3047–3049
131. Pignataro L et al (2009) Combination of a binaphthol-derived phosphite and a C1-symmetric phosphinamine generates heteroleptic catalysts in Rh- and Pd-mediated reactions. Chem Commun 24:3539–3541
132. Birkholz M-N et al (2007) Enantioselective hydrogenation with self-assembling rhodium phosphane catalysts: influence of ligand structure and solvent. Chem Eur J 13:5896–5907
133. Birkholz M-N et al (2007) Enantioselective Pd-catalyzed allylic amination with self-assembling and non-assembling monodentate phosphine ligands. Tetrahedron Asymmetry 18:2055–2060

134. Bayardon J, Sinou D (2008) Enantiopure fluorous 1,2-diaryl-1,2-diaminoethanes; synthesis and applications in asymmetric organometallic catalysis. ARKIVOC 7:26–35
135. Gualandi A et al (2010) Stereoselective synthesis of substituted 1,2-ethylenediaziridines and their use as ligands in palladium-catalyzed asymmetric allylic alkylation. Tetrahedron 66:715–720
136. Giulio Albano V et al (2007) Synthesis, structural characterization, and catalytic activity of chiral diamine and diimine Pd(II)-complexes. Inorg Chimica Acta 360:1000–1008
137. Wojaczynska E, Skarzewski J (2008) Chelating 2-azanorbornyl derivatives as effective nitrogen-nitrogen and nitrogen-chalcogen donating ligands in palladium-catalyzed asymmetric allylic alkylation. Tetrahedron Asymmetry 19:2252–2257
138. Niu J-L et al (2009) Origin of enantioselectivity with heterobidentate sulfide-tertiary amine (sp3) ligands in palladium-catalyzed allylic substitution. Tetrahedron 65:8869–8878
139. Bateman L et al (2008) New chiral diamide ligands: synthesis and application in allylic alkylation. Tetrahedron Asymmetry 19:391–396
140. Bastero A et al (2007) First allylpalladium systems containing chiral imidazolylpyridine ligands - structural studies and catalytic behavior. Eur J Inorg Chem 1:132–139
141. Shibatomi K et al (2009) Development of a new chiral spiro oxazolinylpyridine ligand (Spymox) for asymmetric catalysis. Synlett 241–244
142. Foltz C et al (2007) Using a tripod as a chiral chelating ligand: chemical exchange between equivalent molecular structures in palladium catalysis with 1,1,1-tris(oxazolinyl)ethane ("trisox"). Chem Eur J 13:5994–6008
143. Betz A et al (2008) (N, N) vs. (N, S) chelation of palladium in asymmetric allylic substitution using bis(thiazoline) ligands: A theoretical and experimental study. J Organomet Chem 693:2499–2508
144. Martin E, Dieguez M (2007) Thioether containing ligands for asymmetric allylic substitution reactions. C R Chim 10:188–205
145. Vargas F et al (2008) Modular chiral beta -selenium-, sulfur-, and tellurium amides: synthesis and application in the palladium-catalyzed asymmetric allylic alkylation. Tetrahedron 64:392–398
146. Sehnem JA et al (2008) Modular synthesis of chiral N-protected beta -seleno amines and amides via cleavage of 2-oxazolidinones and application in palladium-catalyzed asymmetric allylic alkylation. Synthesis 1262–1268
147. Zielinska-Blajet M et al (2007) Chiral phenylselenyl derivatives of pyrrolidine and cinchona alkaloids: nitrogen-selenium donating ligands in palladium-catalyzed asymmetric allylic alkylation. Tetrahedron Asymmetry 18:131–136
148. Wojaczynska E, Skarzewski J (2008) Novel C_2-symmetric chiral ligands: enantioselective transformation of cyclic 1,2-diols into 1,2-bis(phenylsulfenyl) and 1,2-bis(phenylselenyl) derivatives. Tetrahedron Asymmetry 19:593–597
149. Robe E et al (2008) Diphosphine sulfides derived from 2,2'-biphosphole: novel chiral S,S ligands for palladium-catalyzed asymmetric allylic substitution. Dalton Trans 2894–2898
150. Wang C-Y et al (2007) Palladium(II) complexes containing a bulky pyridinyl N-heterocyclic carbene ligand: Preparation and reactivity. J Organomet Chem 692:3976–3983
151. Yeagley AA, Chruma JJ (2007) C-C Bond-forming reactions via Pd-mediated decarboxylative α-imino anion generation. Org Lett 9:2879–2882
152. Visentin F, Togni A (2007) Synthesis and characterization of palladium(II) π-allyl complexes with chiral phosphinocarbene ligands. Kinetics and mechanism of allylic amination. Organomet 26:3746–3754
153. Peng HM et al (2008) Synthesis, structures, and solution dynamics of palladium complexes of quinoline-functionalized N-heterocyclic carbenes. Inorg Chem 47:8031–8043
154. Merzouk M et al (2007) Synthesis of chiral iminoalkyl functionalized N-heterocyclic carbenes and their use in asymmetric catalysis. Tetrahedron Lett 48:8914–8917

155. Flahaut A et al (2007) Palladium catalyzed asymmetric allylic alkylation using chelating N-heterocyclic carbene-amino ligands. Tetrahedron Asymmetry 18:229–236
156. Christensen CA, Meldal M (2007) Solid-phase synthesis of a peptide-based P, S-ligand system designed for generation of combinatorial catalyst libraries. J Comb Chem 9:79–85
157. Uozumi Y (2007) Asymmetric allylic substitution of cycloalkenyl esters in water with an amphiphilic resin-supported chiral palladium complex. Pure Appl Chem 79:1481–1489
158. Uozumi Y, Suzuka T (2008) π-Allylic sulfonylation in water with amphiphilic resin-supported palladium-phosphine complexes. Synthesis 12:1960–1964
159. Uozumi Y (2008) Heterogeneous asymmetric catalysis in water with amphiphilic polymer-supported homochiral palladium complexes. Bull Chem Soc Jpn 81:1183–1195
160. Swennenhuis BHG et al (2009) Supported chiral monodentate ligands in rhodium-catalysed asymmetric hydrogenation and palladium-catalysed asymmetric allylic alkylation. Eur J Org Chem 33:5796–5803
161. Jiang Z-D, Meng Z-H (2007) Polymer-supported chiral monodentate phosphoramidites in palladium-catalyzed allylic alkylation reactions. Chinese J Chem 25:542–545
162. Reimann S et al (2007) Enantioselective allylic substitution on Pd/Al_2O_3 modified by chiral diphosphines. J Catal 252:30–38
163. Reimann S et al (2008) A new, efficient heterogeneous Pd catalyst for enantioselective allylic substitution. J Catal 254:79–83
164. Dieguez M et al (2008) Palladium nanoparticles in allylic alkylations and Heck reactions: the molecular nature of the catalyst studied in a membrane reactor. Adv Synth Catal 350:2583–2598
165. Favier I et al (2007) Palladium catalytic species containing chiral phosphites: towards a discrimination between molecular and colloidal catalysts. Adv Synth Catal 349:2459–2469
166. Mukherjee S, List B (2007) Chiral counteranions in asymmetric transition-metal catalysis: highly enantioselective Pd/Bronsted acid-catalyzed direct α-allylation of aldehydes. J Am Chem Soc 129:11336–11337
167. Mang JY et al (2008) Palladium-catalyzed asymmetric allylic alkylation in the presence of chiral cinchonidinium salts. J Korean Chem Soc 52:724–726
168. Lei B-L et al (2009) Kinetic resolution of 2,3-dihydro-2-substituted 4-quinolones by palladium-catalyzed asymmetric allylic alkylation. J Am Chem Soc 131:18250–18251
169. Mino T et al (2008) Kinetic resolution of allylic esters in palladium-catalyzed asymmetric allylic alkylations using C-N bond axially chiral aminophosphine ligands. Tetrahedron Asymmetry 19:2711–2716
170. Jiang X-B et al (2007) SUPRAphos-based palladium catalysts for the kinetic resolution of racemic cyclohexenyl acetate. Chem Commun 22:2287–2289
171. Trost BM, Fandrick DR (2007) Palladium-catalyzed dynamic kinetic asymmetric allylic alkylation with the DPPBA ligands. Aldrichimica Acta 40:59–72

Top Organomet Chem (2012) 38: 155–208
DOI: 10.1007/3418_2011_10

Published online: 14 June 2011

Iridium-Catalyzed Asymmetric Allylic Substitutions

Wen-Bo Liu, Ji-Bao Xia, and Shu-Li You

Abstract Ir-catalyzed asymmetric allylic substitution reactions have been reviewed. This chapter discusses respectively the mechanistic investigation, reaction scope, and synthetic application of Ir-catalyzed allylic substitution reactions. The reaction scope is classified according to different types of nucleophiles such as C, N, O, and S.

Keywords Allylic substitution · Asymmetric catalysis · Enantioselectivity · Iridium · Regioselectivity

Contents

W.-B. Liu, J.-B. Xia and S.-L. You (✉)
State Key Laboratory of Organometallic Chemistry, Shanghai Institute of Organic Chemistry, Chinese Academy of Sciences, 345 Lingling Lu, Shanghai 200032, China
e-mail: slyou@sioc.ac.cn

Abbreviations

Ac	Acetyl
BINOL	Binaphthol
Bn	Benzyl
Boc	*tert*-Butoxycarbonyl
BSA	*N*,*O*-Bis(trimethylsilyl)acetamide
Bz	Benzoyl
Cbz	Benzyloxycarbonyl
cod	1,5-Cyclooctadiene
coe	Cyclooctene
DABCO	1,4-Diazabicyclo[2.2.2]octane
dbcot	Dibenzo[*a*,*e*]cyclooctatetraene
DBU	1,8-Diazabicyclo[5.4.0]undec-7-ene
DCM	Dichloromethane
de	Diastereomeric excess
DMAP	*N*,*N*-Dimethylaminopyridine
DME	1,2-Dimethoxy ethane
DMF	Dimethylformamide
e.s.	Enantiospecificity
ee	Enantiomeric excess
Fmoc	9-Fluorenylmethyloxycarbonyl
LDA	Lithium diisopropylamide
Ms	Methanesulfonyl
MS	Molecular sieves
Ns	Nitrophenylsulfonyl
PS	Proton sponge
RCM	Ring closing metathesis
TBAF	Tetrabutyl ammonium fluoride
TBD	1,5,7-Triazabicyclo-[4.4.0]dec-5-ene

TBDPS	*tert*-Butyldiphenylsilyl
TBS	*tert*-Butyldimethylsilyl
TEA	Triethylamine
Teoc	2-(Trimethylsilyl)ethoxy carbonyl
TES	Triethylsilyl
Tf	Trifluoromethanesulfonyl
THF	Tetrahydrofuran
THT	Tetrahydrothiophene
TIPS	Triisopropylsilyl
TM	Transition metal
TMS	Trimethylsilyl
Troc	2,2,2-Trichlorethoxycarbonyl
Ts	*p*-Toluenesulfonyl

1 Introduction

Transition metal-catalyzed allylic substitutions have been developed as one of the most powerful methods to construct carbon–carbon and carbon–heteroatom bonds [1–3]. Their enantioselective reactions have witnessed wide application in the synthesis of numerous pharmaceutical and natural products [3]. The enantioenriched allylic substitution products could be obtained via transition-metals (TM) such as Pd [1–5], Ir [6–9], W [10], Mo [11–15], Cu [16–22], and Ru [23–25] catalyzed asymmetric allylic substitution reactions, or by Rh- [26–31], Fe- [32–34], and Ni-catalyzed [35–37] double inversion of optically active chiral allylic substrates.

Undoubtedly, palladium complexes have been the most studied catalysts for asymmetric allylic alkylations. However, unsymmetrical allylic substrates, which form an unsymmetrical π-allyl-complex upon oxidative addition, in the presence of palladium catalyst preferentially lead to achiral linear substitution products (Scheme 1). Only in very few cases, the introduction of special ligands [38–45] or substrates [46–50] could deliver good levels of both regio- and enantioselectivity.

Given the ready availability of these unsymmetrical allylic substrates and facile transformation of the branched products, the catalysts that enable regio- and

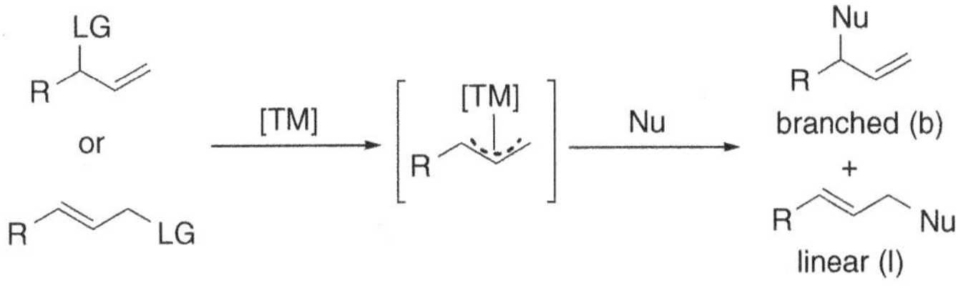

Scheme 1 TM-catalyzed allylic substitutions of monosubstituted allylic substrates

enantioselective allylic substitutions in favor of the branched products would be highly desirable (Scheme 1). Rh-catalysts could give excellent level of regioselectivity, although the number of catalytic enantioselective studies is very limited [28, 29]. The air-sensitivity of W- and Mo-derived catalysts also limited their application despite their excellent performance in selective allylations of unsymmetrical allylic substrates. Comparing all the above-mentioned transition metal catalysts, chiral Ir-complexes have been demonstrated to be highly efficient catalysts for regio- and enantioselective allylic substitutions of this kind of substrates. Since the first report of Ir-catalyzed allylations by Takeuchi et al. [51] and an enantioselective version by Helmchen et al. [6] in 1997, efforts from many research groups have significantly expanded the scope of this reaction. The extensive mechanistic studies by the Hartwig group and others led to the identification of the active catalysts that further broaden the reaction types. The invariably high regio- and enantioselectivity with predictable product configuration of Ir-catalyzed allylic substitutions makes them extremely attractive in organic synthesis. In this chapter, different aspects of Ir-catalyzed allylic substitutions including mechanistic studies, reaction scope, and synthetic applications are discussed, respectively.

2 Mechanistic Studies for Ir-Catalyzed Allylic Substitution Reaction

During the last two decades a wide range of chiral ligands have been developed for asymmetric allylic substitutions, and some of the most frequently used are summarized in Fig. 1.

In 1997, Takeuchi and Kashio for the first time reported the Ir-catalyzed allylic alkylation reaction (Table 1) [51–54]. With the catalyst generated in-situ from $[Ir(cod)Cl]_2$ and $P(OPh)_3$, the reaction of sodium dimethyl malonate with linear (**1**) or branched (**2**) allylic acetate led to the alkylation products favoring the branched isomer (b/l up to 98/2).

In their studies, Takeuchi and Kashio have discovered several general features in Ir-catalyzed allylic alkylations. These include:

1. $[Ir(cod)Cl]_2$ is the most preferred precursor for preparing Ir catalyst.
2. The optimal results are obtained with a 1:2 ratio of $[Ir(cod)Cl]_2$ and a monodentate ligand.
3. Electron-deficient ligands such as $P(OPh)_3$ are optimal.
4. Branched monoallylation products are mainly obtained with linear (*E*)-allylic substrate, while the linear products are favored with linear (*Z*)-allylic substrate maintaining the *Z* geometry of the double bond.

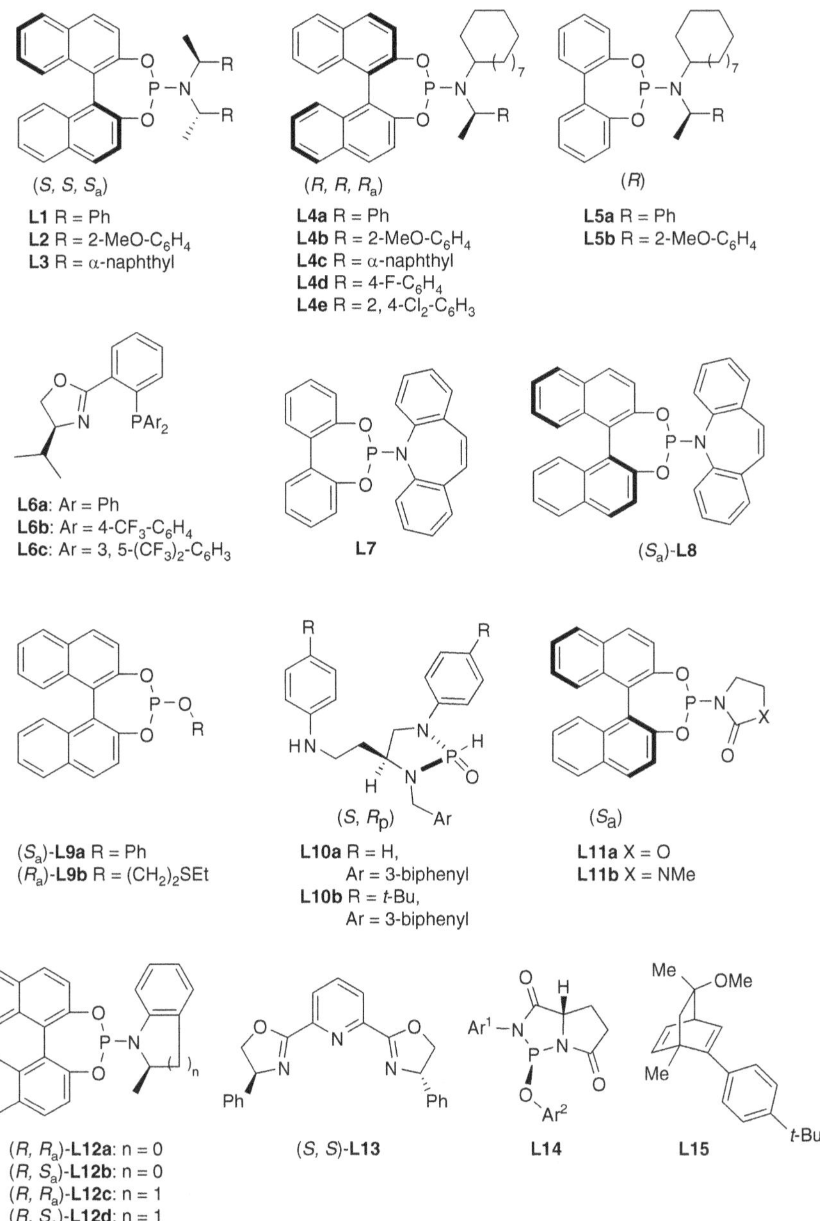

Fig. 1 Representative ligands used in Ir-catalyzed allylic substitution reactions

In the same year, Helmchen's group reported the first Ir-catalyzed asymmetric allylic alkylation (Scheme 2) [6]. Monoaryl-substituted allylic acetates were reacted with sodium dimethyl malonate in the presence of an iridium complex,

Table 1 The first Ir-catalyzed allylic alkylation

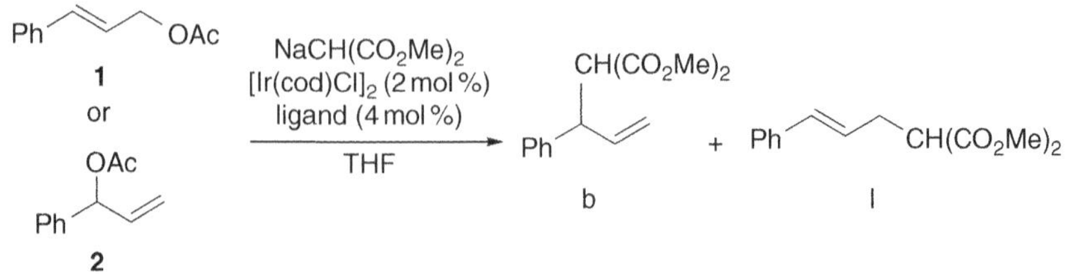

Substrate	Ligand	T (°C)	Time (h)	Yield (%)	b:l
1	–	65	24	89	32:68
1	PPh_3	65	24	59	64:36
1	$P(OPh)_3$	rt	3	98	98:2
2	$P(OPh)_3$	rt	3	99	98:2

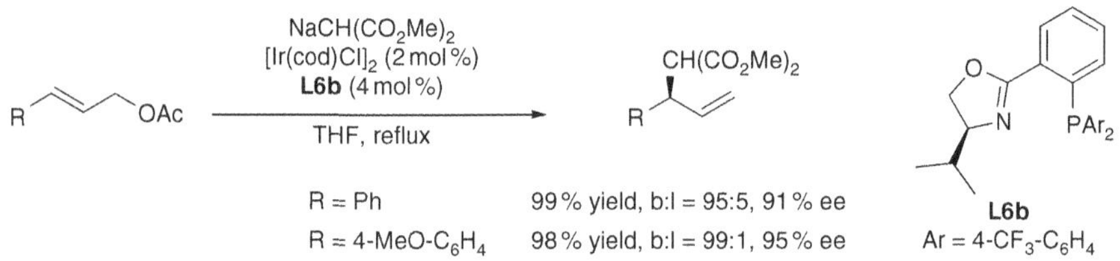

Scheme 2 The first example of an Ir-catalyzed asymmetric allylic alkylation

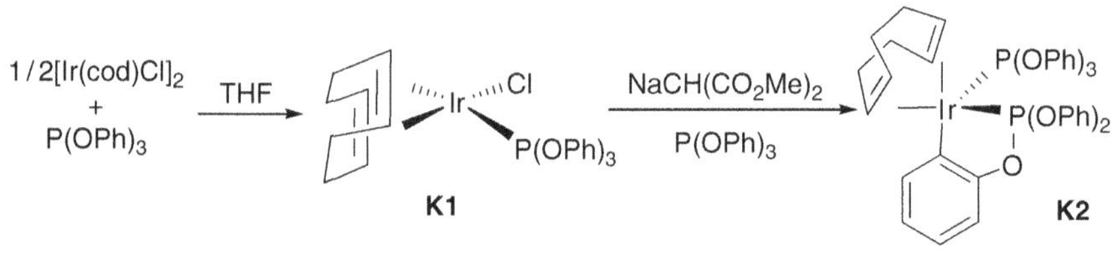

Scheme 3 The cyclometalated species of Ir-complexes with $P(OPh)_3$

generated in situ from $[Ir(cod)Cl]_2$ and a chiral phosphinooxazoline ligand **L6b**. The alkylation products were obtained in good yields with excellent regioselectivities and enantioselectivities.

The initial mechanistic study by Helmchen et al. was carried out with $P(OPh)_3$ as a ligand (Scheme 3) [55]. Mixing $[Ir(cod)Cl]_2$ and 2 equiv of $P(OPh)_3$ yielded the complex **K1**, a coordinatively unsaturated d^8–Ir^I complex [16 valence electrons (VE)s]. However, this complex did not react with typical allylic substrate. The reaction proceeded only upon the addition of $NaCH(CO_2Me)_2$. The nucleophile acting as base assisted the C–H activation of the $P(OPh)_3$ affording an Ir^{III} complex, which eliminated HCl to produce a 16-VE Ir^{I} complex. The subsequent addition of $P(OPh)_3$ led to the coordinatively saturated complex **K2** [56]. As complex **K2** is

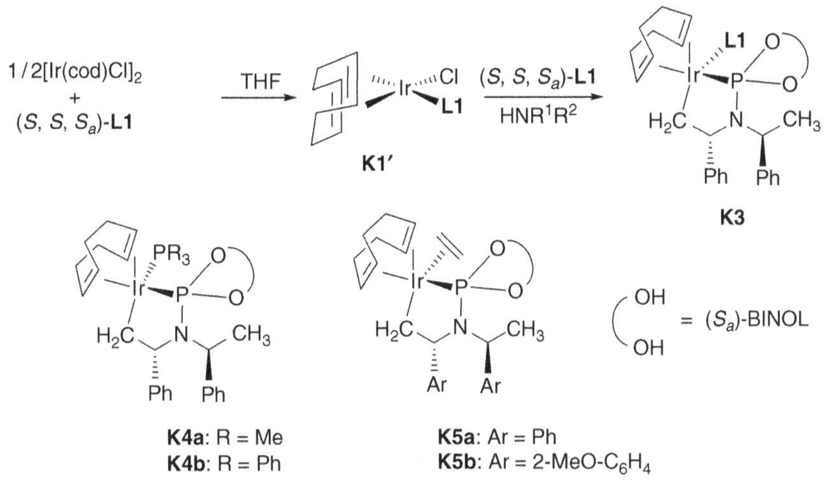

1/2[Ir(cod)Cl]$_2$ + (S, S, S$_a$)-L1 →(THF) K1′ →((S, S, S$_a$)-L1, HNR^{1}R^2) K3

K4a: R = Me
K4b: R = Ph

K5a: Ar = Ph
K5b: Ar = 2-MeO-C$_6$H$_4$

(OH)(OH) = (S$_a$)-BINOL

Scheme 4 The cyclometalated species of Ir-complexes

coordinatively saturated, $P(OPh)_3$ must dissociate in order to obtain a catalytically active 16-VE d^8–Ir^{I} complex. A similar C–H activation was later also found for Ir–phosphoramidite complexes.

The crystal structure of (π-allyl)(PHOX)Ir^{III} complexes was prepared and characterized also by Helmchen's group. The reaction of this complex with sodium dimethyl malonate proceeded by attacking at the central rather than the terminal allylic carbon to yield an iridacyclobutane [57].

In 2002, a Feringa-type phosphoramidite ligand **L1** [58, 59] was used in Ir-catalyzed asymmetric allylations by Ohmura and Hartwig [60]. In the subsequent work by Hartwig's group [61], they found that iridacycle **K3** was formed via a C–H bond activation of the methyl group of the phosphoramidite ligand (Scheme 4). The reactions of **K3** with more strongly coordinating dative ligands such as PMe_3 and PPh_3 via ligand exchange generated complexes **K4a** and **K4b** respectively. Further evidence for the structure of **K3** was obtained by the X-ray diffraction analysis of complex **K4a**. The reactive ethylene complexes **K5a** and **K5b** synthesized by Hartwig et al. were used to improve the efficiency of the allylation process [62].

In 2009, Hartwig and coworkers synthesized and characterized the allyliridium intermediates (Scheme 5, Eq. 1) by treating the cyclometalated iridium complex **K5a** with the corresponding allylic halides in benzene at room temperature [63]. The complexes **K6a** and **K6a′** were synthesized in moderate yields after an anion exchange by adding a silver salt. Both structures of allyliridium intermediates **K6a** and **K6a′** were confirmed by X-ray diffraction analyses. The reactions of **K6a** and **K6a′** with carbon- and heteroatom nucleophiles were also studied. The yields, regioselectivities and enantioselectivities (for **K6a′**) of these stoichiometric reactions were identical to the catalytic reactions of allyl methyl carbonates by Ir-complex **K5a**.

At almost the same time, Helmchen and coworkers reported a more straightforward synthesis of the allyliridium intermediates (Scheme 5, Eq. 2) [64, 65]. They

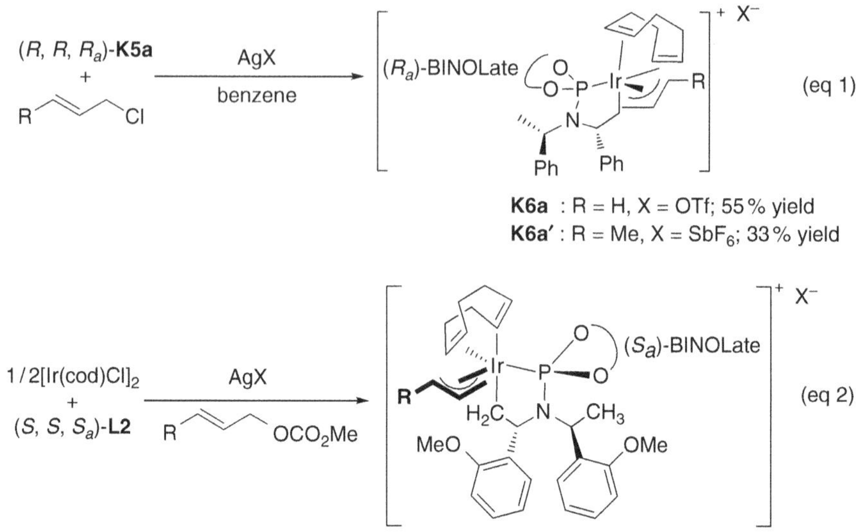

K6a : R = H, X = OTf; 55 % yield
K6a′ : R = Me, X = SbF_6; 33 % yield

Scheme 5 The preparation of allyliridium complexes

found that the allyliridium complex can be directly synthesized in >95% yield using a one-pot procedure according to Scheme 5. These complexes were also characterized by X-ray crystal structure analyses. The determination of the resting state by ^{31}P NMR spectroscopy concluded that the cyclometalation process is reversible.

A catalytic cycle of Ir-catalyzed allylic substitutions is given in Fig. 2. Complex **K1′** is formed by the coordination between $[Ir(cod)Cl]_2$ and a phosphoramidite ligand. Then base induced cyclometalation via a C–H bond activation process in **K1′** leads to the formation of complex **K3**.

This complex is not a directly catalytically active species because it is coordinatively saturated. After the disassociation of ligand, the intermediate **K3′** is generated to mediate the allylic substitution through steps including coordination towards the alkene of the allylic substrate, oxidative addition of the allylic substrate forming the π–allyl complex, substitution by the nucleophile, and dissociation of the substituted products.

3 Ir-Catalyzed Asymmetric Allylic Alkylation Reaction

3.1 Stabilized Enolates

Malonate derivatives are the most frequently used pronucleophiles for the transition metal-catalyzed allylic alkylations. The already mentioned first Ir-catalyzed

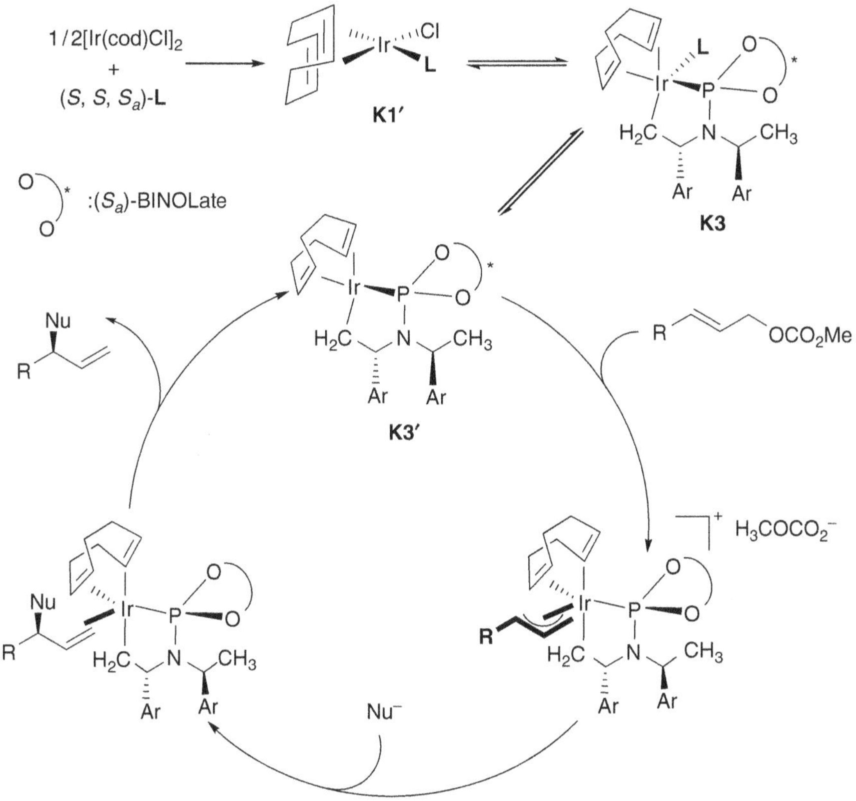

Fig. 2 Catalytic cycle of the Ir-catalyzed allylic substitution reaction

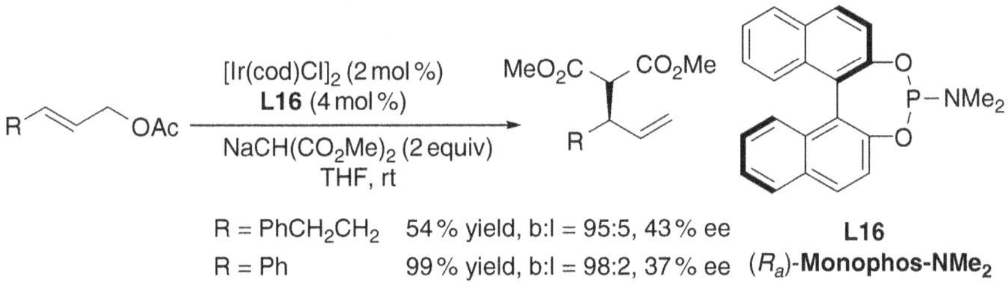

Scheme 6 Ir-(Monophos-NMe_2) complex catalyzed allylic alkylation

enantioselective allylation of dimethyl malonate was reported by Janssen and Helmchen using chiral PHOX ligands **L6** (Scheme 2) [6].

In 1999, Monophos-NMe_2 **L16** was used in the same reaction (Scheme 6) [66]. The alkylation products were obtained with excellent regioselectivity but only moderate enantioselectivity.

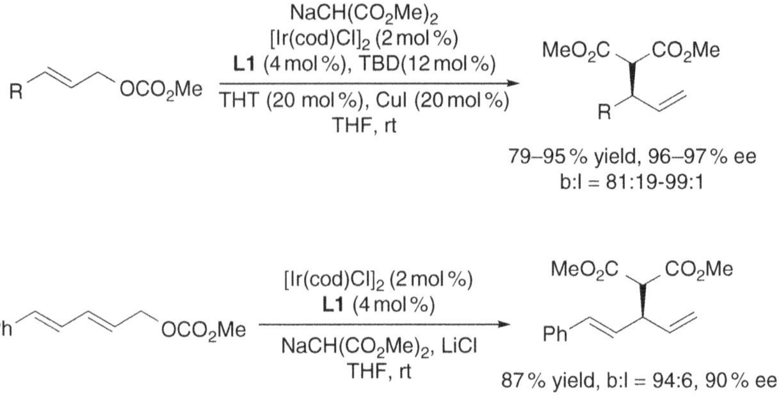

Scheme 7 Ir/**L1** complex catalyzed allylic alkylation with dimethyl malonate

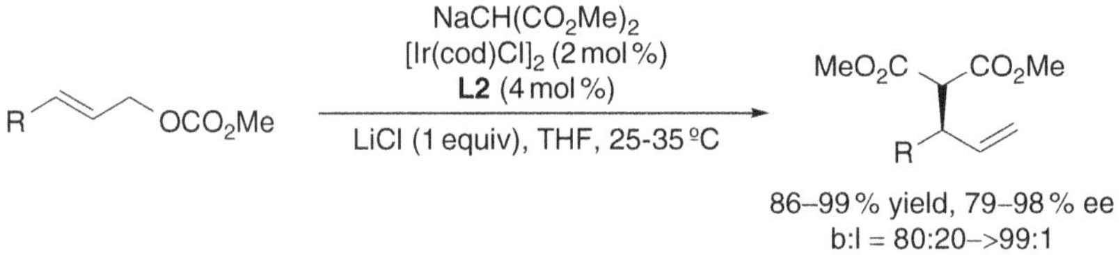

Scheme 8 Ir/**L2** complex catalyzed allylic alkylation with dimethyl malonate

In 2004, Helmchen and coworkers found that high enantioselectivities up to 97% ee were obtained in the Ir-catalyzed asymmetric alkylations with ligand **L1** (Scheme 7) [67]. The reaction rate was dramatically accelerated when both THT (tetrahydrothiophene) and CuI were used as the additives. Notably, the ratio of catalyst to substrate could be lowered down to 0.4% in this reaction. The ee value could be increased slightly when **L2** was used as a ligand under the otherwise same conditions and the catalyst loading could be as low as 0.1 mol% [68]. In the same year, dienyl substrates were also demonstrated to be suitable substrates [69].

Alexakis and coworkers found that ligand **L2**, developed in their own laboratory, was optimal for the Ir-catalyzed allylations in terms of regio- and enantioselectivity (Scheme 8) [70–72]. The selectivity could be dramatically increased by the addition of lithium chloride. This strong salt effect of lithium chloride was also previously noted by Helmchen et al. [66, 69] and Fuji et al. [73]. The kinetic study of various phosphoramidite ligands in the Ir-catalyzed reaction was investigated [72]. It was found that a slight difference in the substitution pattern of the aryl group on the amine moiety of the ligand dramatically alters the activity of the resulting iridium catalyst.

In 2007, a salt-free version of the alkylation reaction was developed by Helmchen and coworkers directly using malonate derivatives as nucleophiles (Scheme 9) [74]. The reaction could be carried out at a 20 mmol scale without loss of ee value. In

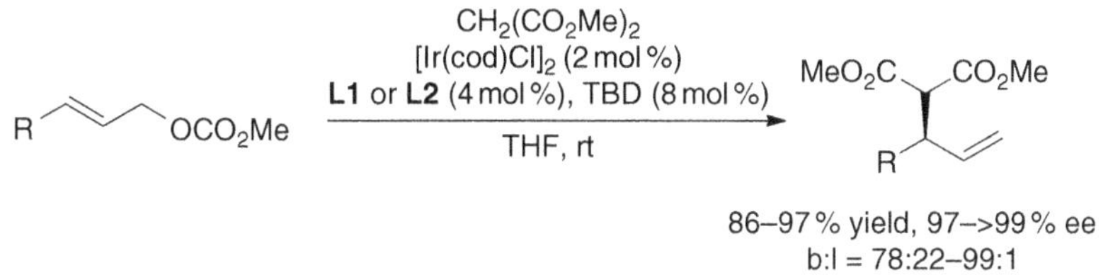

Scheme 9 Salt-free allylic alkylation with dimethyl malonate

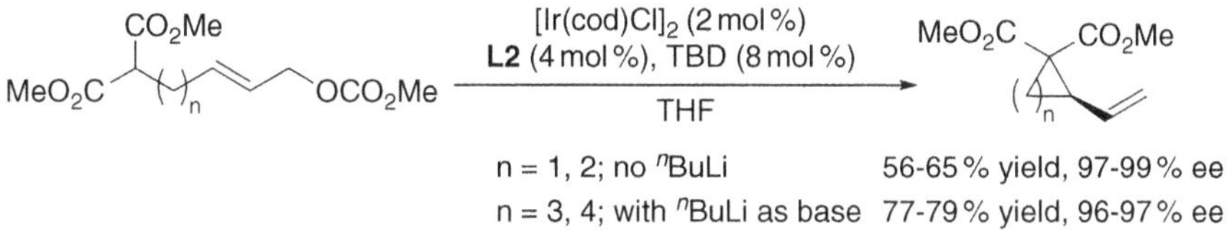

Scheme 10 Ir-catalyzed intramolecular allylic alkylation

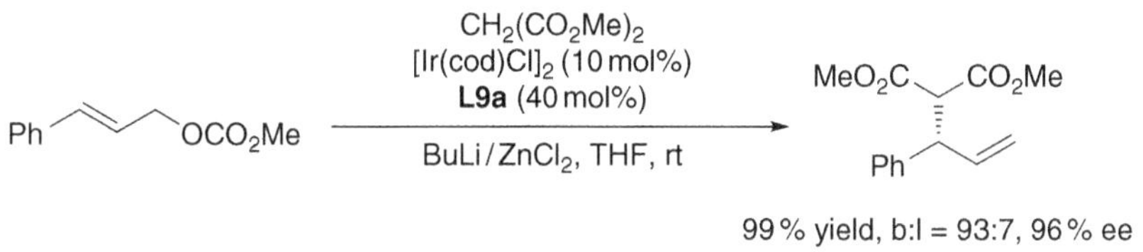

Scheme 11 Ir/**L9a** complex catalyzed allylic alkylation with dimethyl malonate

2008, they also developed a new iridium catalyst system from [Ir(dbcot)Cl]$_2$ and **L2**, which displayed excellent air stability and led to very high regioselectivities [75].

The intramolecular allylic alkylation was also investigated by Helmchen and coworkers (Scheme 10). Cyclopentane and cylcohexane derivatives were obtained with excellent ee's when the malonate anions were prepared using ^{n}BuLi at $-78°$C in order to suppress the noncatalyzed cyclization [68]. The vinylcyclopropane and vinylcylcobutane derivatives were prepared under salt-free conditions [74].

Many other chiral ligands were also investigated. Fuji and coworkers showed that excellent results were obtained with phosphite ligands such as **L9a** (Scheme 11) [73, 76]. The utilization of ^{n}BuLi as base and ZnCl$_2$ as additive was critical for the high ee. The authors hypothesized that the chlorozinc enolate might be responsible for the high enantioselectivity.

Hamada and coworkers synthesized DIAPHOX **L10** and found that allylic alkylation reaction with **L10a** gave excellent yields and selectivities (Scheme 12) [77]. A formal enantioselective synthesis of (−)-paroxetine was achieved in their report.

Chiral phosphoramides with an amide framework were introduced by Takeuchi and coworkers providing good regio- and enantioselectivities (Scheme 13) [78].

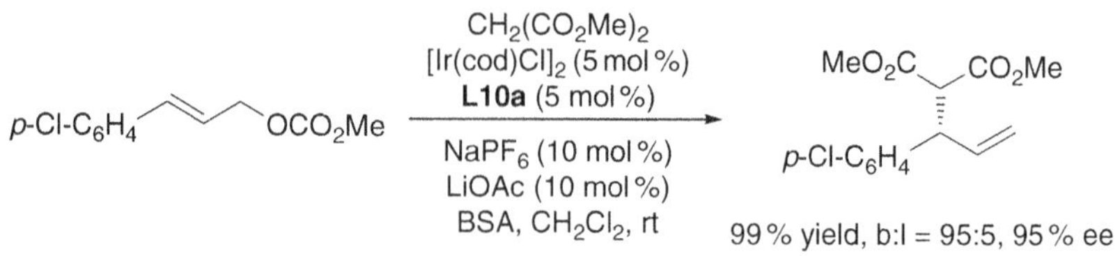

p-Cl-C_6H_4 ... OCO_2Me → $CH_2(CO_2Me)_2$, $[Ir(cod)Cl]_2$ (5 mol %), **L10a** (5 mol %), $NaPF_6$ (10 mol %), LiOAc (10 mol %), BSA, CH_2Cl_2, rt → MeO_2C, CO_2Me, p-Cl-C_6H_4

99 % yield, b:l = 95:5, 95 % ee

Scheme 12 Ir/**L10a** complex catalyzed allylic alkylation with dimethyl malonate

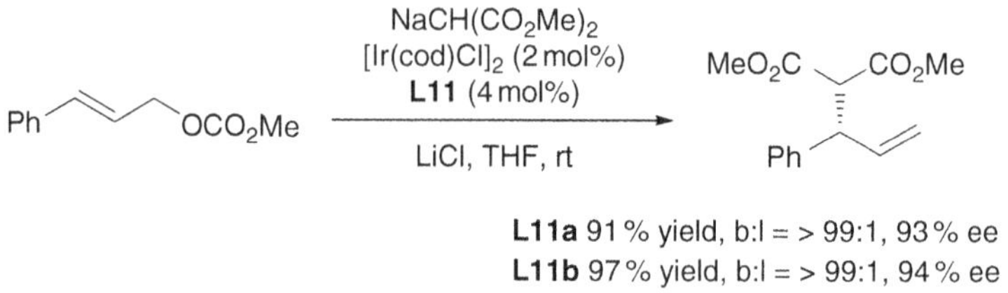

Ph ... OCO_2Me → $NaCH(CO_2Me)_2$, $[Ir(cod)Cl]_2$ (2 mol%), **L11** (4 mol%), LiCl, THF, rt → MeO_2C, CO_2Me, Ph

L11a 91 % yield, b:l = > 99:1, 93 % ee
L11b 97 % yield, b:l = > 99:1, 94 % ee

Scheme 13 Ir/**L11** complex catalyzed allylic alkylation

Ph ... OCO_2Me → $R^1R^2CHNO_2$, $[Ir(cod)Cl]_2$ (2 mol%), **L2** (4 mol%), TBD (8 mol%), THF, rt → R^2, R^1, NO_2, Ph

Cs_2CO_3 as base, R^1 = Me, R^2 = Me 84 % yield, b:l = 96:4, 99 % ee
no base, R^1 = H, R^2 = CO_2Et 90 % yield, b:l = 99:1, 98 % ee

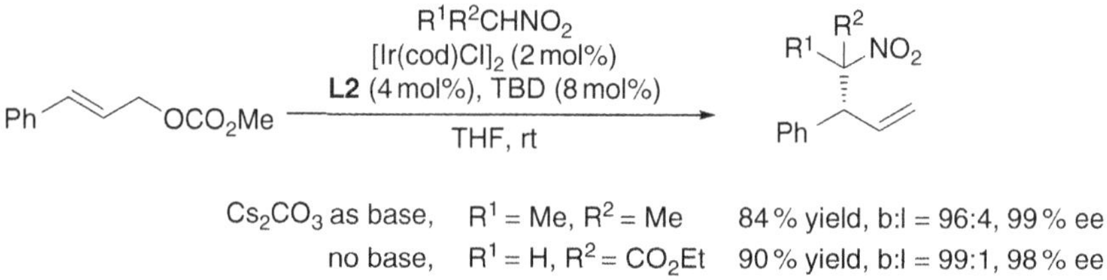

Scheme 14 Ir-catalyzed allylic alkylation with nitro compounds

Substituted malonates, β-keto-esters, and β-amide-esters have also been successfully used as pronucleophiles in Ir-catalyzed asymmetric allylic alkylation reaction, and approximately 1:1 mixtures of diastereomers are generally formed [68, 71, 79].

3.2 Aliphatic Nitro Compounds

Nitro compounds are very attractive since the nitro group of the products allows subsequent versatile transformations. Ir-catalyzed asymmetric allylic alkylations proceeded well with nitroalkane and ethyl nitroacetate as nucleophiles [80]. Excellent regio- and enantioselectivities were obtained using **L2** as chiral ligand (Scheme 14).

3.3 Malononitrile

In addition to aliphatic nitro compound, malononitrile was found to be a suitable nucleophile under salt-free conditions (Scheme 15). Good to excellent regio- and enantioselectivities were obtained using **L2** as chiral ligand [81]. The allylic alkylation products can be easily transformed into methyl esters without loss of ee value.

3.4 Glycine Equivalents

Enantioselective alkylations of glycine derivatives would provide an important class of non-proteinogenic amino acids, which are very useful synthetic intermediates or building blocks for numerous biologically important compounds. In 2003, diphenylimino glycinate was used as pronucleophile in Ir-catalyzed allylations (Scheme 16). Good results were obtained when 3-arylallyl diethyl phosphates were employed with chiral phosphite **L9b** [82, 83]. Interestingly, reverse of the diastereoselectivity was achieved by simply switching the base employed.

In 2008, Eilbracht and coworkers reported that moderate diastereo- but excellent regioselectivities were obtained when **L2** was used as ligand and DABCO as a base (Scheme 17) [84].

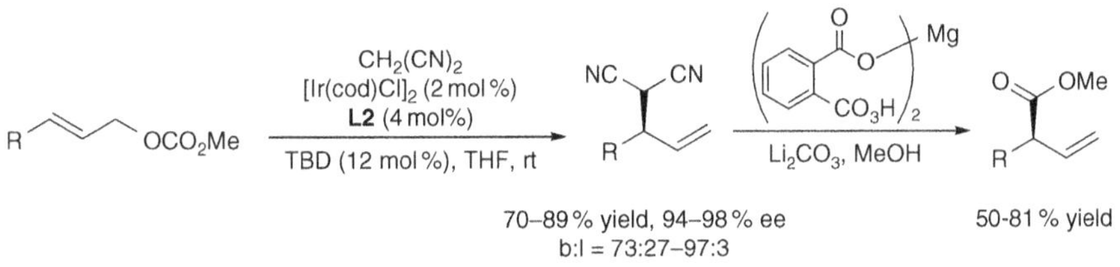

R
OCO2Me
CH2(CN)2
[Ir(cod)Cl]2 (2 mol %)
L2 (4 mol%)
TBD (12 mol %), THF, rt
NC
CN
R
Mg
CO3H
2
Li2CO3, MeOH
O
OMe
R
70–89 % yield, 94–98 % ee
b:l = 73:27–97:3
50-81 % yield

Scheme 15 Ir-catalyzed allylic alkylation with malononitrile

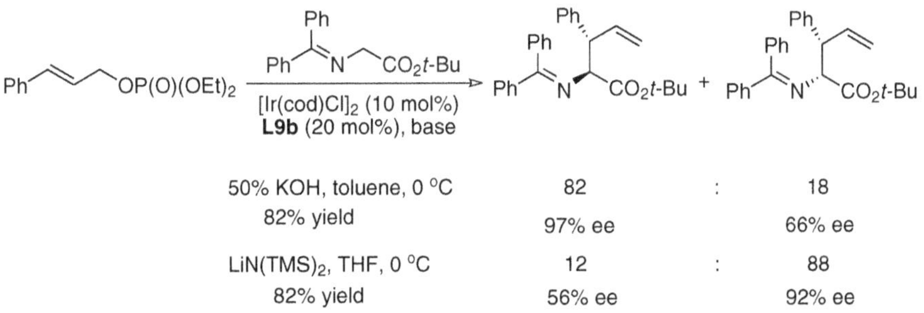

Ph
OP(O)(OEt)2
Ph
Ph
N
CO2t-Bu
[Ir(cod)Cl]2 (10 mol%)
L9b (20 mol%), base
CO2t-Bu
+
50% KOH, toluene, 0 °C
82% yield
82
:
18
97% ee
66% ee
LiN(TMS)2, THF, 0 °C
82% yield
12
:
88
56% ee
92% ee

Scheme 16 Ir-catalyzed allylic alkylation of glycine derivatives with allylic phosphates

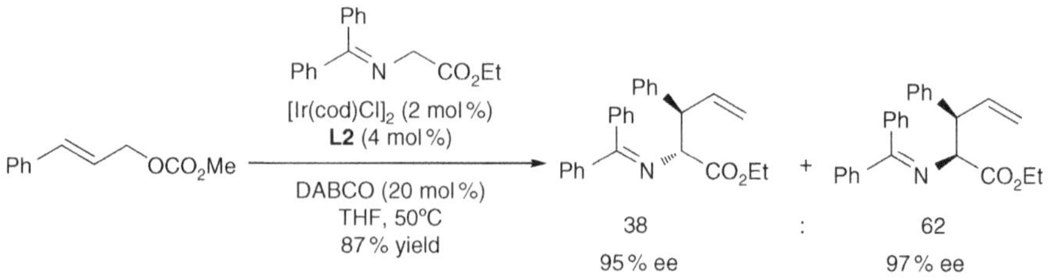

Scheme 17 Ir-catalyzed allylic alkylation of glycine derivatives with allylic carbonates

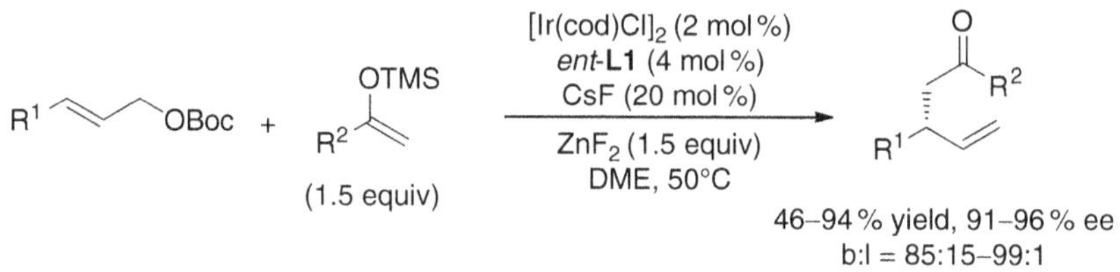

Scheme 18 Ir-catalyzed allylic alkylation with silyl enol ethers

3.5 Ketone Enolates

Non-stabilized ketone enolates have been a great challenge as nucleophiles in transition metal-catalyzed allylic substitution reactions for a long time for several reasons [85–92]: First, the ketone enolate nucleophile must preferentially attack the π-allyl intermediate instead of the carbonyl group, and the latter leads to aldol product. Second, the ketone product must resist re-formation of an enolate followed by reaction with a second allyl electrophile. Third, as a relatively hard nucleophile, the ketone enolate should avoid the attack on the metal center, which normally causes difficulties with respect to enantioselectivity control.

In 2005, Graening and Hartwig reported a highly regio- and enantioselective reaction of silyl enol ethers (Scheme 18) [93]. In the presence of $[Ir(cod)Cl]_2$ (2 mol %), *ent*-**L1** (4 mol%), CsF (20 mol%), and ZnF_2 (1.5 equiv), the allylic alkylations of *tert*-butyl allylic carbonates proceeded well in up to 99:1 regioselectivities, providing the branched products with up to 96% ee.

In 2007, the subsequent work by Weix and Hartwig demonstrated that ketone enamines could be used as well, providing the same homoallylic ketones with similar selectivities (Scheme 19) [94]. Optimal results were obtained using isolated metallacyclic iridium phosphoramidite complex *ent*-**K3** (Scheme 4) as catalyst and isopropyl carbonates as substrates.

About the same time, You and coworkers realized the allylic alkylation of ketone enolates via an Ir-catalyzed asymmetric decarboxylative allylation (Scheme 20) [95]. This protocol features an in-situ generated nucleophile and

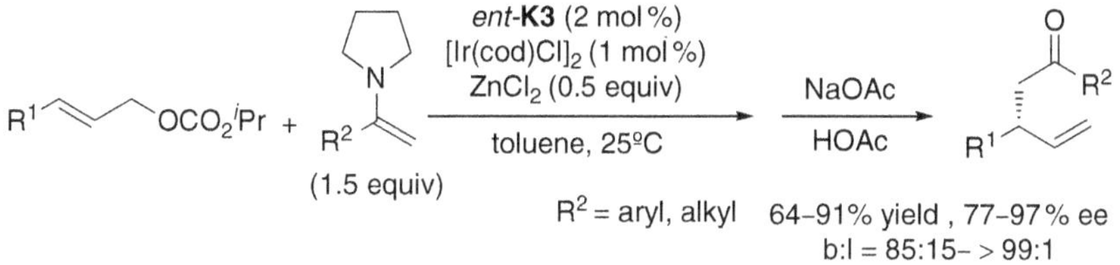

Scheme 19 Asymmetric allylic allylation of enamines

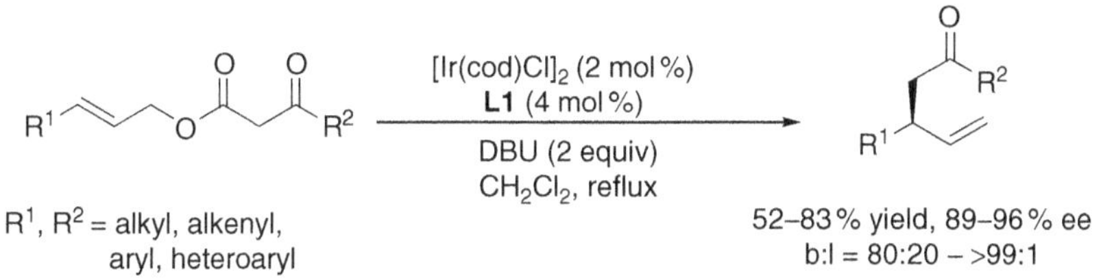

Scheme 20 Decarboxylative allylic alkylation of allyl β-ketocarboxylates

ready availability of the starting materials. The homoallylic ketones were obtained in high regio- and enantioselectivities in the presence of $[Ir(cod)Cl]_2$, **L1** and DBU.

3.6 *Indoles*

Indoles are among the most widely distributed heterocyclic compounds in nature and they serve as the structural core of numerous biologically active natural products and pharmaceuticals. Consequently, the synthesis of enantiomerically pure indole derivatives is highly desirable in both organic synthesis and medicinal chemistry [96]. Recently, the transition metal-catalyzed allylic alkylation has been proved suitable for direct indole functionalization [97–101].

In 2008, You and coworkers reported the Ir-catalyzed allylic alkylation with indoles as nucleophiles [102–104]. They found that $[Ir(cod)Cl]_2$/**L1** is an efficient catalytic system for the highly regio- and enantioselective Friedel–Crafts type allylic alkylation of indoles (Scheme 21) [102]. For various unsymmetric allylic substrates and indoles, this Ir-catalyzed allylic arylation proceeded smoothly with excellent regioselectivities to afford the branched alkylation products with high ee's.

Subsequently, a series of novel phosphoramidite ligands **L12** was synthesized from enantiopure BINOL and 2-methyl 1,2,3,4-tetrahydroquinoline or 2-methylindoline by You and coworkers [103]. These ligands were applied in the above allylic arylation and good results were obtained. These ligands' suitability towards *ortho*-substituted cinnamyl carbonates, which normally afford poor ee's with Feringa

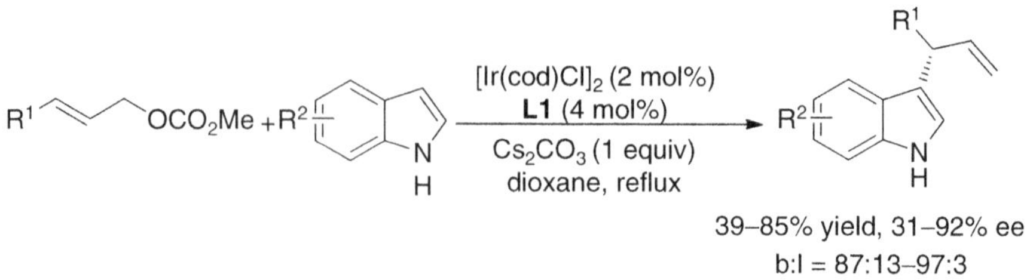

Scheme 21 Asymmetric allylic allylation of indoles

Table 2 Asymmetric allylation of indoles using ligands **L1** and **L12**

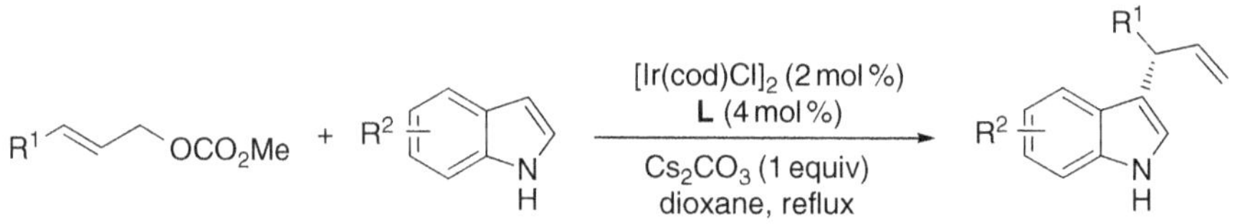

Entry	Ligand	R^1	R^2	Yield	b:l	ee (%)
1	**L1**	2-MeO–C_6H_4	H	84	>99:1	70
2	**L12a**	2-MeO–C_6H_4	H	47	>99:1	90
3	**L1**	2-Cl–C_6H_4	H	63	>99:1	15
4	**L12c**	2-Cl–C_6H_4	H	55	>99:1	79
5	**L1**	2-Br–C_6H_4	H	33	>99:1	4
6	**L12c**	2-Br–C_6H_4	H	41	>99:1	85
7	**L1**	1-Naphthyl	5-MeO	39	>99:1	31
8	**L12c**	1-Naphthyl	5-MeO	85	>99:1	81

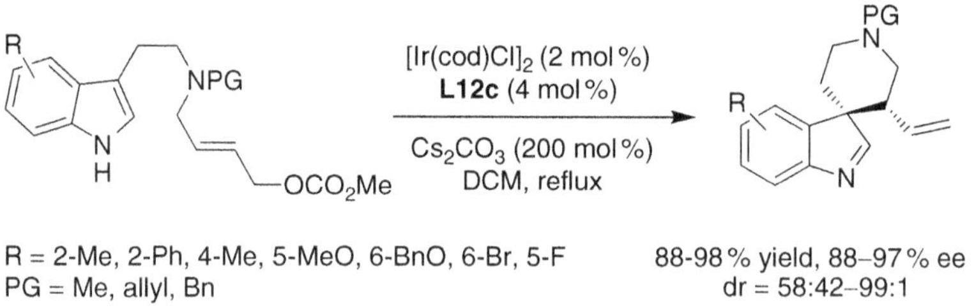

Scheme 22 Ir-catalyzed asymmetric allylic dearomatization of indoles

ligand **L1**, is particularly attractive. As shown in Table 2, a significant improvement of the ee's was observed.

Recently, You and coworkers reported an Ir-catalyzed asymmetric allylic dearomatization reaction [104]. A spiroindolenine substructure was constructed via an Ir-catalyzed intramolecular C-3 allylic alkylation of indoles (Scheme 22).

The allylic dearomatization of indole is only known within a handful of examples in the presence of a palladium catalyst [105–107]. The formation of spiroindolenine is likely caused by the more favorable six-member ring product by reacting at C-3 position over the seven-member ring product at C-2 position. Ligands **L1**, **L2**, and **L3** could catalyze the reaction in excellent *ee*, but with moderate diastereomeric ratios (dr). The iridium complex derived from ligand **L12c**, developed in the You lab [103], with Cs_2CO_3 in DCM, was found to give the best results. Various substituted indoles were probed and afforded the corresponding products in excellent yields, ee's and dr, except 2-substituted indolyl allyl carbonate that gave only a moderate dr.

3.7 Organometallics

Nonstabilized *C*-nucleophiles such as Grignard or organozinc reagents have also been applied in the Ir-catalyzed asymmetric allylic alkylation. Alexakis and coworkers found that good enantioselectivities could be obtained when aryl zinc halides were used as the nucleophiles (Scheme 23) [108, 109]. However, the branched to linear ratio could not be controlled satisfactorily. A formal synthesis of the antidepressant sertraline was carried out as an application of this method.

3.8 Fluorobis(phenylsulfonyl)methane

In 2009, a highly regio- and enantioselective Ir-catalyzed allylation of fluorobis-(phenylsulfonyl)methane (FBSM) has been realized by Zhao, You and their coworkers (Scheme 24) [110]. The fluorine-containing products were obtained with 28–96% yields and 75–95% ee. A gram scale reaction was carried out and the monofluoromethylated products could be transformed to the monofluorinated ibuprofen and naproxen without loss of the enantiomeric purity [111].

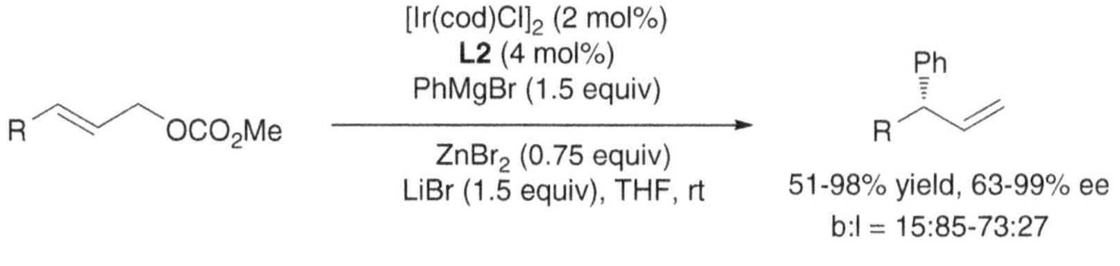

R⁀⁀OCO$_2$Me → [Ir(cod)Cl]$_2$ (2 mol%), **L2** (4 mol%), PhMgBr (1.5 equiv) / ZnBr$_2$ (0.75 equiv), LiBr (1.5 equiv), THF, rt → Ph, R

51-98% yield, 63-99% ee

b:l = 15:85-73:27

Scheme 23 Ir-catalyzed asymmetric allylic alkylation of aryl zinc reagents

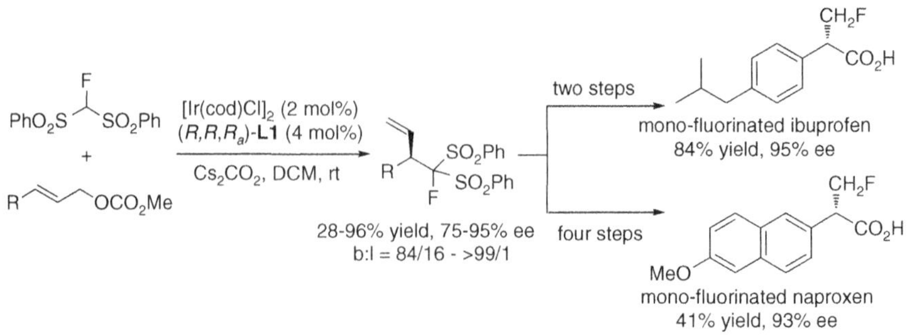

Scheme 24 Preparation of mono-fluorinated ibuprofen family via Ir-catalyzed asymmetric allylation

4 Ir-Catalyzed Asymmetric Allylic Amination Reaction

4.1 Aliphatic and Aryl Amines

Ir-catalyzed asymmetric allylic aminations were for the first time introduced by Hartwig and coworkers [60]. Subsequent efforts from Hartwig and other groups showed that high regio- and enantioselectivities could be obtained from achiral allylic carbonates and acetates with ligand **L1**, affording branched secondary or tertiary allylic amines (Scheme 25) [69, 84, 112, 113]. The nucleophile scope is very general, including aromatic and aliphatic primary amines as well as cyclic and acyclic aliphatic secondary amines. In 2004, Hartwig and coworkers [112] reported highly selective allylations of aromatic amines by conversion of an iridium procatalyst into an active cyclometalated catalyst in the presence of propylamine or DABCO.

Meanwhile, Alexakis and coworkers found that excellent results were obtained in reactions of aromatic and aliphatic primary amines [114]. As depicted in Table 3, with ligand **L2**, the reactions proceeded with increased enantioselectivity and reaction rate, compared to the reactions with ligand **L1**.

A more stable catalyst derived from $[Ir(dbcot)Cl]_2$ and **L2**, introduced by Helmchen and coworkers, was also suitable for the allylic amination and allowed the reaction to be conducted under air with excellent results [75]. The intramolecular reaction was also investigated by Helmchen and coworkers (Scheme 26) [115, 116]. The pyrrolidine, piperidine, and azepane derivatives were obtained in good yields and excellent ee's. The sequential inter- and intramolecular amination of bisallylic carbonates were also investigated and good results were obtained.

Several other chiral ligands were investigated in asymmetric aminations and moderate to good regio- and good enantioselectivities were obtained with chiral Ph-Pybox ligand **L13** [117]. Hamada and coworkers showed that excellent regioselectivities (>99:1) and enantioselectivities (65–95% ee) were obtained employing DIAPHOX **L10b** as a ligand (Scheme 27) [118].

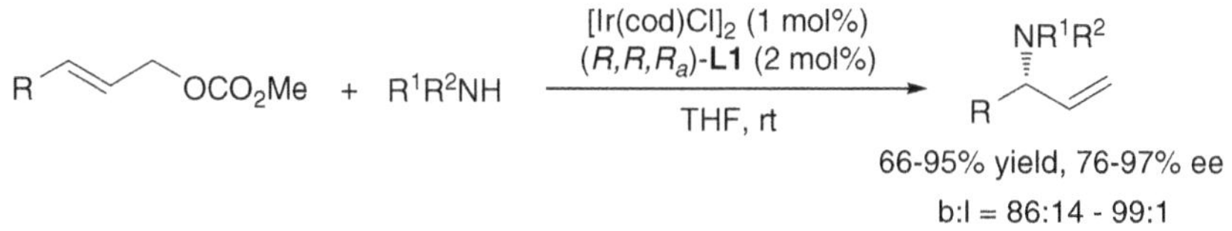

Scheme 25 Ir-catalyzed asymmetric allylic aminations

Table 3 Enantioselective Ir-catalyzed allylic amination with ligands **L1** and **L2**

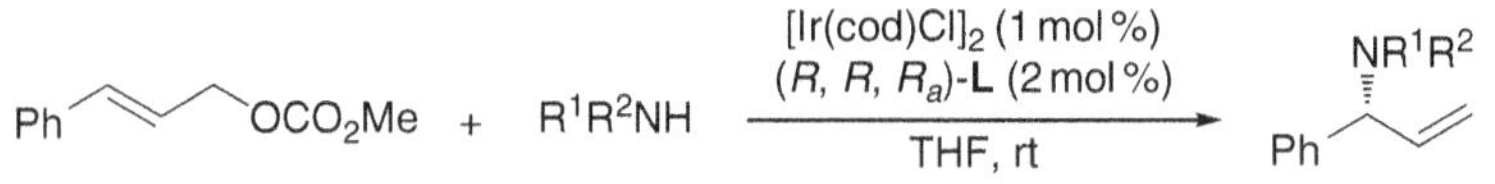

Entry	R^1R^2NH	Ligand[a]	b:l[a]	ee(%)[a]
1	$PhNH_2$	**L2**[**L1**]	98/2[99/1]	97[95]
2	Allylamine	**L2**[**L1**]	99/1[–]	97[97]
3	*n*-Hexylamine	**L2**[**L1**]	98/2[98/2]	98[96]

[a]The value in the square brackets was the results from the study by Ohmura and Hartwig [60]

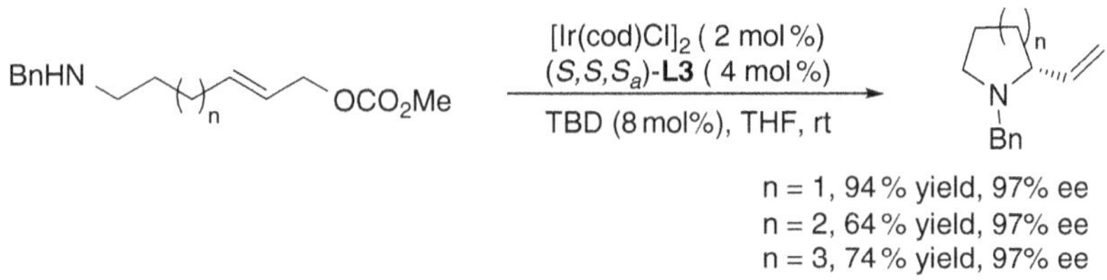

Scheme 26 Ir-catalyzed asymmetric intramolecular allylic aminations

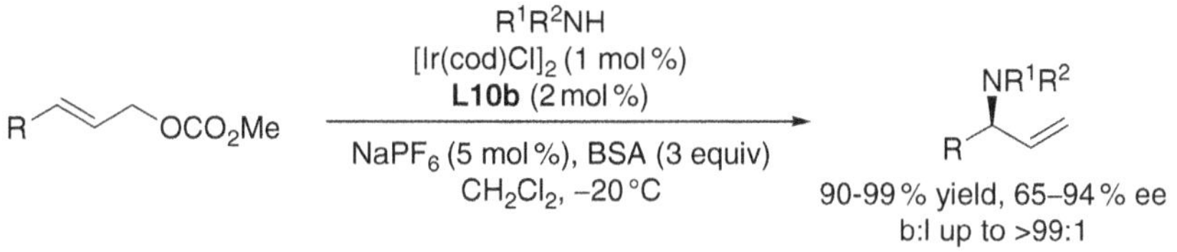

Scheme 27 Asymmetric allylic aminations with DIAPHOX ligand **L10b**

The poor leaving ability of a hydroxyl group causes the fact that the substitution of allylic alcohols typically requires high temperature, neat condition, or an activator. There are few examples of highly enantioselective direct allylation of allylic alcohols. Two procedures have been developed for iridium-catalyzed direct allylic amination of allylic alcohols to form branched allylic amine products with high enantioselectivity by Hartwig et al. [119]. $Nb(OEt)_5$ was found to act as an activator

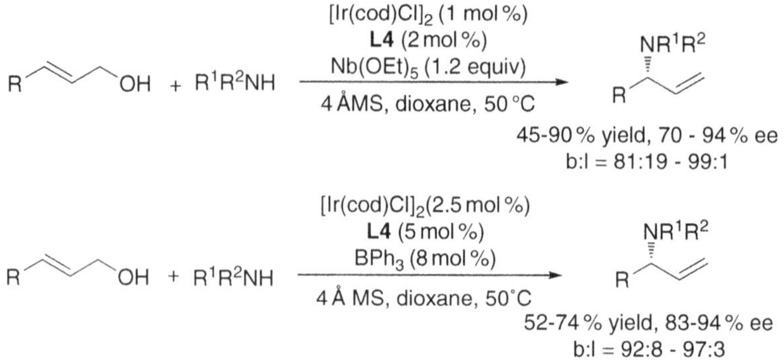

Scheme 28 Ir-catalyzed asymmetric allylic amination with allylic alcohols

of the allylic alcohol in situ, and BPh_3 was also found to act in an analogous manner in catalytic reactions (Scheme 28).

4.2 *Amides*

4.2.1 Sulfonamides

In 2005, sulfonamides were successfully used in asymmetric aminations by Helmchen and coworkers [120]. The reaction with *p*-nosylamide as nucleophile proceeded smoothly without additional base (Table 4). However, the regio- and enantioselectivities could be affected by adding Et_3N as a base. The prepared LiN (CH_2Ph)*p*-Ts was also an effective nucleophile leading to excellent selectivities. Careful studies revealed that the initially formed branched amination product could undergo a rearrangement to give the linear product. This side reaction could be avoided under kinetic control, and excellent regio- and enantioselectivities were obtained with *o*-nosylamides [121].

Recently, You and coworkers reported a highly regio- and enantioselective allylic amination using *N*-tosyl propynylamines (Table 5) [122].

The introduction of an alkyne moiety in the nucleophile allowed a facile access to highly enantioenriched 1,6-enynes, which are very useful building blocks in cycloisomerization reactions (Scheme 29). Subjecting the allylation products towards $PtCl_2$-catalyzed cycloisomerizations led to enantioenriched 3-azabicyclo [4.1.0]heptenes and 3-azabicyclo[3.2.0]heptenes respectively, simply by tuning the substituents on the alkyne. The reactions proceeded to give the cycloisomerization product as a single diastereomer and there was no notable loss of the enantiomeric purity compared with the starting materials.

In 2007, Carreira and coworkers found that the use of sulfamic acid (H_2NSO_3H) as an ammonia equivalent in Ir-catalyzed allylic aminations afforded primary

Table 4 Ir-catalyzed asymmetric allylic amination reaction with nosylamides

R–CH=CH–CH₂OCO₂Me + p-Ns–NH–R′ → [$[Ir(cod)Cl]_2$ (2 mol%), **L2** (4 mol%), TBD (8 mol%); THF, (base), rt] → p-Ns–N(R′)–CH(R)–CH=CH₂

Entry	R	R′	Base	Yield	b:l	ee(%)
1	Ph	H	–	66	32:1	90
2	Ph	H	NEt_3	91	16:1	93
3	PhC_2H_4	Bn	–	93	4:1	84
4	PhC_2H_4	Bn	NEt_3	85	7:1	88
5	PhC_2H_2	Bn	–	50	6:1	90
6	PhC_2H_2	Bn	NEt_3	71	2.5:1	93
7	Ph	Allyl	–	68	19:1	90
8	Ph	Allyl	NEt_3	89	10:1	88
9	3-Pyridyl	Allyl	–	86	13:1	93
10	3-Pyridyl	Allyl	NEt_3	89	4:1	91

Table 5 Ir-catalyzed asymmetric allylic aminations with *N*-tosyl propynylamines

Ph–CH=CH–CH₂OCO₂Me + R–C≡C–CH₂NHTs → [$[Ir(cod)Cl]_2$ (2 mol%), (*S*, *S*, S_a)-**L2** (4 mol%); DABCO, THF, rt] → TsN(CH₂C≡CR)–CH(Ph)–CH=CH₂

Entry	R	Yield	b:l	ee(%)
1	Me	93	95:5	99
2	TMS	83	90:10	99
3	*n*-Bu	84	94:6	97
4	Ph	89	95:5	99

allylic amines [123]. These were obtained in good yields and excellent regioselectivity, especially with secondary allylic alcohols as substrates. An asymmetric example was also carried out and the primary allylic amine was obtained in 70% yield and 70% ee by using **L8** as chiral ligand (Scheme 30, Eq. 1). Recently, aminations of optically active, unactivated secondary alcohols with sulfamic acid were reported by the same group (Scheme 30, Eq. 2) [124]. The addition of LiI (10 mol%) and powdered 4 Å MS was essential for the excellent enantioselectivities. With various aromatic substituents, the primary allylic amines could be isolated directly or protected in situ as benzamides with complete regioselectivity and high enantiospecificity (e.s. 94% to >98%). The aliphatic substrates also gave moderate yields with good stereoselectivity (e.s. 74–96%). [e.s. = $(ee_{product}/ee_{substrate}) \times 100$]

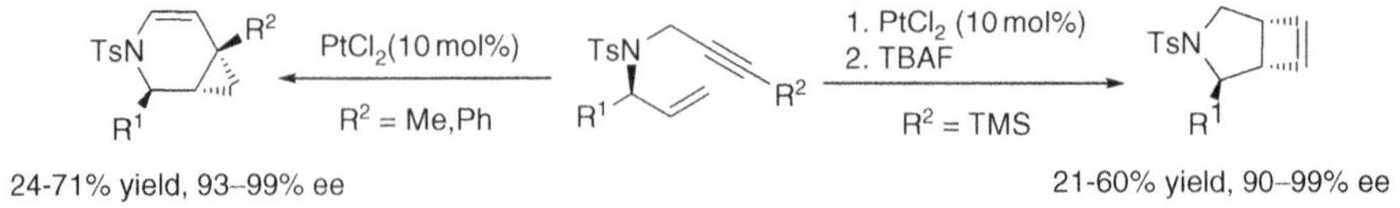

R^2 = Me,Ph: $PtCl_2$(10 mol%)
R^2 = TMS: 1. $PtCl_2$ (10 mol%) 2. TBAF

24-71% yield, 93–99% ee

21-60% yield, 90–99% ee

Scheme 29 Cycloisomerizations of 1,6-enynes

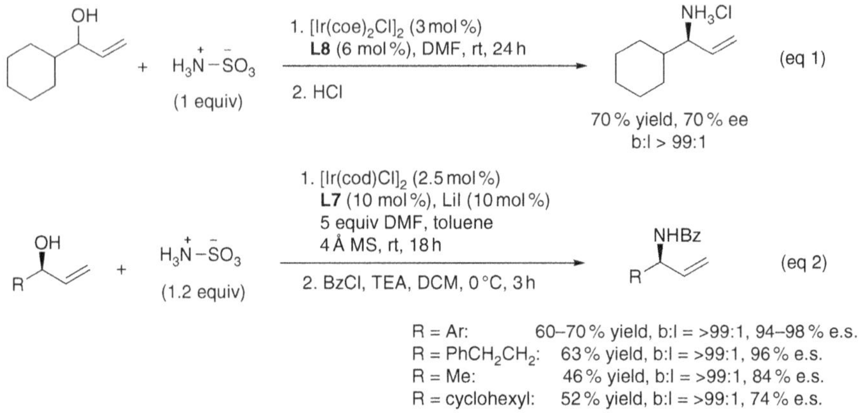

(eq 1): + $H_3\overset{+}{N}-\overset{-}{S}O_3$ (1 equiv); 1. $[Ir(coe)_2Cl]_2$ (3 mol %), **L8** (6 mol %), DMF, rt, 24 h; 2. HCl → NH_3Cl product, 70 % yield, 70 % ee, b:l > 99:1

(eq 2): + $H_3\overset{+}{N}-\overset{-}{S}O_3$ (1.2 equiv); 1. $[Ir(cod)Cl]_2$ (2.5 mol %), **L7** (10 mol %), LiI (10 mol %), 5 equiv DMF, toluene, 4 Å MS, rt, 18 h; 2. BzCl, TEA, DCM, 0 °C, 3 h → NHBz product

R = Ar: 60–70 % yield, b:l = >99:1, 94–98 % e.s.
R = $PhCH_2CH_2$: 63 % yield, b:l = >99:1, 96 % e.s.
R = Me: 46 % yield, b:l = >99:1, 84 % e.s.
R = cyclohexyl: 52 % yield, b:l = >99:1, 74 % e.s.

Scheme 30 Ir-catalyzed allylic aminations of allylic alcohols with sulfamic acid

4.2.2 Carboxamides

Carboxamides were also successfully used as nucleophiles in Ir-catalyzed asymmetric allylic aminations (Table 6) [121, 125–128]. Helmchen and coworkers found that excellent regioselectivity (up to 96:4) and enantioselectivity (up to 98%) were obtained when cyclic phthalimide and succinimide were used as nucleophiles (entries 1–3). (Boc)(CHO)NH was also used successfully and excellent results were obtained (entries 4 and 5). Subsequently, they found that with *tert*-butylbut-2-enoylcarbamate as nucleophile, regioselectivity of up to 98:2 in favor of the branched product and up to 99% ee were achieved (entries 6 and 7).

When $HN(Boc)_2$ was used as nucleophile, Helmchen and coworkers found that the reaction was sluggish under the usual reaction conditions. However, the allylic amination with $(Boc)_2NNa$ occurred much faster and gave excellent results (entries 8 and 9). Hartwig and coworkers also found the beneficial effects by using $(Boc)_2NLi$ as nucleophile (entry 10). There was a slight drop of enantioselectivity when chiral ligand **L5a** or **L5b** was used.

Meanwhile, Singh and Han found that in the presence of $[Ir(cod)Cl]_2$ (2 mol%), **L2** (4 mol%), DBU (20 mol%), the allylic amidation proceeded well to give branched products in high yields and excellent regio- and enantioselectivities

Table 6 Ir-catalyzed asymmetric allylic amination reactions with amides

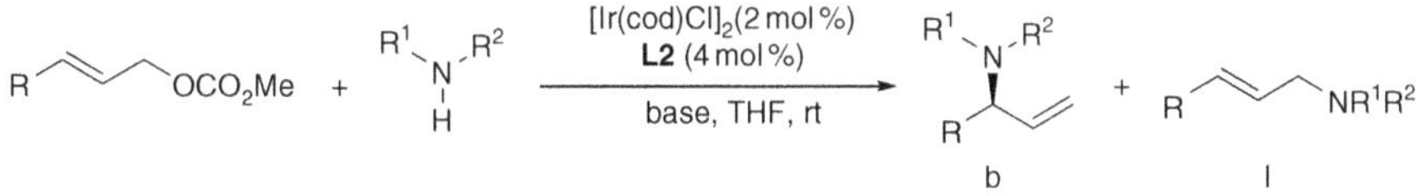

R–CH=CH–CH$_2$OCO$_2$Me + R^{1}N(H)R^2 → [Ir(cod)Cl]$_2$(2 mol%), **L2** (4 mol%), base, THF, rt → b + l (R–CH=CH–CH$_2$NR^{1}R^2)

Entry	Nucleophile	R	t (h)	Yield (%)	b:l	ee (%)	Reference
1	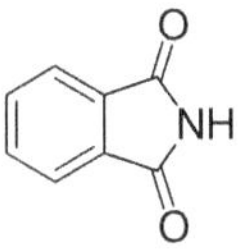 Phthalimide	Ph	2.5	95	96:4	98	[121]
2		n-Pr	2.5	82	94:6	96	[121]
3	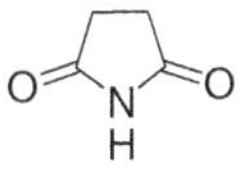 Succinimide	Ph	2.5	87	96:4	98	[121]
4	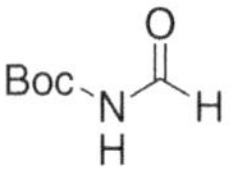 BocNHCHO	Ph	0.7	96	98:2	98.5	[121]
5		n-Pr	1	98	97:3	96	[121]
6	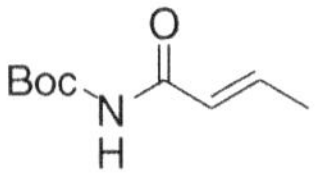 BocNHC(O)CH=CHCH$_3$	Ph	24	64	98:2	95	[125]
7		n-Pr	2	78	98:2	99	[125]
8	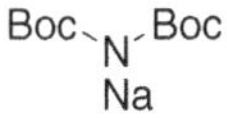 Boc$_2$NNa	Ph	0.7	80	97:3	99	[121]
9		n-Pr	0.7	86	96:4	99	[121]
10	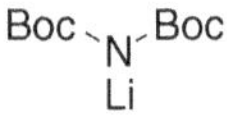 Boc$_2$NLi	Ph	1.5	84	98:2	98	[126]
11	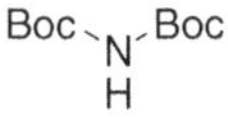 Boc$_2$NH	Ph	1	90	99:1	>99	[127]
12		n-Pr	0.5	93	98:2	96	[127]
13	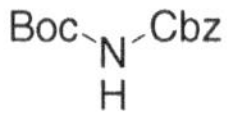 Boc(Cbz)NH	Ph	1	92	99:1	>99	[127]

even with HN(Boc)$_2$ as nucleophile (entries 11 and 12). Excellent results were achieved under these salt-free conditions also for other soft nitrogen nucleophiles (entry 13). Selective/mono deprotections of the different nitrogen protecting groups of the allylation products were also carried out. A series of highly enantiomeric *N*-Boc protected allylic amines were obtained by a one-pot asymmetric allylic amidation/selective deprotection sequence using (Ac)(Boc)NH as nucleophile.

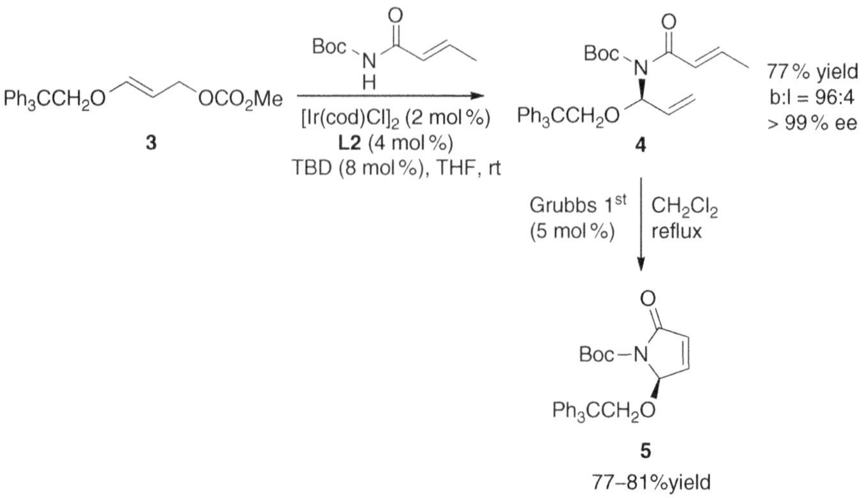

Scheme 31 Synthesis of γ-lactams by allylic amination/ring-closing metathesis reaction

Hartwig and coworkers also demonstrated potassium trifluoroacetamide to be a suitable nucleophile in Ir-catalyzed allylic aminations [126].

When allylic carbonate **3** reacted with *tert*-butylbut-2-enoylcarbamate the amination product **4** was obtained with good yield and excellent selectivity (Scheme 31). With 5 mol% of Grubbs first generation catalyst in dichloromethane, a ring-closing metathesis proceeded smoothly to give lactam **5**, which was a synthetic intermediate for antifungal agent (−)-pramanicine [129].

In 2009, Hartwig and coworkers [130] reported the direct allylic amidation of carbamates and achiral *tert*-butyl allylic carbonates, forming branched, primary allylic amines with high regio- and enantioselectivity (Table 7). This process occurred without base or with 0.5 equiv K_3PO_4 in the presence of a highly reactive metalacyclic iridium catalyst **K5b** containing a labile ethylene ligand (Scheme 4). The reactions of aryl-, heteroaryl-, and alkyl-substituted allylic carbonates with a wide range of carbamates and 2-oxazolidinone proceeded in good yields and high regio- and enantioselectivity.

In 2007, Singh and Han [131] developed an Ir-catalyzed highly regio- and enantioselective decarboxylative allylic amidation. In the presence of [Ir(cod) Cl]$_2$, ligand **L2**, DBU and proton sponge (PS), good yields and excellent regio- and enantioselectivities were obtained (Scheme 32). Here the PS was found to increase the reaction rate by >100-fold.

The reaction mechanism involving an Ir-π-allyl intermediate was confirmed by a crossover experiment (Scheme 33).

To shed light on the timing of decarboxylation, ethyl cinnamyl carbonate was treated with the benzyl carbamate anion under the usual reaction conditions. The fact that no product was observed likely exclude that the benzyl carbamate anion

Table 7 Asymmetric allylic amidation with carbamates

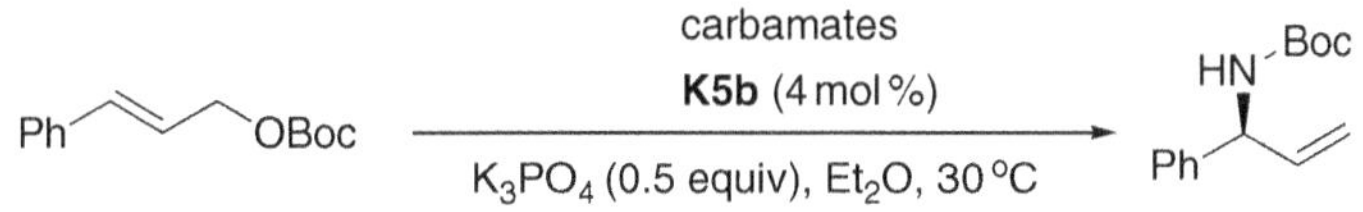

Entry	Nucleophile	R	*t* (h)	Yield (%)	b:l	ee (%)
1[a]	$BocNH_2$	Ph	24	85	94:6	99
2	$FmocNH_2$	Ph	21	57	83:17	94
3	$CbzNH_2$	Ph	9	74	80:20	97
4	$TrocNH_2$	Ph	10	80	96:4	95
5	$TeocNH_2$	Ph	10	73	80:20	98
6	2-Oxazolidinone	Ph	12	72	99:1	99
7[a]	$BocNH_2$	2-Furyl	21	50	83:17	98
8[a]	$BocNH_2$	C_7H_{15}	5	65	85:15	97

[a]The reaction was carried out in THF without K_3PO_4

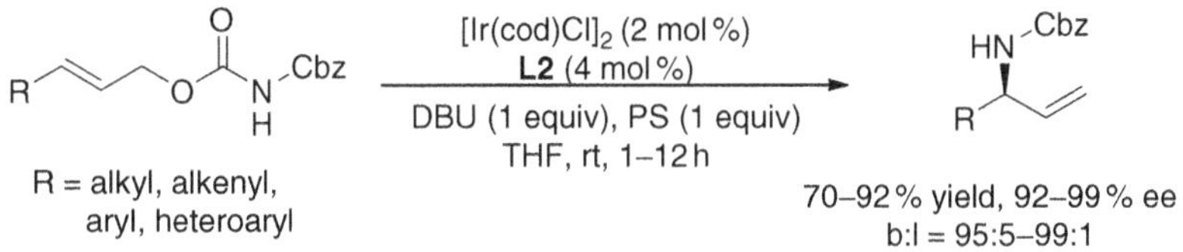

Scheme 32 Asymmetric decarboxylative allylic amidations

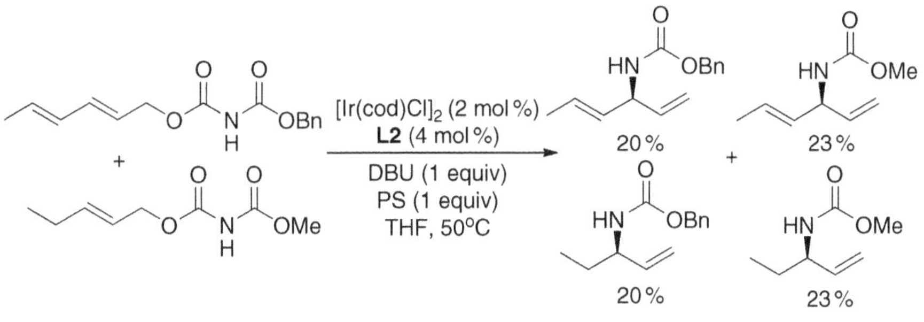

Scheme 33 Crossover experiment

acts as the actual nucleophile to the Ir-π-allyl complex (**II** in Fig. 3). In another word, the allylic amination (**I**) proceeded prior to the decarboxylation process.

Next, Singh and Han [132] showed that Ir-catalyzed decarboxylative allylic amidations of chiral branched benzyl allyl imidodicarboxylates could proceed with complete retention of enantiomeric purity and configuration (Scheme 34).

Fig. 3 Plausible reaction intermediate

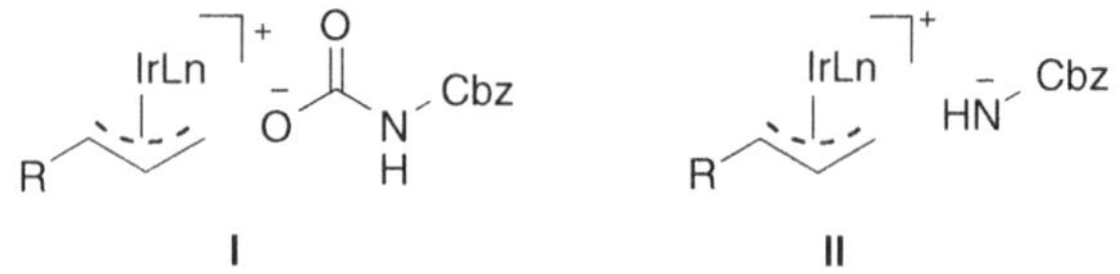

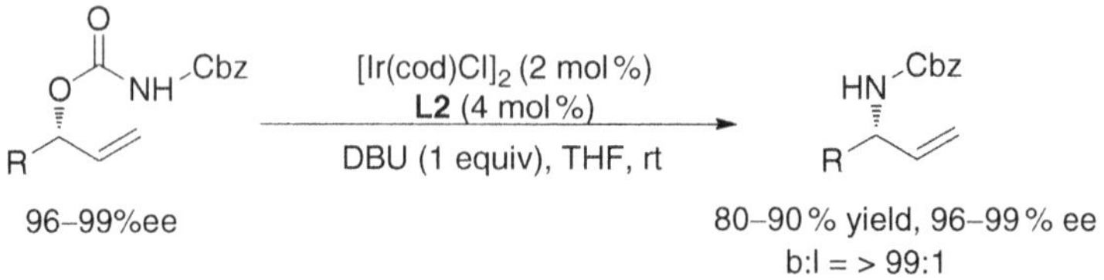

Scheme 34 Decarboxylative allylic amidation of chiral branched benzyl allylic imidodicarboxlate

This is a nice complementary method to the corresponding decarboxylative allylic amidation of achiral linear benzyl allyl imidodicarboxylates.

4.3 *Nitrogen-Containing Heterocycles*

Many heterocycles have been extended as suitable nucleophiles in Ir-catalyzed allylic aminations (Table 8). Highly regio- and enantioselective *N*-allylations of benzimidazoles, imidazoles, and purines have been developed by Stanley and Hartwig [133]. Excellent yields and selectivities were obtained in the presence of ethylene-bound metallacyclic iridium catalysts **K5**.

Subsequently, Stanley and Hartwig [134] explored indole as suitable nitrogen nucleophile (Table 9). With metallacyclic iridium catalyst **K5b**, *N*-allylation of indoles was achieved in high regio- and enantioselectivity. These reactions tolerate a broad range of indoles, bearing electron-withdrawing groups or substituent at 3-position, and allylic carbonates. In addition, *N*-allylation of 7-azaindole was also realized with excellent selectivity. However, allylation of indole, 2-methylindole, and 2-phenylindole occurred selectively at the 3-position under similar reaction conditions [102].

4.4 *Ammonia*

Transition metal-catalyzed direct enantioselective allylations of ammonia are rare but highly desirable because ammonia is inexpensive and such a process avoids the use of protecting groups. Challenges in the direct allylation of ammonia include: (1)

Table 8 Asymmetric allylation of azoles

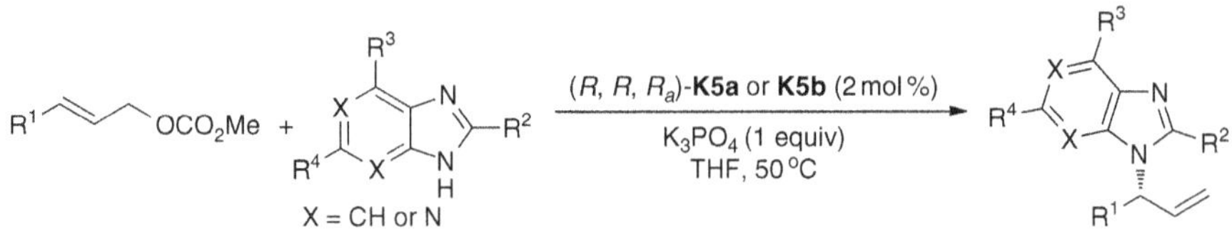

(R, R, R_a)-**K5a** or **K5b** (2 mol%)
K_3PO_4 (1 equiv)
THF, 50 °C

X = CH or N

Entry	Nucleophile	R^1	Ligand	Yield (%)	b:l	ee (%)
1	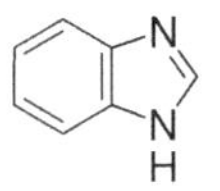	Ph	**K5a**	92	99:1	97
2		2-Naphthyl	**K5a**	90	98.2	97
3		4-$MeOC_6H_4$	**K5a**	90	98:2	97
4		2-$MeOC_6H_4$	**K5b**	93	99:1	80
5		2-Furyl	**K5a**	90	94:6	96
6		*n*-Pr	**K5a**	90	92:8	94
7	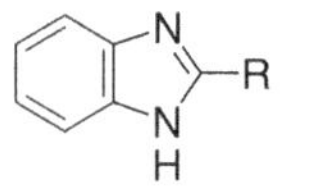 R = Me, Ph, $-CH_2OH$	Ph	**K5b**	75–94	95:5–96:4	96–97
8	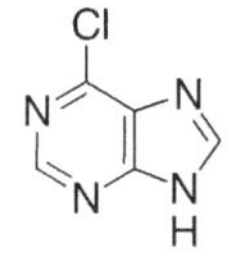	Aryl, alkyl	**K5a** **K5b**	60–91	90:10–98:2	78–96
9	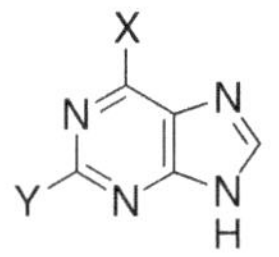 X = Cl, NR_2, SMe; Y = H, NH_2	Aryl, alkyl	**K5a** **K5b**	29–91	92:8–98:2	92–98

ammonia can coordinate towards the metal catalyst, displacing the chiral ligand to generate an achiral catalyst; (2) selective monoallylation of ammonia is difficult to achieve because the allylic amine formed is more nucleophilic than ammonia.

In 2007, Hartwig and coworkers demonstrated that allylation of ammonia could afford the diallylamine as the sole product using a cyclometallated iridium catalyst **K3** (Scheme 35) [126]. The monoallyl amine was hypothesized to be an intermediate in this reaction. Taking advantage of this method and a subsequent ring-closing metathesis (RCM) reaction, chiral 2,5-dinaphthylpyrrolidine was synthesized in high yield and selectivity by Feringa and coworkers [135].

Table 9 Asymmetric allylation of indoles

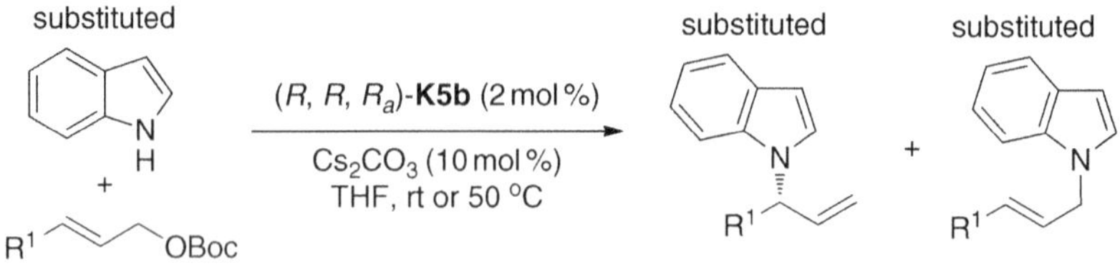

substituted indole + R^1–CH=CH–CH$_2$OBoc → (R, R, R_a)-**K5b** (2 mol %), Cs_2CO_3 (10 mol %), THF, rt or 50 °C → substituted (branched) + substituted (linear)

Entry	Nucleophile	R^1	Yield (%)	b:l	ee (%)
1	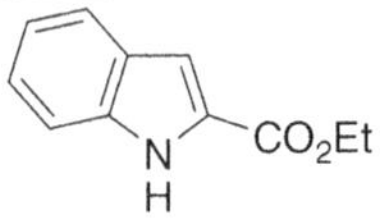 ethyl indole-2-carboxylate (CO_2Et)	Aryl Heteroaryl Alkenyl Alkyl	54–95	77:23–99:1	96–99
2	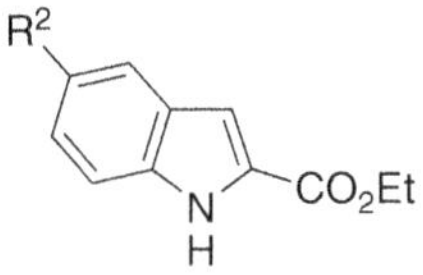 5-R^2-indole-2-CO_2Et R^2 = OMe, F, Cl, NO_2	Ph	84–91	94:6–97:3	99
3	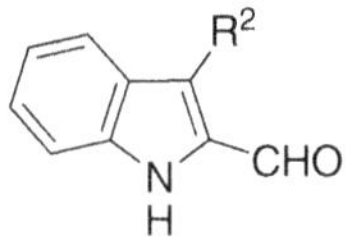 3-R^2-indole-2-CHO R^2 = H, Me	Ph	82–89	98:2	99
4	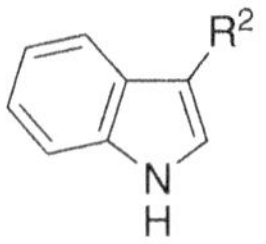 3-R^2-indole R^2 = Ph, CO_2Me CHO, COMe, CN	Ph	21–93	94:6–99:1	96–97
5	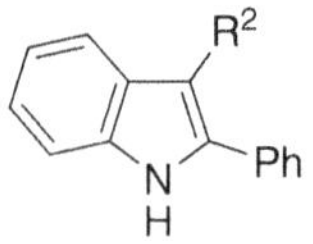 3-R^2-2-Ph-indole R^2 = Me, Ph, CHO	Ph	89–93	96:4–98:2	97–99
6	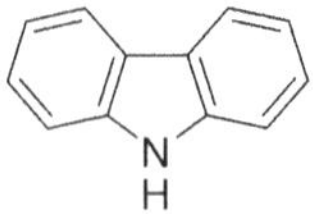 carbazole (NH)	Ph	88	98:2	98
7	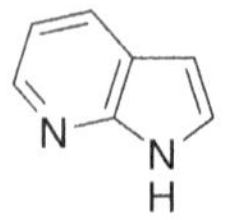 7-azaindole (NH)	Ph	79	91:9	99

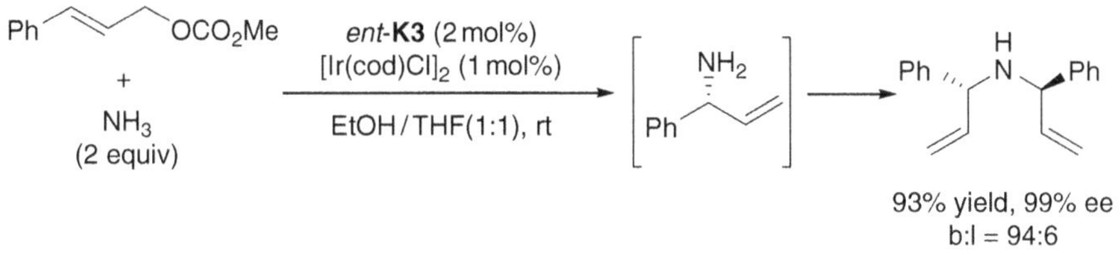

Scheme 35 Ir-catalyzed asymmetric diallylation of ammonia

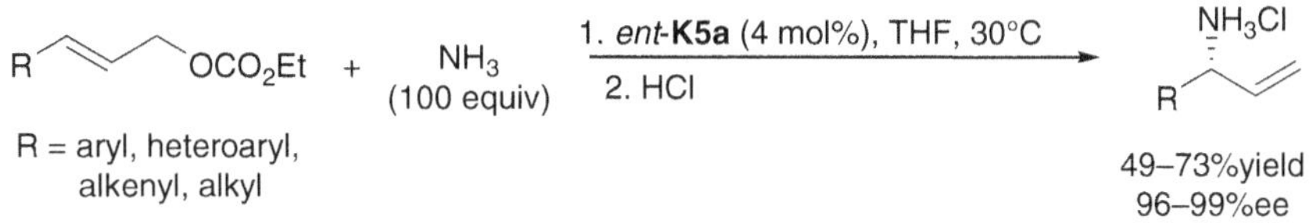

Scheme 36 Ir-catalyzed asymmetric monoallylation of ammonia

In 2009, Ir-catalyzed asymmetric monoallylation of ammonia was achieved also by Hartwig and coworkers [136]. Using metalacyclic complex **K5a** as catalyst, a series of primary allylic amines was obtained in moderate yields but excellent enantioselectivities (Scheme 36). In this monoallylation, a large excess of ammonia was used to avoid the diallylation. The ammonium salts were isolated and characterized after protonation with HCl.

4.5 Guanidines

Protected guanidines were also explored as suitable nucleophiles by Takemoto et al. (Scheme 37) [137]. The monoallylation and double allylation products were achieved using di-Boc-guanidine and tri-Boc-guanidine as substrates, respectively, with Ph-Pybox **L13** as ligand.

5 Ir-Catalyzed Enantioselective Allylic Etherification Reaction

5.1 Phenolates

Ir-catalyzed enantioselective allylic etherifications using phenolates as nucleophiles were introduced by Hartwig and coworkers (Scheme 38) [138]. The reactions were carried out with $[Ir(cod)Cl]_2$/*ent*-**L1** as catalyst. They found that the base has a significant influence on the selectivity. Alkali metal phenolates proved to be superior to ammonium phenolates. Reactions of sodium phenolates with methyl

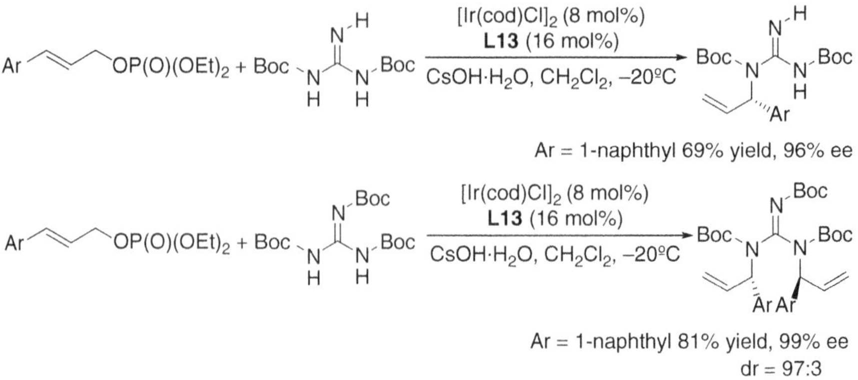

Scheme 37 Enantioselective allylation of guanidines

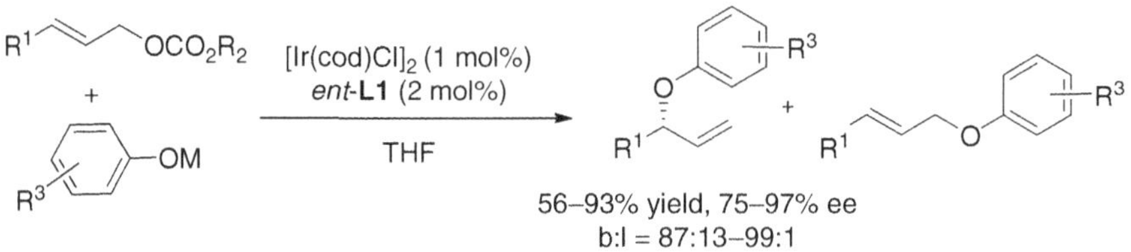

Scheme 38 Asymmetric allylic substitution with alkali phenolates

carbonates formed the alkylation products at room temperature in modest yield because of competing transesterification reactions. With the more hindered and less reactive ethyl carbonates, or using lithium phenolates as nucleophile, the reactions occurred smoothly without significant transesterifications. The reaction outcome was also influenced by the solvents and the best results with respect to rate, regio- and enantioselectivity were obtained in THF.

The reactions with unsubstituted, *ortho*-substituted, or electron-rich phenols gave the branched ethers with good enantioselectivities. In contrast, phenols bearing strong electron-withdrawing groups, such as nitro and cyano, in the *para* position failed to react. The aryl-substituted allyl carbonates showed better results than the aliphatic analogues. *Ortho*-substituted methoxycinnamyl carbonate reacted in high yield and regioselectivity, but the enantioselectivity dropped dramatically (75% ee). Branched allylic carbonates were also tolerated, but gave lower enantioselectivities after full conversion.

The interconversion of branched ether products to linear isomers, catalyzed by the iridium complex, was observed under the standard reaction condition. It has been found that the reaction of LiOPh with cinnamyl methyl carbonate, which could be completed with excellent regio- and enantio-selectivity at 50°C, was performed with lower selectivities when the reaction time was extended significantly. The activated catalyst, formed in situ by treatment of $[Ir(cod)Cl]_2$/*ent*-**L2**

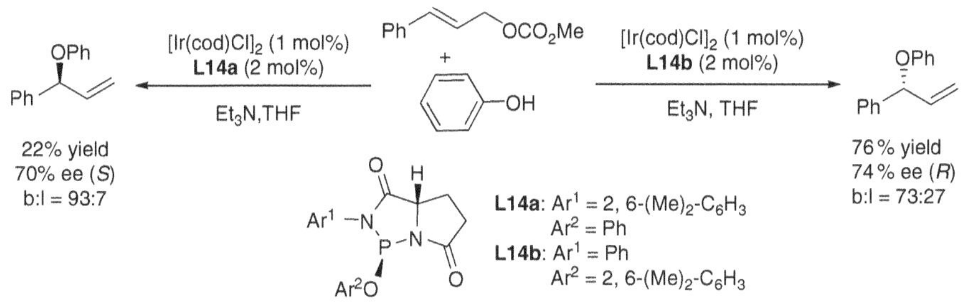

Scheme 39 Asymmetric allylic etherifications using phosphorodiamidite ligands **L14**

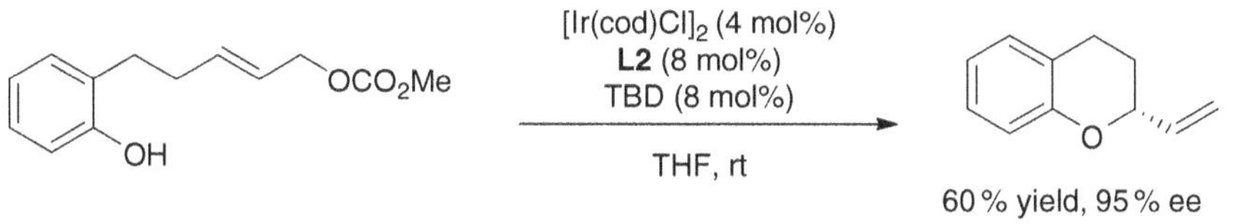

Scheme 40 Asymmetric intramolecular allylic etherification reaction

with DABCO or *n*-propylamine [112], distinctly improved the rate, yields, regioselectivities, and enantioselectivities [139].

By using P-chiral phosphorodiamidites **L14** as ligands, Kimura and Uozumi reported Ir-catalyzed asymmetric allylic etherifications of cinnamyl carbonate with phenol under basic condition, whereby both (*R*)- and (*S*)-configured products could be obtained with good enantioselectivities by simply changing the substituents on the ligand **L14** (Scheme 39) [140].

An intramolecular reaction of a substituted phenol was reported by Helmchen et al. with the activated catalyst derived from $[Ir(cod)Cl]_2$/**L2** to give the chromane derivative with up to 95% ee (Scheme 40) [116].

Terminal branched allylic esters are less substituted at the alkene unit and typically react faster but with dramatically lower enantioselectivity than the corresponding linear esters [55]. It has already been reported that enantiomerically enriched branched allylic esters in the presence of rhodium [26] or iridium complexes [66] yield substitution products with retention of configurations. Some elegant results were obtained using the "memory effect" in kinetic resolutions by treatment of racemic branched allylic esters with half amounts of nucleophiles [71, 141, 142].

For example, Carreira and coworkers used an Ir-complex prepared in situ from $[Ir(coe)_2Cl]_2$ and the chiral [2.2.2]-bicyclooctadiene ligand **L15** as a catalyst (Scheme 41) [141]. Various aryl- and alkyl-substituted allylic carbonates were successfully applied. Both electron-rich and electron-deficient groups were accepted, and the recovered starting carbonates were isolated in 28–46% yields with 80–98% ee.

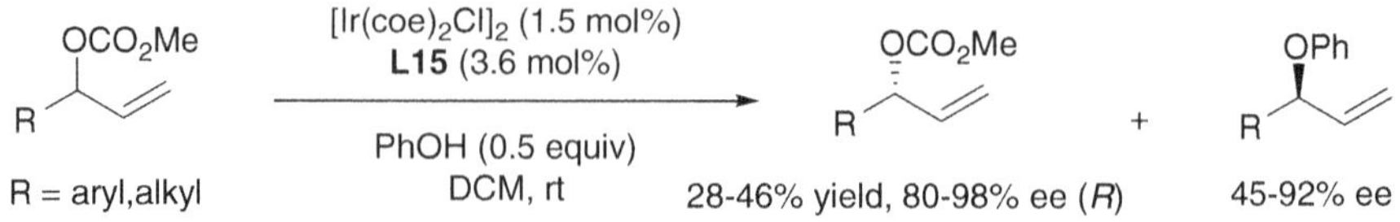

$[Ir(coe)_2Cl]_2$ (1.5 mol%)
L15 (3.6 mol%)
PhOH (0.5 equiv)
DCM, rt

R = aryl,alkyl
28-46% yield, 80-98% ee (*R*)
45-92% ee

Scheme 41 Ir-catalyzed kinetic resolution of allylic carbonates with phenol

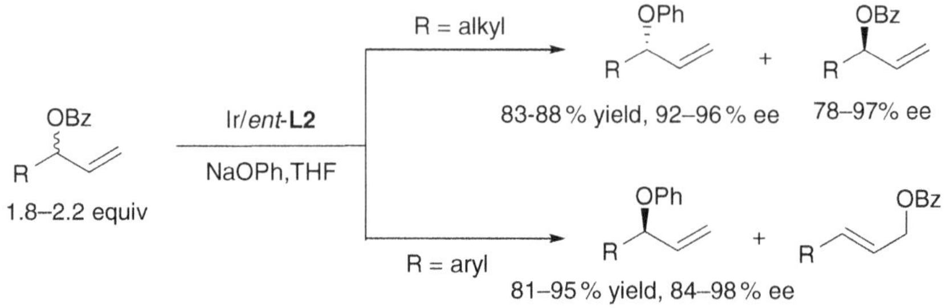

Ir/*ent*-**L2**
NaOPh,THF
1.8–2.2 equiv
R = alkyl
83-88 % yield, 92–96 % ee
78–97% ee
R = aryl
81–95 % yield, 84–98 % ee

Scheme 42 Ir-catalyzed kinetic resolution of allylic benzoates with phenoxide

Recently, Hartwig and coworkers also reported the kinetic resolution of racemic branched allylic benzoates with a series of nucleophiles (Scheme 42) [142]. The substitution products were obtained with high ee's from both aliphatic and aromatic benzoates. The aliphatic benzoates with lower reaction rates left over were isolated in modest to high ee's, depending on the ratio of the starting materials. In the case of the branched aromatic benzoates, the allylic substitutions occurred in competition with an isomerization to the linear isomer. The isomerization of one enantiomer was faster than that of the other. Therefore, the more reactive (*R*)-enantiomer was converted to the linear isomer, and the remaining (*S*)-enantiomer reacted with the nucleophile affording the allylic substitution product with retention of configuration.

5.2 Alkoxides

Enantioselective reactions of aliphatic alkoxides with achiral allylic carbonates in the presence of $[Ir(cod)Cl]_2$/*ent*-**L2** as catalyst were reported by Shu and Hartwig in 2004 (Table 10) [143]. They found that *t*-butyl cinnamyl carbonate provided branched chiral allylic ethers in higher yields than cinnamyl carbonates with smaller alkyl groups, since the latter were accompanied with the competitive transesterification. Copper alkoxides proved to be better suited than zinc alkoxides. In contrast, the reaction with the corresponding alkali metal alkoxides did not result in the formation of the expected ether products. The copper alkoxides derived from both primary and secondary alcohols were well tolerated. With tertiary copper

Table 10 Ir-catalyzed asymmetric allylic substitutions with alkoxides

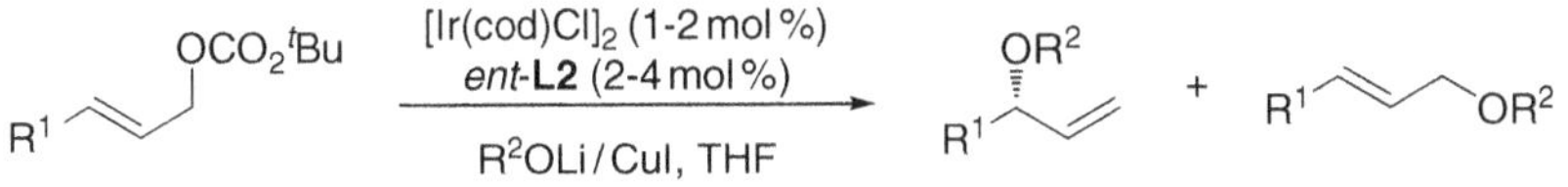

R^1-CH=CH-CH$_2$-$OCO_2{}^tBu$ → [$[Ir(cod)Cl]_2$ (1-2 mol %), *ent*-**L2** (2-4 mol %), R^2OLi/CuI, THF] → R^1-CH(OR^2)-CH=CH$_2$ + R^1-CH=CH-CH$_2$-OR^2

Entry	R^1	R^2	Yield (%)	b:l	ee (%)
1	Ph	Bn	92	99:1	94
2	Me	Bn	80	95:5	97
3	Ph	i-Pr_2CH	86	99:1	96
4	Ph	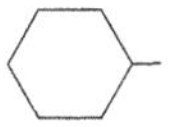 cyclohexyl	79	97:3	94
5	p-$NO_2C_6H_4$	cyclohexyl	59	96:4	86
6	p-$MeOC_6H_4$	cyclohexyl	86	99:1	95
7	Ph	BocN (piperidin-4-yl)	70	98:2	95
8	Me	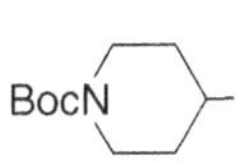 BocN (piperidin-4-yl)	56	97:3	94
9	n-Pr	BocN (piperidin-4-yl)	66	92:8	93
10	Ph	t-Bu	80	96:4	63

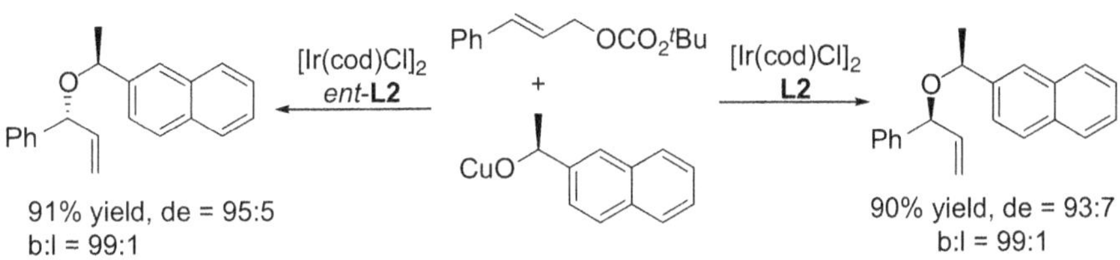

Ph-CH=CH-CH$_2$-$OCO_2{}^tBu$ + CuO-CH(CH$_3$)-(2-naphthyl)

[$[Ir(cod)Cl]_2$, *ent*-**L2**] → 91% yield, de = 95:5, b:l = 99:1

[$[Ir(cod)Cl]_2$, **L2**] → 90% yield, de = 93:7, b:l = 99:1

Scheme 43 Asymmetric allylic etherification reaction with chiral alkoxide

alkoxides, the regioselectivities and yields were excellent but the enantioselectivities dropped. The etherification with primary and secondary alkoxides encompassed both aryl and alkyl-substituted allylic carbonates.

The allylic etherification of a chiral alkoxide, derived from (*S*)-1-(2-naphthyl) ethanol, in the presence of the enantiomer of ligand **L2** (*ent*-**L2**) formed the opposite diastereomer of the allylic ether with excellent yield, regio- and diastereoselectivity (Scheme 43). These results clearly indicated that the enantioselective control was dominated by the chiral ligand and not the substrate, and the reaction was not constrained by a match and mismatch of the substrate and catalyst chirality.

Roberts and Lee reported an efficient Ir-catalyzed allylic etherification of aliphatic alcohols employing stoichiometric zinc alkoxides (Scheme 44) [144]. The

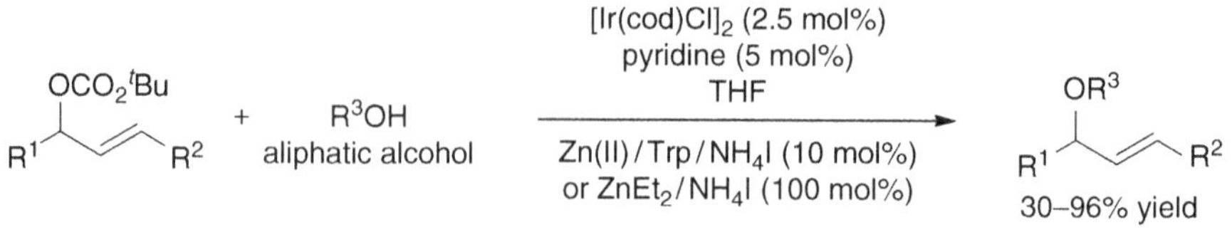

Scheme 44 Allylic etherification reaction via bimetallic catalysis

reactions occurred with complete regiospecificity and tolerated a wide scope of substrates.

5.3 *Hydroxylamine Derivatives*

Hydroxylamine derivatives are attractive nucleophiles for allylic substitutions because both nitrogen and oxygen atoms can react to give the *N*-allylated and *O*-allylated products respectively, depending on which of the reactive centers is protected (For a review, see [145]). Hydroxamic acids with an electron-withdrawing group on the nitrogen acted as oxygen nucleophile in asymmetric allylations with arylallyl phosphates by using the iridium complex of chiral pybox ligand **L13** (Table 11) [146]. Under optimal reaction conditions, using $PhCF_3/H_2O$ as solvent and barium hydroxide as base, moderate to high regio- and enantioselectivities were obtained.

The reaction of oximes with allyl phosphates was catalyzed by the iridium/pybox system at −20°C to give the *O*-substituted products, which could be easily converted into their corresponding alcohols [117]. The best results were obtained with $Ba(OH)_2{\cdot}H_2O$ as base and only arylallyl phosphates have been reported as suitable substrates (Table 12).

5.4 *Silanolates*

Silanols are considerably more acidic than the corresponding alcohols [147] and have been used as nucleophiles in Pd-catalyzed ring opening of vinyl epoxides [148, 149]. Enantioselective allylation of potassium triethylsilanolate with the iridium/**L1** catalyst was reported by Carreira et al. (Table 13) [150]. The alcoholate exchange was observed between the cinnamyl carbonates with small alkyl groups and the reaction in CH_2Cl_2 gave the best results. Various silanolates were utilized forming the products in excellent ee's.

The chiral allylic alcohols (The first Ir-catalyzed enantioselective allylic hydroxylation to prepare branched allylic alcohols using bicarbonate as nucleophile was reported recently, see [151]) were obtained by cleavage of the silyl ether under mild conditions in a one-pot procedure [150], and a wide variety of allylic substrates were tolerated (Table 14).

Table 11 Asymmetric allylations of hydroxamic acid

R^1R^2NOH + Ar–CH=CH–CH$_2$–OP(O)(OEt)$_2$ → [Ir(cod)Cl]$_2$ (6 mol %), **L13** (12 mol %), PhCF$_3$/H$_2$O (2:1, v/v), Ba(OH)$_2$·H$_2$O → branched O–NR^{1}R^2 product + Ar–CH=CH–CH$_2$–O–NR^{1}R^2

Entry	Ar	R[1]	R[2]	Yield (%)	b:l	ee (%)
1	Ph	Ph	Bz	70	80:20	87
2	Ph	Ph	Ac	75	77:23	82
3	Ph	Ph	Cbz	63	64:36	68
4	Ph	Bn	Bn	No reaction		
5	Ph	Bn	Ac	70	76:24	82
6	Ph	Bn	Bz	68	80:20	87
7	4-F–C_6H_4	Ph	Bz	61	72:28	62
8	4-Me–C_6H_4	Ph	Bz	67	79:21	72
9	1-Naphthyl	Ph	Bz	60	94:6	78

Table 12 Asymmetric allylations of oximes

PhCH=N–OH + Ar–CH=CH–CH$_2$–OP(O)(OEt)$_2$ → [Ir(cod)Cl]$_2$ (4-8 mol %), **L13** (8-16 mol %), Ba(OH)$_2$·H$_2$O, CH$_2$Cl$_2$, –20 °C → branched O-allyl oxime + Ar–CH=CH–CH$_2$–O–N=CHPh

Entry	Ar	Yield (%)	b:l	ee (%)
1	Ph	87	90:10	95
2	4-F–C_6H_4	89	69:31	90
3	3-Cl–C_6H_4	86	76:24	92
4	4-Me–C_6H_4	84	83:17	90
5	4-MeO–C_6H_4	35	84:16	70
6	1-Naphthyl	83	94:6	90
7	2-Naphthyl	81	83:17	89

5.5 *Aliphatic Alcohols and Silanols*

Generally, aliphatic alcohols are difficult to be used directly for allylic substitution because oxygen is a relatively hard nucleophile. Thus, metal alkoxides are often used in order to soften the oxygen. Ueno and Hartwig recently realized the Ir-catalyzed allylation of aliphatic primary, secondary, and tertiary alcohols in the presence of potassium phosphate [152]. The addition of 1-phenyl propyne (20 mol%) was found to be essential to suppress the olefin isomerization of the

Table 13 Asymmetric allylations of silanoates

ROK + Ph⁄⁄OCO$_2{}^t$Bu → [Ir(cod)Cl]$_2$ (3 mol %), **L1** (6 mol %), CH$_2$Cl$_2$, rt → branched OR product + Ph⁄⁄OR

Entry	R	Yield (%)	b:l	ee (%)
1	Me_3Si	30	n.d.	97
2	Et_3Si	90	99:1	94
3	t-$BuMe_2Si$	79	97:3	98
4	i-Pr_3Si	64	86:4	99

Table 14 Asymmetric allylations of K-silanolates

R⁄⁄OCO$_2{}^t$Bu → 1. [Ir(cod)Cl]$_2$ (3 mol %) / **L1** (6 mol %), TESOK (200 mol %), CH$_2$Cl$_2$, rt; 2. 30 % aq. NaOH, MeOH or ^{n}Bu$_4$NF → branched OH product (R)

Entry	R	Yield (%)	ee (%)
1	alkyl	56	95
2	aryl	64–88	92–98
3	heteroaryl	50–67	97–99
4	alkenyl	65	97

product (Scheme 45). Various allylic acetates and alcohols were tolerated with excellent enantioselectivities and regioselectivities.

Notably, *tert*-butyldimethylsilanol under the optimal reaction conditions underwent smoothly in 85% yield and 98% ee [150]. The decarboxylative etherification reaction with higher catalyst loading was also realized but in a moderate yield and a slightly lower ee as shown in Scheme 46.

6 Ir-Catalyzed Asymmetric Allylic Substitution of Other Nucleophiles

6.1 *Sulfur Nucleophiles*

The utilization of sulfur nucleophiles in transition metal-catalyzed asymmetric reaction [153–155] is very challenging due to the deactivation of the metal catalyst by sulfur. Ueda and Hartwig [156] reported the use of sodium sulfinate as a

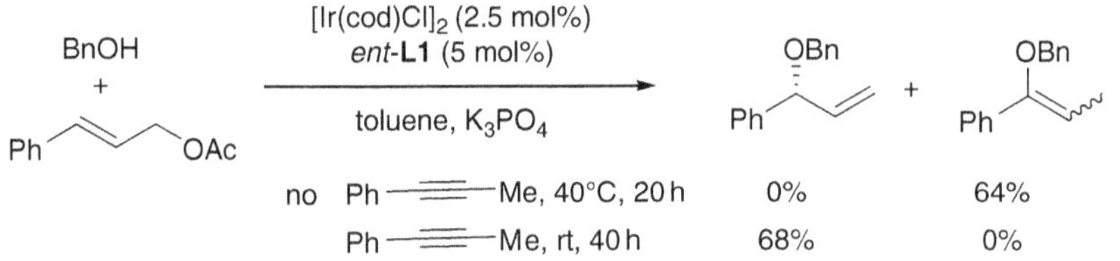

BnOH + Ph⌒⌒OAc

$[Ir(cod)Cl]_2$ (2.5 mol%), *ent*-**L1** (5 mol%)

toluene, K_3PO_4

	branched (OBn)	linear (OBn)
no Ph—≡—Me, 40°C, 20 h	0%	64%
Ph—≡—Me, rt, 40 h	68%	0%

Scheme 45 Ir-catalyzed allylic etherification reaction with aliphatic alcohols

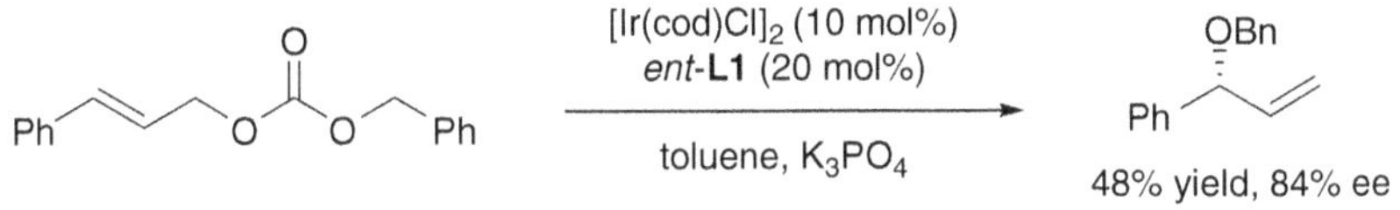

$[Ir(cod)Cl]_2$ (10 mol%), *ent*-**L1** (20 mol%)

toluene, K_3PO_4

48% yield, 84% ee

Scheme 46 Decarboxylative allylic etherification reaction

nucleophile in Ir-catalyzed allylations to form branched allylic sulfones (Table 15). With ligand **L1**, the reactions between various sodium sulfinates and achiral allylic carbonates occurred in good yields, with high selectivities for the branched isomer, and high enantioselectivities.

At the same time, You and coworkers also investigated an intramolecular type Ir-catalyzed allylic substitution using sulfinates as nucleophiles [157]. With DBU as the base, the isomerization of the allyl sulfones occurred quickly giving rise to the trisubstituted vinyl sulfones in high yields as pure *E* isomer (Scheme 47).

Subsequently, You and coworkers [158] reported another atom-economical carbon-sulfur bond formation via intramolecular allylic substitution of *O*-allyl carbamothioates (Table 16). With $[Ir(cod)Cl]_2$/**L1**, enantioenriched *S*-allyl carbamothioates were obtained in high yields with up to 95% ee and moderate to excellent regioselectivities. A crossover experiment was carried out and the observation of the crossover products likely excluded the rearrangement pathway. The products derived from the same *O*-allyl carbamothioate were favorably formed suggest the existence of coordination between the Ir-allyl complex and sulfur-containing nucleophile.

More recently, selective allylations of sodium thiophenolate were realized with CsF as an additive by Zhao et al. [159], producing allyl phenyl sulfides with up to 99% ee.

6.2 *Ir-Catalyzed Allylic Vinylation Reaction*

In general, the carbon nucleophiles used in transition metal-catalyzed allylic substitutions are limited to soft carbon nucleophiles, their conjugated acids have p*K*a

Table 15 Ir-catalyzed allylic substitution of sodium sulfinates

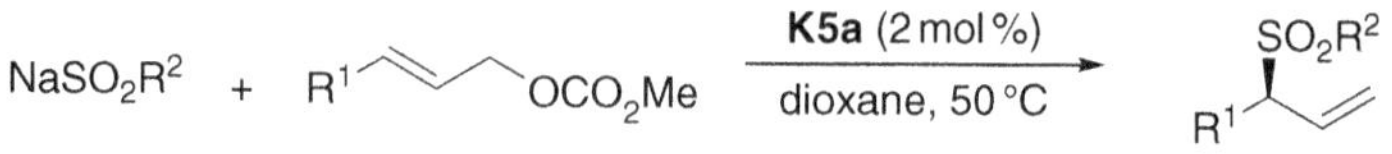

$NaSO_2R^2$ + R^1–CH=CH–CH$_2$OCO$_2$Me → (**K5a** (2 mol%), dioxane, 50 °C) → R^1CH(SO$_2$R^2)CH=CH$_2$

Entry	R^1	R^2	Yield (%)	b:l	ee (%)
1	Ph	Ph	92	97:3	94
2	4-MeOC$_6$H$_4$	Ph	95	99:1	89
3	4-MeOC$_6$H$_4$	Me	95	>99:1	92
5	4-MeOC$_6$H$_4$	*i*-Pr	99	>99:1	93
5	4-MeOC$_6$H$_4$	4-ClC$_6$H$_4$	85	95:5	86
6	Ph	4-MeC$_6$H$_4$	95	94:6	92
7	Me	4-MeC$_6$H$_4$	92	99:1	94
8	*i*-Pr	4-MeC$_6$H$_4$	86	99:1	97

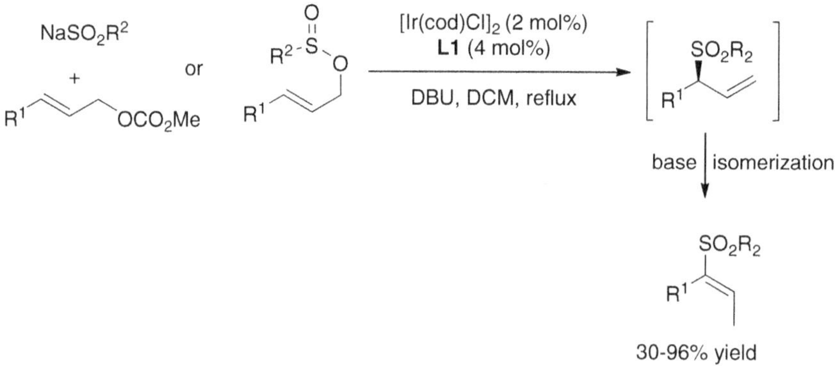

$NaSO_2R^2$ + R^1–CH=CH–CH$_2$OCO$_2$Me or R^2–S(=O)–O–CH$_2$CH=CH–R^1 → ([Ir(cod)Cl]$_2$ (2 mol%), **L1** (4 mol%), DBU, DCM, reflux) → [R^1CH(SO$_2$R$_2$)CH=CH$_2$] → (base isomerization) → R^1C(SO$_2$R$_2$)=CH–CH$_3$

30-96% yield

Scheme 47 Ir-catalyzed allylic substitution of sulfinates and isomerization reactions

Table 16 Ir-catalyzed intramolecular allylic substitution of *O*-allyl carbamothioates

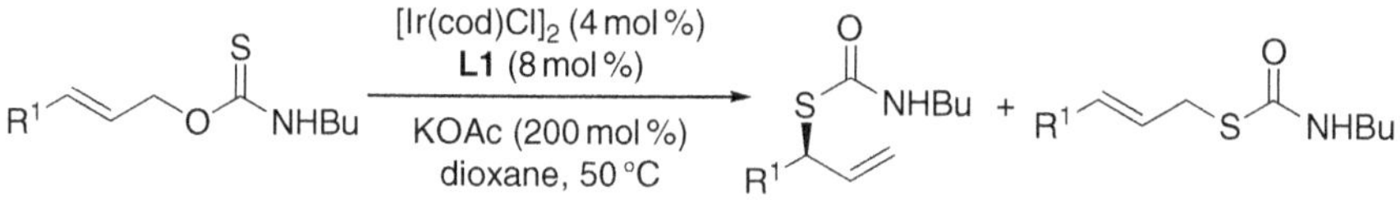

R^1–CH=CH–CH$_2$–O–C(=S)–NHBu → ([Ir(cod)Cl]$_2$ (4 mol%), **L1** (8 mol%), KOAc (200 mol%), dioxane, 50 °C) → R^1CH(CH=CH$_2$)–S–C(=O)–NHBu + R^1–CH=CH–CH$_2$–S–C(=O)–NHBu

Entry	R^1	Yield (%)	b:l	ee (%)
1	Ph	90	81:19	90
2	4-CF$_3$C$_6$H$_4$	68	74:26	90
3	4-BrC$_6$H$_4$	80	62:38	85
5	4-MeC$_6$H$_4$	52	93:7	93
5	4-MeOC$_6$H$_4$	61	61:39	64
6	*n*-Pr	79	61:39	89

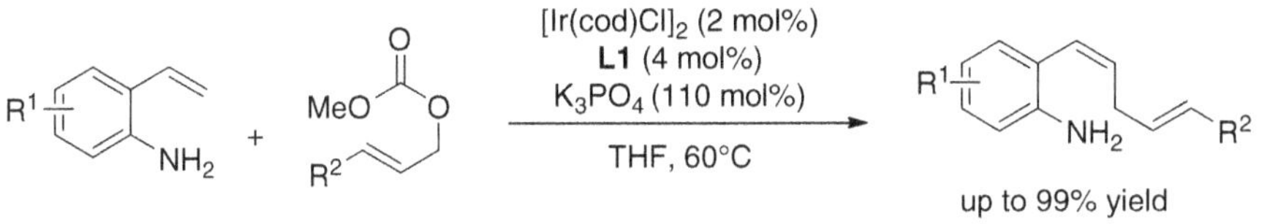

Scheme 48 Ir-catalyzed allylic alkylation of alkenes

Table 17 Tandem allylic vinylation/intramolecular allylic amination reaction

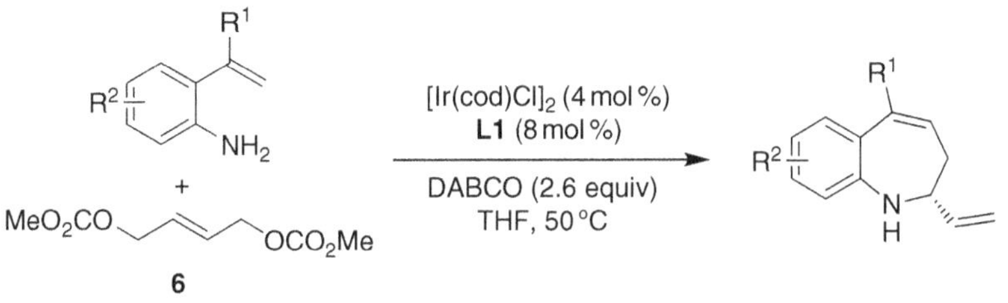

Entry	R^1	R^2	Yield (%)	ee (%)
1	H	H	95	91
2	H	5-MeO	89	91
3	H	4-Me	86	91
5	H	5-Br	75	90
5	Me	H	92	94
6	Me	5-Cl	75	90
7	Ph	4-MeO	81	90

$<$ 25. Ir-catalyzed direct allylic alkylations of alkenes was not known until 2009, when You and coworkers reported an Ir-catalyzed vinylation of allylic carbonates with *ortho*-amino styrene derivatives (Scheme 48) [160]. This protocol afforded *Z*, *E* skipped dienes in up to 99% yield with the exclusive formation of a *cis* double bond. The proposed mechanism involved an Ir-catalyzed amine-assisted vinyl C–H activation process.

Based on this allylic vinylation reaction, You and coworkers further developed an Ir-catalyzed tandem allylic vinylation/intramolecular allylic amination reaction [161]. With dicarbonate **6** as a substrate, the product of the vinylation contains a free amine group and an allylic carbonate, and the intramolecular asymmetric allylic amination can take place under the same catalytic system. This hypothesis worked well, and Ir-catalyzed tandem process was realized to afford 1-benzazepine derivatives with high enantioselectivity (Table 17).

7 Applications of Ir-Catalyzed Asymmetric Allylic Substitutions

The regiospecificity of Ir-catalyzed allylic substitutions affords the branched products bearing a terminal alkene that could find broad synthetic application in organic synthesis. In many cases, this allylic substitution has been combined with a RCM due to the characteristic terminal alkene in the products. It should be mentioned that this strategy had previously been used in conjunction with allylic substitutions catalyzed by other transition-metals [31, 162–166].

The synthesis of the prostaglandin analogue TEI-9826 was reported by Helmchen and coworkers based on an Ir-catalyzed allylation of a Weinreb type malonic amide **7** (Scheme 49) [79, 167]. The allylation product **8**, which could smoothly be converted into an enone **9**, was obtained with high ee using **L1** or **L2** as ligand. The corresponding substituted cyclopent-2-enone **10** was obtained by RCM. A subsequent aldol condensation was carried out according to a known procedure [168] affording TEI-9826 in 45% overall yield starting from cyclopentenone **10**.

The synthesis of an α,β-unsaturated γ-lactam **12**, a useful building block, was accomplished by an enantioselective allylic amination using imide **11** as a pronucleophile to afford a diene, which served as substrate for the RCM [113, 125]. The lactam **12** was successfully used in the synthesis of a Baclofen derivative, a potential monoamine oxidase inhibitor (Scheme 50) [169, 170].

The compounds listed in Fig. 4 [80, 108, 169–171] have also been synthesized via the combination of Ir-catalyzed allylic substitution and a Ru-catalyzed olefin metathesis protocol.

An elegant enantioselective total synthesis of (−)-α-kainic acid was recently reported by Helmchen and coworkers (Scheme 51) [172]. A key step for the preparation of the enantiomeric pure synthon **13** was based on an asymmetric allylic amination with a propargylic amine or *N,N*-diacylamine. The (−)-α-kainic acid was synthesized by subsequent steps involving a diastereoselective intramolecular Pauson–Khand reaction, Baeyer–Villiger reaction, reduction of the resultant lactone and direct Jones oxidation of a silyl ether.

The antiepilepsy drug (*S*)-vigabatrin was prepared in five steps with 51% overall yield from vinyloxirane. The Ir-catalyzed asymmetric allylic amination with allyl carbonate **14** as a key step was used to form the chiral carbon center (Scheme 52) [47, 128].

Helmchen's group [173, 174] has also developed the stereoselective synthesis of 2,6-disubstituted piperidines based on an Ir-catalyzed allylic cyclization as a configurational switch. This approach was successfully applied to the total syntheses of dendrobate alkaloid (+)-241 D and its C6-epimer, a spruce alkaloid, and the prosopis alkaloids (+)-prosopinine, (+)-prosophylline, (+)-prosopine, and (+)-6-*epi*-prosopine (Scheme 53).

The *N*-Boc protected chiral allylic amine **15**, obtained via Ir-catalyzed allylic amination, was used as the key intermediate for the asymmetric synthesis of (−)-cytoxazone (Scheme 54). The cytokinin modulator was achieved in

53% overall yield in five steps starting form (*E*)-ethyl 3-(4-methoxyphenyl)allyl carbonate [127, 175].

Several other target molecules that could be accessed via an Ir-catalyzed allylic substitution and subsequent diversified conversions of the terminal alkenes are depicted in Fig. 5 [77, 131–133, 167, 176–178].

8 Conclusions and Perspectives

Since the first report in 1997, Ir-catalyzed asymmetric allylic substitutions have been developed into a synthetically robust and practical method. With various applicable catalysts, represented by the $[Ir(cod)Cl]_2$/phosphoramidite complexes, the reactions proceeded with a variety of *C*-, *N*-, *O*-, and *S*- nucleophiles to afford a wide range of branched substitution products with excellent regioselectivity and enantioselectivity. This Ir-catalytic system featuring the invariably high selectivity with predictable configuration of products is particularly attractive. Their synthetic applications have been shown successfully for a broad range of biologically active compounds of interest in medicinal chemistry. In addition, the significant extension of the nucleophiles has resulted in many new transformations. We believe that more novel transformations will be discovered with the continuous efforts from research groups in this field. On the other hand, the catalytic efficiency of the known Ir-catalytic system is still rather low in the industrial viewpoint. This can be attributed to the instability of catalysts, product inhibition of the catalysts, and many other unknown factors. Given the rising price of Ir-catalysts, developing highly efficient catalytic systems is going to be the key that eventually leads to the industrial application of these reactions.

9 Experimental Section

9.1 *General Procedures for Iridium-Catalyzed Enantioselective Allylic Aminations*

Method A [112]: In a drybox, $[Ir(cod)Cl]_2$ (46.9 mg, 0.070 mmol), **L1** (75.5 mg, 0.140 mmol) were dissolved in 0.3 mL THF and 0.3 mL propylamine and heated to 50°C for 20 min [36]. Afterwards all volatiles were removed. The yellow solid was dissolved in 3.5 mL THF to get a stock solution. In a drybox, aniline (1.20 mmol) was added to 0.5 mL of the stock solution of the catalyst (0.010 mmol) in a screw capped vial equipped with a stirring bar. The vial was sealed with a cap containing a PTFE septum and removed from the drybox. Allyl carbonate (1.00 mmol) was added with a syringe and the reaction stirred at room temperature until the

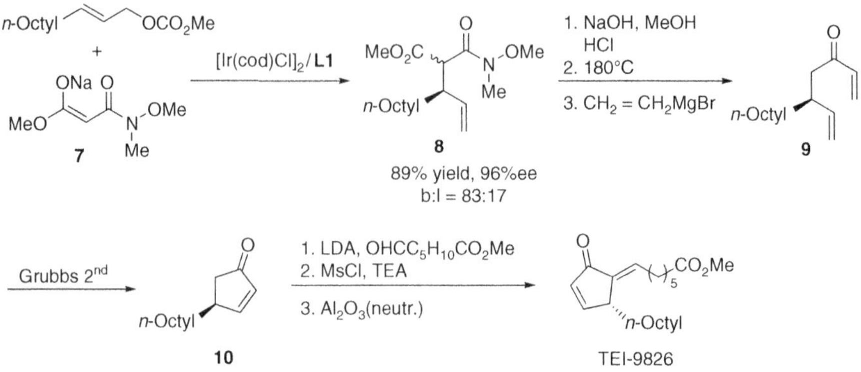

Scheme 49 Synthesis of the prostaglandin analogue TEI-9826

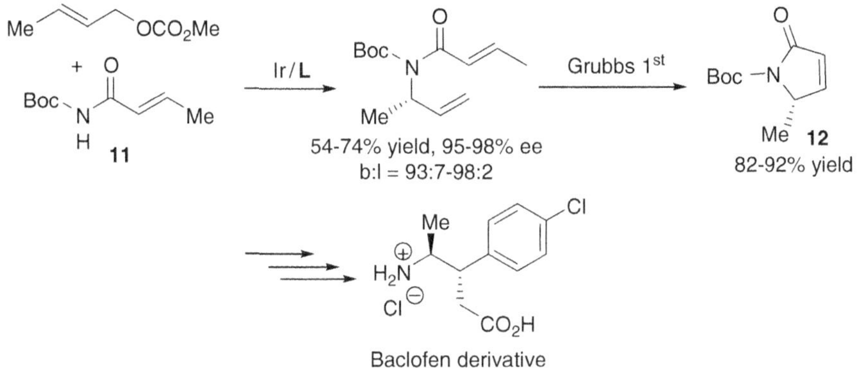

Scheme 50 Synthesis of Baclofen derivative

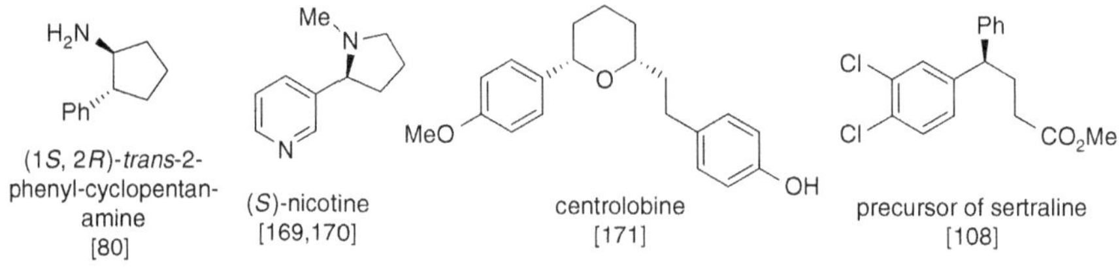

(1*S*, 2*R*)-*trans*-2-phenyl-cyclopentan-amine [80]

(*S*)-nicotine [169,170]

centrolobine [171]

precursor of sertraline [108]

Fig. 4 Compounds synthesized via Ir-catalyzed allylic substitution and olefin metathesis

carbonate was fully converted, as determined by GC and TLC. The volatile materials were evaporated. The ratio of regioisomers was determined by ^{1}H NMR analysis of the residue crude mixture. The crude reaction mixture was purified by flash column chromatography on silica gel to give desired product.

Method B [67, 112]: In a drybox, $[Ir(cod)Cl]_2$ (3.4 mg, 0.005 mmol) and (R, R, R_a)-**L2** (6.4 mg, 0.010 mmol) were dissolved in 0.5 mL of THF in a screw-

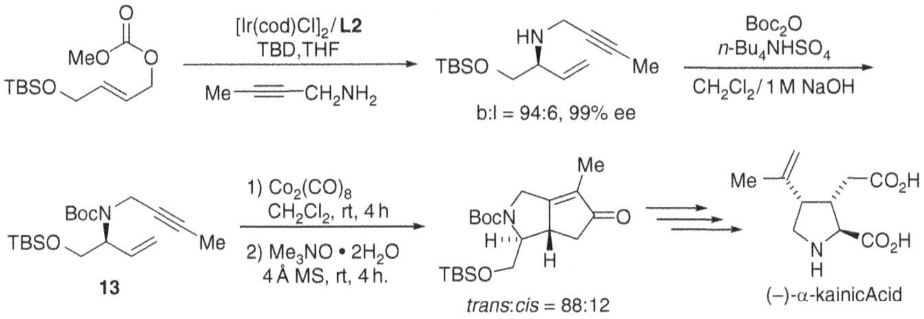

Scheme 51 Total synthesis of (−)-α-kainic acid

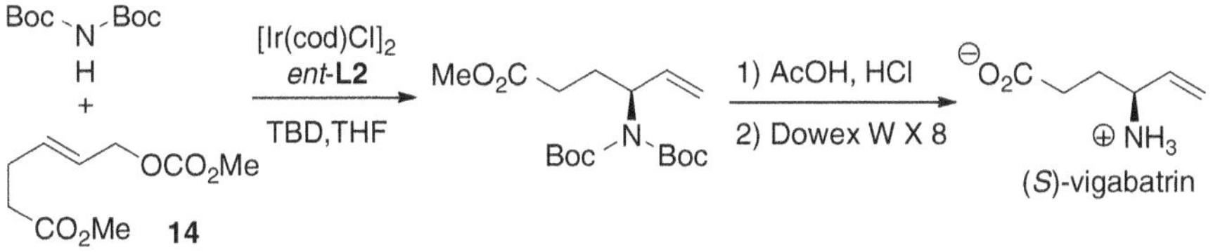

Scheme 52 Preparation of (*S*)-vigabatrin

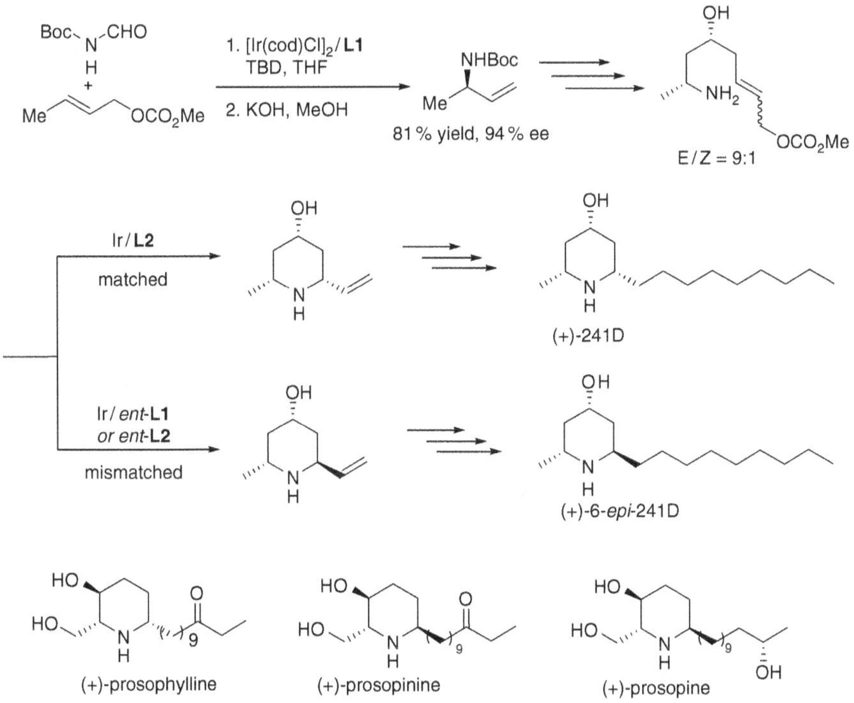

Scheme 53 Total synthesis of (+)-241D and related piperidine alkaloids

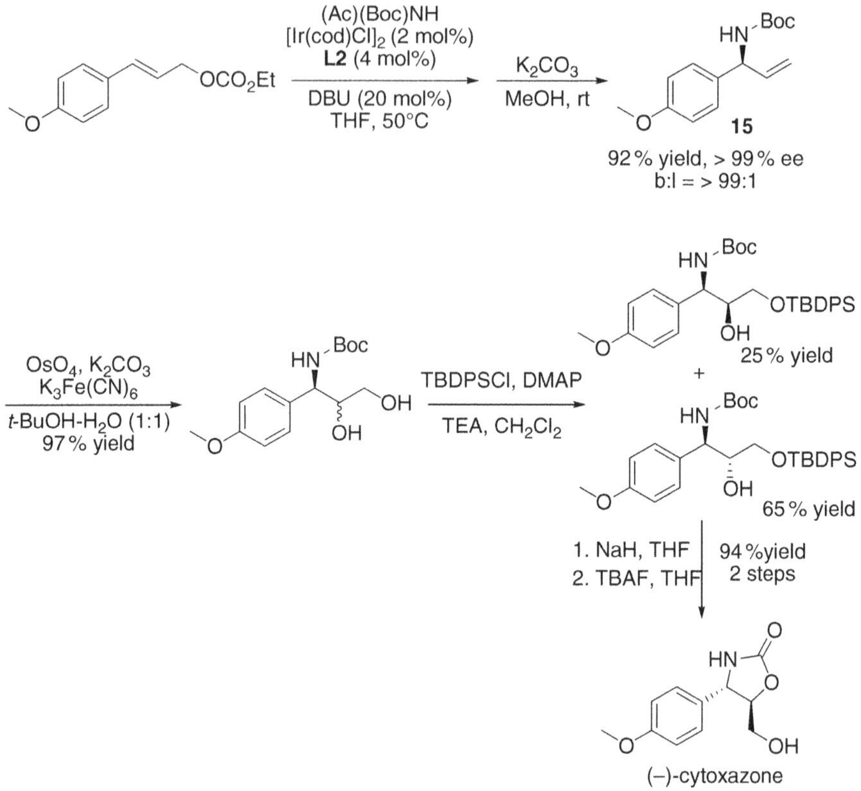

Scheme 54 Asymmetric synthesis of (−)-cytoxazone from chiral allylic amide **15**

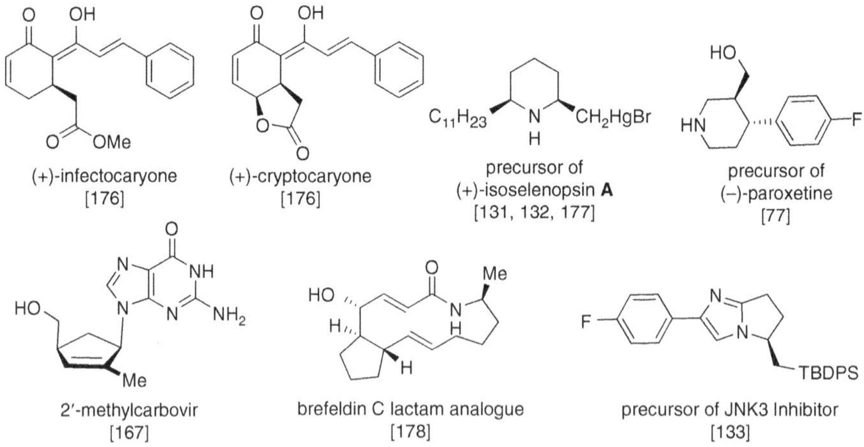

Fig. 5 Selected compounds synthesized via Ir-catalyzed allylic substitution

capped vial, DABCO [36] (11.2 mg, 0.100 mmol for cinnamyl methyl carbonate, 5.6 mg, 0.050 mmol for other aromatic and aliphatic methyl carbonates), or anhydrous TBD [11] (2.8 mg, 0.02 mmol) and a small magnetic stir bar were added. The vial was sealed with a cap containing a PTFE septum and removed from the drybox. Aniline (1.20 mmol) and methyl allyl carbonate (1.00 mmol) were added by syringe and the reaction was continued at room temperature for the specified time. After the reaction was complete as monitored by GC and TLC, the volatile materials were evaporated. The ratio of regioisomers was determined by ^{1}H NMR analysis of the residue crude mixture. The mixture was then purified by flash column chromatography on silica gel to give desired product.

Acknowledgments Financial support was provided by Chinese Academy of Sciences, National Natural Science Foundation of China (20872159, 20821002), and National Basic Research Program of China (973 Program 2009CB825300). We thank Prof. Wei Zhang for assisting in the preparation of the manuscript.

References

1. Trost BM, Van Vranken DL (1996) Asymmetric transition metal-catalyzed allylic alkylations. Chem Rev 96:395
2. Lu Z, Ma S (2008) Metal-catalyzed enantioselective allylation in asymmetric synthesis. Angew Chem Int Ed 47:258
3. Trost BM, Crawley ML (2003) Asymmetric transition-metal-catalyzed allylic alkylations: applications in total synthesis. Chem Rev 103:2921
4. Trost BM, Strege PE (1977) Asymmetric induction in catalytic allylic alkylation. J Am Chem Soc 99:1649
5. Diéguez M, Pàmies O (2010) Biaryl phosphites: new efficient adaptative ligands for Pd-catalysed asymmetric allylic substitution reactions. Acc Chem Res 43:312
6. Janssen JP, Helmchen G (1997) First enantioselective alkylations of monosubstituted allylic acetates catalyzed by chiral iridium complexes. Tetrahedron Lett 38:8025
7. Helmchen G, Dahnz A, Dübon P, Schelwies M, Weihofen R (2007) Iridium-catalysed asymmetricallylic substitutions. Chem Commun 675
8. Helmchen G (2009) Iridium-catalyzed asymmetric allylic substitutions. In: Oro LA, Claver C (eds) Iridium complexes in organic synthesis. Wiley-VCH, Weinheim, Germany, p 211
9. Hartwig JF, Stanley LM (2010) Mechanistically driven development of iridium catalysts for asymmetric allylic substitution. Acc Chem Res 43:1461
10. Lloyd-Jones GC, Pfaltz A (1995) Chiral phosphanodihydrooxazoles in asymmetric catalysis: tungsten-catalyzed allylic substitution. Angew Chem Int Ed 34:462
11. Trost BM, Hachiya I (1998) Asymmetric molybdenum-catalyzed alkylations. J Am Chem Soc 120:1104
12. Trost BM, Zhang Y (2007) Mo-catalyzed regio-, diastereo-, and enantioselective allylic alkylation of 3-aryloxindoles. J Am Chem Soc 129:14548
13. Trost BM, Zhang Y (2010) Catalytic double stereoinduction in asymmetric allylic alkylation of oxindoles. Chem Eur J 16:296
14. Belda O, Moberg C (2004) Molybdenum-catalyzed asymmetric allylic alkylations. Acc Chem Res 37:159
15. Malkov AV, Gouriou L, Lloyd-Jones GC, Stary I, Langer V, Spoor P, Vinader V, Kočovský P (2006) Asymmetric allylic substitution catalyzed by C_1-symmetrical complexes

of molybdenum: structural requirements of the ligand and the stereochemical course of the reaction. Chem Eur J 12:6910
16. Malda H, van Zijl AW, Arnold LA, Feringa BL (2001) Enantioselective copper-catalyzed allylic alkylation with dialkylzincs using phosphoramidite ligands. Org Lett 3:1169
17. Fananas-Mastral M, Feringa BL (2010) Copper-catalyzed regio- and enantioselective synthesis of chiral enol acetates and beta.-substituted aldehydes. J Am Chem Soc 132:13152
18. Van Veldhuizen JJ, Campbell JE, Giudici RE, Hoveyda AH (2005) A readily available chiral Ag-based N-heterocyclic carbene complex for use in efficient and highly enantioselective Ru-catalyzed olefin metathesis and Cu-catalyzed allylic alkylation reactions. J Am Chem Soc 127:6877
19. Yoshikai N, Zhang SL, Nakamura E (2008) Origin of the regio- and stereoselectivity of allylic substitution of organocopper reagents. J Am Chem Soc 130:12862
20. Selim KB, Matsumoto Y, Yamada K, Tomioka K (2009) Efficient chiral N-heterocyclic carbene/copper(I)-catalyzed asymmetric allylic arylation with aryl Grignard reagents. Angew Chem Int Ed 48:8733
21. Langloisa JB, Alexakis A (2010) Copper-catalyzed asymmetric allylic alkylation of racemic cyclic substrates: application of dynamic kinetic asymmetric transformation (DYKAT). Adv Synth Catal 352:447
22. Fang F, Zhang H, Xie F, Yang G, Zhang W (2010) Highly enantioselective copper-catalyzed allylic alkylation with atropos phosphoramidites bearing a D2-symmetric biphenyl backbone. Tetrahedron 66:3593
23. Matsushima Y, Onitsuka K, Kondo T, Mitsudo T, Takahashi S (2001) Asymmetric catalysis of planar-chiral cyclopentadienylruthenium complexes in allylic amination and alkylation. J Am Chem Soc 123:10405
24. Onitsuka K, Matsushima Y, Takahashi S (2005) Kinetic resolution of allyl carbonates in asymmetric allylic alkylation catalyzed by planar-chiral cyclopentadienyl-ruthenium complexes. Organometallics 24:6472
25. Onitsuka K, Okuda H, Sasai H (2008) Regio- and enantioselective O-allylation of phenol and alcohol catalyzed by a planar-chiral cyclopentadienyl ruthenium complex. Angew Chem Int Ed 47:1454
26. Evans PA, Nelson JD (1998) Conservation of absolute configuration in the acyclic rhodium catalyzed allylic alkylation reaction: evidence for an *enyl* ($\sigma + \pi$) organorhodium intermediate. J Am Chem Soc 120:5581
27. Evans PA, Kennedy LJ (2001) Regioselective rhodium-catalyzed allylic linchpin cross-coupling reactions: diastereospecific construction of *anti*-1,3-carbon stereogenic centers and C2-symmetrical fragments. J Am Chem Soc 123:1234
28. Selvakumar K, Valentini M, Pregosin PS, Albinati A (1999) Chiral phosphito-thioether complexes of palladium(0). comments on the Pd, Rh, and Ir regio- and enantioselective allylic alkylations of PhCH:CHCH(OAc)R, R = H, Me, Et. Organometallics 18:4591
29. Hayashi T, Okada A, Suzuka T, Kawatsura M (2003) High enantioselectivity in rhodium-catalyzed allylic alkylation of 1-substituted 2-propenyl acetates. Org Lett 5:1713
30. Kazmaier U, Stolz D (2006) Regio- and stereoselective rhodium-catalyzed allylic alkylations of chelated enolates. Angew Chem Int Ed 45:3072
31. Leahy DK, Evans PA (2005) Rhodium(I)-catalyzed allylic substitution reactions and their applications to target directed synthesis. In: Evans PA (ed) Modern rhodium-catalyzed organic reactions. Wiley, New York, p 191
32. Plietker B (2006) A highly regioselective salt-free iron-catalyzed allylic alkylation. Angew Chem Int Ed 45:1469
33. Plietker B (2006) Regioselective iron-catalyzed allylic amination. Angew Chem Int Ed 45:6053
34. Plietker B, Dieskau A, Möws K, Jatsch A (2008) Ligand-dependent mechanistic dichotomy in iron-catalyzed allylic substitutions: σ-allyl versus π-allyl mechanism. Angew Chem Int Ed 47:198

35. Yatsumonji Y, Ishida Y, Tsubouchi A, Takeda T (2007) Nickel(0) triethyl phosphite complex-catalyzed allylic substitution with retention of regio- and stereochemistry. Org Lett 9:4603
36. Chung KG, Miyake Y, Uemura S (2000) Nickel(0)-catalyzed asymmetric cross-coupling reactions of allylic compounds with arylboronic acids. J Chem Soc Perkin Trans 1 15
37. Didiuk MT, Morken JP, Hoveyda AH (1995) Directed regio- and stereoselective nickel-catalyzed addition of alkyl grignard reagents to allylic ethers. J Am Chem Soc 117:7273
38. Prétôt R, Pfaltz A (1998) New ligands for regio- and enantiocontrol in Pd-catalyzed allylic alkylations. Angew Chem Int Ed 37:323
39. Hayashi T, Kawatsura M, Uozumi Y (1998) Retention of regiochemistry of allylic esters in palladium-catalyzed allylic alkylation in the presence of a MOP ligand. J Am Chem Soc 120:1681
40. You SL, Zhu XZ, Luo YM, Hou XL, Dai LX (2001) Highly regio- and enantioselective Pd-catalyzed allylic alkylation and amination of monosubstituted allylic acetates with novel ferrocene P, N-ligands. J Am Chem Soc 123:7471
41. Zheng WH, Sun N, Hou XL (2005) Highly regio- and enantioselective palladium-catalyzed allylic alkylation and amination of dienyl esters with 1,1′-P, N-ferrocene ligands. Org Lett 7:5151
42. Zheng WH, Zheng BH, Zhang Y, Hou XL (2007) Highly regio-, diastereo-, and enantioselective Pd-catalyzed allylic alkylation of acyclic ketone enolates with monosubstituted allyl substrates. J Am Chem Soc 129:7718
43. Liu W, Chen D, Zhu XZ, Wan XL, Hou XL (2009) Highly diastereo- and enantioselective Pd-catalyzed cyclopropanation of acyclic amides with substituted allyl carbonates. J Am Chem Soc 131:8734
44. Fang P, Ding CH, Hou XL, Dai LX (2010) Palladium-catalyzed regio- and enantio-selective allylic substitution reaction of monosubstituted allyl substrates with benzyl alcohols. Tetrahedron Asymmetr 21:1176
45. Chen JP, Ding CH, Liu W, Hou XL, Dai LX (2010) Palladium-catalyzed regio-, diastereo-, and enantioselective allylic alkylation of acylsilanes with monosubstituted allyl substrates. J Am Chem Soc 132:15493
46. Trost BM, Toste FD (1999) Regio- and enantioselective allylic alkylation of an unsymmetrical substrate: a working model. J Am Chem Soc 121:4545
47. Trost BM, Bunt RC, Lemoine RC, Calkins TL (2000) Dynamic kinetic asymmetric transformation of diene monoepoxides: a practical asymmetric synthesis of vinylglycinol, vigabatrin, and ethambutol. J Am Chem Soc 122:5968
48. Krafft ME, Sugiura M, Abboud KA (2001) Novel use of ring strain to control regioselectivity: alkene-directed palladium-catalyzed allylation. J Am Chem Soc 123:9174
49. Cook GR, Yu H, Sankaranarayanan S, Shanker PS (2003) Hydrogen bond directed highly regioselective palladium-catalyzed allylic substitution. J Am Chem Soc 125:5115
50. Trost BM, Osipov M, Dong GB (2010) Palladium-catalyzed dynamic kinetic asymmetric transformations of vinyl aziridines with nitrogen heterocycles: rapid access to biologically active pyrroles and indoles. J Am Chem Soc 132:15800
51. Takeuchi R, Kashio M (1997) Highly selective allylic alkylation with a carbon nucleophile at the more substituted allylic terminus catalyzed by an iridium complex: an efficient method for constructing quaternary carbon centers. Angew Chem Int Ed 36:263
52. Takeuchi R, Kashio M (1998) Iridium complex-catalyzed allylic alkylation of allylic esters and allylic alcohols: unique regio- and stereoselectivity. J Am Chem Soc 120:8647
53. Takeuchi R (2002) Iridium complex-catalyzed highly selective organic synthesis. Synlett 1954
54. Takeuchi R, Kezuka S (2006) Iridium-catalyzed formation of carbon-carbon and carbon-heteroatom bonds. Synthesis 3349

55. Bartels B, García-Yebra C, Rominger F, Helmchen G (2002) Iridium-catalysed allylic substitution: stereochemical aspects and isolation of Ir^{III} complexes related to the catalytic cycle. Eur J Inorg Chem 2569
56. Bedford RB, Castillòn S, Chaloner PA, Claver C, Fernandez E, Hitchcock PB, Ruiz A (1996) Iridium complexes of orthometalated triaryl phosphites: synthesis, structure, reactivity, and use as imine hydrogenation catalysts. Organometallics 15:3990
57. García-Yebra C, Janssen JP, Rominger F, Helmchen G (2004) Asymmetric iridium(I)-catalyzed allylic alkylation of monosubstituted allylic substrates with phosphinooxazolines as ligands. Isolation, characterization, and reactivity of chiral (allyl)iridium(III) complexes. Organometallics 23:5459
58. Feringa BL (2000) Phosphoramidites: marvellous ligands in catalytic asymmetric conjugate addition. Acc Chem Res 33:346
59. Teichert JF, Feringa BL (2010) Phosphoramidites: privileged ligands in asymmetric catalysis. Angew Chem Int Ed 49:2486
60. Ohmura T, Hartwig JF (2002) Regio- and enantioselective allylic amination of achiral allylic esters catalyzed by an iridium-phosphoramidite complex. J Am Chem Soc 124:15164
61. Kiener CA, Shu C, Incarvito C, Hartwig JF (2003) Identification of an activated catalyst in the iridium-catalyzed allylic amination and etherification. Increased rates, scope, and selectivity. J Am Chem Soc 125:14272
62. Markovic D, Hartwig JF (2007) Resting state and kinetic studies on the asymmetric allylic substitutions catalyzed by iridium-phosphoramidite complexes. J Am Chem Soc 129:11680
63. Madrahimov ST, Markovic D, Hartwig JF (2009) The allyl intermediate in regioselective and enantioselective iridium-catalyzed asymmetric allylic substitution reaction. J Am Chem Soc 131:7228
64. Spiess S, Raskatov JA, Gnamm C, Brödner K, Helmchen G (2009) Ir-catalyzed asymmetric allylic substitutions with (phosphoramidite)Ir complexes – resting states, synthesis, and characterization of catalytically active (π-allyl)Ir complexes. Chem Eur J 15:11087
65. Raskatov JA, Spiess S, Gnamm C, Brödner K, Rominger F, Helmchen G (2010) Ir-catalysed asymmetric allylic substitutions with cyclometalated (phosphoramidite)Ir complexes – resting states, catalytically active (π-allyl)Ir complexes and computational exploration. Chem Eur J 16:6601
66. Bartels B, Helmchen G (1999) Ir-catalysed allylic substitution: mechanistic aspects and asymmetric synthesis with phosphorus amidites as ligands. Chem Commun 741
67. Lipowsky G, Miller N, Helmchen G (2004) Regio- and enantioselective iridium-catalyzed allylic alkylation with in situ activated P, C-chelate complexes. Angew Chem Int Ed 43:4595
68. Streiff S, Welter C, Schelwies M, Lipowsky G, Miller N, Helmchen G (2005) Carbocycles *via* enantioselective inter- and intramolecular iridium-catalysed allylic alkylations. Chem Commun 2957
69. Lipowsky G, Helmchen G (2004) Regio- and enantioselective iridium-catalysed allylic aminations and alkylations of dienyl esters. Chem Commun 116
70. Alexakis A, Polet D (2004) Very efficient phosphoramidite ligand for asymmetric iridium-catalyzed allylic alkylation. Org Lett 6:3529
71. Polet D, Alexakis A, Tissot-Croset K, Corminboeuf C, Ditrich K (2006) Phosphoramidite ligands in iridium-catalyzed allylic substitution. Chem Eur J 12:3596
72. Polet D, Alexakis A (2005) Kinetic study of various phosphoramidite ligands in the iridium-catalyzed allylic substitution. Org Lett 7:1621
73. Fuji K, Kinoshita N, Kawabata T, Tanaka K (1999) Enantioselective allylic substitution catalyzed by an iridium complex: remarkable effects of the counter cation. Chem Commun 2289
74. Gnamm C, Förster S, Miller N, Brödner K, Helmchen G (2007) Enantioselective iridium-catalyzed allylic alkylations; improvements and applications based on salt-free reaction conditions. Synlett 790

75. Spiess S, Welter C, Franck G, Taquet J-P, Helmchen G (2008) Iridium-catalyzed asymmetric allylic substitutions – very high regioselectivity and air stability with a catalyst derived from dibenzo[a, e]cyclooctatetraene and a phosphoramidite. Angew Chem Int Ed 47:7652
76. Kinoshitta N, Marx KH, Tanaka K, Tsubaki K, Kawabata T, Yoshikai N, Nakamura E, Fuji K (2004) Enantioselective allylic substitution of cinnamyl esters catalyzed by iridium-chiral aryl phosphite complex: conspicuous change in the mechanistic spectrum by a countercation and solvent. J Org Chem 69:7960
77. Nemoto T, Sakamoto T, Fukuyama T, Hamada Y (2007) Ir-catalyzed asymmetric allylic alkylation using chiral diaminophosphine oxides: DIAPHOXs. Formal enantioselective synthesis of (−)-paroxetine. Tetrahedron Lett 48:4977
78. Onodera G, Watabe K, Matsubara M, Oda K, Kezuka S, Takeuchi R (2008) Iridium-catalyzed enantioselective allylic alkylation using chiral phosphoramidite ligand bearing an amide moiety. Adv Synth Catal 350:2725
79. Schelwies M, Dübon P, Helmchen G (2006) Enantioselective modular synthesis of 2,4-disubstituted cyclopentenones by iridium-catalyzed allylic alkylation. Angew Chem Int Ed 45:2466
80. Dahnz A, Helmchen G (2006) Iridium-catalyzed enantioselective allylic substitutions with aliphatic nitro compounds as prenucleophiles. Synlett 697
81. Förster S, Tverskoy O, Helmchen G (2008) Malononitrile as acylanion equivalent. Synlett 2803
82. Kanayama T, Yoshida K, Miyabe H, Takemoto Y (2003) Enantio- and diastereoselective Ir-catalyzed allylic substitutions for asymmetric synthesis of amino acid derivatives. Angew Chem Int Ed 42:2054
83. Kanayama T, Yoshida K, Miyabe H, Kimachi T, Takemoto Y (2003) Synthesis of β-substituted β-amino acids with use of iridium-catalyzed asymmetric allylic substitution. J Org Chem 68:6197
84. Bondzic B, Farwick A, Liebich J, Eilbracht P (2008) Application of iridium catalyzed allylic substitution reactions in the synthesis of branched tryptamines and homologues *via* tandem hydroformylation–fischer indole synthesis. Org Biomol Chem 6:3723
85. Trost BM, Schroeder GM (1999) Palladium-catalyzed asymmetric alkylation of ketone enolates. J Am Chem Soc 121:6759
86. Braun M, Laicher F, Meier T (2000) Diastereoselective and enantioselective palladium-catalyzed allylic substitution with nonstabilized ketone enolates. Angew Chem Int Ed 39:3494
87. You SL, Hou XL, Dai LX, Zhu XZ (2001) Highly efficient ligands for palladium-catalyzed asymmetric alkylation of ketone enolates. Org Lett 3:149
88. Trost BM, Schroeder GM (2005) Palladium-catalyzed asymmetric allylic alkylation of ketone enolates. Chem Eur J 11:174
89. Yan XX, Liang CG, Zhang Y, Hong W, Cao BX, Dai LX, Hou XL (2005) Highly enantioselective Pd-catalyzed allylic alkylations of acyclic ketones. Angew Chem Int Ed 44:6544
90. Bélanger E, Cantin K, Messe O, Tremblay M, Paquin JF (2007) Enantioselective Pd-catalyzed allylation reaction of fluorinated silyl enol ethers. J Am Chem Soc 129:1034
91. Braun M, Meier T (2006) Tsuji–Trost allylic alkylation with ketone enolates. Angew Chem Int Ed 45:6952
92. Braun M, Meier T (2006) New developments in stereoselective palladium-catalyzed allylic alkylations of preformed enolates. Synlett 661
93. Graening T, Hartwig JF (2005) Iridium-catalyzed regio- and enantioselective allylation of ketone enolates. J Am Chem Soc 127:17192
94. Weix DJ, Hartwig JF (2007) Regioselective and enantioselective iridium-catalyzed allylation of enamines. J Am Chem Soc 129:7720
95. He H, Zheng X-J, Li Y, Dai L-X, You S-L (2007) Ir-catalyzed regio- and enantioselective decarboxylative allylic alkylations. Org Lett 9:4339

96. Zeng M, You SL (2010) Asymmetric friedel-crafts alkylation of indoles. The control of enantio- and regioselectivity. Synlett 1289
97. Malkov AV, Davis SL, Baxendale IR, Mitchell WL, Kočovský P (1999) Molybdenum(II)-catalyzed allylation of electron-rich aromatics and heteroaromatics. J Org Chem 64:2751
98. Bandini M, Melloni A, Umani-Ronchi A (2004) New versatile Pd-catalyzed alkylation of indoles via nucleophilic allylic substitution: controlling the regioselectivity. Org Lett 6:3199
99. Bandini M, Melloni A, Piccinelli F, Sinisi R, Tommasi S, Umani-Ronchi A (2006) Highly enantioselective synthesis of tetrahydro-β-carbolines and tetrahydro-γ-carbolines *via* Pd-catalyzed intramolecular allylic alkylation. J Am Chem Soc 128:1424
100. Ma S, Yu S, Peng Z, Guo H (2006) Palladium-catalyzed functionalization of indoles with 2-acetoxymethyl-substituted electron-deficient alkenes. J Org Chem 71:9865
101. Cheung HY, Yu WY, Lam FL, Au-Yeung TTL, Zhou Z, Chan TH, Chan ASC (2007) Enantioselective Pd-catalyzed allylic alkylation of indoles by a new class of chiral ferrocenyl P/S ligands. Org Lett 9:4295
102. Liu WB, He H, Dai LX, You SL (2008) Ir-catalyzed regio- and enantioselective Friedel-Crafts-type allylic alkylation of indoles. Org Lett 10:1815
103. Liu WB, He H, Dai LX, You SL (2009) Synthesis of 2-methylindoline- and 2-methyl-1,2,3,4-tetrahydroquinoline-derived phosphoramidites and their applications in iridium-catalyzed allylic alkylation of indoles. Synthesis 10:2076
104. Wu QF, He H, Liu WB, You SL (2010) Enantioselective construction of spiroindolenines by Ir-catalyzed allylic alkylation reactions. J Am Chem Soc 132:11418
105. Kimura M, Futamata M, Mukai R, Tamaru Y (2005) Pd-catalyzed C3-selective allylation of indoles with allyl alcohols promoted by triethylborane. J Am Chem Soc 127:4592
106. Trost BM, Quancard J (2006) Palladium-catalyzed enantioselective C-3 allylation of 3-substituted-1H-indoles using trialkylboranes. J Am Chem Soc 128:6314, and references therein
107. Kagawa N, Malerich JP, Rawal VH (2008) Palladium-catalyzed β-allylation of 2,3-disubstituted indoles. Org Lett 10:2381
108. Alexakis A, Hajjaji SE, Polet D, Rathgeb X (2007) Iridium-catalyzed asymmetric allylic substitution with aryl zinc reagents. Org Lett 9:3393
109. Polet D, Rathgeb X, Falciola CA, Langlois J-B, Hajjaji SE, Alexakis A (2009) Enantioselective iridium-catalyzed allylic arylation. Chem Eur J 15:1205
110. Liu WB, Zheng SC, He H, Zhao XM, Dai LX, You SL (2009) Iridium-catalyzed regio- and enantioselective allylic alkylation of Fluorobis(phenylsulfonyl)methane. Chem Commun 6604
111. Fukuzumi T, Shibata N, Sugiura M, Yasui H, Nakamura S, Toru T (2006) Fluorobis (phenylsulfonyl)methane: a fluoromethide equivalent and palladium-catalyzed enantioselective allylic monofluoromethylation. Angew Chem Int Ed 45:4973
112. Shu C, Leitner A, Hartwig JF (2004) Enantioselective allylation of aromatic amines after in situ generation of an activated cyclometalated iridium catalyst. Angew Chem Int Ed 43:4797
113. Lee JH, Shin S, Kang J, S-g L (2007) Ir-catalyzed allylic amination/ring-closing metathesis: a new route to enantioselective synthesis of cyclicβ-amino alcohol derivatives. J Org Chem 72:7443
114. Tissot-Croset K, Polet D, Alexakis A (2004) A highly effective phosphoramidite ligand for asymmetric allylic substitution. Angew Chem Int Ed 43:2426
115. Welter C, Koch O, Lipowsky G, Helmchen G (2004) First intramolecular enantioselective iridium-catalysed allylic aminations. Chem Commun 896
116. Welter C, Dahnz A, Brunner B, Streiff S, Dübon P, Helmchen G (2005) Highly enantioselective syntheses of heterocycles *via* intramolecular Ir-catalyzed allylic amination and etherification. Org Lett 7:1239
117. Miyabe H, Matsumura A, Moriyama K, Takemoto Y (2004) Utility of the iridium complex of the pybox ligand in regio- and enantioselective allylic substitution. Org Lett 6:4631

118. Nemoto T, Sakamoto T, Matsumoto T, Hamada Y (2006) Ir-catalyzed asymmetric allylic amination using chiral diaminophosphine oxides. Tetrahedron Lett 47:8737
119. Yamashita Y, Gopalarathnam A, Hartwig JF (2007) Iridium-catalyzed, asymmetric amination of allylic alcohols activated by Lewis acids. J Am Chem Soc 129:7508
120. Weihofen R, Dahnz A, Tverskoy O, Helmchen G (2005) Highly enantioselective iridium-catalysed allylic aminations with anionic N-nucleophiles. Chem Commun 3541
121. Weihofen R, Tverskoy O, Helmchen G (2006) Salt-free iridium-catalyzed asymmetric allylic aminations with N, N-diacylamines and ortho-nosylamide as ammonia equivalents. Angew Chem Int Ed 45:5546
122. Xia JB, Liu WB, Wang TM, You SL (2010) Enantioselective synthesis of 3-azabicyclo [4.1.0]heptenes and 3-azabicyclo[3.2.0]heptenes by Ir-catalyzed asymmetric allylic amination of *N*-tosyl propynylamine and Pt-catalyzed cycloisomerization. Chem Eur J 16:6442
123. Defieber C, Ariger MA, Moriel P, Carreira EM (2007) Iridium-catalyzed synthesis of primary allylic amines from allylic alcohols: sulfamic acid as ammonia equivalent. Angew Chem Int Ed 46:3139
124. Roggen M, Carreira EM (2010) Stereospecific substitution of allylic alcohols to give optically active primary allylic amines: unique reactivity of a (P, alkene)Ir complex modulated by iodide. J Am Chem Soc 132:11917
125. Spiess S, Berthold C, Weihofen R, Helmchen G (2007) Synthesis of α, β-unsaturated γ-lactams via asymmetric iridium-catalysed allylic substitution. Org Biomol Chem 5:2357
126. Pouy MJ, Leitner A, Weix DJ, Ueno S, Hartwig JF (2007) Enantioselective iridium-catalyzed allylic amination of ammonia and convenient ammonia surrogates. Org Lett 9:3949
127. Singh OV, Han H (2007) Iridium(I)-catalyzed regio- and enantioselective allylic amidation. Tetrahedron Lett 48:7094
128. Gnamm C, Franck G, Miller N, Stork T, Brödner K, Helmchen G (2008) Enantioselective iridium-catalyzed allylic aminations of allylic carbonates with functionalized side chains. Asymmetric total synthesis of (*S*)-Vigabatrin. Synthesis 3331
129. Barrett AGM, Head J, Smith ML, Stock NS, White AJP, Williams DJ (1999) Fleming-tamao oxidation and masked hydroxyl functionality: total synthesis of (+)-pramanicin and structural elucidation of the antifungal natural product (−)-pramanicin. J Org Chem 64:6005
130. Weix DJ, Marković D, Ueda M, Hartwig JF (2009) Direct, intermolecular, enantioselective, iridium-catalyzed allylation of carbamates to form carbamate-protected, branched allylic amines. Org Lett 11:2944
131. Singh OV, Han H (2007) Iridium(I)-catalyzed regio- and enantioselective decarboxylative allylic amidation of substituted allyl benzyl imidodicarbonates. J Am Chem Soc 129:774
132. Singh OV, Han H (2007) Iridium(I)-catalyzed stereospecific decarboxylative allylic amidation of chiral branched benzyl allyl imidodicarboxylates. Org Lett 9:4801
133. Stanley LM, Hartwig JF (2009) Regio- and enantioselective N-allylations of imidazole, benzimidazole, and purine heterocycles catalyzed by single-component metallacyclic iridium complexes. J Am Chem Soc 131:8971
134. Stanley LM, Hartwig JF (2009) Iridium-catalyzed regio- and enantioselective N-allylation of indoles. Angew Chem Int Ed 48:7841
135. Teichert JF, Ferriga BL (2010) Catalytic asymmetric synthesis of 2,5-dinaphthylpyrrolidine. Synthesis 1200
136. Pouy MJ, Stanley LM, Hartwig JF (2009) Enantioselective, iridium-catalyzed monoallylation of ammonia. J Am Chem Soc 131:11312
137. Miyabe H, Yoshida K, Reddy VK, Takemoto Y (2009) Palladium- or iridium-catalyzed allylic substitution of guanidines: convenient and direct modification of guanidines. J Org Chem 74:305
138. López F, Ohmura T, Hartwig JF (2003) Regio- and enantioselective iridium-catalyzed intermolecular allylic etherification of achiral allylic carbonates with phenoxides. J Am Chem Soc 125:3426

139. Leitner A, Shu C, Hartwig JF (2005) Effects of catalyst activation and ligand steric properties on the enantioselective allylation of amines and phenoxides. Org Lett 7:1093
140. Kimura M, Uozumi Y (2007) Development of new P-chiral phosphorodiamidite ligands having a pyrrolo[1,2-c]diazaphosphol-1-one unit and their application to regio- and enantioselective iridium-catalyzed allylic etherification. J Org Chem 72:707
141. Fischer C, Defieber C, Suzuki T, Carreira EM (2004) Readily available [2.2.2]-bicyclooctadienes as new chiral ligands for Ir(I): catalytic, kinetic resolution of allyl carbonates. J Am Chem Soc 126:1628
142. Stanley LM, Bai C, Ueda M, Hartwig JF (2010) Iridium-catalyzed kinetic asymmetric transformations of racemic allylic benzoates. J Am Chem Soc 132:8918
143. Shu C, Hartwig JF (2004) Iridium-catalyzed intermolecular allylic etherification with aliphatic alkoxides: asymmetric synthesis of dihydropyrans and dihydrofurans. Angew Chem Int Ed 43:4794
144. Roberts J, Lee C (2005) Allylic etherification *via* Ir(I)/Zn(II) bimetallic catalysis. Org Lett 7:2679
145. Miyabe H, Takemoto Y (2005) Regio- and stereocontrolled palladium- or iridium-catalyzed allylation. Synlett 1641
146. Miyabe H, Yoshida K, Yamauchi M, Takemoto Y (2005) Hydroxylamines as oxygen atom nucleophiles in transition-metal-catalyzed allylic substitution. J Org Chem 70:2148
147. Soderquist JA, Vaquer J, Diaz MJ, Rane AM, Bordwell FG, Zhang SZ (1996) Triisopropylsilanol: a new type of phase transfer catalyst for dehydrohalogenation. Tetrahedron Lett 37:2561
148. Trost BM, Ito N, Greenspan PD (1993) Triphenylsilanol as a water surrogate for regioselective Pd catalyzed allylations. Tetrahedron Lett 34:1421
149. Geissler H, Kim JH, Greeves N (1994) A total synthesis of (+)-2′*S*, 3′*R*-zoapatanol. Angew Chem Int Ed 33:2182
150. Lyothier I, Defieber C, Carreira EM (2006) Iridium-catalyzed enantioselective synthesis of allylic alcohols: silanolates as hydroxide equivalents. Angew Chem Int Ed 45:6204
151. Gärtner M, Mader S, Seehafer K, Helmchen G (2011) Enantio- and regioselective iridium-catalyzed allylic hydroxylation. J Am Chem Soc 133:2072
152. Ueno S, Hartwig JF (2008) Direct, iridium-catalyzed enantioselective and regioselective allylic etherification with aliphatic alcohols. Angew Chem Int Ed 47:1928
153. Trost BM, Crawley ML, Lee CB (2000) α-Acetoxysulfones as "Chiral Aldehyde" equivalents. J Am Chem Soc 122:6120
154. Felpin FX, Landais Y (2005) Practical Pd/C-mediated allylic substitution in water. J Org Chem 70:6441
155. Chandrasekhar S, Jagadeshwar V, Saritha B, Narsihmulu C (2005) Palladium-triethylborane-triggered direct and regioselective conversion of allylic alcohols to allyl phenyl sulfones. J Org Chem 70:6506
156. Ueda M, Hartwig JF (2010) Iridium-catalyzed, regio- and enantioselective allylic substitution with aromatic and aliphatic sulfinates. Org Lett 12:92
157. Xu QL, Dai LX, You SL (2010) Tandem Ir-catalyzed allylic substitution reaction of allyl sulfinates and isomerization. Org Lett 12:800
158. Xu QL, Liu WB, Dai LX, You SL (2010) Iridium-catalyzed enantioselective allylic substitution of O-allyl carbamothioates. J Org Chem 75:4615
159. Zheng SC, Gao N, Liu W, Liu DG, Zhao XM, Cohen T (2010) Regio- and enantioselective iridium-catalyzed allylation of thiophenol: synthesis of enantiopure allyl phenyl sulfides. Org Lett 12:4454
160. He H, Liu WB, Dai LX, You SL (2009) Ir-catalyzed cross-coupling of styrene derivatives with allylic carbonates: free amine assisted vinyl C-H bond activation. J Am Chem Soc 131:8346

161. He H, Liu WB, Dai LX, You SL (2010) Enantioselective synthesis of 2,3-dihydro-1*H*-benzo [*b*]azepines: iridium-catalyzed tandem allylic vinylation/amination reaction. Angew Chem Int Ed 49:1496
162. Evans PA, Robinson JE (1999) Regioselective Rh-catalyzed allylic amination/ring-closing metathesis approach to monocyclic azacycles: diastereospecific construction of 2,5-disubstituted pyrrolines. Org Lett 1:1929
163. Evans PA, Leahy DK (2003) Recent developments in rhodium-catalyzed allylic substitution and carbocyclization reactions. Chemtracts 16:567
164. Evans PA, Leahy DK, Andrews WJ, Uraguchi D (2004) Stereodivergent construction of cyclic ethers by a regioselective and enantiospecific rhodium-catalyzed allylic etherification: total synthesis of gaur acid. Angew Chem Int Ed 43:4788
165. Trost BM, Jiang C (2003) Pd-catalyzed asymmetric allylic alkylation. a short route to the cyclopentyl core of viridenomycin. Org Lett 5:1563
166. Alexakis A, Polet D (2002) Tandem copper-catalyzed enantioselective allylation-metathesis. Org Lett 4:4147
167. Dübon P, Schelwies M, Helmchen G (2008) Preparation of 2,4-disubstituted cyclopentenones by enantioselective iridium-catalyzed allylic alkylation: synthesis of 2′-methylcarbovir and *TEI-9826*. Chem Eur J 14:6722
168. Kobayashi Y, Murugesh MG, Nakano M, Takahisa E, Usmani SB, Ainai T (2002) A new method for installation of aryl and alkenyl groups onto a cyclopentene ring and synthesis of prostaglandins. J Org Chem 67:7110
169. Welter C, Moreno RM, Streiff S, Helmchen G (2005) Enantioselective synthesis of (+)(*R*)- and (−)(*S*)-nicotine based on Ir-catalysed allylic amination. Org Biomol Chem 3:3266
170. Dübon P, Farwick A, Helmchen G (2009) Enantioselective syntheses of 2-substituted pyrrolidines from allylamines by domino hydroformylation-condensation: short syntheses of (*S*)-nicotine and the alkaloid 225C. Synlett 9:1413
171. Böhrsch V, Blechert S (2006) A concise synthesis of (−)-centrolobine *via* a diastereoselective ring rearrangement metathesis–isomerisation sequence. Chem Commun 1968
172. Farwick A, Helmchen G (2010) Enantioselective total synthesis of (−)-α-kainic acid. Org Lett 12:1108
173. Gnamm C, Krauter CM, Brödner K, Helmchen G (2009) Stereoselective synthesis of 2,6-disubstituted piperidines using the iridium-catalyzed allylic cyclization as configurational switch: asymmetric total synthesis of (+)-241 D and related piperidine alkaloids. Chem Eur J 15:2050
174. Gnamm C, Krauter CM, Brödner K, Helmchen G (2009) A configurational switch based on iridium-catalyzed allylic cyclization: application in asymmetric total syntheses of prosopis, dendrobate, and spruce alkaloids. Chem Eur J 15:10514
175. Kakeya H, Morishita M, Kobinata K, Osono M, Osada H (1998) Isolation and biological activity of a novel cytokine modulator, cytoxazone. J Antibiot 51:1126
176. Franck G, Brödner K, Helmchen G (2010) Enantioselective modular synthesis of cyclohexenones: total syntheses of (+)-crypto- and (+)-infectocaryone. Org Lett 12:3886
177. Singh OV, Han H (2004) Tandem overman rearrangement and intramolecular amidomercuration reactions. Stereocontrolled synthesis of *cis*- and *trans*-2,6-dialkylpiperidines. Org Lett 6:3067
178. Förster S, Helmchen G (2008) Stereoselective synthesis of a lactam analogue of brefeldin C. Synlett 831

Top Organomet Chem (2012) 38: 209–234
DOI: 10.1007/3418_2011_11

Published online: 14 June 2011

Molybdenum-Catalyzed and Tungsten-Catalyzed Enantioselective Allylic Substitutions

Christina Moberg

Abstract Asymmetric allylic substitutions catalyzed by molybdenum and tungsten complexes provide branched chiral products from unsymmetrically substituted allylic reagents. Highly selective chiral ligands are available for both types of reactions, but for the tungsten-catalyzed substitutions, enantioselective reactions are only possible starting from achiral linear allylic substrates. A variety of stabilized carbanions can be used as nucleophiles. The molybdenum-catalyzed reaction has been applied to the synthesis of several biologically active compounds.

Keywords Allylation · Bispyridylamide · Molybdenum · Phosphinooxazoline · Regioselectivity · Tungsten

Contents

C. Moberg
Organic Chemistry, KTH School of Chemical Science and Engineering, 100 44 Stockholm, Sweden
e-mail: kimo@kth.se

Abbreviations

b	Branched
Bipy	Bipyridine
Boc	*tert*-Butoxycarbonyl
l	Linear

1 Introduction

The introduction of molybdenum catalysts for allylic alkylations by Trost and Lautens in 1982 [1] and by tungsten catalysts by Trost and Hung the year after [2] offered useful alternatives to the previously predominantly used palladium catalysts. Molybdenum- and tungsten-catalyzed allylic alkylations proceed with a regioselectivity complementary to that of the palladium-catalyzed process. Thus, monosubstituted allylic substrates result in chiral branched products whereas the palladium-catalyzed reactions commonly yield linear products (Scheme 1). Although the potential of the new processes for enantioselective transformations of unsymmetrically substituted allylic substrates was realized early, it took more than a decade to find suitable chiral catalysts.

The molybdenum-catalyzed reactions are characterized by rapidly equilibrating allyl complexes. For this reason, the reactions lead to essentially the same product distribution regardless of which regio- or stereoisomer of the substrate that is used, and the reactions may lead to high regio- and stereoselectivity also from mixtures of isomers. The reactions have been used as key steps in enantioselective syntheses of several biologically active compounds.

The allyl complexes with tungsten undergo slower isomerization than those containing molybdenum. Whereas high levels of asymmetric induction has been achieved in reactions of both linear and branched substrates catalyzed by

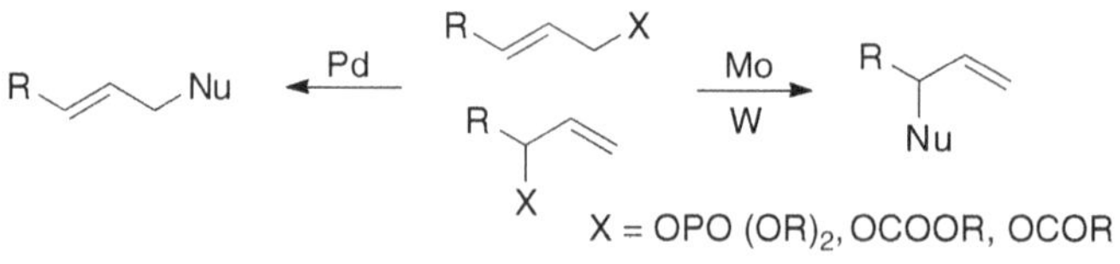

R Nu Pd R X R X Mo W R Nu

X = OPO $(OR)_2$, OCOOR, OCOR

Scheme 1 Transition metal catalyzed allylic alkylations

molybdenum complexes, reactions of branched racemic substrates in the presence of tungsten catalysts result in racemic products.

Iridium complexes react with the same regiochemistry as molybdenum and tungsten complexes. They have wider applicability since they are compatible with a wider range of nucleophiles, and use of molybdenum catalysts has therefore been somewhat limited. However, the latter are often the catalysts of choice due to the lower price of molybdenum and because convenient reaction conditions are available.

This chapter describes the scope and limitations of the catalytic processes, chiral catalysts for enantioselective reactions, and the mechanism and stereochemistry whereby the reactions proceed. The reactions are illustrated by a few examples of synthetic applications. A previous review covering molybdenum-catalyzed substitutions is available [3] and reactions using molybdenum and tungsten have been surveyed together with reactions employing other metals [4–6].

2 Reaction Conditions

Molybdenum and, in particular, tungsten catalysts are generally less reactive than palladium catalysts in allylic alkylations, but the reactivity varies with the nature of the ligand used. For both types of reactions, efficient chiral ligands are available. Bispyridylamides are the ligands of choice in molybdenum-catalyzed reactions, and for tungsten-catalyzed reactions phosphinooxazoline ligands are most successful. Rather high catalyst loadings are usually necessary. The molybdenum-catalyzed reactions can be run in air with lower catalyst loadings using microwave irradiation, and are complete within 5–6 min under these conditions. In tungsten-catalyzed reactions, phosphines have been shown to poison the catalyst [2]. The order of addition of the reagents may be crucial in the latter type of reactions [7].

2.1 *Molybdenum and Tungsten Sources*

$M(CO)_3(EtCN)_3$ and $M(CO)_3(\eta^6\text{-}C_7H_8)$ (C_7H_8 = cycloheptatriene), M = Mo or W, are the most commonly used catalyst precursors. They are prepared from $M(CO)_6$ and propionitrile [8] or cycloheptatriene [9, 10]. The hexacarbonyl complexes do not catalyze the reactions under normal conditions, but stable and crystalline $Mo(CO)_6$ can be used as precatalyst in microwave-mediated reactions [11] and under reflux in toluene [12] or when the complex is preheated in presence of ligand [13].

2.2 *Electrophiles*

Allylic phosphates are more reactive than allylic carbonates, which in turn are more reactive than allylic carboxylates. Molybdenum-catalyzed reactions can be performed with all three types of electrophiles, although carboxylates may require high temperatures. In tungsten-catalyzed reactions, allylic substrates with more reactive leaving groups are needed, and allylic phosphates are therefore the substrates of choice. Branched allylic substrates are more reactive than their linear analogs. In tungsten-catalyzed reactions of (1-phenyl)-2-propenyl and (3-phenyl)-2-propenyl carbonates, competition experiments showed that the branched carbonate was approximately six times more reactive than the linear isomer [14]. An electron-withdrawing group, such as CF_3, in *para*-position of the aromatic ring increased the reactivity to a small extent.

2.3 *Nucleophiles*

The range of nucleophiles that has been used in molybdenum- and tungsten-catalyzed allylic alkylations is limited to stabilized carbon nucleophiles, and is thus narrower than in the palladium- and iridium-catalyzed processes. Malonate esters are good nucleophiles in molybdenum-catalyzed allylations. Diketones and β-ketoesters react poorly, but a variety of stabilized enolates exhibit high reactivity as nucleophiles. In tungsten-catalyzed allylic alkylation, acetylenic esters and ketoesters [7] as well as sulfone-stabilized anions [2] can be used in addition to malonates, but no examples of enantioselective processes using these types of nucleophiles have been reported.

3 Regiochemistry

The ratio of regioisomers is a consequence of steric and electronic factors and depends on the catalyst, the reactivity and steric properties of the nucleophile, the steric demands and the charge distribution in the electrophile, and the reaction conditions. In both molybdenum- and tungsten-catalyzed reactions, the regioselectivity is lower with alkyl-substituted than with aryl-substituted allylic electrophiles.

In tungsten-catalyzed reactions, the regioselectivity was found to be independent of the location of the leaving group, as shown by the formation of equal product mixtures from linear and branched carbonates [14], thus suggesting a common intermediate. There is a correlation between regioselectivity and charge location in the electrophile [14–16]. The influence of electronic and steric parameters on the regiochemistry of tungsten-catalyzed processes was studied using 1- and 3-aryl-substituted electrophiles (**1**,**2**) with different groups in *para*-position (Fig. 1). The regioselectivities observed in the reaction with sodium malonate were found to correlate with Swain–Lupton parameters, and to be consistent with the effect of charge location [14]. Particularly high regioselectivity, 65.7:1 in favor or the branched isomer, was observed in the reaction with a naphthyl derivative **3**.

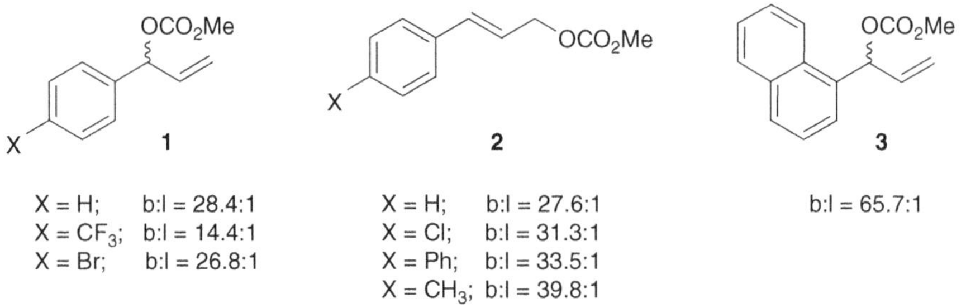

1	**2**	**3**
X = H; b:l = 28.4:1	X = H; b:l = 27.6:1	b:l = 65.7:1
X = CF_3; b:l = 14.4:1	X = Cl; b:l = 31.3:1	
X = Br; b:l = 26.8:1	X = Ph; b:l = 33.5:1	
	X = CH_3; b:l = 39.8:1	

Fig. 1 Regioselectivity in W-catalyzed allylations of malonate

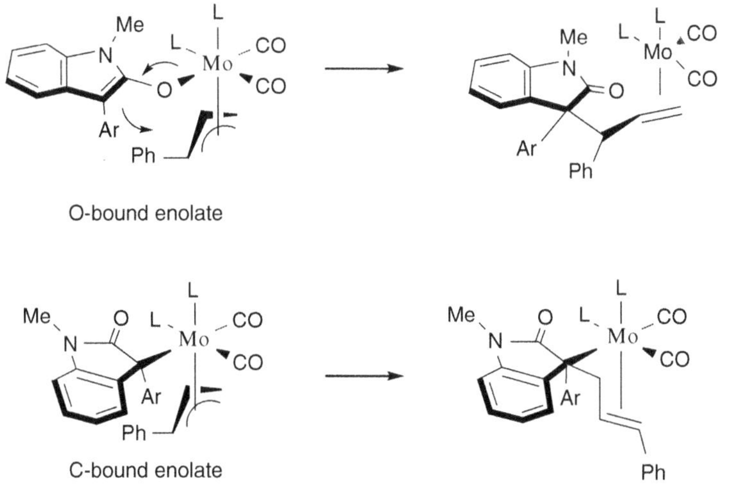

O-bound enolate

C-bound enolate

Scheme 2 Mechanistic proposal for origin of regioselectivity in allylations of oxindoles

A mechanistic explanation for the formation of the different regioisomers was offered for molybdenum-catalyzed allylations. In reactions with oxindoles as nucleophiles it was suggested that an *O*-bound enolate reacts to give the branched product via a Claisen-like rearrangement whereas a *C*-bound enolate leads to the linear product via direct reductive elimination (Scheme 2) [17, 18].

4 Chemoselectivity

As a consequence of the higher reactivity of branched compared to linear allylic substrates in tungsten-catalyzed allylations, the secondary carbonate function was selectively replaced in dicarbonate **4** [7] (Scheme 3). For the same reason, linanyl methyl carbonate reacts readily, while geranyl methyl carbonate provides only trace

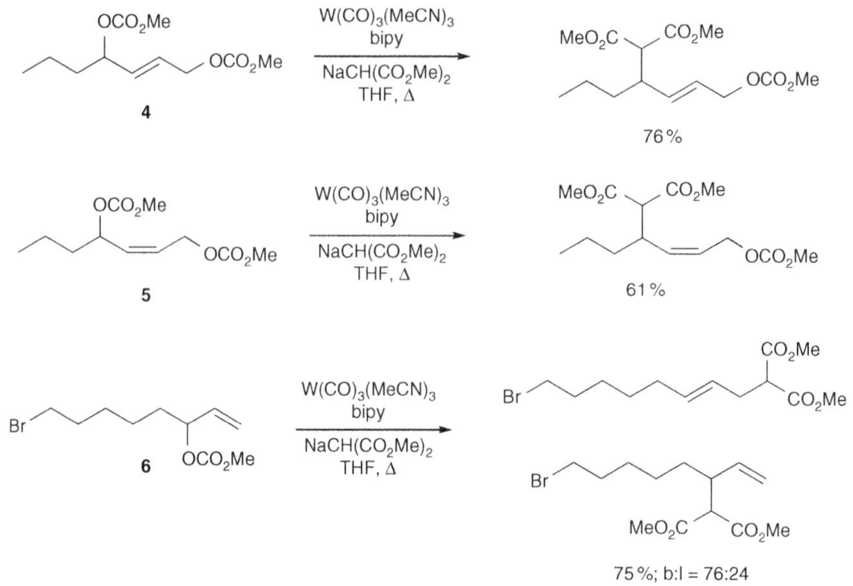

Scheme 3 W-catalyzed chemoselective allylic alkylations

amounts of product, even after prolonged reaction time. Due to slow isomerization of allyl intermediates in tungsten-catalyzed reactions, retention of the configuration of the double bond is observed [7, 16], and allylic substrates differing in the configuration of the olefinic bond thus provide different products (compare **4** and the *Z*-isomer **5**).

In the bromocarbonate **6**, selective substitution of the allylic carbonate function occurs [2], although replacement of the bromide is observed under noncatalytic conditions. Molybdenum catalysts react with the same selectivity [19]. The molybdenum catalysts are also compatible with allyl and vinyl silanes [20].

5 Stereochemistry

5.1 *Tungsten-Catalyzed Reactions*

The stereochemistry of the tungsten-catalyzed process was elucidated via substitution of *rac*-3-carbomethoxy-5-cyclohexenecarbonate **7** with malonate, which gave the isomer **8** formed by retention of stereochemistry as the major product (Scheme 4). The formation of the minor isomer **9** was shown not to be due to isomerization of starting material or product [2].

The use of racemic substrates did not allow conclusion regarding any η^3–η^1–η^3 isomerization of the intermediate allyl complexes. Such isomerization was, however,

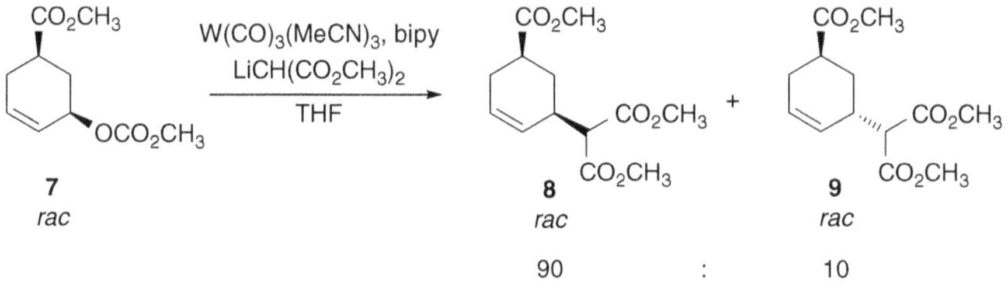

Scheme 4 Stereoselective W-catalyzed allylic alkylations

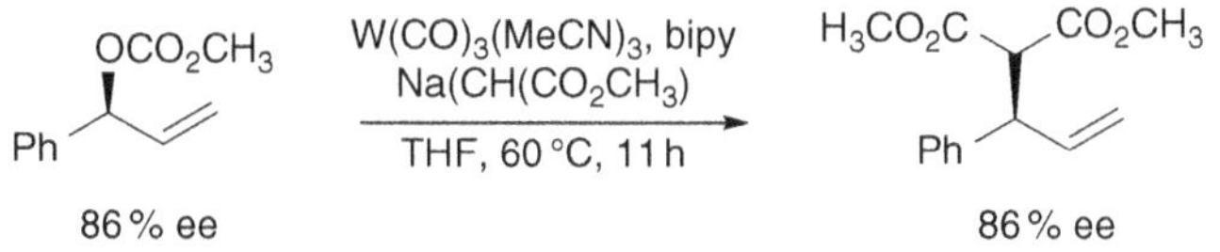

Scheme 5 Stereospecific W-catalyzed allylic alkylations

precluded from the results of a reaction using of an enantiomerically enriched carbonate [14]. It was found that the branched isomer, obtained along with a small amount of the linear isomer, had the same enantiomeric purity as the substrate, and that the reaction was completely *syn*-stereoselective (Scheme 5).

5.2 Molybdenum-Catalyzed Reactions

The molybdenum-catalyzed reactions have been studied in greater detail. Although inversion of configuration has been observed under stoichiometric conditions [21], overall retention of configuration is the rule under catalytic conditions [1, 19], but in contrast to the palladium-catalyzed reaction, molybdenum-catalyzed reactions proceed via a retention–retention pathway of stereochemistry, as unambiguously demonstrated by employing D-labeled substrates (see below) [22–24].

Use of chiral racemic substrates in combination with chiral ligands results in diasteromeric allyl complexes (**10** and **11**, Scheme 6). Equilibration of the diastereomeric η^3-allyl complexes, which mainly proceeds via η^1-allyl intermediates, is required to obtain enantioenriched compounds. This is an example of a dynamic kinetic asymmetric transformation, DYKAT [25].

Since diastereomeric allyl complexes are formed in the presence of chiral ligands, the substrate enantiomers may react with different rates, resulting in kinetic resolution (see Sect. 8.1.2). The faster reacting enantiomer, (*S*)-**12** in Scheme 7, reacts with higher enantioselectivity and with higher branched to linear ratio than its enantiomer [26]. From studies of D-labeled substrates it was demonstrated that the matched substrate, which is the faster reacting enantiomer, reacts with net retention of configuration and with retention of the configuration of the double

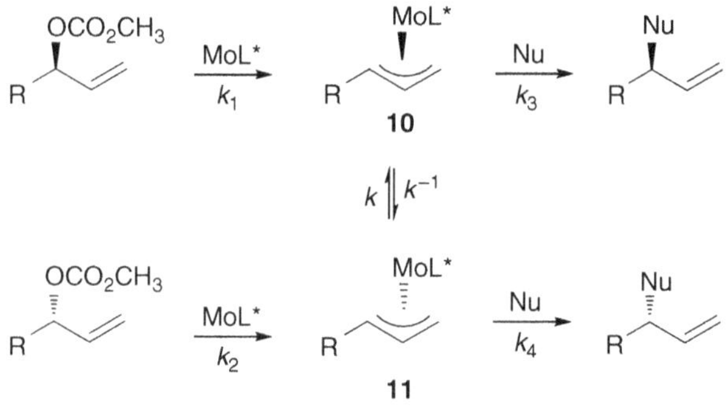

Scheme 6 Dynamic kinetic asymmetric allylic alkylations

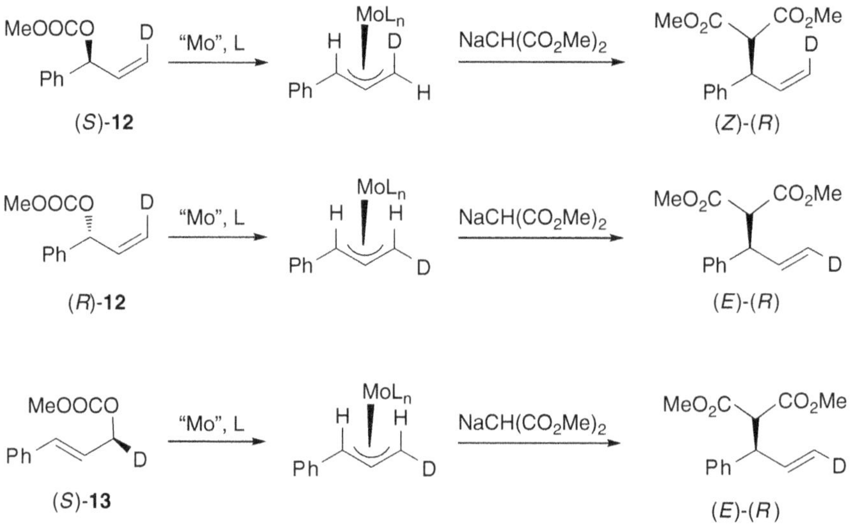

Scheme 7 Allylic alkylations via a retention–retention mechanism

bond. The mismatched substrate, in this case (*R*)-**12**, also reacts with net retention, but with isomerization of the allyl complex, thus resulting in apparent inversion. The linear substrate, (*S*)-**13**, also reacts via a retention–retention pathway (Scheme 7) [22, 24].

6 Enantioselective Reactions

6.1 Tungsten-Catalyzed Reactions

The first examples of enantioselective tungsten-catalyzed allylic alkylations were presented by Lloyd-Jones and Pfaltz in 1995 [27]. They found that

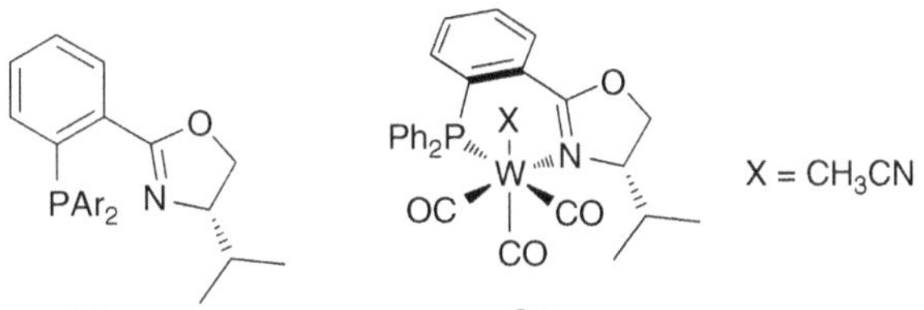

Fig. 2 Phosphinoxazoline ligand **L1** and its W-complex **C1**

Table 1 Allylation of malonate catalyzed by complex **C1**

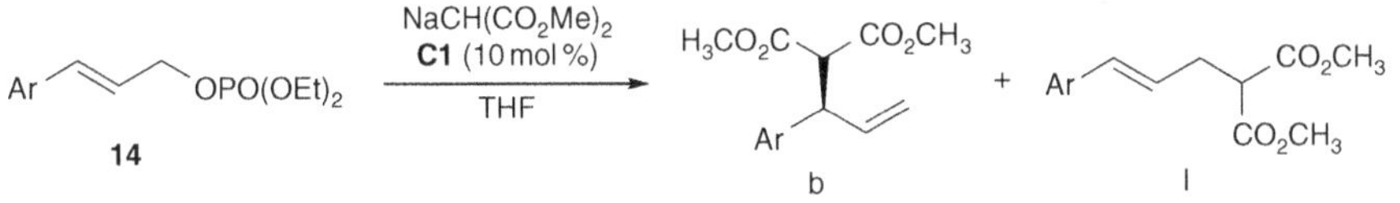

Entry	Substrate	Ar	Temp (°C)	Time (h)	Yield	b:l	ee (%) of b
1	(*E*)-**14a**	Ph	−10	71	89	74:26	96 (*R*)
2	(*E*)-**14b**	4-MeC_6H_4	−13	208	92	80:20	95 (*R*)
3	(*E*)-**14b**	4-MeC_6H_4	25	136	86	80:20	94 (*R*)
4	(*E*)-**14c**	4-ClC_6H_4	−13	214	95	80:20	89 (*R*)
5	(*E*)-**14d**	4-PhC_6H_4	−13	228	98	79:21	90 (*R*)

phosphinoxazolines (Fig. 2, **L1**), which previously had been employed as ligands in palladium-catalyzed allylic alkylations as well as in a variety of other enantioselective metal-catalyzed processes [28], resulted in highly enantioenriched products from linear allylic phosphates. Reaction of 3-phenyl-2-propenyl diethyl phosphate **14** with dimethyl sodiomalonate in the presence of ligand **L1** and $W(CO)_3(MeCN)_3$ or $W(CO)_3(\eta^6\text{-}C_7H_8)$ gave 95% of a 72:28 mixture of the branched and linear products, the former with 96% ee, at −10°C. Use of preformed tungsten complex **C1** [29] led to similar results (Table 1, entry 1). Other aromatic substrates were transformed to products with equally high regio- and enantioselectivity (entries 2–5), while an aliphatic substrate, *E*-2-butenyl diethyl phosphate, reacted with reverse regioselectivity (27:73 branched to linear ratio) and moderate enantioselectivity (65% ee of branched product) when **C1** was used as catalyst at 25°C. Tungsten complexes of phosphinooxazolines with other substituents in the oxazoline ring gave inferior results.

As a consequence of slow isomerization of the intermediate allyl complexes, racemic products were obtained when racemic branched substrates were used as substrates [14]. Due to the lack of *E*–*Z* olefin isomerization, the opposite product enantiomer was obtained from the *Z*-isomer (*Z*)-**14a**, but the enantioselectivity was poor (Scheme 8) [30].

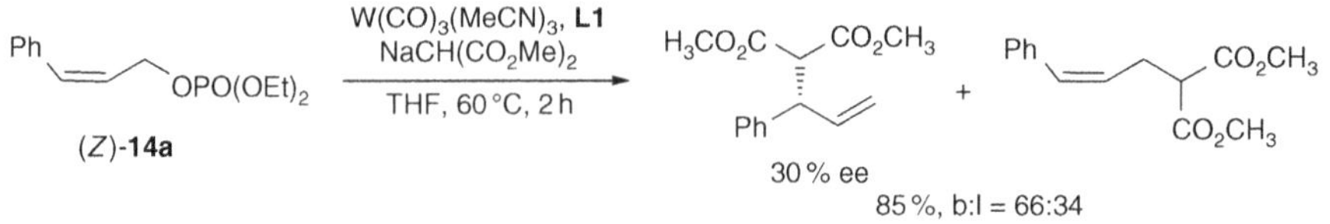

Scheme 8 Allylic alkylation with (*Z*)-**14a**

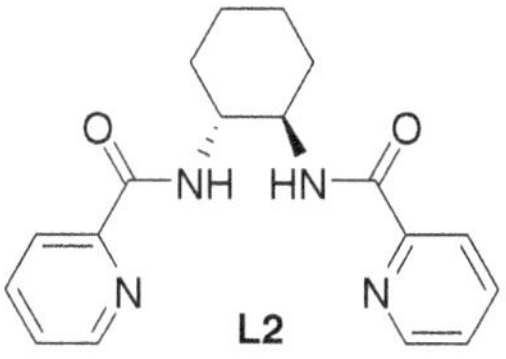

Fig. 3 Chiral bispyridylamide ligand **L2**

Shortly after the presentation of the first enantioselective tungsten-catalyzed allylic alkylations, Trost and Hachiya found that a complex prepared from W(CO)$_3$(EtCN)$_3$ and chiral bispyridylamide **L2** (Fig. 3) [31] catalyzed the reaction of cinnamyl carbonate with dimethyl sodiomalonate. When 15 mol% catalyst was used, high regioselectivity (49:1 branched to linear ratio) and high enantioselectivity (98% ee) were observed, but the yield was unsatisfactory (55%) [32].

6.2 *Molybdenum-Catalyzed Reactions*

6.2.1 Bispyridylamides

In contrast to phosphinooxazoline **L1**, which was found not to serve as a useful ligand for molybdenum-catalyzed allylic alkylations, bispyridylamide **L2** proved to afford excellent results when used in combination with Mo(CO)$_3$(EtCN)$_3$ and is still the ligand of choice for the molybdenum-catalyzed reaction. It can be conveniently prepared, even in large scale [33], from the diamine and a pyridine carboxylic acid derivative, and both enantiomers are commercially available.

Both linear (**15**) and branched allylic carbonates (**16**) with aromatic and heteroaromatic substituents can be used as substrates in the reaction [32]. In general, linear substrates **15** lead to products with higher branched to linear ratio and higher enantiomeric excess (Table 2, entries 1 and 2). The reactions were run in THF at room temperature or under reflux using 15 mol% catalyst. At room temperature higher regioselectivity was observed (49:1 as compared to 32:1 at reflux temperature for cinnamyl carbonate), but the yields were somewhat lower.

Ligand **L2** has also been successfully used in reactions with polyenes [34]. High regioselectivities as well as high enantioselectivities were observed (Scheme 9).

Table 2 Allylation of malonates catalyzed by bispyridylamide ligand **L2**

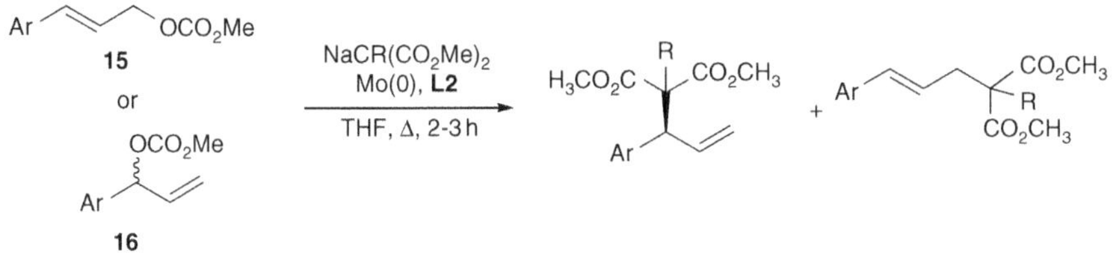

Entry	Substrate	Ar	R	Yield (%)	b:l	ee (%) of b
1	**15a**	Ph	H	88	32:1	99
2	**16a**	Ph	H	70	13:1	92
3	**16b**	2-Thienyl	H	78	19:1	88
4	**15c**	2-Furyl	Me	71	32:1	97
5	**16c**	2-Furyl	Me	65	32:1	87
6	**16d**	2-Pyridyl	Me	71	5:1	94

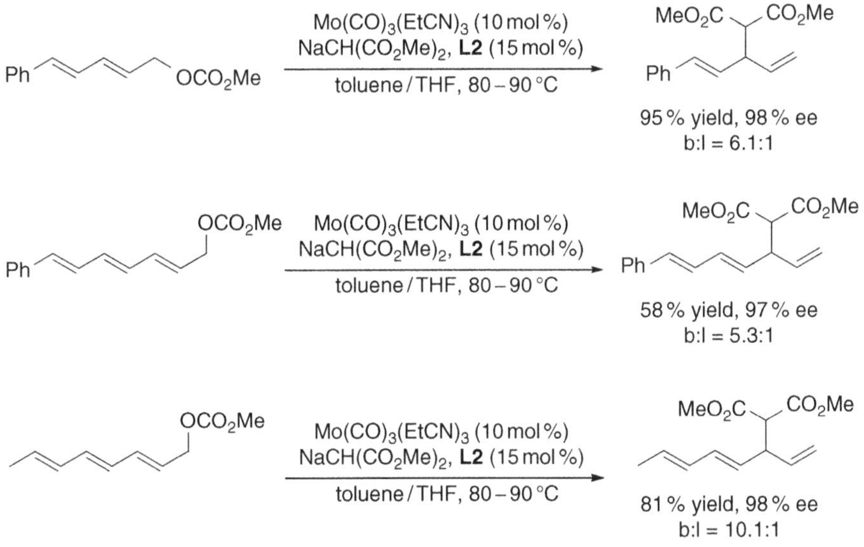

Scheme 9 Regio- and enantioselective molybdenum-catalyzed alkylations of polyenyl esters

A variety of modified bispyridylamides **L3** have been employed as ligands in the reaction. Bispyridylamide derivatives with π-donating substituents in the 4-positions of the pyridine rings resulted in even higher regioselectivity and equally high enantioselectivity [35, 36]. Under microwave conditions in the presence of Mo $(CO)_6$ the products were obtained within 5–6 min (Table 3). Although these conditions result in somewhat reduced selectivity (compare reactions using parent ligand **L2**, entry 1 of Tables 2 and 3), the reactions are simple to perform since inert conditions are not required and since a stable and crystalline molybdenum source

Table 3 Mo-catalyzed allylations using substituted bispyridylamides **L3**

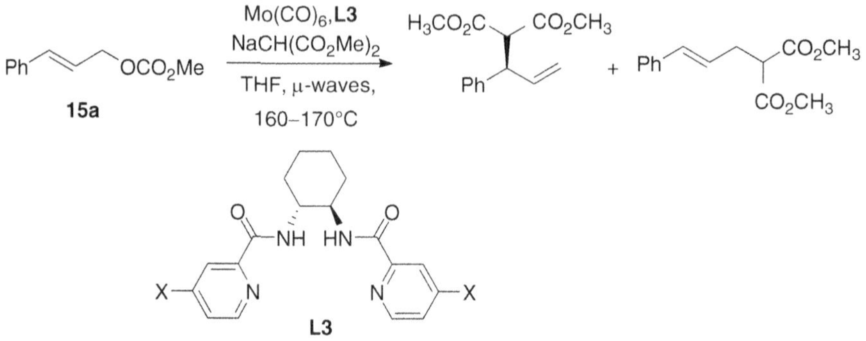

Entry	Ligand	X	Yield (GC, %)	b:l	ee (%) of b
1	**L2**	H	82	19:1	98
2	**L3-1**	OCH_3	>95	41:1	>99
3	**L3-2**	Cl	89	74:1	96
4	**L3-3**	$N(CH_2)_4$	91	88:1	96

Table 4 Microwave-mediated allylations

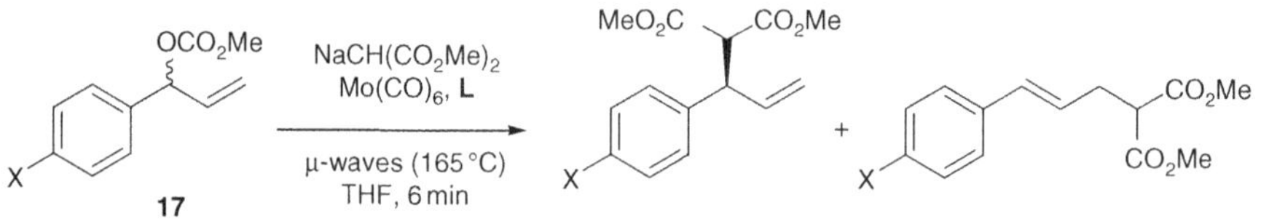

Entry	Substrate	X	Ligand	Yield (%)	b:l	ee
1	**17a (15a)**	H	**L2**	76	13:1	96
2	**17a (15a)**	H	**L3-2**	89	69:1	86
3	**17b**	Cl	**L2**	78	26:1	96
4	**17b**	Cl	**L3-2**	70	34:1	74
5	**17c**	CF_3	**L2**	48	10:1	98
6	**17c**	CF_3	**L3-2**	52	20:1	90

can be used; at lower temperatures a more reactive, and more unstable, molybdenum source is required.

Ligand **L3-2** was also used in reactions of some branched *para*-substituted allylic carbonates **17** and the results compared to those obtained using parent ligand **L2** under the same conditions (Table 4). Higher regioselectivity but lower enantioselectivity were observed using the chloro-substituted ligand [36].

The reaction with the *p*-chloro-substituted substrate **17b** was applied as a key step in the synthesis of (*R*)-baclofen, a derivative of γ-aminobutyric acid (Scheme 10) [37]. (*R*)-Baclofen is the main inhibitory neurotransmitter in the central nervous system. It activates $GABA_B$ receptors and is used for the treatment of spastic

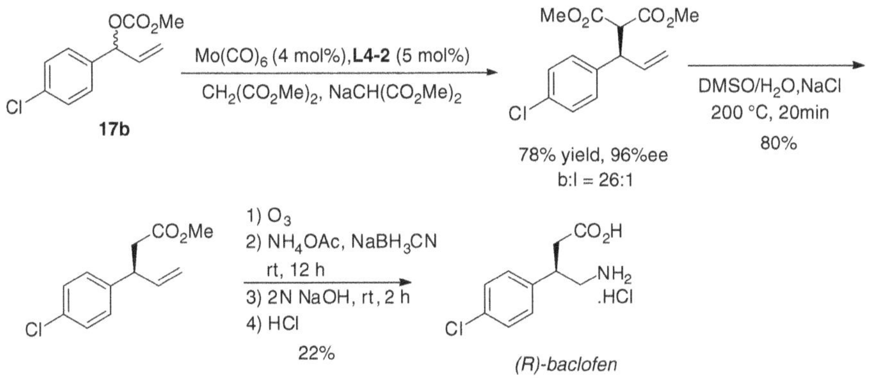

Scheme 10 Synthesis of baclofen via Mo-catalyzed enantioselective allylation

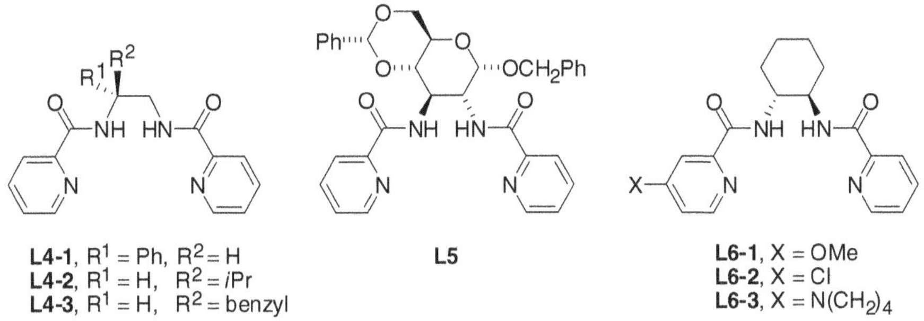

Fig. 4 Bispyridylamides with C1 symmetry

movement [38]. The compound is presently in clinical trials for treatment of fragile X and autism.

Twofold rotational symmetry in the ligands is not a prerequisite for successful results (Fig. 4). Monosubstituted diamines were used for the preparation of C_1-symmetric ligands **L4** with a single stereocenter [24, 39]. In allylations of cinnamyl carbonates and a few heteroaromatic derivatives, the regio- and stereoselectivities were high (Table 5, entries 1–3), although slightly lower than those observed with ligand **L2** (Table 2). Unsymmetrical bispyridylmethanes were also obtained by the use of diamines prepared from carbohydrates [40]. The best performance was shown by a ligand derived from glucose (**L5**). Bispyridylamides **L6** containing one pyridine ring with a π-donating substituent in 4-position led to products with very high regio- and enantioselectivities (Table 5, entries 5–7) [37].

Replacement of one of the pyridine nuclei with a phenyl ring also provided a ligand which led to high regio- and stereoselectivity, actually higher than observed with the parent ligand **L2**. Use of the unsymmetrical ligand in the reaction of **15a** with dimethyl malonate gave 90% of a product with a branched to linear ratio of 60:1 and an ee of the branched product of 99% [41]. Replacement of one pyridine ring by a *tert*-buyl group also resulted in a selective ligand, while a ligand

Table 5 Mo-catalyzed allylations using C_1-symmetric ligands **L4–L6**

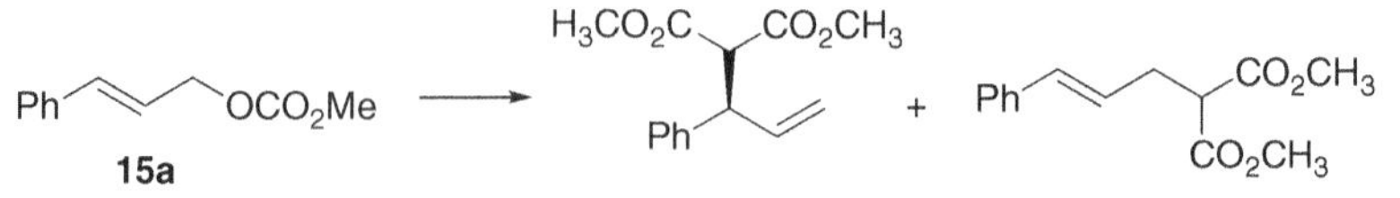

Entry	Ligand	Conditions	Yield (%)	b:l	ee (%) of b
1	**L4-1**	60°C, 4 h	65	8:1	92 (*S*)
2	**L4-2**	60°C, 12 h	68	32:1	98 (*R*)
3	**L4-3**	60°C, 4 h	69	13:1	89 (*R*)
4	**L5**	μ-Waves, 6 min	90	49:1	99 (*R*)
5	**L6-1**	μ-Waves, 6 min	90	98:1	97 (*R*)
6	**L6-2**	μ-Waves, 6 min	89	74:1	97 (*R*)
7	**L6-3**	μ-Waves, 6 min	89	75:1	99 (*R*)

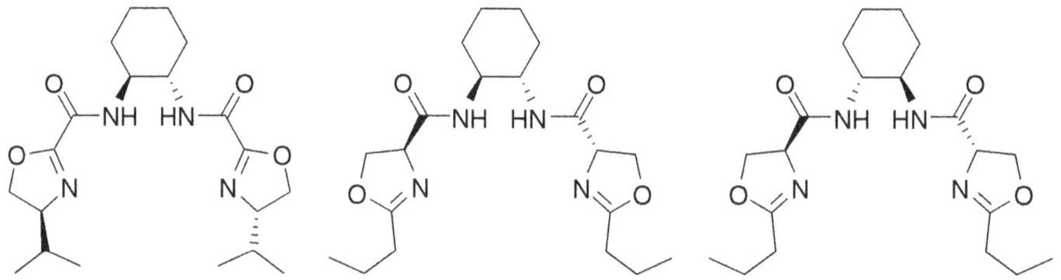

Fig. 5 Bisdihydrooxazole ligands

containing a quinoline ring probably was too sterically hindered, as indicated by a low yield of the product.

Several ligands have also been prepared where one or both amide functions were replaced by ester groups. A ligand with one ester group generated a poor catalyst [24], and replacement of both amide groups destroyed the catalytic activity [41]. Methylation of the amide groups also resulted in a ligand with very poor activity [41].

6.2.2 Bisdihydrooxazoleamides

C_2-Symmetric ligands containing dihydrooxazole rings in place of pyridine rings were prepared by Pfaltz and co-workers and assessed in molybdenum-catalyzed allylic alkylations employing $Mo(CO)_3(EtCN)_3$ as catalyst precursor [42, 43]. The presence of four stereocenters allows for wide structural variations. Among the best and generally useful ligands were **L7** and **L8** (Fig. 5). The reactions using these ligands were slower than those employing ligand **L2**, and the branched to linear ratios were generally lower (Table 6). As expected, allylic acetates were less

Table 6 Mo-catalyzed allylations of aliphatic substrates using bisdihydrooxazoleamides

R–CH=CH–CH$_2$–OCO$_2$Me **15a, 18** or R–CH(OCO$_2$Me)–CH=CH$_2$ **19** → [Mo(CO)$_3$(EtCN)$_3$, **L**, CH$_2$(CO$_2$Me)$_2$, NaH; THF, 70 °C] → (H$_3$CO$_2$C)(CO$_2$CH$_3$)CH–CH(R)–CH=CH$_2$ + R–CH=CH–CH$_2$–CH(CO$_2$CH$_3$)CO$_2$CH$_3$

Entry	Substrate	R	Ligand	Time (d)	Yield (%)	b:l	ee (%) of b
1	**15a**	Ph	**L7**	0.5	86	14:1	99 (*R*)
2	**15a**	Ph	**L8-1**	1	83	6:1	98 (*R*)
3	**18a**	Me	**L2**	1	85	5:1	94 (*R*)
4	**18a**	Me	**L7**	1	88	1.5:1	94 (*R*)
5	**18a**	Me	**L8-1**	1	81	9:1	97 (*R*)
6	**18a**	Me	**L8-2**	2	80	11:1	96 (*S*)
7	**19a**	Me	**L8-1**	1	83	5:1	80 (*R*)
8	**18b**	Pr	**L2**	1.5	80	8:1	98 (+)
9	**18b**	Pr	**L8-1**	2	69	2:1	96 (+)
10	**18b**	Pr	**L8-2**	1.5	84	8:1	98 (−)
11	**18c**	OPh	**L8-1**	2	79	>20:1	98 (−)

reactive than allylic carbonates, the former electrophiles requiring a reaction temperature of 110°C in toluene. The regio- and enantioselectivities were, however, similar to those observed in reactions with allylic carbonates. Different diastereomers of the ligands gave different results. The C_2-symmetric diastereomer of **L7**, for example, was considerable less reactive and less selective than **L7**.

Reactions with cinnamyl carbonate **15a** and several allylic carbonates with aliphatic substituents, **18** and **19**, were studied (Table 6). The selectivities were equal to or higher than those observed using ligand **L2**.

7 Double Stereoinduction

Use of prochiral nucleophiles in combination with prochiral or racemic electrophiles introduces additional complexity since diastereomeric products may be obtained as a result of enantiocontrol of both the electrophile and the nucleophile. A variety of stabilized prochiral enolates were found to exhibit high reactivity as nucleophiles. Thus, glycine ester **20** served as an excellent nucleophile in the reaction with cinnamyl phosphate **14a** (Scheme 11) [44]. The product was obtained with excellent enantioselectivity (98% ee) and diastereoselectivity (dr 20:1) but modest regioselectivity (2:1 in favor of the branched product).

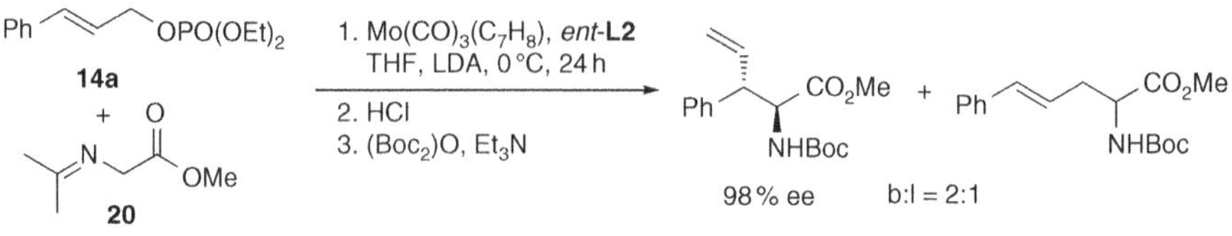

Scheme 11 Enantioselective allylation of glycine esters **20**

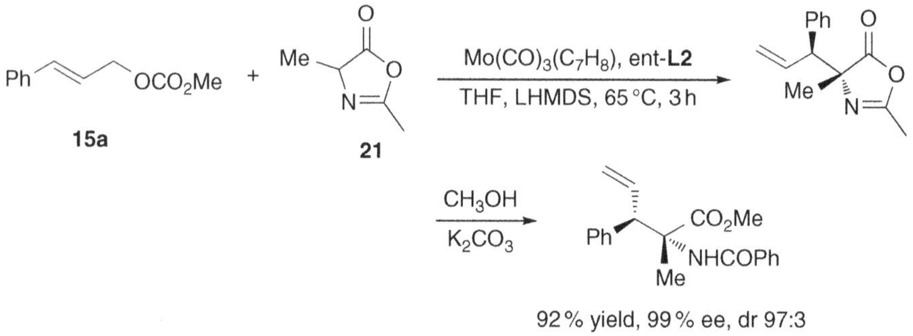

Scheme 12 Enantioselective allylation of azlactones

Azlactones provided products with equally high enantio- and diastereoselectivity and, in contrast to **20**, also high regioselectivity (Scheme 12) [44]. Allylic substrates with carbonate or phosphate as leaving group reacted with the enolate of α-substituted azlactones **21**, prepared using lithium hexamethyldisilamide, to give only the branched product; use of the sodium and potassium enolates resulted in formation of some linear product. The product was transformed to a quaternary amino acid by treatment with basic methanol. The two steps could be performed in one pot, resulting in 92% yield of product with 99% ee and a diastereomeric ratio of 97:3. Allylic carbonates with a variety of aryl substituents were combined with substituted azlactones to yield quaternary amino acids in yields of 76–92% and excellent enantiomeric and diastereomeric purity. The reaction complements the analogous palladium-catalyzed process, which gives products with the alternative regiochemistry [45].

5*H*-Alkyl-2-phenyl-oxazol-4-ones (oxalactimes) **22** reacted analogously to provide high yields of products, which could be transformed to highly enantioenriched α-hydroxyamides, and further to α-hydroxycarboxylic acids (Table 7) [46].

3,3-Disubstituted oxindoles, with quaternary carbon centers, are important synthetic targets since many compounds with such structures have significant biological properties. 3,3′-Dialkyloxindoles are accessible via molybdenum-catalyzed allylation of the anions of 3′-alkyloxindoles **23** by use of ligand **L2** [47]. The base has a large effect on the enantioselectivity and like in the reactions with azlactones and oxalactimes, nucleophiles with lithium as countercation provided the best results. Thus, reactions of a variety of 3-substituted and 3,5-disubstituted

Table 7 Allylations of oxalactimes

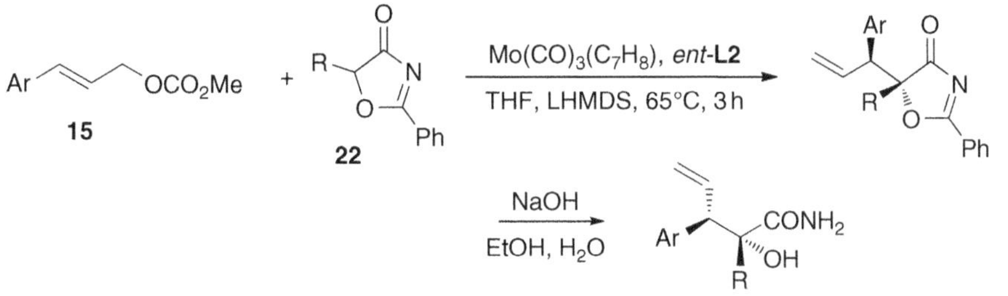

Entry	Substrate	Ar	Oxalactime	R	yield	b:l	dr	ee (%) of b
1	**15a**	Ph	**22a**	Me	91	99:1	11.5:1	>99
2	**15b**	3-Thienyl	**22b**	*i*-Pentenyl	89	14:1	10:1	99
3	**16e**[a]	2-Br-C_6H_4[a]	**22a**	Me	78	27:1	24:1	99

[a]Branched substrate was used

Table 8 Allylations of oxindoles

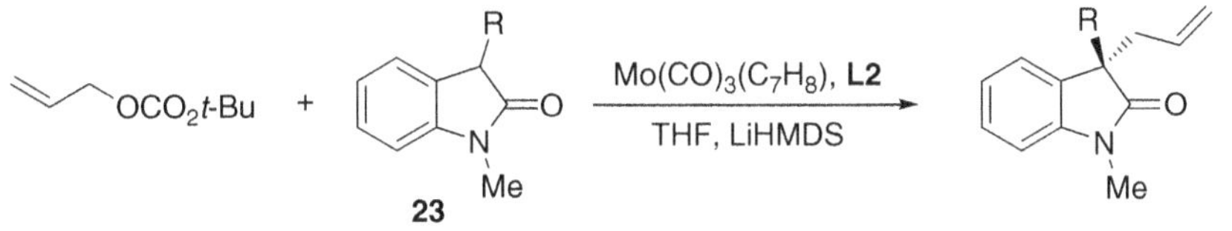

Entry	Substrate	R	ee (%)	yield (%)
1	**23a**	Me	81	99
2	**23b**	Bn	93	95
3	**23c**	*i*-Pr	91	96
4	**23d**	CH_2CN	93	99

oxindoles with allyl carbonate gave highly enantioenriched products (Table 8) [47]. The reaction was highly chemoselective, as shown for example by the lack of α-alkylation of the nitrile in substrate **23d** (entry 4).

Double stereocontrol was achieved in reactions where aryl- and alkyloxindoles were reacted with *t*-butyl cinnamic carbonate (Scheme 13) [17, 18]. Compounds with Boc-protected nitrogen provided products with modest regio- and diastereoselectivity whereas *N*-alkyl derivatives reacted with higher selectivity. The steric properties of the protecting group were not important, however, and nucleophiles with *N*-methyl, *N*-benzyl, and *N*-methoxymethyl functions led to similar results. The 3-methyl substituted oxindole reacted with poor selectivity. In contrast, the 3-phenyl derivative **23e** gave the products with high selectivity. The electronic properties of the aryl group were shown to be important. Decreased regio- and

Ph–CH=CH–CH2–OCO2t-Bu + 23e → Mo(CO)3(C7H8), L2 / THF, NaOt-Bu, RT, 8 h

88% yield, 92% ee
b:l = 18:1, 8:1 dr

Scheme 13 Stereoselective allylations of oxindoles

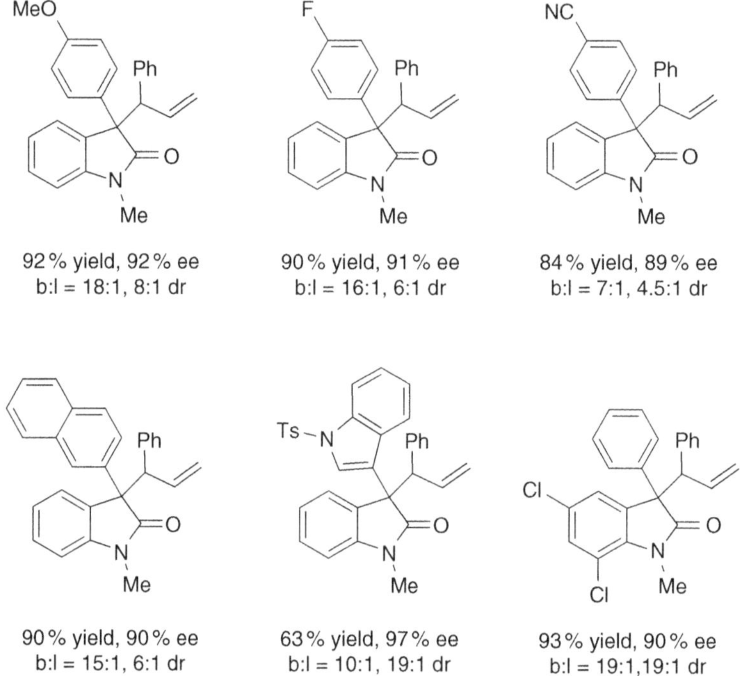

Fig. 6 Stereoselective allylations of oxindoles

diastereoselectivity were observed for compounds carrying aryl groups with electron-withdrawing groups in *para*-position, whereas electron-donating groups had little effect on the selectivity (Fig. 6). The parent ligand **L2** was used in most reactions, but improved selectivity, in particular enantioselectivity, was observed with the more electron-rich *p*-methoxy-substituted derivative **L3-1**.

The steric properties of the nucleophile had a major effect on the outcome of the reaction. Bulky nucleophiles gave products with higher diastereoselectivity (Fig. 6).

Sterically less bulky nucleophiles gave predominantly linear product. Thus, *N*-methyl-5-(2-thienyl)-oxindole gave the linear product in 90% yield, whereas *N*-methyl-5-[2-(3-methyl)-thienyl]-oxindole led to 95% product with a branched

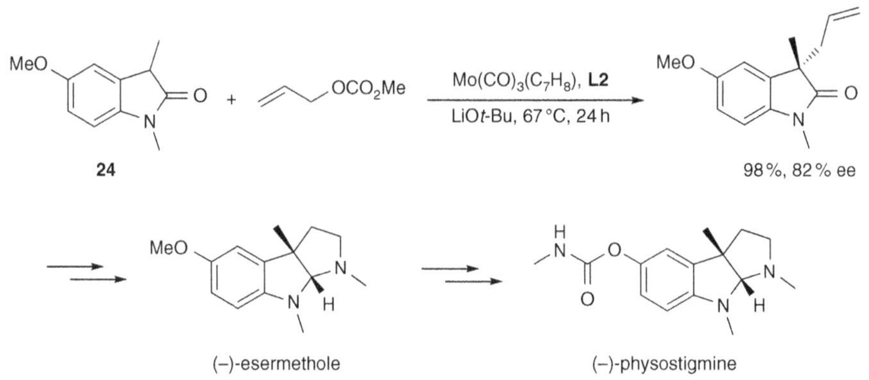

Scheme 14 Stereoselective allylations of oxindoles in natural product synthesis

to linear ratio of 11:1 (19:1 dr, 92% ee). Reaction of an alkyloxindol **24** was used as a key step in the syntheses of (−)-esermethole and (−)-physostigmine (Scheme 14) [47].

8 Mechanism

8.1 Molybdenum-Catalyzed Reactions

The mechanism of the molybdenum-catalyzed process has been investigated in some detail. The structures of complexes involved in the catalytic cycle have been studied using monopyridine ligand **L9** [48, 49]. The proposed catalytic cycle (Scheme 15) is based on results from X-ray crystallography and NMR spectroscopy [41] studies, and from DFT computations [50].

The ligand **L9** first reacts with the Mo(0) precursor (in this case Mo(norbornadiene)$(CO)_4$) to form a neutral, unsymmetrical complex (**C9**). The linear or branched allylic substrate reacts with the Mo(0) complex in an oxidative addition with formation of an 18-electron η^3-allyl Mo(II) complex (**C9-1**). The reaction involves two equivalents of **C9** and one equivalent of the allyl carbonate, and gives MeOH, CO_2, free ligand **L9**, and $Mo(CO)_6$ along with **C9-1** [51]. The structure of **C9-1** was determined by X-ray crystallography and shown to have close to octahedral coordination geometry. The weakly coordinating amide oxygen is then replaced by CO (**C9-2**), followed by coordination of the nucleophile to the metal (**C9-3**). Final reductive elimination leads to the observed product and **C9-4**, which is the resting state of the catalyst. The observed stereochemistry of the product is in accordance with pre-coordination of the nucleophile to the metal and attack at the allyl group from the side of the metal.

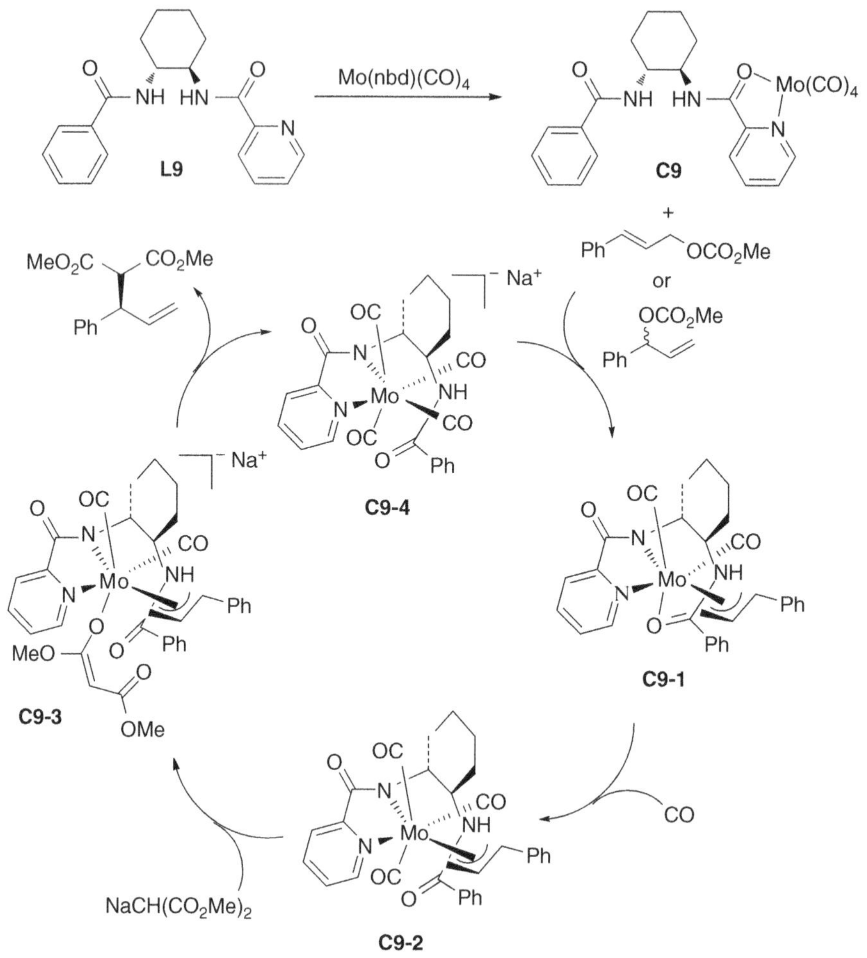

Scheme 15 Mechanistic proposal for enantioselective Mo-catalyzed allylic alkylations

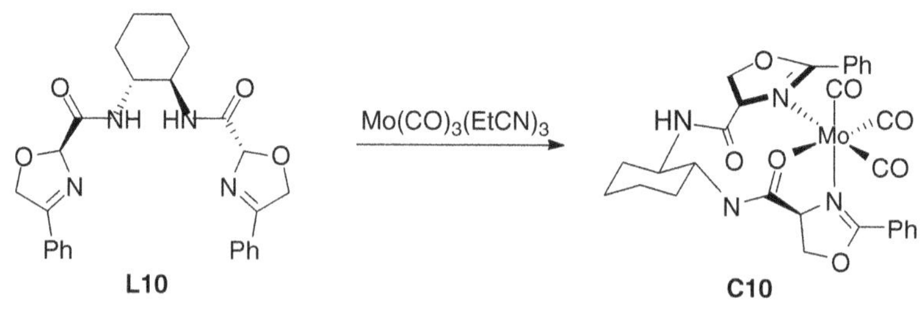

Scheme 16 Bidendate coordination of ligand **L10**

CO plays a crucial role in the reaction by activating the complex **C9-1** for nucleophilic attack; addition of malonate to **C9-1** in the absence of CO does not yield any product [51].

The structures of the complexes involved in the catalytic cycle comprising ligand **L2** have not been studied, but it has been shown that a bisdihydrooxazoleamide **L10** coordinates via both oxazole nitrogen atoms (**C10**, Scheme 16) [43]. Ligand **L2** most probably coordinates in an analogous fashion [52].

8.1.1 Memory Effects

In case equilibration of the diastereomeric allyl complexes is slower than nucleophilic attack, memory effects are observed, leading to lower enantiomeric ratios from branched substrates than observed with linear substrates.

The product obtained in the initial part of the reaction of dimethyl malonate with (1-phenyl)-2-propenyl methyl carbonate in the presence of *ent*-**L2** has a high enantiomeric ratio, which, due to a slower reaction of the mismatched substrate, decreases as the reaction proceeds (from 98 to 87% overall ee). The branched to linear ratio also decreases from 35:1 to 25:1. In contrast, the enantiomeric excess of the product obtained from the linear substrate is constant over time (97% ee) [26].

Memory effects are generally small [53], but vary with the solvent as well as with the ligand [24]. Change of solvent may thus help to improve an unsatisfactory enantioselectivity. The effect decreases with decreasing concentration of the nucleophile [49]. A more reactive nucleophile, dimethyl methylmalonate, gives rise to a larger memory effect [19] consistent with the assumption that the memory effect originates from equilibration being slow in comparison to nucleophilic attack.

8.1.2 Kinetic Resolution

When chiral racemic substrates are used in combination with chiral ligands, kinetic resolution is occasionally observed ($k_1 > k_2$). Relative rates of the reaction of the enantiomers in the range of 8–13 have been observed [26]. The difference in reaction rates of the two enantiomers is occasionally high enough to permit the isolation of highly enantioenriched substrate [54].

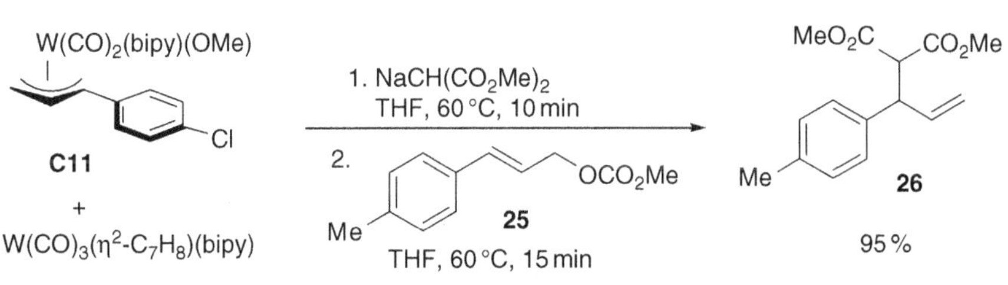

$W(CO)_2(bipy)(OMe)$ **C11** + $W(CO)_3(\eta^2\text{-}C_7H_8)(bipy)$

1. $NaCH(CO_2Me)_2$, THF, 60 °C, 10 min
2. **25**, THF, 60 °C, 15 min

26 95 %

Scheme 17 Tungsten-bipyridine catalyzed alkylation

8.2 Tungsten-Catalyzed Reactions

The tungsten-catalyzed process does not follow the conventional route for metal-catalyzed allylic alkylations, as demonstrated by the following observations: Treatment of complex **C11**, obtained along with $W(CO)_3(\eta^2\text{-}C_7H_8)(bipy)$ from (3-(*p*-chlorophenyl))-2-propenyl carbonate, bipy and $W(CO)_3(\eta^6\text{-}C_7H_8)$, with (3-(*p*-methylphenyl))-2-propenyl carbonate **25** gave the *p*-methyl-substituted product **26** (Scheme 17). The reversed order of addition of the allylic carbonates gave the *p*-chloro-substituted product [14]. These observations are evidently not in accordance with product formation taking place via nucleophilic attack at an allyl complex such as **C11**.

9 Synthetic Applications

The catalytic reactions described in this review have been illustrated by a number of synthetic applications [(−)-esermethole, (−)-physostigmine, and (*R*)-baclofen]. A few additional applications are described here.

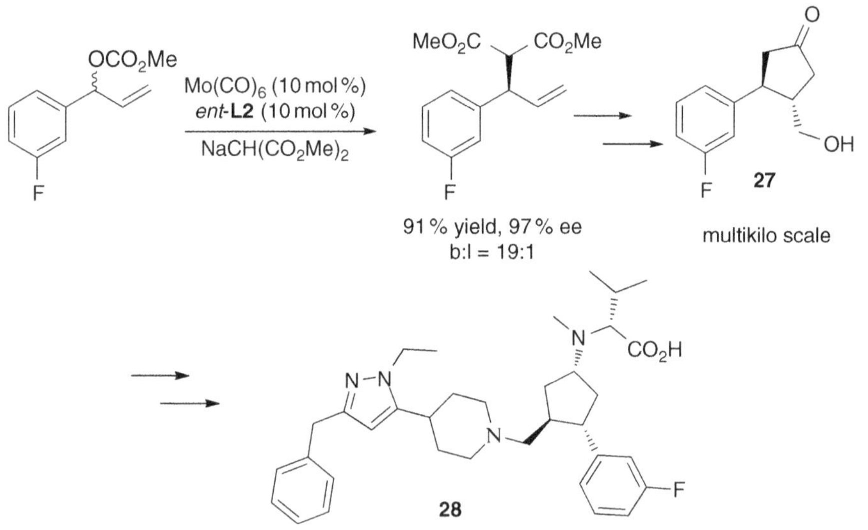

Scheme 18 Molybdenum-catalyzed allylation in the synthesis of anti-HIV drug candidate **28**

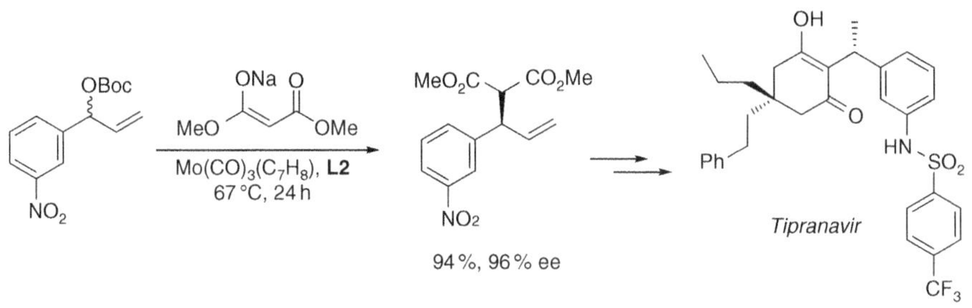

Scheme 19 Molybdenum-catalyzed allylation in the synthesis of tipranavir

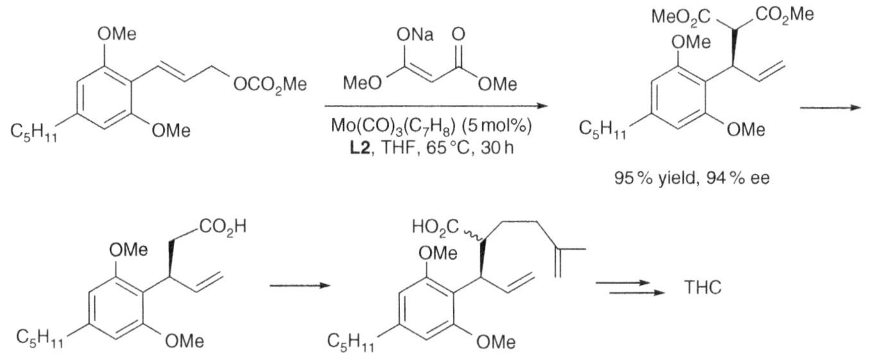

Scheme 20 Molybdenum-catalyzed allylation in the synthesis of (−)-Δ^9-*trans*-tetrahydrocannabinol

(3*S*,4*S*)-3-(3-Fluorophenyl)-4-(hydroxymethyl)cyclopentanone, **27** [55], identified as a key intermediate in the synthesis of the anti-HIV drug candidate **28**, was prepared via enantioselective molybdenum-catalyzed allylation (Scheme 18) [52].

Tipranavir, a nonpeptide HIV protease inhibitor was synthesized using molybdenum-catalyzed alkylation as a key step (Scheme 19) [56].

(−)-Δ^9-*Trans*-tetrahydrocannabinol (THC), finally, was prepared employing the same methodology (Scheme 20) [57]. These examples demonstrate that aromatic allylic carbonates with a variety of substitution patterns can be successfully used in the catalytic reaction.

10 Conclusions

Molybdenum-catalyzed allylic substitutions proceed with a reactivity complimentary to that of the more commonly used palladium catalysts. Readily available catalysts are known whereby products with high enantio- and regioselectivity can be obtained. Via the combination of prochiral nucleophiles and prochiral or racemic electrophiles, high diastereoselectivity can also be achieved. Tungsten catalysts give products with the same regiochemistry, but enantioselective reactions are possible only starting from linear, monosubstituted allylic derivatives.

References

1. Trost BM, Lautens M (1982) Molybdenum catalysts for allylic alkylation. J Am Chem Soc 104:5543
2. Trost BM, Hung M-H (1983) Tungsten-catalyzed allylic alkylations. New avenues for selectivity. J Am Chem Soc 105:7757
3. Belda O, Moberg C (2004) Molybdenum-catalyzed asymmetric allylic alkylations. Acc Chem Res 37:159
4. Pfaltz A, Lautens M (1999) In: Jacobsen EN, Pfaltz A, Yamamoto H (eds) Comprehensive asymmetric catalysis. Springer, Berlin, p 833
5. Trost BM, Crawley ML (2003) Asymmetric transition-metal-catalyzed allylic alkylations: applications in total synthesis. Chem Rev 103:2921
6. Lu Z, Ma S (2008) Metal-catalyzed enantioselective allylation in asymmetric synthesis. Angew Chem Int Ed 47:258
7. Trost BM, Tometzki GB, Hung M-H (1987) Unusual chemoselectivity using difunctional allylic alkylating agents. J Am Chem Soc 109:2176
8. Kubas GJ, Van Der Sluys LS (1990) Tricarbonyltris(nitrile) complexes of Cr, Mo, and W. Inorg Synth 28:29
9. Cotton FA, McCleverty JA, White JE (1990) Tricarbonyl(cycloheptatriene)molybdenum(0). Inorg Synth 28:45
10. Kubas GJ (1983) Preparation and use of $W(CO)_3(NCR)_3$ (R = Et, Pr) as improved starting materials for synthesis of tricarbonyl(η^6-cycloheptatriene)tungsten and other substituted carbonyl complexes. Inorg Chem 22:692
11. Kaiser N-FK, Bremberg U, Larhed M, Moberg C, Hallberg A (2000) Fast, convenient, and efficient molybdenum-catalyzed asymmetric allylic alkylation under noninert conditions: an example of microwave promoted fast chemistry. Angew Chem Int Ed 39:3596
12. Trost BM, Lautens M, Hung M-H, Carmichael CS (1984) Tandem alkylation-cycloadditions. Control by transition-metal templates. J Am Chem Soc 106:7641
13. Palucki M, Um JM, Conlon DA, Yasuda N, Hughes DL, Mao B, Wang J, Reider PJ (2001) Molybdenum-catalyzed asymmetric allylic alkylation reactions using $Mo(CO)_6$ as precatalyst. Adv Synth Catal 343:46
14. Lehmann J, Lloyd-Jones GC (1995) Regiocontrol and stereoselectivity in tungsten-bipyridine catalysed allylic alkylation. Tetrahedron 51:8863
15. Trost BM, Hung M-H (1984) On the regiochemistry of metal-catalyzed allylic alkylation: a model. J Am Chem Soc 106:6837
16. Frisell H, Åkermark B (1995) Influence of different 4,7-substituted 1,10-phenanthroline ligands on reactivity and regio- and stereocontrol in tungsten-catalyzed allylic alkylations. Organometallics 14:561
17. Trost BM, Zhang Y (2007) Mo-catalyzed regio-, diastereo-, and enantioselective allylic alkylation of 3-aryloxindoles. J Am Chem Soc 129:14548
18. Trost BM, Zhang Y (2010) Catalytic double stereoinduction in asymmetric allylic alkylation of oxindoles. Chem Eur J 16:296
19. Trost BM, Lautens M (1987) Chemoselectivity and stereocontrol in molybdenum-catalyzed allylic alkylations. J Am Chem Soc 109:1469
20. Trost BM, Lautens M (1983) Regiochemical control in the molybdenum-catalyzed reactions of trimethylsilyl- and ester-substituted allylic acetates. Organometallics 2:1687

21. Faller JW, Linebarrier D (1988) Reversal of stereochemical path in allylic alkylations promoted by palladium and molybdenum complexes. Organometallics 7:1670
22. Lloyd-Jones GC, Krska SW, Hughes DL, Gouriou L, Bonnet VD, Jack K, Sun Y, Reamer RA (2004) Conclusive evidence for a retention-retention pathway for the molybdenum-catalyzed asymmetric alkylation. J Am Chem Soc 126:702
23. Lloyd-Jones GC, Muños MP (2007) Isotopic labelling in the study of organic and organometallic mechanism and structure: an account. J Label Cmpd Radiopharm 50:1072
24. Malkov AV, Gouriou L, Lloyd-Jones GC, Starý I, Langer V, Spoor P, Vinader V, Kočovský P (2006) Asymmetric allylic substitution catalyzed by C_1-symmetrical complexes of molybdenum: structural requirements of the ligand and the stereochemical course of the reaction. Chem Eur J 12:6910
25. Trost BM (2002) Pd asymmetric allylic alkylation (AAA). A powerful synthetic tool. Chem Pharm Bull 50:1
26. Hughes DL, Palucki M, Yasuda N, Reamer RA, Reider PJ (2002) Solvent-dependent dynamic kinetic asymmetric transformation/kinetic resolution in molybdenum-catalyzed asymmetric allylic alkylations. J Org Chem 67:2762
27. Lloyd-Jones GC, Pfaltz A (1995) Chiral phosphanodihydrooxazoles in asymmetric catalysis: tungsten-catalyzed allylic substitution. Angew Chem Int Ed Engl 34:462
28. Helmchen G, Pfaltz A (2000) Phosphinooxazolines-a new class of versatile, modular P, N-ligands for asymmetric catalysis. Acc Chem Res 33:336
29. Lloyd-Jones GC, Pfaltz A (1995) Synthesis and structure of low-valent tungsten complexes bearing chiral oxazoline-derived ligands. Z Naturforsch 50b:361
30. Prétôt R, Lloyd-Jones GC, Pfaltz A (1998) Enantio- and regiocontrol in palladium- and tungsten-catalyzed allylic substitutions. Pure Appl Chem 70:1035
31. Barnes DJ, Chapman RL, Vagg RS, Watton EC (1978) Synthesis of novel bis(amides) by means of triphenyl phosphite intermediates. J Chem Eng Data 23:349
32. Trost BM, Hachiya I (1998) Asymmetric molybdenum-catalyzed alkylations. J Am Chem Soc 120:1104
33. Conlon DA, Yasuda N (2001) Practical synthesis of chiral N,Nc-Bis(2°-pyridinecarboxamide)-1,2-cyclohexane ligands. Adv Synth Catal 343:137
34. Trost BM, Hildbrand S, Dogra K (1999) Regio- and enantioselective molybdenum-catalyzed alkylations of polyenyl esters. J Am Chem Soc 121:10416
35. Belda O, Kaiser N-F, Bremberg U, Larhed M, Hallberg A, Moberg C (2000) Highly stereo- and regioselective allylations catalyzed by Mo-pyridylamide complexes. Electronic and steric effects of the ligand. J Org Chem 65:5868
36. Belda O, Moberg C (2002) Substituted pyridylamide ligands in microwave-accelerated Mo (0)-catalysed allylic alkylations. Synthesis 1601
37. Belda O, Lundgren S, Moberg C (2003) Recoverable resin-supported pyridylamide ligand for microwave-accelerated molybdenum-catalyzed asymmetric allylic alkylations: enantioselective synthesis of baclofen. Org Lett 5:2275
38. Ong J, Kerr DIB (2005) Clinical potential of $GABA_B$ receptor modulators. CNS Drug Rev 11:317
39. Malkov AV, Spoor P, Vinader V, Kočovský P (2001) Asymmetric molybdenum(0)-catalyzed allylic substitution. Tetrahedron Lett 42:509
40. Del Litto R, Benessere V, Ruffo F, Moberg C (2009) Carbohydrate-based pyridine-2-carboxamides for Mo-catalyzed asymmetric allylic alkylations. Eur J Org Chem 1352
41. Trost BM, Dogra K, Hachiya I, Emura T, Hughes DL, Krska S, Reamer RA, Palucki M, Yasuda N, Reider PJ (2002) Designed ligands as probes for the catalytic binding mode in Mo-catalyzed asymmetric allylic alkylation. Angew Chem Int Ed 41:1929
42. Glorius F, Pfaltz A (1999) Enantioselective molybdenum-catalyzed allylic alkylation using chiral bisoxazoline ligands. Org Lett 1:141
43. Glorius F, Neuburger M, Pfaltz A (2001) Highly enantio- and regioselective allylic alkylations catalyzed by chiral [bis(dihydrooxazole)]molybdenum complexes. Helv Chim Acta 84:3178

44. Trost BM, Dogra K (2002) Synthesis of novel quaternary amino acids using molybdenum-catalyzed asymmetric allylic alkylation. J Am Chem Soc 124:7256
45. Trost BM, Ariza X (1997) Catalytic asymmetric alkylation of nucleophiles: asymmetric synthesis of α-alkylated amino acids. Angew Chem Int Ed Engl 36:2635
46. Trost BM, Dogra K, Franzini M (2004) 5H-Oxazol-4-ones as building blocks for asymmetric synthesis of -hydroxycarboxylic acid derivatives. J Am Chem Soc 126:1944
47. Trost BM, Zhang Y (2006) Molybdenum-catalyzed asymmetric allylation of 3-alkyloxindoles: application to the formal total synthesis of (−)-physostigmine. J Am Chem Soc 128:4590
48. Hughes DL, Lloyd-Jones GC, Krska SW, Gouriou L, Bonnet VD, Jack K, Sun Y, Mathre DJ, Reamer RA (2004) Mechanistic studies of the molybdenum-catalyzed asymmetric alkylation reaction. Proc Natl Acad Sci USA 101:5379
49. Krska SW, Hughes DL, Reamer RA, Mathre DJ, Palucki M, Yasuda N, Sun Y, Trost BM (2004) New insights into the mechanism of molybdenum-catalyzed asymmetric alkylation. Pure Appl Chem 76:625
50. Luft JAR, Yu Z-X, Hughes DL, Lloyd-Jones GC, Krska SW, Houk KN (2006) On the stability of the π-allyl intermediate in molybdenum-catalyzed asymmetric alkylations. Tetrahedron: Asymmetry 17:716
51. Krska SW, Hughes DL, Reamer RA, Mathre DJ, Sun Y, Trost BM (2002) The unusual role of CO transfer in molybdenum-catalyzed asymmetric alkylations. J Am Chem Soc 124:12656
52. Kočovský P, Malkov AV, Vyskočil Š, Lloyd-Jones GC (1999) Transition metal catalysis in organic synthesis – reflections, chirality and new vistas. Pure Appl Chem 71:1425
53. Palucki M, Um JM, Yasuda N, Conlon DA, Tsay F-R, Hartner FW, Hsiao Y, Marcune B, Karady S, Hughes DL, Dormer PG, Reider PJ (2002) Development of a new and practical route to chiral 3,4-disubstituted cyclopentanones: asymmetric alkylation and intramolecular cyclopropanation as key C–C bond-forming steps. J Org Chem 67:5508
54. Faller JW, Sarantopoulos N (2004) Retention of configuration and regiochemistry in allylic alkylations via the memory effect. Organometallics 23:2179
55. Conlon DA, Jensen MS, Palucki M, Yasuda N, Um JM, Yang C, Hartner FW, Tsay F-R, Hsiao Y, Pye P, Rivera NR, Hughes DL (2005) Stereoselective synthesis of an anti-HIV drug candidate. Chirality 17:149
56. Trost BM, Andersen NG (2002) Utilization of molybdenum- and palladium-catalyzed dynamic kinetic asymmetric transformations for the preparation of tertiary and quaternary stereogenic centers: a concise synthesis of tipranavir. J Am Chem Soc 124:14320
57. Trost BM, Dogra K (2007) Synthesis of (−)-Δ^9-trans-tetrahydrocannabinol: stereocontrol via Mo-catalyzed asymmetric allylic alkylation reaction. Org Lett 9:861

Top Organomet Chem (2012) 38: 235–268
DOI: 10.1007/3418_2011_12

Published online: 3 July 2011

Copper-catalyzed Enantioselective Allylic Substitution

Jean-Baptiste Langlois and Alexandre Alexakis

Abstract The efficiency of organocopper reagents in the displacement of allylic leaving groups has been well established during the past five decades. In sharp contrast, catalytic asymmetric version of this reaction using a chiral catalyst is a more recent field of research. This chapter presents an overview of tremendous studies towards the development of an "ideally" active catalyst achieving high regio- and enantioselectivities. The comparative reactivity and generality of peptides, phosphorus, as well as *N*-heterocyclic carbenes based catalysts are discussed in the first part. Then, relevant scope and synthetic applications are reviewed. Noteworthily, this chapter is restricted to C–C bond formation processes, excluding C–B and C–Si bond formations.

Keywords Copper catalysis · Asymmetric catalysis · Allylic substitution · C–C bond formation · Organometallic reagents

Contents

J.-B. Langlois and A. Alexakis (✉)
Department of Organic Chemistry, University of Geneva, 30 Quai Ernest Ansermet, 1211 Geneva 4, Switzerland
e-mail: alexandre.alexakis@unige.ch

1 Introduction

During the last decades, the demand for fine-chemicals has considerably increased. For instance, the development of new potent pharmaceuticals is of major interest in our society. These expectations required the construction of stereodefined chiral centers, which are often the basis of the activity of the molecule. In response to this, organic chemists need powerful tools, which are cheap, easy-to-use, and able to provide the desired product in high level of selectivity. In this view, asymmetric processes catalyzed by various transition metals have proven their efficacy [1, 2]. Copper is among the cheapest and most readily available catalyst. Used in combination with organometallic reagents (Li, Mg, Zn, Al), it is one of the best methods to form an allylic C–C bond [3–7].

This reaction has been thoroughly studied by the research groups of Bäckvall [8–12], Goering [13–27], Nakamura [28–31] and others. After four decades of extensive work, a significant knowledge has been acquired, allowing for the establishment of the following mechanism (Scheme 1). A copper (I) species forms a π-complex

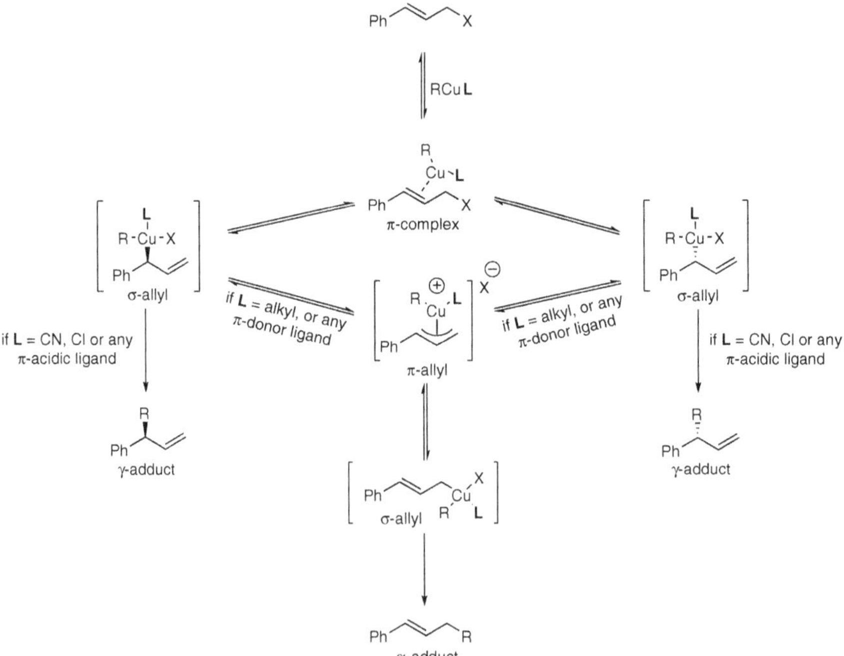

Scheme 1 Proposed reaction mechanism

with the substrate, which favors the oxidative addition at the γ-position. The newly formed isomeric σ-allylcopper (III) complexes might react under two distinct pathways. If ligand **L** is an electron acceptor (CN or any π-acidic ligand) the reductive elimination step will be promoted, affording products of γ-alkylation. However, if ligand **L** is an electron donor (alkyl or any π-donor ligand), the reductive elimination step will be retarded allowing for the formation of a π-allylcopper (III) complex. Such a complex might lead to the formation of the product of α-alkylation after π–σ equilibrations. It is worth to note that even though the oxidative addition occurred in γ-position, a mixture of both regioisomers might be obtained at the end of the reaction. The control of this regioselectivity as well as the enantioselectivity of the γ-adduct are fascinating challenges for organic chemists. In this view, very efficient diastereoselective processes have been developed giving rise to tremendous applications in total synthesis [32, 33]. All these studies have been carefully reviewed and it will not be the purpose of this chapter. We will focus our attention on enantioselective processes of copper-catalyzed allylic substitution [33–43]. Indeed, owing to the predominant role of **L** in this process, significant efforts have been dedicated to the development and the design of suitable chiral ligands that enable the control of both the regio- and the enantioselectivities of the reaction. In the first part, we will detail all the different catalytic systems that have been reported in the literature. Then, a large overview of the reaction scope will be presented.

2 Development of Enantioselective Processes

2.1 Arenethiolatocopper (I)

The first report on enantioselective copper-catalyzed allylic substitutions came in 1995 [44–47]. Bäckvall used a copper(I)-complex of arenethiolate ligand **L1**, developed by van Koten, as chiral catalyst for the alkylation of (*E*)-3-cyclohexyl-2-propenylacetate (**1**) with *n*-BuMgI. A perfect γ-selectivity was observed as well

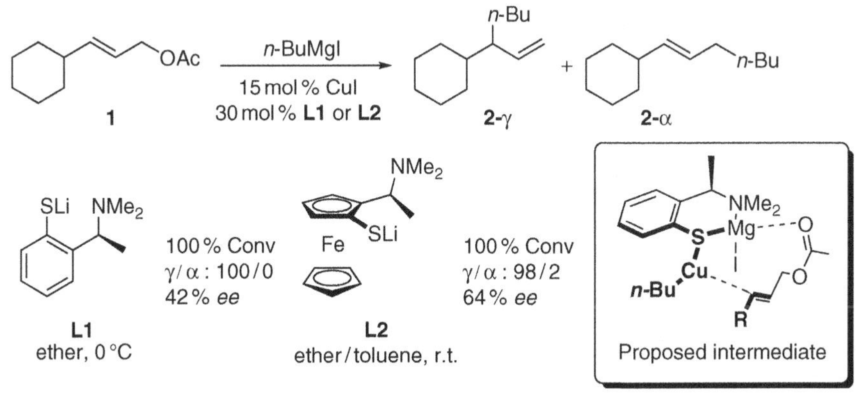

OAc
n-BuMgI
15 mol % CuI
30 mol % **L1** or **L2**
1
n-Bu
2-γ
+
n-Bu
2-α
SLi NMe$_2$
L1
ether, 0 °C
100 % Conv
γ/α : 100/0
42 % *ee*
NMe$_2$
Fe SLi
L2
ether/toluene, r.t.
100 % Conv
γ/α : 98/2
64 % *ee*
NMe$_2$
S Mg O
Cu
n-Bu
O
R
Proposed intermediate

Scheme 2 Arenethiolatocopper (I), Bäckvall and van Koten 1995

as an enantiomeric induction of 42% (Scheme 2). This value was improved to 64% *ee* using a new set of conditions and the ferrocenyl-based second generation ligand **L2**. The proposed reaction intermediate is depicted in Scheme 2. The copper forms a π-complex with the olefin favoring the γ-substitution. The magnesium atom is coordinated to the carbonyl oxygen of the acetate allowing for an electrophilic activation of the starting material.

2.2 Primary and Secondary Chiral Amines

In 1999, Knochel and Dübner described a highly enantioselective reaction using primary ferrocenyl amine **L3** as chiral ligand (Scheme 3) [48, 49]. Even though this process was limited to the addition of hindered dialkylzinc reagents such as $(neo\text{-pentyl})_2Zn$, 82% *ee* was obtained along with a γ-selectivity of 95% for the reaction of cinnamyl chloride **3**. Thereafter, a fine-tuning of the chiral ligand revealed the bulky amine **L4** as optimal catalyst and 96% *ee* was attained.

Inspired by this work, the group of Woodward developed in 2005 a process employing the chiral secondary amine **L5** [50]. They applied it to the alkylation of Baylis–Hillman adducts with diethylzinc in the presence of 5 mol% of CuTC (copper (I) thiophenecarboxylate). Enantioselectivities ranging from 76% to 90% *ee* were obtained for a large range of substrates (Scheme 4). Of note is the use of methylaluminoxane (MAO; $[\text{-Al(Me)O}]_n$), which acts as a strong halide scavenger in order to displace the zinc-Schlenk equilibrium through the formation of Et_2Zn (by trapping $ZnCl_2$) rather than EtZnCl. The exact role of the amine remained unclear but different assumptions can be invoked. Indeed, the chiral amine might simply act as a copper ligand but also as a chiral leaving group displacing the allylic chloride.

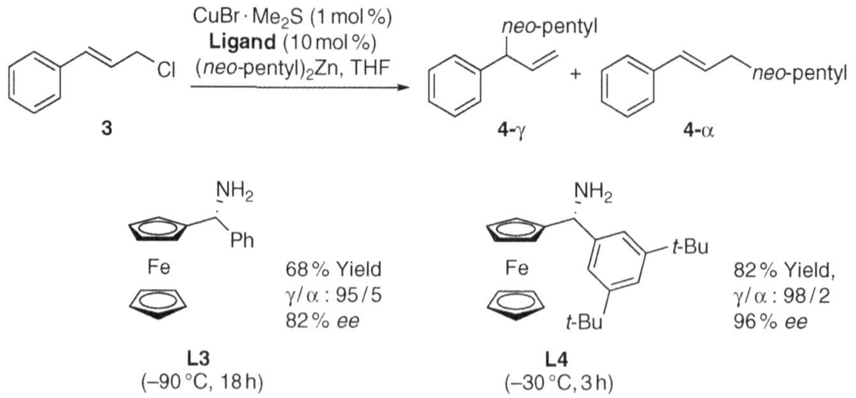

CuBr · Me_2S (1 mol %)
Ligand (10 mol %)
$(neo\text{-pentyl})_2Zn$, THF

Cl → neo-pentyl + neo-pentyl

3 **4-γ** **4-α**

NH_2 Ph Fe — 68 % Yield, γ/α : 95/5, 82 % *ee*

L3 (–90 °C, 18 h)

NH_2 t-Bu Fe t-Bu — 82 % Yield, γ/α : 98/2, 96 % *ee*

L4 (–30 °C, 3 h)

Scheme 3 Chiral primary amines, Knochel 1999

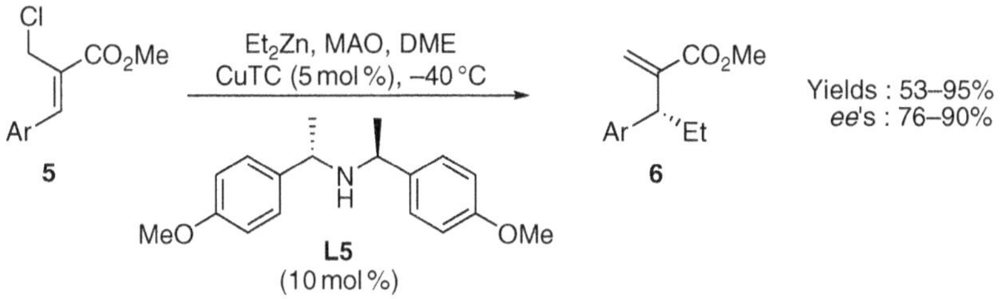

Scheme 4 Chiral secondary amines, Woodward 2005

2.3 Binol-Based Catalyst

In 2000, Woodward and Gladiali described a binol-based catalyst **L6** for the asymmetric alkylation of Baylis–Hillman adducts (Scheme 5) [51]. This represents the unique example of the use of chiral BINOL in copper-catalyzed allylic alkylation (*ee*'s up to 64%).

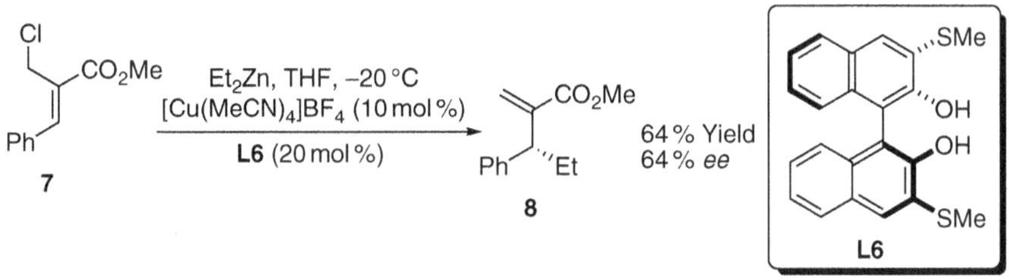

Scheme 5 Binol-based catalyst, Woodward 2000

2.4 Chiral Phosphite Ligands

After an extensive study of phosphorus-based ligands, Alexakis and co-workers reported in 2001 the simple phosphite ligand **L7** incorporating a chiral TADDOL backbone and a *N*-methylephedrine moiety (Scheme 6) [52, 53]. In combination with 1 mol% of CuCN, this ligand allowed for the addition of simple Grignard reagents to cinnamyl chloride **3** in good enantioselectivity around 73%. Replacing the copper source by CuTC afforded alkylation product **9**-γ in 82% *ee* and 94/6 regioselectivity. It is worth noting that without ligand, CuTC promoted the formation of **9**-α, suggesting that the use of phosphorus ligand totally changes the properties of the organocopper species.

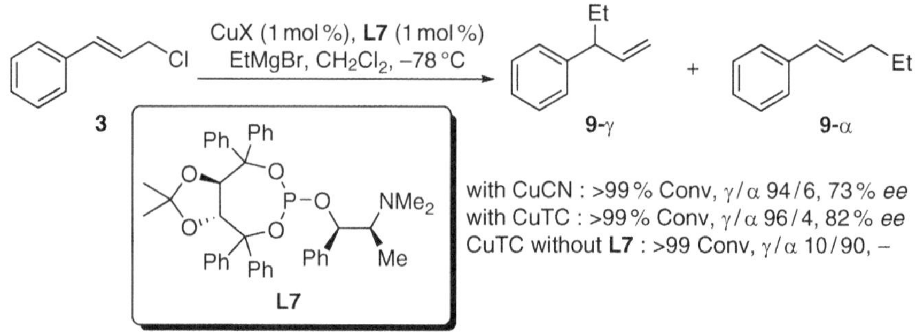

CuX (1 mol%), **L7** (1 mol%)
EtMgBr, CH_2Cl_2, −78 °C

with CuCN : >99% Conv, γ/α 94/6, 73% *ee*
with CuTC : >99% Conv, γ/α 96/4, 82% *ee*
CuTC without **L7** : >99 Conv, γ/α 10/90, –

Scheme 6 Chiral phosphite ligand, Alexakis 2001

2.5 Chiral Phosphoramidite Ligands

Parallel to this study, the group of Feringa successfully introduced a phosphoramidite ligand composed of a BINOL backbone and a *C2*-symmetric amine (eq 1, Scheme 7) [54]. They demonstrated that both axial and central chiralities of this ligand were required to get high asymmetric induction. The promising results obtained for the addition of diethylzinc, combined with the high modularity of this family of ligand, rapidly attracted the attention of several research groups. Thus, a wide range of phosphoramidites were prepared and evaluated in this reaction. For instance, Zhou and co-workers adapted their original spirobindane-7,7-diol to the Feringa's ligand (**L9**, eq 2, Scheme 7) [55]. Evaluated in similar conditions than the one previously developed by Feringa, 71% *ee* and a γ/α ratio of 88/12 were attained. Then, after an impressive screening of phosphoramidites, Feringa improved these results to 86% *ee* and 93/7, using a hydrogenated analogue of **L8** (**L10**, eq 3, Scheme 7) [56].

In the same time, the group of Alexakis has also participated to this "phosphoramidite rush" but focused its attention to the reactivity of organomagnesium reagent. In 2002, they replaced the BINOL backbone of the Feringa's ligand by a biphenol moiety and applied this new ligand **L11** to the alkylation of substrate **3** (Scheme 8) [53]. Enantioselectivities of 79% and 83% were respectively obtained for the introduction of an ethyl and an *iso*propyl group. This level of enantioinduction was kept whatever the degree and the nature of the substitution of the aryl group initially present on the substrate. Later on, Alexakis disclosed a BINOL-based ligand **L12** possessing an electron-donating group (*o*-MeO) on the amine moiety [57–59]. This ligand, now commercially available, proved to be highly effective, providing *ee*'s up to 96% as well as high regioselectivity in the model reaction. Furthermore, 96% *ee* and good regioselectivity were obtained for the challenging allylic methylation of substrate **3**.

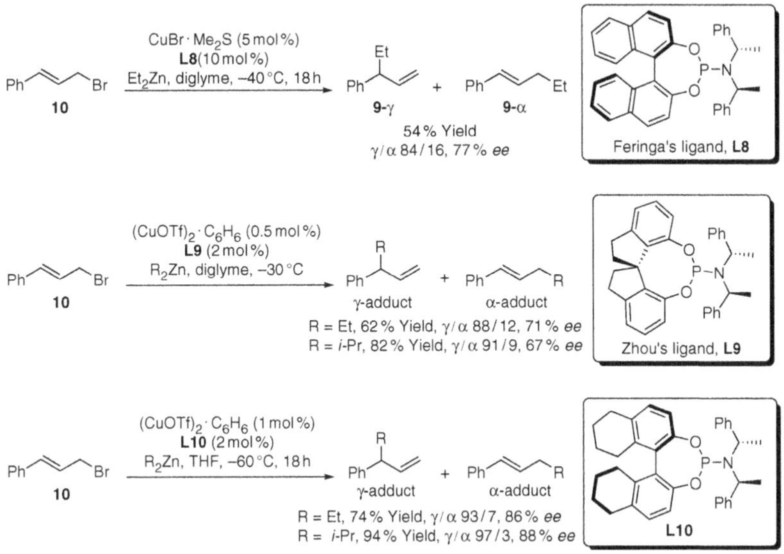

Ph⁀⁀Br (10) — CuBr·Me_2S (5 mol%), **L8** (10 mol%), Et_2Zn, diglyme, –40 °C, 18 h → **9-γ** + **9-α**
54% Yield
γ/α 84/16, 77% *ee*
Feringa's ligand, **L8** (eq1)

Ph⁀⁀Br (10) — $(CuOTf)_2 \cdot C_6H_6$ (0.5 mol%), **L9** (2 mol%), R_2Zn, diglyme, –30 °C → γ-adduct + α-adduct
R = Et, 62% Yield, γ/α 88/12, 71% *ee*
R = *i*-Pr, 82% Yield, γ/α 91/9, 67% *ee*
Zhou's ligand, **L9** (eq2)

Ph⁀⁀Br (10) — $(CuOTf)_2 \cdot C_6H_6$ (1 mol%), **L10** (2 mol%), R_2Zn, THF, –60 °C, 18 h → γ-adduct + α-adduct
R = Et, 74% Yield, γ/α 93/7, 86% *ee*
R = *i*-Pr, 94% Yield, γ/α 97/3, 88% *ee*
L10 (eq3)

Scheme 7 Phosphoramidites and organozinc reagents, Feringa and Zhou 2001–2004

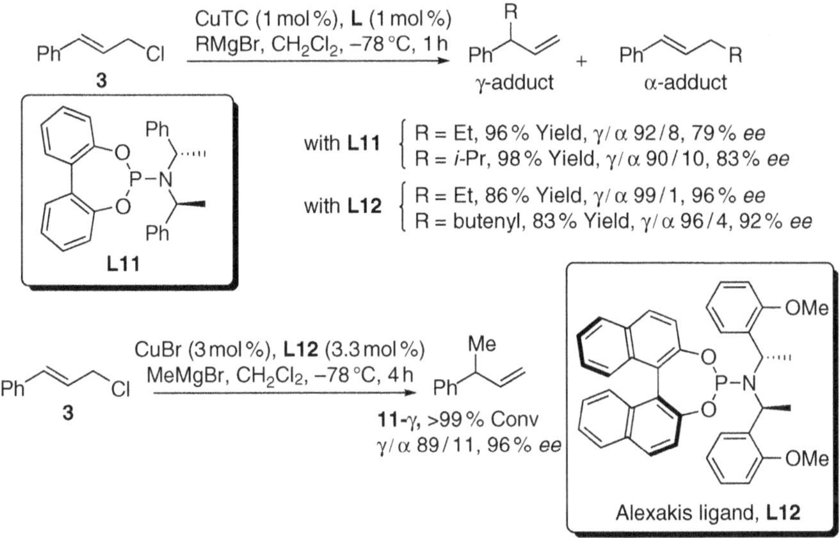

Scheme 8 Phosphoramidites and Grignard reagents, Alexakis 2002–2004

2.6 Peptide-Based Ligands

Efforts have been directed toward the design of new chiral ligands that would be easy to prepare and combine high level of reactivity and selectivity. For this purpose, Hoveyda developed in 2001 a readily modular and non-*C*2-symmetric

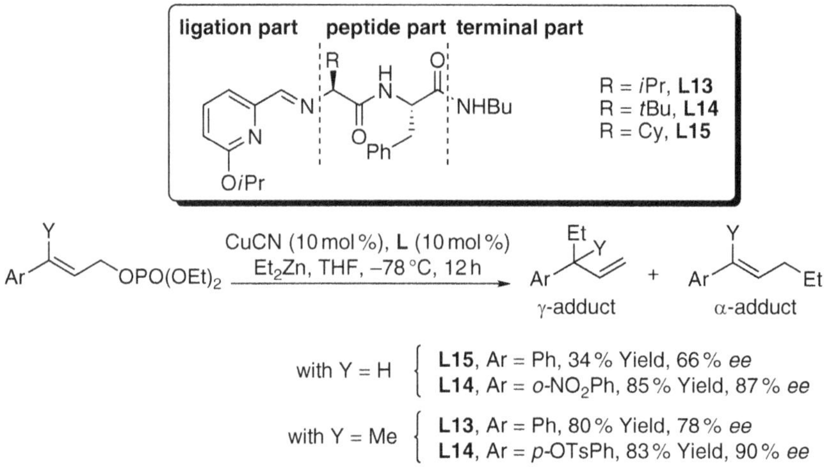

Scheme 9 First generation of peptide-based ligands, Hoveyda 2001

peptide-based family of ligands (**L13–L15**) (Scheme 9) [60]. These catalysts contained three distinct parts: (1) the ligation part, possessing a Schiff base and a pyridine moiety which constitutes a second point of coordination, (2) a chain of two amino acids forming the peptide unit needed for the asymmetric induction, (3) the terminal part which must be a *N*-butylamine to obtain good results.

This combination led to an interesting level of enantioselectivity and high regioselectivity for the alkylation of cinnamyl phosphates with diethylzinc. Noteworthy, better *ee*'s were obtained with electron-deficient substrates (87% *ee* for *o*-NO_2Ph, Scheme 9). Two years later, some modifications of the ligand structure (**L16**) gave rise to significant improvement of the enantioselectivity to 96% *ee* (Scheme 10) ([61], for a review see [62]). For the first time, the scope of the reaction was large, proceeding with the same efficacy for various alkyl, alkenyl, alkynyl, and aryl allylic phosphates. Moreover, the impressive potential of this class of ligands was highlighted by the formation of quaternary centers. This reaction is still very challenging and so far only few processes enable the generation of highly enantioenriched quaternary centers. In this case, a regioselectivity of 30/1 and an enantioinduction up to 92% were attained. A key reaction intermediate was proposed in order to explain such results. The alkylcopper species is bound to the oxygen of the naphthol unit while being coordinated by the Schiff base rigidifying the system. Then, the substrate approaches from the top face and is stabilized by the formation of a π-complex as well as the coordination of the oxygen atom of the terminal amide group. The chirality of the internal amino acid induced preferential approach of the substrate from a single enantiotopic face due to steric interactions with the substituents of the olefin (Scheme 10).

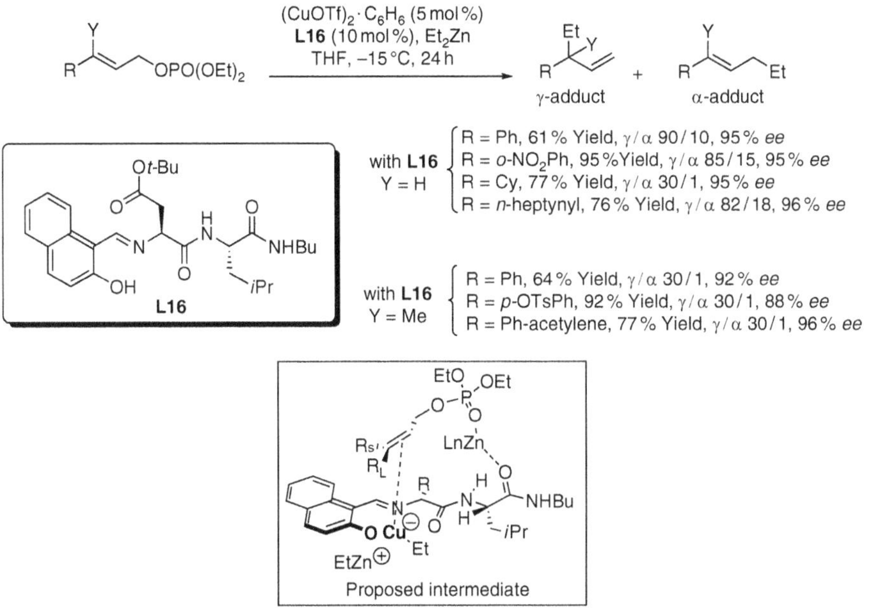

Scheme 10 Second generation of peptide-based ligands, Hoveyda 2004

2.7 Phosphine-Based Ligands

Parallel to this study, the group of Gennari described a series of aminosulfonamide phosphine ligands such as **L17** [63]. Although they present some similarities with the ligands previously reported by Hoveyda, some specific features must be noted. The pyridine moiety was replaced by a diphenylphosphine, the Schiff base by an amide and the terminal amide by a chiral sulfonamide group. Based on the same concept than the ligand of Hoveyda, the great advantages of this family are their high modularity and availability. Promising results were obtained in terms of activity in the model reaction using allylic phosphate **12** and diethylzinc (γ/α ratio of 90/10 and up to 40% *ee*, Scheme 11).

In 2006, Feringa and co-workers reported the use of commercially available Taniaphos ligand **L18** in the alkylation of various allylic bromides by Grignard reagents (Scheme 12) [64]. The wide reaction scope combined with the high selectivities contribute to make this process particularly relevant in this field. The reaction occurred smoothly for aryl and alkyl allylic bromides. The challenging introduction of a methyl group proved to be efficient, reaching almost perfect regioselectivity and 98% *ee*.

More recently, Tomioka described the amidophosphine ligand **L19**, readily prepared from (*S*)-proline (Scheme 13) [65]. A good asymmetric induction (91% *ee*) was obtained for the alkylation of cinnamyl bromide (**10**), albeit with a moderate regioselectivity (γ/α ratio of 65/35). Interestingly, this catalytic system

$(CuOTf)_2 . C_6H_6$(5 mol %)
L17 (10 mol %), Et_2Zn
THF, −78 to −55 °C

Ph–CH=CH–CH$_2$–OPO(OEt)$_2$ **12** → **9-γ** + **9-α**

L17

93 % Conv
γ/α 90/10, 40 % *ee*

Scheme 11 Aminosulfonamide phosphine ligand, Gennari 2002

CuBr · Me_2S (1 mol %)
L18 (1.1 mol %), R′MgBr
CH_2Cl_2, −78 °C, 12 h

γ-adduct + α-adduct

L18

R = Ph, R′ = Et, 92 % Yield, γ/α 81/19, 95 % *ee*
R = Napht, R′ = Et, 86 % Yield, γ/α 87/13, 90 % *ee*
R = Ph, R′ = Me, 91 % Yield, γ/α 97/3, 98 % *ee*
R = $BnOCH_2$, R′ = Me, 93 % Yield, γ/α 100/0, 92 % *ee*

Scheme 12 Taniaphos ligand, Feringa 2006

CuX, L19, R′MgBr
CH_2Cl_2, −78 °C

γ-adduct + α-adduct

L19

with $[Cu(MeCN)_4]BF_4$ (5 mol %)
L19 (6 mol %), R = Ph, R′ = Et
95 % Yield, γ/α 62/38, 91 % ee

with CuTC (2 mol %)
L19 (4.4 mol %), R = Cy, R′ = Ph
99 % Yield, γ/α 76/24, 81 % ee

Scheme 13 Amidophosphine ligand, Tomioka 2008

also allowed for the arylation of various allylic substrates with comparable efficiency (up to 81% *ee*). This represented the first example of copper-catalyzed asymmetric allylic arylation.

After many years of theoretical mechanistic studies of the present reaction [5], Nakamura postulated that, to be effective, a chiral ligand should coordinate the copper but also the Lewis acidic metal atom of the organometallic reagent. To this end, he introduced in 2009 the aminohydroxyphosphines such as **L20** in the ethylation reaction of cinnamyl phosphate derivatives (Scheme 14) [66]. Careful investigations

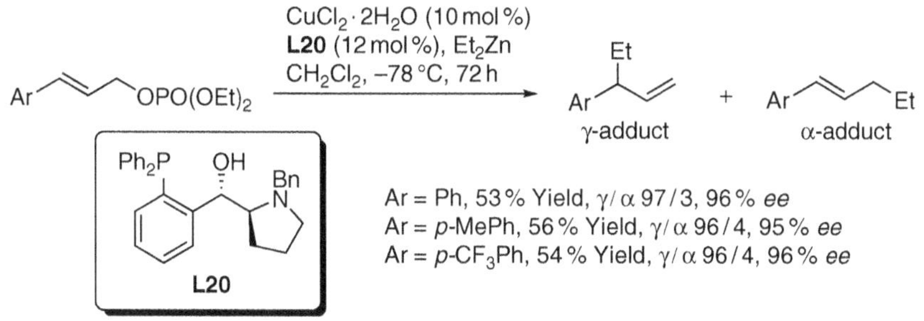

Scheme 14 Aminohydroxyphosphine ligand, Nakamura 2009

identified $CuCl_2 \cdot 2\ H_2O$ as the best copper source in dichloromethane at −78°C. It is worth to note that even though a copper (II) salt is used, the active catalyst is assumed to be a copper (I) species, stemming from an in situ reduction of the initial catalyst by the organometallic reagent. Finally, the high selectivities obtained were rationalized by a computational modeling study using DFT calculations.

2.8 N-Heterocyclic Carbenes

N-heterocyclic carbenes (NHC) have received a broad attention during the last decade [67–77]. Particular features such as: (1) strong σ-donor properties, (2) high modularity, (3) possibility to bring the chirality closer to the reaction site compared to phosphine ligands, contribute to make these ligands interesting candidates for asymmetric catalysis. As such, the groups of Okamoto and Hoveyda pioneered their use in copper-catalyzed asymmetric allylic substitution. In 2004, Okamoto and co-workers employed the NHC complex **C21**, previously described by Alexakis in copper-catalyzed conjugate addition, to perform the alkylation of allylic pyridyl ether (Scheme 15) [78–80]. It is worth to note that this original leaving group as

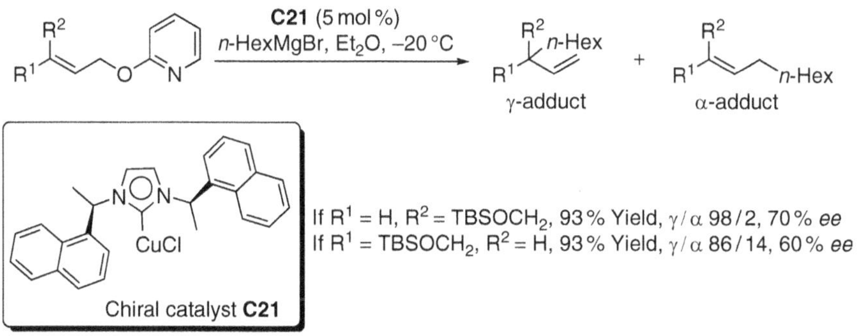

Scheme 15 Monodentate NHC, Okamoto 2004

well as the *Z*-configuration of the olefin are crucial parameters to obtain good results.

Parallel to this study, Hoveyda developed a first generation of bidentate chiral NHC ligands, assuming that strong binding of the metal center is required for an efficient asymmetric induction [81]. These ligands are used as silver dimeric complexes **C22** allowing for easier handling and better *in situ* transmetalation to copper (Scheme 16). High enantioselectivity and almost perfect regioselectivity were observed in the model reaction of allylic phosphates with diethylzinc. As for the peptide ligands, the formation of quaternary centers was achieved. Interestingly, the treatment of the silver complex with $CuCl_2 \cdot 2\ H_2O$ led to the formation of the corresponding air-stable dimeric NHC-Cu(II) complex **C23**. The latter species being catalytically potent, it seems to indicate that a copper species is effectively involved in the reaction mechanism.

One year later, the same group developed a new generation of NHC containing a biphenyl moiety instead of the chiral binaphthyl [82]. The source of chirality was installed on the NHC backbone by use of an asymmetric diamine (**C24**, Scheme 17). This new set of catalysts was significantly more effective in terms of scope and enantioselectivity. Thus, the introduction of methyl, ethyl, *n*-butyl and even bulky *iso*propyl group was achieved with *ee*'s up to 98% for the formation of both tertiary and quaternary centers. Furthermore, the low catalyst loading needed (0.5 mol%) as well as the nature of the copper source (cheap and air-stable $CuCl_2 \cdot 2\ H_2O$) definitively positioned this method as one of the most efficient in this field.

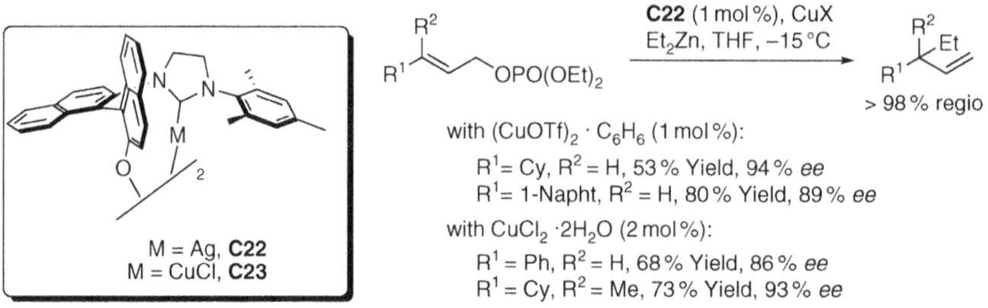

M = Ag, **C22**
M = CuCl, **C23**

C22 (1 mol%), CuX
Et_2Zn, THF, −15 °C

> 98% regio

with $(CuOTf)_2 \cdot C_6H_6$ (1 mol%):
R^1 = Cy, R^2 = H, 53% Yield, 94% *ee*
R^1 = 1-Napht, R^2 = H, 80% Yield, 89% *ee*
with $CuCl_2 \cdot 2H_2O$ (2 mol%):
R^1 = Ph, R^2 = H, 68% Yield, 86% *ee*
R^1 = Cy, R^2 = Me, 73% Yield, 93% *ee*

Scheme 16 First generation of bidentate NHC ligand, Hoveyda 2004

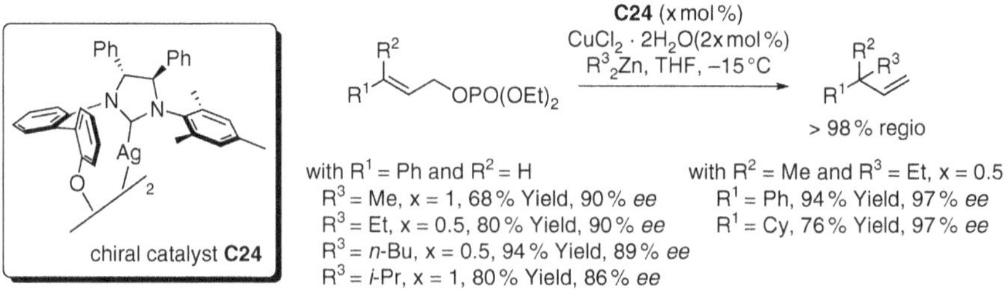

chiral catalyst **C24**

C24 (x mol%)
$CuCl_2 \cdot 2H_2O$(2x mol%)
$R^3{}_2Zn$, THF, −15 °C

> 98% regio

with R^1 = Ph and R^2 = H
R^3 = Me, x = 1, 68% Yield, 90% *ee*
R^3 = Et, x = 0.5, 80% Yield, 90% *ee*
R^3 = *n*-Bu, x = 0.5, 94% Yield, 89% *ee*
R^3 = *i*-Pr, x = 1, 80% Yield, 86% *ee*

with R^2 = Me and R^3 = Et, x = 0.5
R^1 = Ph, 94% Yield, 97% *ee*
R^1 = Cy, 76% Yield, 97% *ee*

Scheme 17 Second generation of bidentate NHC ligand, Hoveyda 2005

Later on, Hoveyda replaced the phenol group by a sulfonate (**C25**, Scheme 18) [83]. This slight modification allowed for the asymmetric introduction of vinylaluminium reagents. This challenging reaction is to date the unique highly enantioselective method to introduce an alkenyl group in copper-catalyzed allylic alkylation. The vinylic coupling partners were prepared by hydroalumination of terminal alkynes with DIBAL-H (i-Bu_2AlH) and directly used in the copper-catalyzed reaction. Chiral dienes were afforded in perfect regioselectivity and enantioselectivities ranging for 77–98%.

The same group has also defined suitable conditions for the hydroalumination of silyl-protected alkynes such as **13**, leading selectively to the *E*- or *Z*-vinylaluminium reagents (**14** and **15** respectively, Scheme 19) [84]. Using the conditions previously described, the alkenylation products were afforded in up to 98% *ee* and stereoselectivity. Noteworthy, in the case of the *Z*-vinylaluminium reagent **16** a dimethylsilyl hydride unit was used instead of trimethylsilyl due to a stronger preference for the transfer of the vinyl unit (vs i-Bu group stemming from the

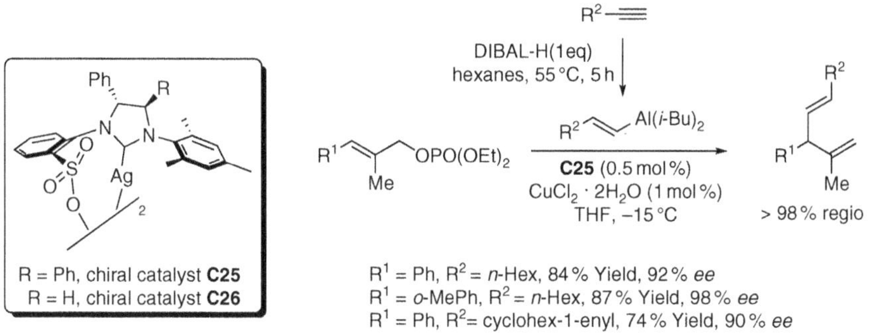

Scheme 18 Asymmetric allylic alkenylation, Hoveyda 2007

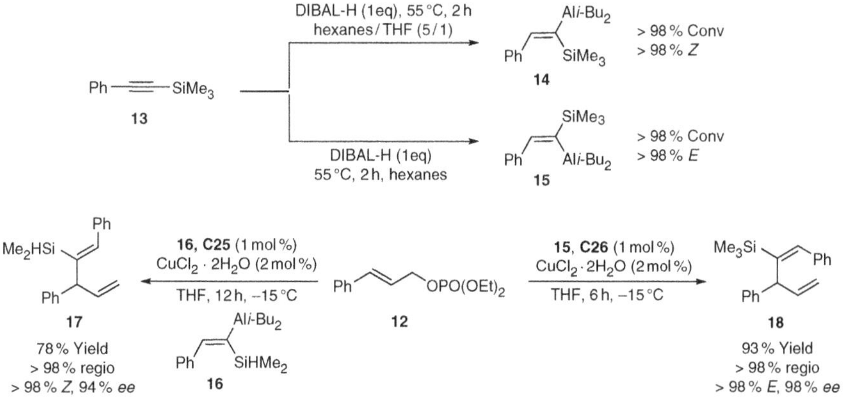

Scheme 19 Synthesis of pure *E*- or *Z*-alkenes, Hoveyda 2010

DIBAL). The deprotected alkene products were obtained after a clean protodesilylation in a mixture of TFA/$CHCl_3$ at 0°C for 15 h.

The group of Tomioka has recently used NHC's to solve another synthetic challenge, the allylic arylation [85]. Indeed, using the *C2*-symmetric NHC-copper complex **C27**, they performed the addition of various aryl Grignard reagents to cinnamyl bromide derivatives (Scheme 20). High selectivities were attained whatever the electronic properties of the substrate. Interestingly, for all the processes mentioned in this section, better reactivities and selectivities were achieved using a preformed catalyst, which is a general trend in carbene–copper chemistry.

A notable exception has been reported during the preparation of this chapter by Mauduit and co-workers (Scheme 21) [86]. They introduced hydroxyalkyl NHC ligand **L28**, which should be formed in situ from imidazolinium salt **Im** by addition of catalytic amount of *n*-butyllithium. Using this readily available chiral ligand, the alkylation of cinnamylphosphate **12** was achieved with good results. Interestingly, the formation of a silver complex of this ligand displayed lower activity and selectivities.

Having established effective catalytic systems, chemists turned their attention to overcome the challenging reactivity of different substrates and/or to expand the reaction scope in view of significant applications in total synthesis. For that purpose, new substrates have to be designed, not only for the copper-catalyzed alkylation step but also for subsequent transformations.

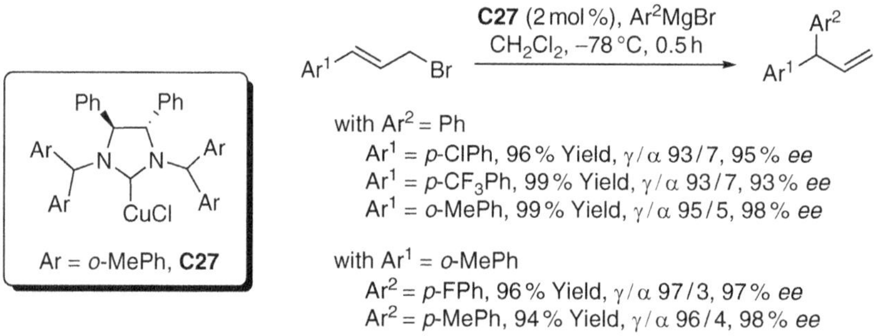

C27 (2 mol %), Ar2MgBr
CH2Cl2, −78 °C, 0.5 h
Ar1
Br
Ar2
Ar1
Ph
Ph
Ar
Ar
Ar
Ar
N
N
CuCl
Ar = o-MePh, C27
with Ar2 = Ph
Ar1 = p-ClPh, 96 % Yield, γ/α 93/7, 95 % ee
Ar1 = p-CF3Ph, 99 % Yield, γ/α 93/7, 93 % ee
Ar1 = o-MePh, 99 % Yield, γ/α 95/5, 98 % ee
with Ar1 = o-MePh
Ar2 = p-FPh, 96 % Yield, γ/α 97/3, 97 % ee
Ar2 = p-MePh, 94 % Yield, γ/α 96/4, 98 % ee

Scheme 20 Asymmetric allylic arylation, Tomioka 2009

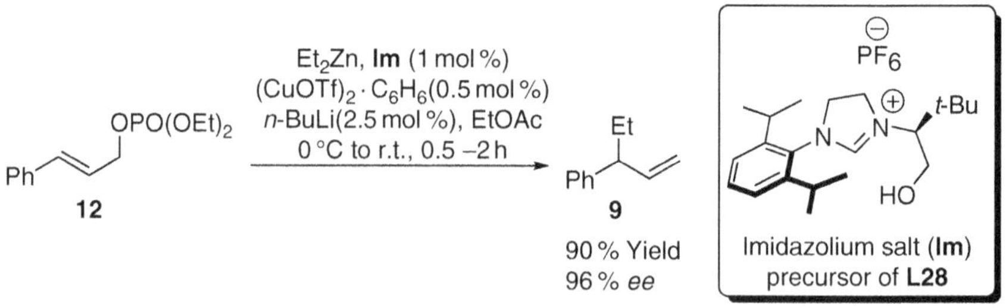

Et2Zn, Im (1 mol %)
(CuOTf)2 · C6H6(0.5 mol %)
n-BuLi(2.5 mol %), EtOAc
0 °C to r.t., 0.5 –2 h
OPO(OEt)2
Ph
12
Et
Ph
9
90 % Yield
96 % ee
PF6
t-Bu
N
N
HO
Imidazolium salt (Im)
precursor of L28

Scheme 21 Hydroxyalkyl NHC ligand, Mauduit 2010

3 Scope of the Reaction

3.1 Highly Functionalized Substrates

3.1.1 (*E*)-Methyl 4-Diethylphosphate-But-2-Enoate

In 2003, Hoveyda and co-workers described the alkylation of substrates bearing an ester group in γ-position to the leaving group providing α-alkyl-β,γ-unsaturated chiral esters [87, 88]. Using the peptide-based second generation of catalysts (**L29**), outstanding regio- and enantiocontrols were achieved (eq 1, Scheme 22). The synthetic utility of the resulting products was demonstrated with the synthesis of (*R*)-(−)-elenic acid **20**, which is an inhibitor of topoisomerase II. A fine-tuning of the chiral ligand allowed for the formation of quaternary centers in *ee*'s up to 98% (eq 2, Scheme 22). Impressively, quaternary centers containing a phenyl group could be prepared with equal efficiency using **L31**.

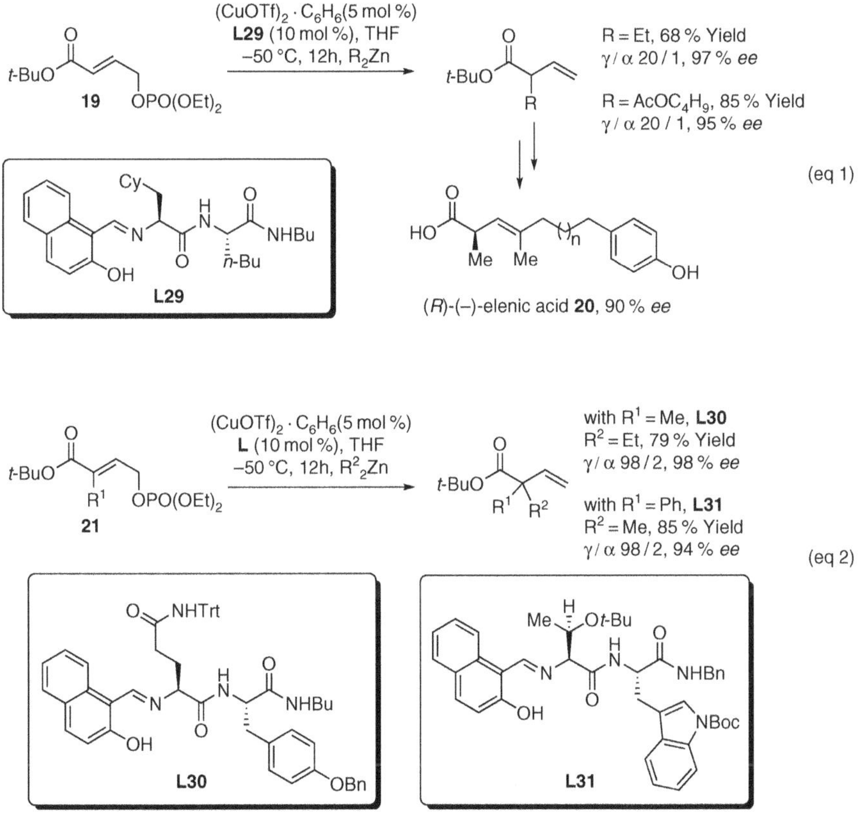

Scheme 22 Formal asymmetric α-alkylation of esters, Hoveyda 2003

3.1.2 Preparation of Allylboronates and Allylsilanes

The group of Hall published in 2007 an interesting preparation of α-chiral allylboronates by copper-catalyzed allylic alkylation [89]. They notably described the use of new phosphoramidite **L32** for the asymmetric alkylation of 3-chloropropenylboronate **22**, giving rise to the formation of various enantioenriched allylboronates (up to 94% *ee*, Scheme 23). Allylboronate **23** was subsequently trapped in situ with various aldehydes yielding the corresponding homoallylic alcohols as single isomers.

At the same time, Hoveyda applied his chiral NHC-$(CuOTf)_2{\cdot}C_6H_6$ systems (Scheme 16) to the substitution of vinylsilane **24** (Scheme 24) [90]. The first generation of NHC catalysts (**C22**) allowed for the formation of tertiary and quaternary alkyl-substituted allylsilanes in high selectivities. However, the second and the third generation of NHC catalysts (**C24** and **C25**) were required to respectively form tertiary and quaternary aryl-substituted allylsilanes.

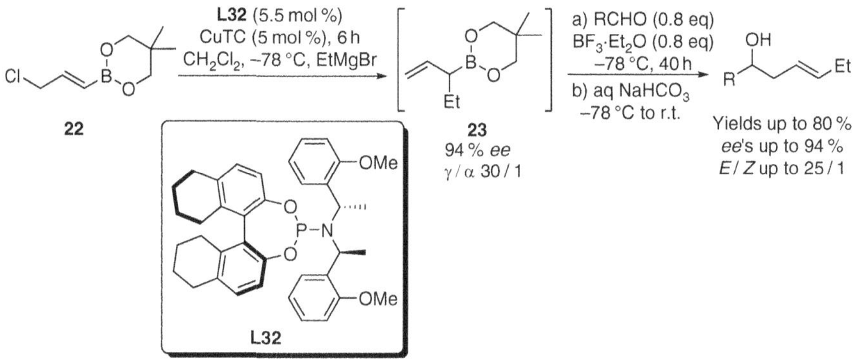

Scheme 23 Preparation of allylboronates, Hall 2007

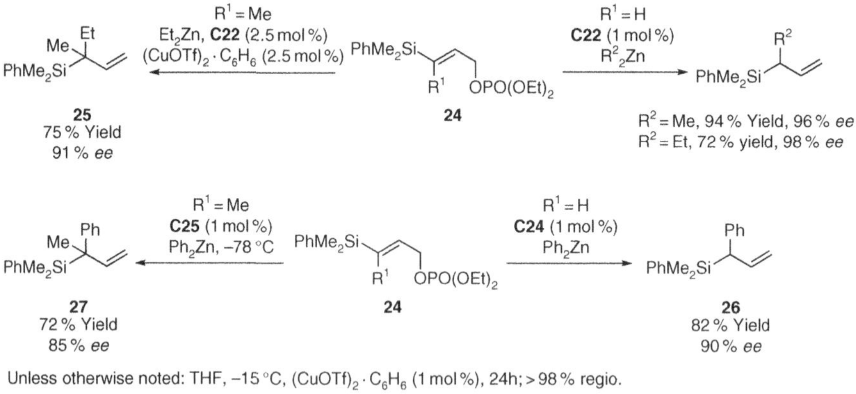

Unless otherwise noted: THF, –15 °C, $(CuOTf)_2 \cdot C_6H_6$ (1 mol %), 24h; > 98 % regio.

Scheme 24 Preparation of various allylsilanes, Hoveyda 2007

3.1.3 β-Substituted Acyclic and Cyclic Subtrates

It is well known in the literature that β-substituted substrates afforded poorly selective alkylation reactions, certainly due to additional steric constraints. Nonetheless, Alexakis reported in 2006 that, using Feringa's phosphoramidite (*S*,*S*,*S*)-**L33** in combination with CuTC and Grignard reagents, this type of substrates could be alkylated in good regioselectivity and *ee*'s up to 97% (eq 1, Scheme 25) [91, 92]. The addition of a 3-butenylMgBr followed by a ring closing metathesis using the Grubbs second generation catalyst gave synthetically interesting cyclopentene **30**. Furthermore, this research group introduced the endocyclic (1-chloromethyl)cycloalkene in this reaction. The higher stability of the product bearing an *exo* double bond will favor the attack in γ-position resulting in the formation of a stereogenic center. Using similar conditions, the products were obtained in excellent selectivities ranging from 95% to 99.6% *ee* (eq 2, Scheme 25). A beneficial effect was noted for Grignard reagent owning a double bond on the alkyl chain. The role of this unsaturation is not fully understood even though it is conceivable that π–π or π-cation interactions might be involved in the transition state allowing for a better stereocontrol. The synthetic potential of a chiral adduct bearing an *exo* double bond has been illustrated by Renaud with the preparation of (−)-indolizidine 167B [93].

This unexpected interaction has been recently used to access enantioenriched cyclic substrates via a one-pot Cu-catalyzed allylic alkylation/ring closing

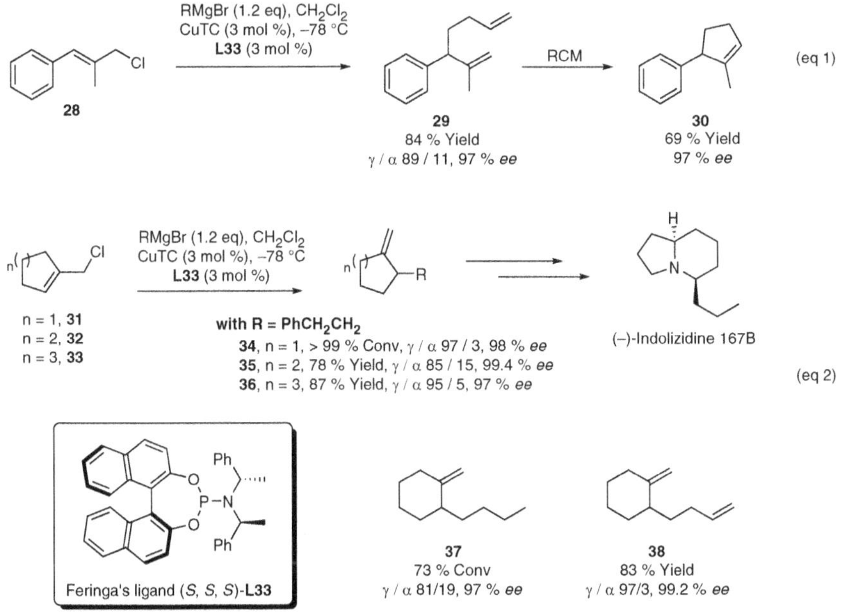

RMgBr (1.2 eq), CH_2Cl_2
CuTC (3 mol %), –78 °C
L33 (3 mol %)

28

29
84 % Yield
γ / α 89 / 11, 97 % *ee*

RCM

30
69 % Yield
97 % *ee*

(eq 1)

RMgBr (1.2 eq), CH_2Cl_2
CuTC (3 mol %), –78 °C
L33 (3 mol %)

n = 1, **31**
n = 2, **32**
n = 3, **33**

with R = $PhCH_2CH_2$
34, n = 1, > 99 % Conv, γ / α 97 / 3, 98 % *ee*
35, n = 2, 78 % Yield, γ / α 85 / 15, 99.4 % *ee*
36, n = 3, 87 % Yield, γ / α 95 / 5, 97 % *ee*

(–)-Indolizidine 167B

(eq 2)

Feringa's ligand (*S, S, S*)-**L33**

37
73 % Conv
γ / α 81/19, 97 % *ee*

38
83 % Yield
γ / α 97/3, 99.2 % *ee*

Scheme 25 β-Substituted substrates, Alexakis 2006

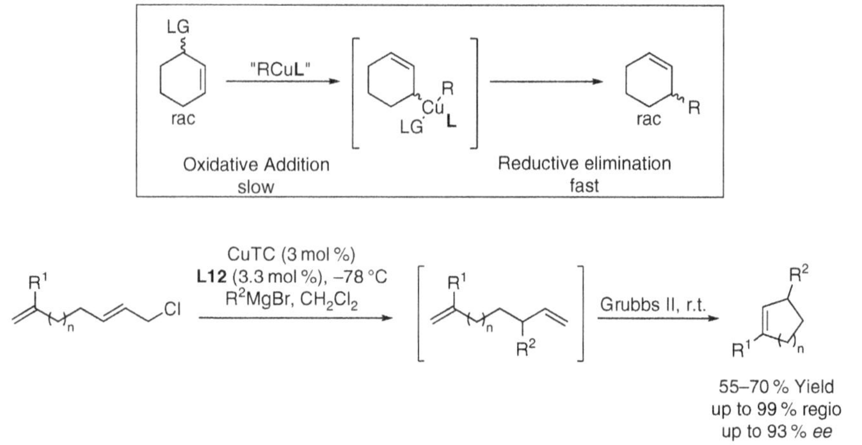

Scheme 26 Formal alkylation of cyclic substrates, Alexakis 2010

metathesis protocol (Scheme 26) [94]. Cyclic substrates are especially challenging due to the lack of σ–π isomerization involved in the reaction mechanism (see Scheme 1). Indeed, if the reductive elimination step is fast, a racemic cyclic substrate will lead to a racemic product. The alternative proposed by Alexakis was to use a ω-ethylenic allylic substrate, which after the alkylation step underwent a ring closing metathesis by treatment with the Grubbs second generation catalyst affording the desired enantioenriched cyclic product. Promising results were obtained for differently substituted five, six and seven-membered rings.

3.1.4 1,4-Disubstituted But-2-Enes

In 2007, Alexakis and Falciola disclosed a new class of symmetrical substrates containing two possible reaction sites (Scheme 27) [95, 96]. The substitution in γ-position of 1,4-dibromo-but-2-ene **39** provided a homoallylic bromide, which was inert under the reaction conditions. Such functionality could be a powerful tool for further derivatization of the alkylation products. The perfect γ-regioselectivity was rationalized by a possible stabilization of the empty *d* orbitals of the

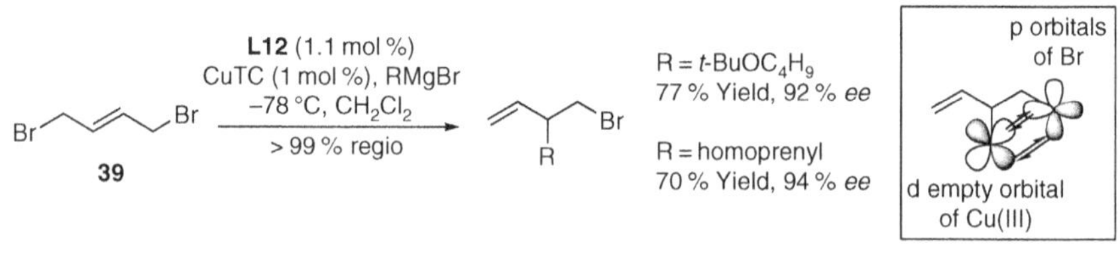

Scheme 27 1,4-Dibromout-2-ene, Alexakis 2007

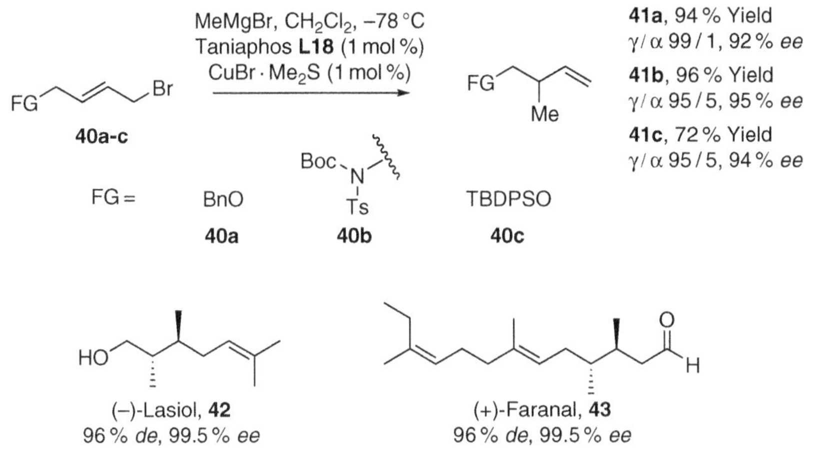

MeMgBr, CH_2Cl_2, –78 °C
Taniaphos **L18** (1 mol %)
CuBr · Me_2S (1 mol %)

40a-c

FG = BnO **40a**; Boc(Ts)N **40b**; TBDPSO **40c**

41a, 94 % Yield, γ/α 99/1, 92 % *ee*
41b, 96 % Yield, γ/α 95/5, 95 % *ee*
41c, 72 % Yield, γ/α 95/5, 94 % *ee*

(–)-Lasiol, **42**
96 % *de*, 99.5 % *ee*

(+)-Faranal, **43**
96 % *de*, 99.5 % *ee*

Scheme 28 Bifunctional chiral building blocks, Feringa 2007

transition metal, present in γ-position, by the *p* orbitals of the halide. This process appeared to be efficient for the introduction of various alkyl Grignard reagents reaching 94% *ee*.

The same year, Feringa replaced the second bromide group by a less reactive hydroxyl or amine protected unit and submitted these new substrates to his previously described conditions (Scheme 28) [97]. The high versatility of the corresponding products was smartly demonstrated by the enantioselective syntheses of ant pheromones (–)-Lasiol **42** and (+)-Faranal **43**.

3.1.5 3-Bromopropenyl Ester

Feringa developed a process for the formation of benzoyl-protected secondary allylic alcohols via asymmetric alkylation of 3-bromopropenylester **44** (Scheme 29) [98]. Outstanding regio and enantioselectivity were attained for the introduction of various alkyl groups.

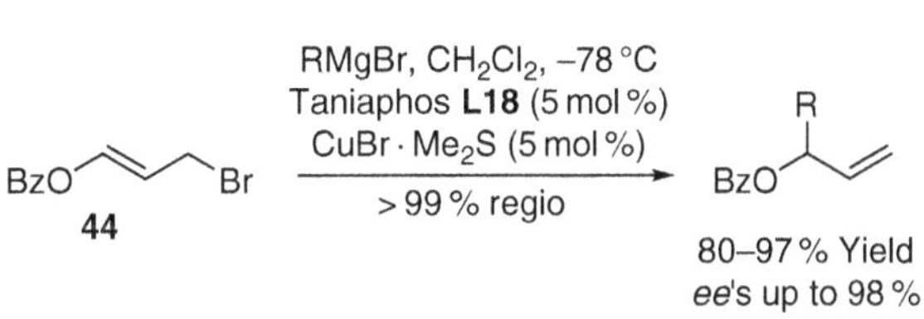

Scheme 29 Formation of benzoyl-protected chiral allylic alcohols, Feringa 2006

44

RMgBr, CH_2Cl_2, –78 °C
Taniaphos **L18** (5 mol %)
CuBr · Me_2S (5 mol %)
> 99 % regio

80–97 % Yield
ee's up to 98 %

3.2 *Desymmetrization of* meso *Allylic Substrates*

A *meso* compound possesses an even number of stereogenic centers as well as a plane of symmetry. Desymmetrization reaction of such type of achiral substrates is of particular interest because it could lead to the obtention of several real stereocenters in only one step. Most relevant examples will be detailed in this section (vide infra).

3.2.1 Cyclic Allylic Bis-Diethylphosphates

In 2003, the group of Gennari introduced the *meso* cyclic allylic bis-diethylphosphates **45a–c** as potential candidates for copper-catalyzed desymmetrization reactions (eq 1, Scheme 30) [99, 100]. Using their aminosulfonamide phenols **L34** and **L35** as chiral ligands, they performed the enantioselective alkylation of substrates owning rings of various sizes (5, 6 and 7-membered rings). Interestingly, products of invertive and retentive substitutions were obtained in the case of six-membered ring reaction partners (**46b**). This pair of diastereoisomers was recovered in their racemic form. Later on, in collaboration with Feringa, the same group replaced their ligands by phosphoramidites **L8** (eq 2, Scheme 30) [101]. In this case, high asymmetric induction was observed for a wide range of cyclic substrates. Noteworthily, both diastereoisomers of the Feringa's ligand **L8** and **L33** showed comparable efficiency.

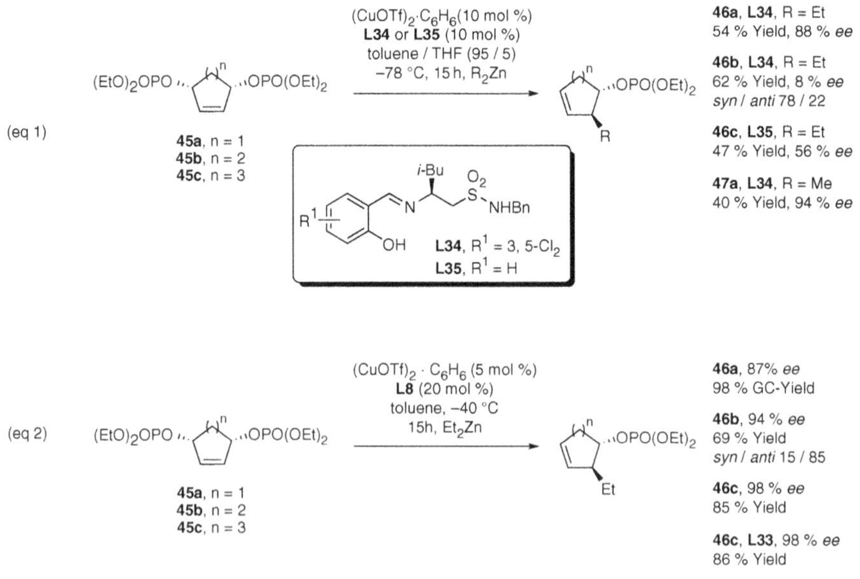

(eq 1)
$(EtO)_2OPO$ ... $OPO(OEt)_2$
45a, n = 1
45b, n = 2
45c, n = 3

$(CuOTf)_2{\cdot}C_6H_6$(10 mol %)
L34 or **L35** (10 mol %)
toluene / THF (95 / 5)
−78 °C, 15 h, R_2Zn

$OPO(OEt)_2$, R

i-Bu, O_2, S, N, NHBn, R^1, OH
L34, R^1 = 3, 5-Cl_2
L35, R^1 = H

46a, **L34**, R = Et
54 % Yield, 88 % *ee*

46b, **L34**, R = Et
62 % Yield, 8 % *ee*
syn / *anti* 78 / 22

46c, **L35**, R = Et
47 % Yield, 56 % *ee*

47a, **L34**, R = Me
40 % Yield, 94 % *ee*

(eq 2)
$(EtO)_2OPO$ ... $OPO(OEt)_2$
45a, n = 1
45b, n = 2
45c, n = 3

$(CuOTf)_2 \cdot C_6H_6$ (5 mol %)
L8 (20 mol %)
toluene, −40 °C
15h, Et_2Zn

$OPO(OEt)_2$, Et

46a, 87% *ee*
98 % GC-Yield

46b, 94 % *ee*
69 % Yield
syn / *anti* 15 / 85

46c, 98 % *ee*
85 % Yield

46c, **L33**, 98 % *ee*
86 % Yield

Scheme 30 Cyclic allylic bis-diethylphosphates, Gennari and Feringa 2003

3.2.2 Symmetric Allylic Epoxides

During their investigations in kinetic resolution of allylic epoxides (see section 3.3.1), Pineschi and Feringa developed interesting symmetrical substrates such as methylidene epoxides **47** [102]. The S_N2' attack of the organocopper species to one or the other enantiotopic double bond will afford the chiral dienols **48** in good yields and selectivities (eq 1, Scheme 31). Thereafter, Pineschi successfully applied the same conditions to the desymmetrization of 1,3,5,7-cyclooctatetraene mono-epoxide **49** (eq 2, Scheme 31) [103].

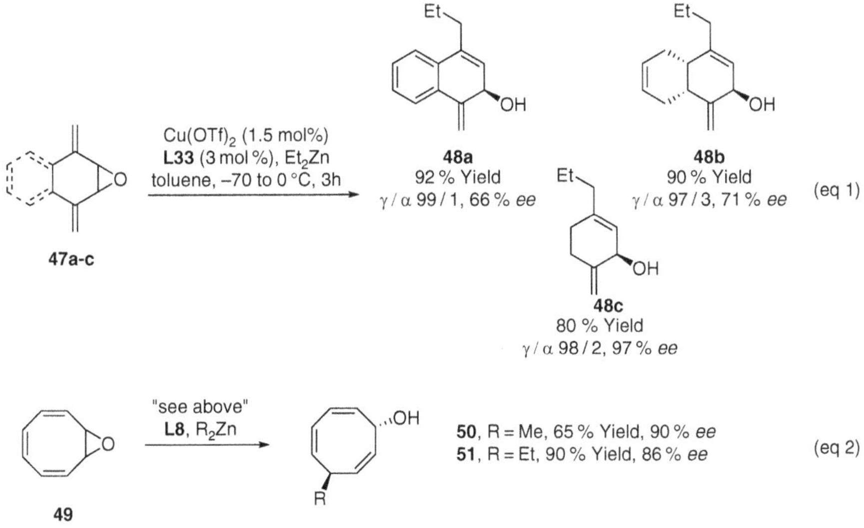

Cu(OTf)$_2$ (1.5 mol%)
L33 (3 mol %), Et$_2$Zn
toluene, −70 to 0 °C, 3h

47a-c

48a
92 % Yield
γ / α 99 / 1, 66 % *ee*

48b
90 % Yield
γ / α 97 / 3, 71 % *ee*

(eq 1)

48c
80 % Yield
γ / α 98 / 2, 97 % *ee*

"see above"
L8, R$_2$Zn

49

50, R = Me, 65 % Yield, 90 % *ee*
51, R = Et, 90 % Yield, 86 % *ee*

(eq 2)

Scheme 31 Symmetric monoepoxides, Pineschi and Feringa 2000–2003

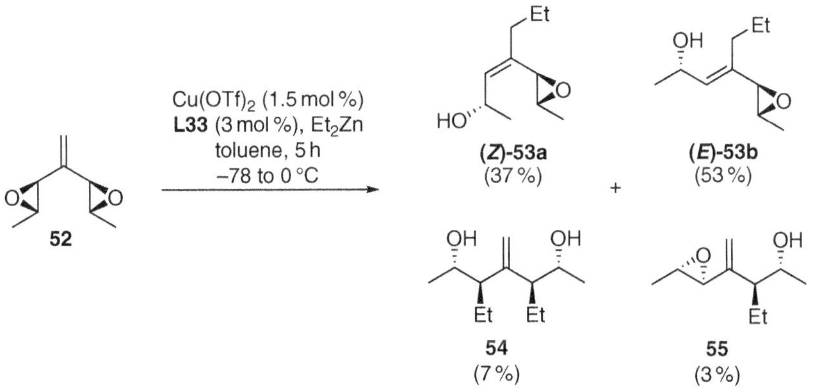

Cu(OTf)$_2$ (1.5 mol %)
L33 (3 mol %), Et$_2$Zn
toluene, 5 h
−78 to 0 °C

52

(Z)-53a
(37 %)

(E)-53b
(53 %)

54
(7 %)

55
(3 %)

Scheme 32 Vinyl diepoxides, Pineschi 2003

Later on, the same group reversed its strategy and replaced the enantiotopic double bonds by two enantiotopic epoxides (Scheme 32) [104]. Thus, substrate **52** was submitted to the reaction conditions leading to a complex mixture of isomers (**53**–**55**). Interestingly, 90% of the isolated products resulted from an S_N2' addition. Both isomers of the 6:4 *E*/*Z*-mixture were obtained in approximately 50% *ee*, clearly indicating that the *s-cis*/*s-trans* conformers did not affect the enantio-determining step.

3.2.3 Oxabicyclic Alkenes

Pineschi and Feringa reported in 2002 a fascinating application of their method to the ring-opening reaction of oxabenzonorbornadienes **56** (Scheme 33) [105]. They observed variable enantioselectivities depending on the electronic properties of the starting material. Electron-deficient substrates provided slightly less enantioenriched benzylic alcohols. It can be assumed that two different mechanisms might be in competition in the present reaction, leading to different stereoisomeric adducts. An *anti* S_N2' reaction could be envisaged, resulting from the stabilization of the copper species by the σ* orbital of the C–O bond of the epoxide leading

$Cu(OTf)_2$ (3 mol %)
L8 (7 mol %), $R^2{}_2Zn$
toluene, r.t.

56 → **57** + **58**

Yield up to 90 %, *ee*'s up to 99 %, **57**/**58** up to 99/1
R^1 = 6, 7-F_2, 6, 7-OMe_2, 5, 8-Me_2, 5, 8-OMe_2
R^2 = Et, Me, *n*-Bu

Scheme 33 Oxabicyclic alkenes, Pineschi and Feringa 2002

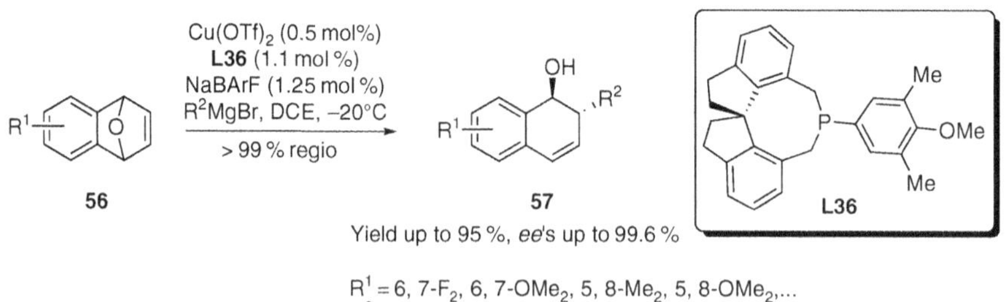

R^1 = 6, 7-F_2, 6, 7-OMe_2, 5, 8-Me_2, 5, 8-OMe_2,...
R^2 = Et, *n*-Bu, *i*-Pr, *i*-Bu, *t*-Bu

Scheme 34 Oxabicyclic alkenes, Zhou 2008

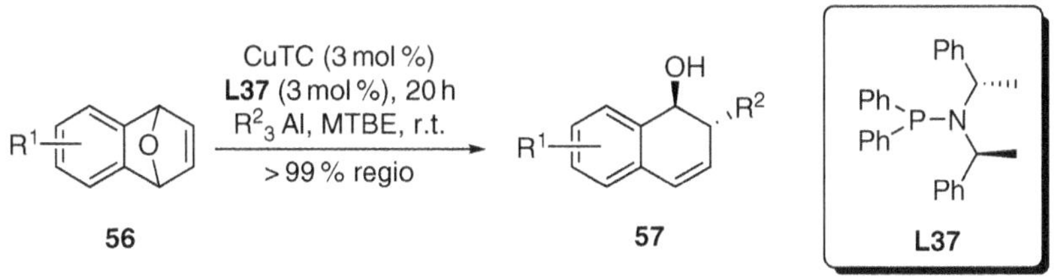

CuTC (3 mol %)
L37 (3 mol %), 20 h
$R^2{}_3$ Al, MTBE, r.t.
> 99 % regio

56 **57** **L37**

Yield up to 95 %, *ee*'s up to 94 %

R^1 = 6, 7-F_2, 6, 7-OMe_2, 5, 8-Me_2, 5, 8-OMe_2,...
R^2 = Me, Et, *n*-Bu, *n*-Pr, *i*-Bu

Scheme 35 Oxabenzonorbornadienes, Alexakis 2009

to the *anti* adduct **57**. Or a *syn* S_N2' process might be considered, resulting from a stabilization of the copper by a *p* orbital of the oxygen atom yielding the *syn* adduct **58**. Therefore, even if the regiocontrol is generally high in this type of reaction, the stereocontrol remained a challenge.

In 2008, Zhou and co-workers improved the stereoselectivity of the reaction using an *in situ* prepared cationic copper(I) catalyst bearing two chiral spiro phosphines **L36** (Scheme 34) [106]. With a new set of conditions in hand, a wide range of substrates was investigated affording benzylic alcohols under complete stereocontrol in favor of the *anti* adduct **57** with enantioselectivities up to 99.6%.

Later on, Alexakis introduced trialkylaluminium and the SimplePhos ligand **L37** in this reaction (Scheme 35) [107, 108]. A complete stereocontrol and high asymmetric inductions were observed for the same range of substrates.

3.2.4 Polycyclic Hydrazines

In 2005, Pineschi slightly modified his protocol by replacing the diorganozinc by trialkylaluminium reagents (eq 1, Scheme 36) [109]. This optimization allowed for the ring-opening of *meso* polycyclic hydrazine **59** in 80% *ee*. Thereafter, Micouin and Alexakis disclosed that trimethylaluminium modified the phosphoramidite ligand ***ent*-L33** under the reaction conditions (dichloromethane, 0°C) leading to the formation of an aminophosphine ligand **L38** (eq 2, Scheme 36) [110]. Variation of the substitution pattern of the ligand allowed to improve the ee-value to 85% with the naphthylamine-derived ligand **L39**, and after replacing the dimethylphosphine unit by a diphenylphosphine (**L40**) even 94% *ee* could be obtained (eq 3. Scheme 36) [111, 112].

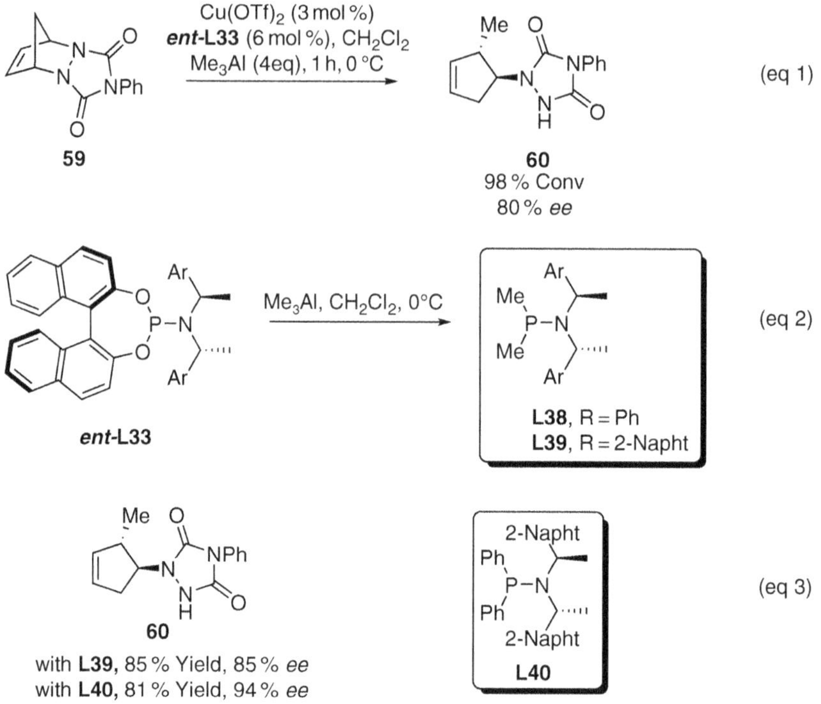

Scheme 36 Polycyclic hydrazines, Pineschi, Alexakis and Micouin 2005–2010

3.2.5 Symmetric Dienol

Hoveyda published in 2007 the use of a substrate **61** possessing two possible reaction sites (Scheme 37). [113] A double asymmetric allylic alkylation of **61**

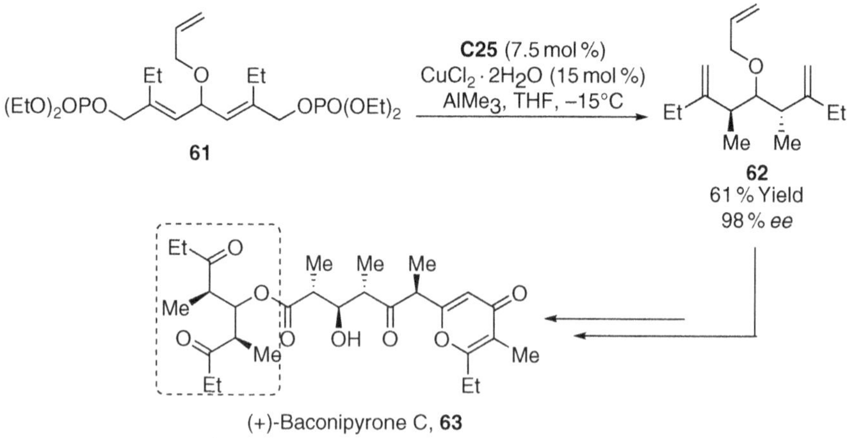

Scheme 37 Symmetric dienol, Hoveyda 2007

provided versatile chiral 1,6-diene **62**, which have been subsequently engaged in the synthesis of (+)-Baconipyrone C.

3.3 Kinetic Resolution Processes

The kinetic resolution is a powerful tool in organic synthesis. This process, based on the difference in reactivity between two enantiomers, can lead to the formation of a highly enantioenriched product starting from a racemic substrate. Unfortunately, a maximum yield of 50% can be expected for an enantiopure product. We will see in this section that the second enantiomer of the starting material is not necessarily unreactive (for a review on this section: [114]).

3.3.1 Cycloalkene Monoepoxides

The first example of a kinetic resolution in copper-catalyzed allylic alkylations, decribed in 1998, was the result of a fruitful collaboration of Pineschi and Feringa [115]. They applied phosporamidite ligand **L8** to the alkylation of cyclic 1,3-diene monoepoxides **64a–c**. Using the aforementioned conditions, good regiocontrol as well as excellent enantioselectivities were obtained (eq 1, Scheme 38). Noteworthy, analogous allylic aziridines were opened with comparable efficacy [116]. Eight years later, the Alexakis group turned its attention to this reaction and developed

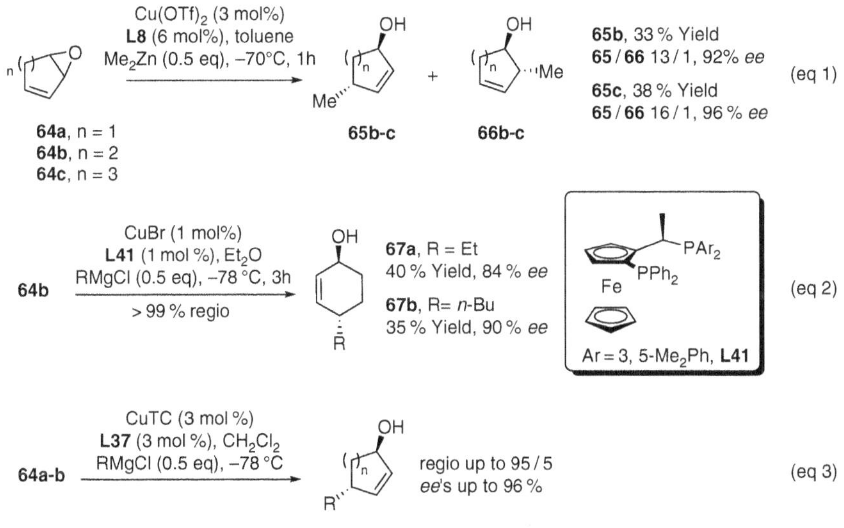

$Cu(OTf)_2$ (3 mol%), **L8** (6 mol%), toluene, Me_2Zn (0.5 eq), −70°C, 1h

64a, n = 1; **64b**, n = 2; **64c**, n = 3 → **65b-c** + **66b-c**

65b, 33 % Yield, **65** / **66** 13 / 1, 92% *ee*; **65c**, 38 % Yield, **65** / **66** 16 / 1, 96 % *ee* (eq 1)

64b → CuBr (1 mol%), **L41** (1 mol %), Et_2O, RMgCl (0.5 eq), −78 °C, 3h, > 99 % regio

67a, R = Et, 40 % Yield, 84 % *ee*; **67b**, R= *n*-Bu, 35 % Yield, 90 % *ee*

Ar = 3, 5-Me_2Ph, **L41** (eq 2)

64a-b → CuTC (3 mol %), **L37** (3 mol %), CH_2Cl_2, RMgCl (0.5 eq), −78 °C

regio up to 95 / 5, *ee*'s up to 96 % (eq 3)

Scheme 38 Cyclic 1,3-diene monoepoxides, Pineschi and Alexakis 1998–2008

a combination of Josiphos type ligands (**L41**) and Grignard reagents leading to a perfectly regioselective reaction [117]. Selectivity factors (*s*) ranging from 25 to 42 were attained with this catalytic system (eq 2, Scheme 38) (For information about kinetic resolution and selectivity factor, see: [118]). Later on, the same group enlarged the scope of the reaction with the introduction of a wide range of alkyl groups using SimplePhos **L37** [119].

3.3.2 Regiodivergent Kinetic Resolution

In 2001, Pineschi and Feringa described a new elegant process in copper-catalyzed asymmetric allylic alkylation: the regiodivergent kinetic resolution [120–122]. This process allows for the selective conversion of both enantiomers of a racemic substrate into two distinct regioisomers. Cyclic allylic epoxides were optimal candidates to demonstrate the utility of this reaction. Thus, substrate **64c** proved to be a good reaction partner affording 50% of the allylic alcohol **65c** in 90% *ee* and 50% of homoallylic alcohol **66c** in 95% *ee* (eq 1, Scheme 39). Unfortunately, the reaction was very sensitive to the ringsize of the starting material and lower selectivities were obtained for five, six, and eight-membered rings. A substrate with an *exo* double bond was also considered. In this case, the reaction was limited to cyclohexyl derivatives leading to the regioisomers in *ee*'s around 94% (eq 2, Scheme 39). Such results might indicate that the *s-trans* or *s-cis* conformation of the starting material played an important role in the regio- and enantiodetermining step.

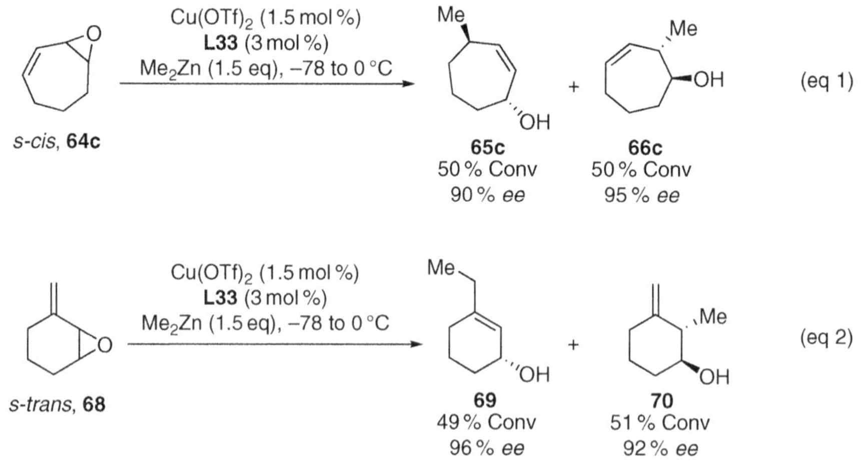

$Cu(OTf)_2$ (1.5 mol %), **L33** (3 mol %), Me_2Zn (1.5 eq), –78 to 0 °C

s-cis, **64c** → **65c** (50 % Conv, 90 % *ee*) + **66c** (50 % Conv, 95 % *ee*) (eq 1)

s-trans, **68** → **69** (49 % Conv, 96 % *ee*) + **70** (51 % Conv, 92 % *ee*) (eq 2)

Scheme 39 Regiodivergent kinetic resolution, Pineschi and Feringa 2001

3.4 Dynamic Kinetic Asymmetric Transformation

Trost has developed, in 1994, the Dynamic Kinetic Asymmetric Transformation (DYKAT) in palladium-catalyzed allylic alkylation (First report: [123], For a review see: [124]). Based on the formation of a *meso* π-allyl intermediate, it allows for the conversion of both enantiomers of a racemic substrate in a single enantioenriched product. Such a process was unknown in copper chemistry particularly because the systematic formation of a π-allyl intermediate was not proved. Indeed, numerous catalytic systems were based on a selective oxidative addition on one enantiotopic face of a prochiral substrate leading to the formation of a σ-allylcopper complex, which undergoes a rapid reductive elimination to give the desired enantioenriched γ-adduct (Scheme 1). However, in 2009, Alexakis described that using the suitable leaving group (bromide), the alkylation of a cyclohexenyl

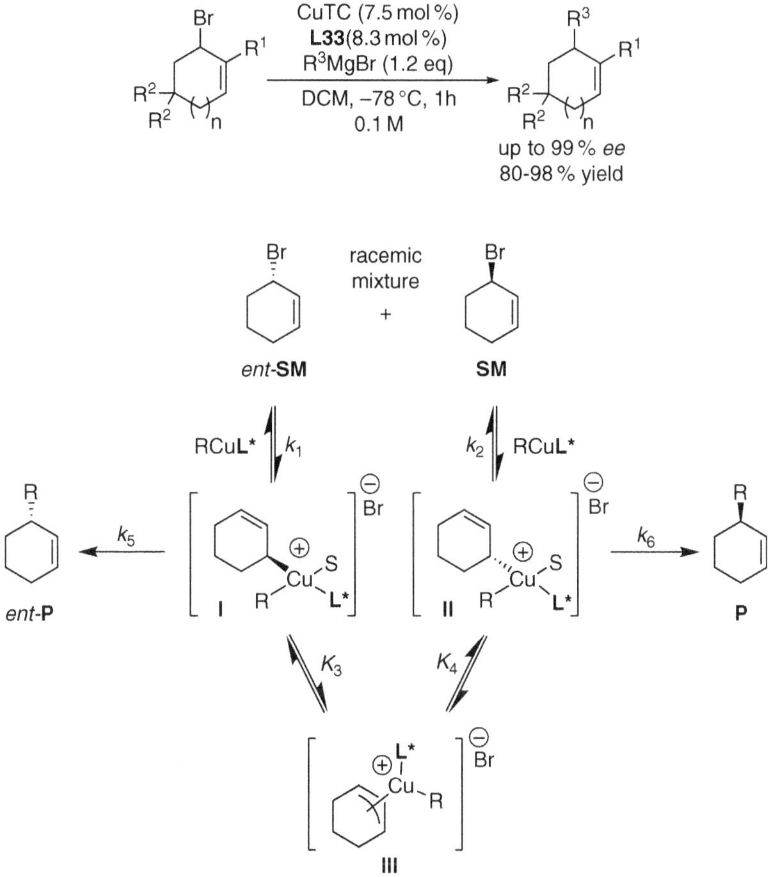

Scheme 40 Dynamic kinetic asymmetric transformation, Alexakis 2009. *S* molecule of solvent, *k* relative rate constant, *K* equilibrium constant

derivative could be quantitative and above all highly enantioselective (*ee*'s up to 99%) [125, 126]. This observation was generalized to differently decorated six-membered rings and for the addition of a large range of primary alkyl Grignard reagents. A thorough study of the reaction outcome led to the proposition of the following mechanism. Both enantiomers of the starting material underwent an oxidative addition leading to the formation of diastereoisomeric σ-allylcopper complexes **I** and **II** (Scheme 40). These intermediates might be in rapid equilibrium via the *meso* π-allyl intermediate **III**. Then, if the rate of reductive elimination k_5 and k_6 are different, all the system will be displaced through the formation of the enantiomer of the product having the lowest free energy of formation.

4 Conclusion and Outlook

The copper-catalyzed asymmetric allylic alkylation has been particularly developed during the past two decades. The conjugated efforts of a relatively restricted amount of research groups have led to a considerable extension of the field of application of the reaction. Thus, a wide array of interesting chiral synthons could be readily obtained in high optical purity ranging from simple alkenes to more elaborated allylsilanes or allylboronates. The development of a large variety of catalytic systems using different organometallic reagents (Mg, Zn, Al) as well as different chiral ligands (phosphine, phosphoramidite, dipeptide, NHC) has allowed to override several synthetic challenges such as allylic methylation, alkenylation or arylation in highly selective manner. However, the mechanism of this reaction is still not fully understood, certainly due to the extreme reactivity of the Cu(III) species, which renders elusive their isolation and/or characterization. The improvement of analytical and computational techniques will help organic chemists to get more insights into this reaction mechanism, which will certainly coincide with new extension of the reaction scope.

References

1. Ojima I (2000) Catalytic asymmetric synthesis, 2nd edn. Wiley, New York
2. Jacobsen EN, Pfaltz A, Yamamoto H (1999) Comprehensive asymmetric catalysis I-III. Springer, Berlin
3. Takeuchi R (2002) Iridium complex-catalyzed highly selective organic synthesis. Synlett 1954–1965
4. Trost BM, Crawey ML (2003) Asymmetric transition-metal-catalyzed allylic alkylations: applications in total synthesis. Chem Rev 103:2921–2944
5. Trost BM (2004) Asymmetric allylic alkylation, an enabling methodology. J Org Chem 69:5813–5837
6. Miyabe H, Takemoto Y (2005) Regio- and stereocontrolled palladium- or iridium-catalyzed allylation. Synlett 1641–1655

7. Lu Z, Ma SM (2008) Metal-catalyzed enantioselective allylation in asymmetric synthesis. Angew Chem Int Ed 47:258–297
8. Bäckvall JE, Sellen M, Grant B (1990) Regiocontrol in copper-catalyzed Grignard reactions with allylic substrates. J Am Chem Soc 112:6615–6621
9. Bäckvall JE, Persson ESM, Bombrun A (1994) Regiocontrol in copper-catalyzed cross coupling of allylic chlorides with aryl Grignard reagents. J Org Chem 59:4126–4130
10. Persson ESM, Bäckvall JE (1995) The effect of solvent and Copper(I) precursor on the regioselectivity in the cross-coupling reaction of allylic acetates with preformed mono- and dibutylcuprates. Acta Chem Scand 49:899–906
11. Persson ESM, Vanklaveren M, Grove DM, Bäckvall JE, Vankoten G (1995) *ortho*-chelating arenethiolatocopper(I) complexes as versatile catalysts in the regioselective cross-coupling of allylic derivatives with *n*BuMgI – an example of reversed reactivity of leaving groups. Chem Eur J 1:351–359
12. Karlstrom ASE, Bäckvall JE (2001) Experimental evidence supporting a CuIII intermediate in cross-coupling reactions of allylic esters with diallylcuprate species. Chem Eur J 7:1981–1989
13. Goering HL, Kantner SS (1981) Alkylation of allylic derivatives. Regio- and stereochemistry of alkylation of allylic alcohols by the Murahashi method. J Org Chem 46:2144–2148
14. Goering HL, Kantner SS, Tseng CC (1983) Alkylation of allylic derivatives. 4. On the mechanism of alkylation of allylic N-phenylcarbamates with lithium dialkylcuprates. J Org Chem 48:715–721
15. Goering HL, Kantner SS (1983) Alkylation of allylic derivatives. 5. Loss of double-bond configuration associated with.alpha.-alkylation of allylic carboxylates with dialkylcuprates. J Org Chem 48:721–724
16. Goering HL, Singleton VD (1983) Alkylation of allylic derivatives. 6. Regiochemistry of alkylation of allylic acetates with dialkylcuprates. J Org Chem 48:1531–1533
17. Goering HL, Tseng CC (1983) Alkylation of allylic derivatives. 7. Stereochemistry of alkylation of the isomeric trans-.alpha., gamma.-methyl(phenyl)allyl acetates with lithium dialkylcuprates and alkylcyanocuprates. J Org Chem 48:3986–3990
18. Goering HL, Kantner SS (1984) Alkylation of allylic derivatives. 8. Regio- and stereochemistry of alkylation of allylic carboxylates with lithium methylcyanocuprate. J Org Chem 49:422–426
19. Goering HL, Tseng CC (1985) Alkylation of allylic derivatives. 9. On the stereochemistry of alkylation of acyclic allylic alcohols by the Murahashi method. J Org Chem 50:1597–1599
20. Goering HL, Kantner SS, Seitz EP (1985) Alkylation of allylic derivatives. 10. Relative rates of reactions of allylic carboxylates with lithium dimethylcuprate. J Org Chem 50:5495–5499
21. Tseng CC, Paisley SD, Goering HL (1986) Alkylation of allylic derivatives. 11. Copper(I)-catalyzed cross coupling of allylic carboxylates with Grignard reagents. J Org Chem 51: 2884–2891
22. Tseng CC, Yen SJ, Goering HL (1986) Alkylation of allylic derivatives. 12. Stereochemistry of copper(I)-catalyzed cross coupling of allylic carboxylates with Grignard reagents. J Org Chem 51:2892–2895
23. Underiner TL, Goering HL (1988) Alkylation of allylic derivatives. 13. Cross-coupling reactions of the isomeric 2,3,4,4a,5,6-hexahydro-2-naphthalenyl carboxylates with organocopper and Grignard reagents. J Org Chem 53:1140–1146
24. Underiner TL, Paisley SD, Schmitter J, Lesheski L, Goering HL (1989) Alkylation of allylic derivatives. 14. Relationship of double-bond configuration between reactant and product for cross-coupling reactions of Z-allylic carboxylates with organocopper reagents. J Org Chem 54:2369–2374
25. Underiner TL, Goerinng HL (1989) Cross coupling of allylic derivatives. 15. Regio- and stereospecfic cross-coupling reactions of dienyl allylic N-phenylcarbamates with phenylcopper reagents. J Org Chem 54:3239–3240

26. Underiner TL, Goering HL (1990) Cross coupling of allylic derivatives. 16. Regiochemistry of cross coupling of the isomeric (E, E)-3,5-heptadien-2-yl and (E, E)-2,5-heptadien-4-yl pivalates with organocopper and Grignard reagents. J Org Chem 55:2757–2761
27. Underiner TL, Goering HL (1991) Alkylation of allylic derivatives. 17. Cross-coupling reactions of diallylic pivalates with butyl- and phenylcopper reagents. J Org Chem 56: 2563–2572
28. Mori S, Nakamura E, Morokuma K (2000) Mechanism of SN2 alkylation reactions of lithium organocuprate clusters with alkyl halides and epoxides. Solvent Effects, BF3 effects, and trans-diaxial epoxide opening. J Am Chem Soc 122:7294–7307
29. Yamanaka M, Kato S, Nakamura E (2004) Mechanism and regioselectivity of reductive elimination of π-allylcopper (III) intermediates. J Am Chem Soc 126:6287–6293
30. Norinder J, Bäckvall JE, Yoshikai N, Nakamura E (2006) Unusual homocoupling in the reaction of diorganocuprates with an allylic halide. Organometallics 25:2129–2132
31. Yoshikai N, Zhang SL, Nakamura B (2008) Origin of the regio- and stereoselectivity of allylic substitution of organocopper reagents. J Am Chem Soc 130:12862–12863
32. Breit B, Demel P (2002) In: Krause N (ed) Modern organocopper chemistry. Wiley-VCH, Weinheim, p 210
33. Spino C (2009) In: Rappoport Z, Marek I (eds) The chemistry of organocopper compounds. Wiley, Hoboken, p 603
34. Christoffers J, Mann A (2001) Enantioselective construction of quaternary stereocenters. Angew Chem Int Ed 40:4591–4597
35. Karlström ASE, Bäckvall JE (2002) In: Krause N (ed) Modern organocopper chemistry. Wiley-VCH, Weinheim, p 259
36. Christoffers J, Baro A (2005) Stereoselective construction of quaternary stereocenters. Adv Synth Catal 347:1473–1482
37. Kar A, Argade NP (2005) A concise account of recent SN2′ Grignard coupling reactions in organic synthesis. Synthesis 2995–3022
38. Yorimitsu H, Oshima K (2005) Recent progress in asymmetric allylic substitutions catalyzed by chiral copper complexes. Angew Chem Int Ed 44:4435–4439
39. Alexakis A, Malan C, Lea L, Tissot-Croset K, Polet D, Falciola C (2006) The Copper-catalyzed asymmetric allylic substitution. Chimia 60:124–130
40. Geurts K, Fletcher SP, van Zijl AW, Minnaard AJ, Feringa BL (2008) Copper-catalyzed asymmetric allylic substitution reactions with organozinc and Grignard reagents. Pure Appl Chem 80:1025–1037
41. Harutyunyan SR, den Hartog T, Geurts K, Minnaard AJ, Feringa BL (2008) Catalytic asymmetric conjugate addition and allylic alkylation with Grignard reagents. Chem Rev 108:2824–2852
42. Alexakis A, Bäckvall JE, Krause N, Pamies O (2008) Enantioselective copper-catalyzed conjugate addition and allylic substitution reactions. Chem Rev 108:2796–2823
43. Falciola CA, Alexakis A (2008) Copper-catalyzed asymmetric allylic alkylation. Eur J Org Chem 3765–3780
44. Vanklaveren M, Persson ESM, Delvillar A, Grove DM, Bäckvall JE, van Koten G (1995) Chiral arenethiolatocopper(I) catalyzed substitution reactions of acyclic allylic substrates with Grignard reagents. Tetrahedron Lett 36:3059–3062
45. Meuzelaar GJ, Karlstrom ASE, van Klaveren M, Persson ESM, del Villar A, van Koten G (2000) Asymmetric Induction in the arenethiolatocopper(I)-catalyzed substitution reaction of Grignard reagents with allylic substrates. Tetrahedron 56:2895–2903
46. Karlstrom ASE, Huerta FF, Meuzelaar GJ, Bäckvall JE (2001) Ferrocenyl thiolates as ligands in the enantioselective copper-catalyzed substitution of allylic acetates with Grignard reagents. Synlett 923–926
47. Cotton HK, Norinder J, Bäckvall JE (2006) Screening of ligands in the asymmetric metallocenethiolatocopper(I)-catalyzed allylic substitution with Grignard reagents. Tetrahedron 62: 5632–5640

48. Dübner F, Knockel P (1999) Copper(I)-catalyzed enantioselective substitution of allyl chlorides with diorganozinc compounds. Angew Chem Int Ed 38:379–381
49. Dübner F, Knochel P (2000) Highly enantioselective copper-catalyzed substitution of allylic chlorides with diorganozincs. Tetrahedron Lett 41:9233–9237
50. Goldsmith PJ, Teat SJ, Woodward S (2005) Enantioselective preparation of β, β-disubstituted α-methylenepropionates by MAO promotion of the zinc Schlenk equilibrium. Angew Chem Int Ed 44:2235–2237
51. Borner C, Gimeno J, Gladiali S, Goldsmith PJ, Ramazzotti D, Woodward S (2000) Asymmetric chemo- and regiospecific addition of organozinc reagents to Baylis–Hillman derived allylic electrophiles. Chem Commun 2433–2434
52. Alexakis A, Malan C, Lea L, Benhaim C, Fournioux X (2001) Enantioselective copper-catalyzed SN2′ substitution with Grignard reagents. Synlett 927–930
53. Alexakis A, Croset K (2002) Tandem copper-catalyzed enantioselective allylation–metathesis. Org Lett 4:4147–4149
54. Malda H, van Zijl AW, Arnold LA, Feringa BL (2001) Enantioselective copper-catalyzed allylic alkylation with dialkylzincs using phosphoramidite ligands. Org Lett 3:1169–1171
55. Shi WJ, Wang LX, Fu Y, Zhu SF, Zhou QL (2003) Highly regioselective asymmetric copper-catalyzed allylic alkylation with dialkylzincs using monodentate chiral spiro phosphoramidite and phosphite ligands. Tetrahedron Asymmetr 14:3867–3872
56. Van Zijl AW, Arnold LA, Minnaard AJ, Feringa BL (2004) Highly enantioselective copper-catalyzed allylic alkylation with phosphoramidite ligands. Adv Synth Catal 346:413–420
57. Tissot-Croset K, Polet D, Alexakis A (2004) A highly effective phosphoramidite ligand for asymmetric allylic substitution. Angew Chem Int Ed 43:2426–2428
58. Tissot-Croset K, Polet D, Gille S, Hawner C, Alexakis A (2004) Synthesis and use of a phosphoramidite ligand for the copper-catalyzed enantioselective allylic substitution. Tandem allylic substitution/ring-closing metathesis. Synthesis 2586–2590
59. Tissot-Croset K, Alexakis A (2004) Copper catalyzed enantioselective allylic substitution by MeMgX. Tetrahedron Lett 45:7375–7378
60. Luchaco-Cullis CA, Mizutani H, Murphy KE, Hoveyda AH (2001) Modular pyridinyl peptide ligands in asymmetric catalysis: enantioselective synthesis of quaternary carbon atoms through copper-catalyzed allylic substitutions. Angew Chem Int Ed 40:1456–1460
61. Kacprzynski MA, Hoveyda AH (2004) Cu-catalyzed asymmetric allylic alkylations of aromatic and aliphatic phosphates with alkylzinc reagents. An effective method for enantioselective synthesis of tertiary and quaternary carbons. J Am Chem Soc 126:10676–10681
62. Hoveyda AH, Hird AW, Kacprzynski MA (2004) Small peptides as ligands for catalytic asymmetric alkylations of olefins. Rational design of catalysts or of searches that lead to them? Chem Commun 1779–1785
63. Ongeri S, Piarulli U, Roux M, Monti C, Gennari C (2002) Synthesis and screening of new chiral ligands for the copper-catalysed enantioselective allylic substitution. Helv Chim Acta 85:3388–3399
64. Lopez F, van Zijl AW, Minnaard AJ, Feringa BL (2006) Highly enantioselective Cu-catalysed allylic substitutions with Grignard reagents. Chem Commun 409–411
65. Selim KB, Yamada K, Tomioka K (2008) Copper-catalyzed asymmetric allylic substitution with aryl and ethyl Grignard reagents. Chem Commun 5140–5142
66. Yoshikai N, Miura K, Nakamura E (2009) Enantioselective copper-catalyzed allylic substitution reaction with aminohydroxyphosphine ligand. Adv Synth Catal 351:1014–1018
67. Bourissou D, Guerret O, Gabbaie FP, Bertrand G (2000) Stable carbenes. Chem Rev 100: 39–92
68. César V, Bellemin-Laponnaz S, Gade LH (2004) Chiral N-heterocyclic carbenes as stereo-directing ligands in asymmetric catalysis. Chem Soc Rev 33:619–636
69. Crudden CM, Allen DP (2004) Stability and reactivity of N-heterocyclic carbene complexes. Coord Chem Rev 248:2247–2273

70. Crabtree RH (2005) NHC ligands versus cyclopentadienyls and phosphines as spectator ligands in organometallic catalysis. J Organomet Chem 690:5451–5457
71. Dible BR, Sigman MS (2006) Steric effects in the aerobic oxidation of π-allylnickel(II) complexes with N-heterocyclic carbenes. Inorg Chem 45:8430–8441
72. Clavier H, Nolan SP (2007) N-heterocyclic carbenes: advances in transition metal-mediated transformations and organocatalysis. Chem B Org Chem 103:193–222
73. Kuehl O (2007) The chemistry of functionalised N-heterocyclic carbenes. Chem Soc Rev 36:592–607
74. Liddle ST, Edworthy IS, Arnold PL (2007) Anionic tethered N-heterocyclic carbene chemistry. Chem Soc Rev 36:1732–1744
75. Diez-Gonzalez S, Nolan SP (2007) Stereoelectronic parameters associated with N-heterocyclic carbene (NHC) ligands: a quest for understanding. Coord Chem Rev 251:874–883
76. Marion N, Nolan SP (2008) N-heterocyclic carbenes in gold catalysis. Chem Soc Rev 37: 1776–1782
77. Kelly RA III, Clavier H, Giudice S, Scott NM, Stevens ED, Bordner J, Samardjiev I, Hoff CD, Cavallo L, Nolan SP (2008) Determination of N-heterocyclic carbene (NHC) steric and electronic parameters using the [(NHC)Ir(CO)2Cl] system. Organometallics 27:202–210
78. Tominaga S, Oi Y, Kato T, An DK, Okamoto S (2004) γ-selective allylic substitution reaction with Grignard reagents catalyzed by copper N-heterocyclic carbene complexes and its application to enantioselective synthesis. Tetrahedron Lett 45:5585–5588
79. Okamoto S, Tominaga S, Saino N, Kase K, Shimoda K (2005) Allylic substitution reactions with Grignard reagents catalyzed by imidazolium and 4,5-dihydroimidazolium carbene–CuCl complexes. J Organomet Chem 690:6001–6007
80. Alexakis A, Winn CL, Guillen F, Pytkowicz J, Roland S, Mangeney P (2003) Asymmetric synthesis with *N*-heterocyclic carbenes. Application to the copper-catalyzed conjugate addition. Adv Synth Catal 345:345–348
81. Larsen AO, Leu W, Oberhuber CN, Campbell JE, Hoveyda AH (2004) Bidentate NHC-based chiral ligands for efficient Cu-catalyzed enantioselective allylic alkylations: structure and activity of an air-stable chiral Cu complex. J Am Chem Soc 126:11130–11131
82. Van Veldhuizen JJ, Campbell JE, Giudici RE, Hoveyda AH (2005) A readily available chiral Ag-based N-heterocyclic carbene complex for use in efficient and highly enantioselective Ru-catalyzed olefin metathesis and Cu-catalyzed allylic alkylation reactions. J Am Chem Soc 127:6877–6882
83. Lee Y, Akiyama K, Gillingham DG, Brown MK, Hoveyda AH (2008) Highly site- and enantioselective Cu-catalyzed allylic alkylation reactions with easily accessible vinylaluminum reagents. J Am Chem Soc 130:446–447
84. Akiyama K, Gao F, Hoveyda AH (2010) Stereoisomerically pure trisubstituted vinylaluminum reagents and their utility in copper-catalyzed enantioselective synthesis of 1,4-dienes containing *Z* or *E* alkenes. Angew Chem Int Ed 49:419–423
85. Selim KB, Matsumoto Y, Yamada K, Tomioka K (2009) Efficient chiral N-heterocyclic carbene/copper(I)-catalyzed asymmetric allylic arylation with aryl Grignard reagents. Angew Chem Int Ed 48:8733–8735
86. Jennequin T, Wencel-Delord J, Rix D, Daubignard J, Crévisy C, Mauduit M (2010) Chelating hydroxyalkyl NHC as efficient chiral ligands for room-temperature copper-catalyzed asymmetric allylic alkylation. Synlett 1661–1665
87. Murphy KE, Hoveyda AH (2003) Enantioselective synthesis of α-alkyl-β, γ-unsaturated esters through efficient Cu-catalyzed allylic alkylations. J Am Chem Soc 125:4690–4691
88. Murphy KE, Hoveyda AH (2005) Catalytic enantioselective synthesis of quaternary all-carbon stereogenic centers. Preparation of α, α′-disubstituted β, γ-unsaturated esters through Cu-catalyzed asymmetric allylic alkylations. Org Lett 7:1255–1258
89. Carosi L, Hall DG (2007) Catalytic enantioselective preparation of α-substituted allylboronates: One-pot addition to functionalized aldehydes and a route to chiral allylic trifluoroborate reagents. Angew Chem Int Ed 46:5913–5915

90. Kacprzynski MA, May TL, Kazane SA, Hoveyda AH (2007) Enantioselective synthesis of allylsilanes bearing tertiary and quaternary Si-substituted carbons through Cu-catalyzed allylic alkylations with alkylzinc and arylzinc reagents. Angew Chem Int Ed 46:4554–4558
91. Falciola CA, Tissot-Croset K, Alexakis A (2006) β-disubstituted allylic chlorides: substrates for the Cu-catalyzed asymmetric SN2′ reaction. Angew Chem Int Ed 45:5995–5998
92. Falciola CA, Tissot-Croset K, Reyneri H, Alexakis A (2008) β- and γ-disubstituted olefins: substrates for copper-catalyzed asymmetric allylic substitution. Adv Synth Catal 350: 1090–1100
93. Kapat A, Nyfeler E, Giuffredi GT, Renaud P (2009) Intramolecular Schmidt reaction involving primary azidoalcohols under nonacidic conditions: synthesis of indolizidine (−)-167B. J Am Chem Soc 131:17746
94. Giacomina F, Riat D, Alexakis A (2010) ω-ethylenic allylic substrates as alternatives to cyclic substrates in copper- and iridium-catalyzed asymmetric allylic alkylation. Org Lett 12:1156–1159
95. Falciola CA, Alexakis A (2007) 1,4-dichloro- and 1,4-dibromo-2-butenes as substrates for Cu-catalyzed asymmetric allylic substitution. Angew Chem Int Ed 46:2619–2622
96. Falciola CA, Alexakis A (2008) High diversity on simple substrates: 1,4-dihalo-2-butenes and other difunctionalized allylic halides for copper-catalyzed SN2′ reactions. Chem Eur J 14:10615–10627
97. Van Zijl AW, Szymanski W, Lopez F, Minnaard AJ, Feringa BL (2008) Catalytic enantioselective synthesis of vicinal dialkyl arrays. J Org Chem 73:6994–7002
98. Geurts K, Fletcher SP, Feringa BL (2006) Copper catalyzed asymmetric synthesis of chiral allylic esters. J Am Chem Soc 128:15572–15573
99. Piarulli U, Daubos P, Claverie C, Roux M, Gennari C (2003) A catalytic and enantioselective desymmetrization of *meso* cyclic allylic bisdiethylphosphates with organozinc reagents. Angew Chem Int Ed 42:234–236
100. Piarulli U, Daubos P, Claverie C, Monti C, Gennari C (2005) Copper-catalysed, enantioselective desymmetrisation of *meso* cyclic allylic bis(diethyl phosphates) with organozinc reagents. Eur J Org Chem 895–906
101. Piarulli U, Claverie C, Daubos P, Gennari C, Minnaard AJ, Feringa BL (2003) Copper phosphoramidite-catalyzed enantioselective desymmetrization of meso-cyclic allylic bisdiethyl phosphates. Org Lett 5:4493–4496
102. Bertozzi F, Crotti P, Macchia F, Pineschi M, Arnold A, Feringa BL (2000) A new catalytic and enantioselective desymmetrization of symmetrical methylidene cycloalkene oxides. Org Lett 2:933–936
103. Del Moro F, Crotti P, Di Bussolo V, Macchia F, Pineschi M (2003) Catalytic enantioselective desymmetrization of COT-monoepoxide. Maximum deviation from coplanarity for an SN2′-cuprate alkylation. Org Lett 5:1971–1974
104. Bertozzi, F, Crotti P, Del Moro F, Di Bussolo V, Macchia F, Pineschi M (2003) Copper-catalysed addition of organometallic reagents to vinyl diepoxides – a novel route to oxa-bridged systems and to substituted allylic alcohols. Eur J Org Chem 1264–1270
105. Bertozzi F, Pineschi M, Macchia F, Arnold LA, Minnaard AJ, Feringa BL (2002) Copper phosphoramidite catalyzed enantioselective ring-opening of oxabicyclic alkenes: remarkable reversal of stereocontrol. Org Lett 4:2703–2705
106. Zhang W, Zhu SF, Qiao XC, Zhou QL (2008) Highly enantioselective copper-catalyzed ring opening of oxabicyclic alkenes with Grignard reagents. Chem Asian J 3:2105–2111
107. Millet R, Bernardez T, Palais L, Alexakis A (2009) Copper-catalyzed desymmetrization of oxabenzonorbornadienes with aluminum reagents. Tetrahedron Lett 50:3474–3477
108. Millet R, Gremaud L, Bernardez T, Palais L, Alexakis A (2009) Copper-catalyzed asymmetric ring-opening reaction of oxabenzonorbornadienes with Grignard and aluminum reagents. Synthesis 2101–2112

109. Pineschi M, Del Moro F, Crotti P, Macchia F (2005) Catalytic asymmetric ring opening of 2,3-substituted norbornenes with organometallic reagents: A new formal aza functionalization of cyclopentadiene. Org Lett 7:3605–3607
110. Bournaud C, Falciola C, Lecourt T, Rosset S, Alexakis A, Micouin L (2006) On the use of phosphoramidite ligands in copper-catalyzed asymmetric transformations with trialkylaluminum reagents. Org Lett 8:3581–3584
111. Palais L, Mikhel IS, Bournaud C, Micouin L, Falciola CA, Vuagnoux-d'Augustin M, Rosset S, Bernardinelli G, Alexakis A (2007) SimplePhos monodentate ligands: synthesis and application in copper-catalyzed reactions. Angew Chem Int Ed 46:7462–7465
112. Palais L, Bournaud C, Micouin L, Alexakis A (2010) Copper-catalysed ring opening of polycyclic *meso*-hydrazines with trialkylaluminium reagents and simplephos ligands. Chem Eur J 16:2567–2573
113. Gillingham DG, Hoveyda AH (2007) Chiral N-heterocyclic carbenes in natural product synthesis: application of Ru-catalyzed asymmetric ring-opening/cross-metathesis and Cu-catalyzed allylic alkylation to total synthesis of baconipyrone C. Angew Chem Int Ed 46:3860–3864
114. Pineschi M (2004) Copper-catalyzed enantioselective allylic alkylation ring-opening reactions of small-ring heterocycles with hard alkyl metals. New J Chem 28:657–665
115. Badalassi F, Crotti P, Macchia F, Pineschi M, Arnold A, Feringa BL (1998) Catalytic enantioselective carbon–carbon bond formation by addition of dialkylzinc reagents to cyclic 1,3-diene monoepoxides. Tetrahedron Lett 39:7795–7798
116. Gini F, Del Moro F, Macchia F, Pineschi M (2003) Regio- and enantioselective copper-catalyzed addition of dialkylzinc reagents to cyclic 2-alkenyl aziridines. Tetrahedron Lett 44:8559–8562
117. Millet R, Alexakis A (2007) Copper-catalyzed kinetic resolution of 1,3-cyclohexadiene monoepoxide with Grignard reagents. Synlett 435–438
118. Kagan HB, Fiaud JC (1988) Kinetic resolution. Top Stereochem 18:249–330
119. Millet R, Alexakis A (2008) SimplePhos as efficient ligand for the copper-catalyzed kinetic resolution of cyclic vinyloxiranes with grignard reagents. Synlett 1797–1800
120. Bertozzi F, Crotti P, Macchia F, Pineschi M, Feringa BL (2001) Highly enantioselective regiodivergent and catalytic parallel kinetic resolution. Angew Chem Int Ed 40:930–932
121. Pineschi M, Del Moro F, Crotti P, Di Bussolo V, Macchia F (2004) Catalytic regiodivergent kinetic resolution of allylic epoxides: a new entry to allylic and homoallylic alcohols with high optical purity. J Org Chem 69:2099–2105
122. Pineschi M, Del Moro F, Crotti P, Di Bussolo V, Macchia F (2005) Copper-catalyzed highly enantioselective synthesis of cyclic allylic and homoallylic alcohols with dialkylzinc reagents. Synthesis 334–337
123. Trost BM, Bunt RC (1994) Asymmetric induction in allylic alkylations of 3-(acyloxy) cycloalkenes. J Am Chem Soc 116:4089–4090
124. Trost BM, Fandrick DR (2007) Palladium-catalyzed dynamic kinetic asymmetric allylic alkylation with the DPPBA ligands. Adrichim Acta 40:59–72
125. Langlois JB, Alexakis A (2009) Dynamic kinetic asymmetric transformation in copper catalyzed allylic alkylation. Chem Commun 3868–3870
126. Langlois JB, Alexakis A (2010) Copper-catalyzed asymmetric allylic alkylation of racemic cyclic substrates: application of dynamic kinetic asymmetric transformation (DYKAT). Adv Synth Catal 352:447–457

Top Organomet Chem (2012) 38: 269–320
DOI: 10.1007/3418_2011_15

Published online: 24 September 2011

Allylic Substitutions Catalyzed by Miscellaneous Metals

Jeanne-Marie Begouin, Johannes E.M.N. Klein, Daniel Weickmann, and Bernd Plietker

Abstract Allylic substitutions catalyzed by miscellaneous metals have recently been uncovered as useful alternatives to the established corresponding transition metal catalyzed transformations. In particular, the interesting regioselectivity course of the allylic substitutions is of synthetic interest. In this chapter, we summarize the most recent findings in the field of group 8–10 metal (Fe, Ru, Co, Rh, Ni, Pt) catalyzed substitutions with a strong emphasis on the substrate range and the regioselectivity.

Keywords Allylic Substitution · Group 8-10 · Miscellaneous Metals · Regioselectivity · Substrate Range

Contents

J.-M. Begouin, J.E.M.N. Klein, D. Weickmann, and B. Plietker (✉)
Institute of Organic Chemistry, University of Stuttgart, Pfaffenwaldring 55, 70569 Stuttgart, Germany
e-mail: bernd.Plietker@oc.uni-stuttgart.de

Abbreviations

Ac	Acetyl
acac	Acetylacetonate
Alk	Alkyl
Ar	Aryl
bmim	Butylmethylimidazolium
Boc	*tert*-Butyloxycarbonyl
cat	Catalytic
cod	Cyclooctadiene
Cp	Cyclopentadienyl
Cp*	Pentamethylcyclopentadienyl
DCE	1,2-Dichloroethane
DCM	Dichloromethane
DMA	*N,N*-Dimethylacetamide
DME	1,2-Dimethoxyethane
DMF	*N,N*-Dimethylformamide
DMM	Dimethylmalonate
DPM	Bis(diphenylphosphino)methan
dppe	Bis(diphenylphosphino)ethane
dppf	Bis(diphenylphosphino)ferrocene
dr	Diastereomeric ratio
en	1,2-Ethylenediamine
equiv	Equivalent(s)
Et	Ethyl
FG	Functional Group
h	Hour(s)
i-Bu	*iso*-Butyl
i-Pr	*iso*-Propyl
LG	Leaving group
Mbs	*N*-4-Methoxybenzenesulfonyl toluidin
Me	Methyl
Mes	Mesityl
min	Minute(s)
mol	Mole(s)
MTBE	*tert*-Butylmethylether
NHC	*N*-Heterocyclic carbene
NMP	*N*-Methyl-pyrrolidinon
Nu	Nucleophile
Pent	Pentyl
Ph	Phenyl
Pin	Pinacol
pip	Piperidinium
PMP	*p*-Methoxyphenyl

rt	Room temperature
S_N	Nucleophilic Substitution
TBAF	Tetrabutylammonium fluoride
t-Bu	*tert*-Butyl
Tfa	Trifluoroacetic acid
THF	Tetrahydrofuran
tmeda	*N,N,N′,N′*-Tetramethyl-1,2-ethylenediamine
TMS	Trimethylsilyl
TsOH	Toluenesulfonic acid

1 Introduction

The combination of a polarized C–X-bond and an allylic π-bond opposes fundamental mechanistic changes in substitution chemistry. Due to the electron-withdrawing nature of the leaving group, the allyl moiety might be regarded as an a^1–a^3-system. Hence, a control of the regioselective course is necessary in order to obtain a synthetically useful catalytic method. However, independent of the nature of the catalyst, this synthetic requirement is difficult to achieve. Employing Lewis-acidic metal complexes the nucleophilic substitution might follow an S_N2-type mechanism (Scheme 1), in which the metal activates the leaving group. Depending on the nature of the leaving group and the Lewis acidic or basic character of the catalyst, a concerted stereo- and regioselective S_N2- or S_N2'-type reaction might take place. In the presence of strongly acidic catalysts and/or good leaving groups,

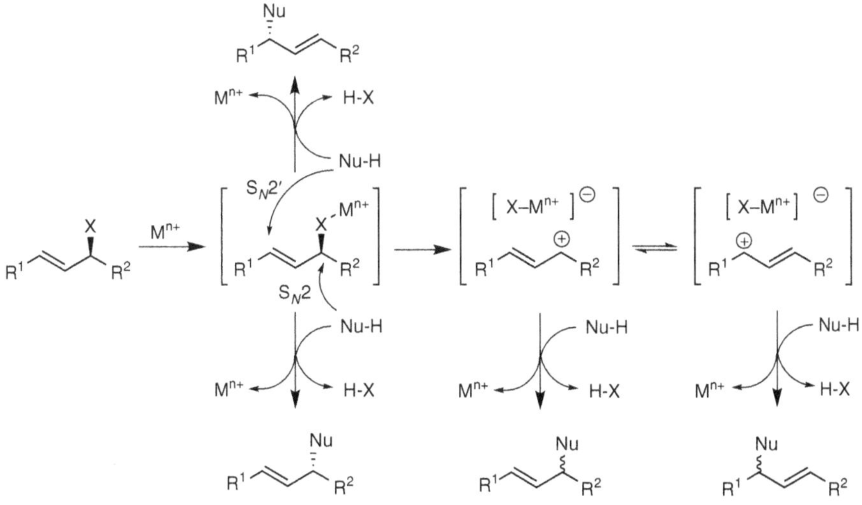

Scheme 1 Regioselectivities in allylic substitutions catalyzed by Lewis-acidic complexes

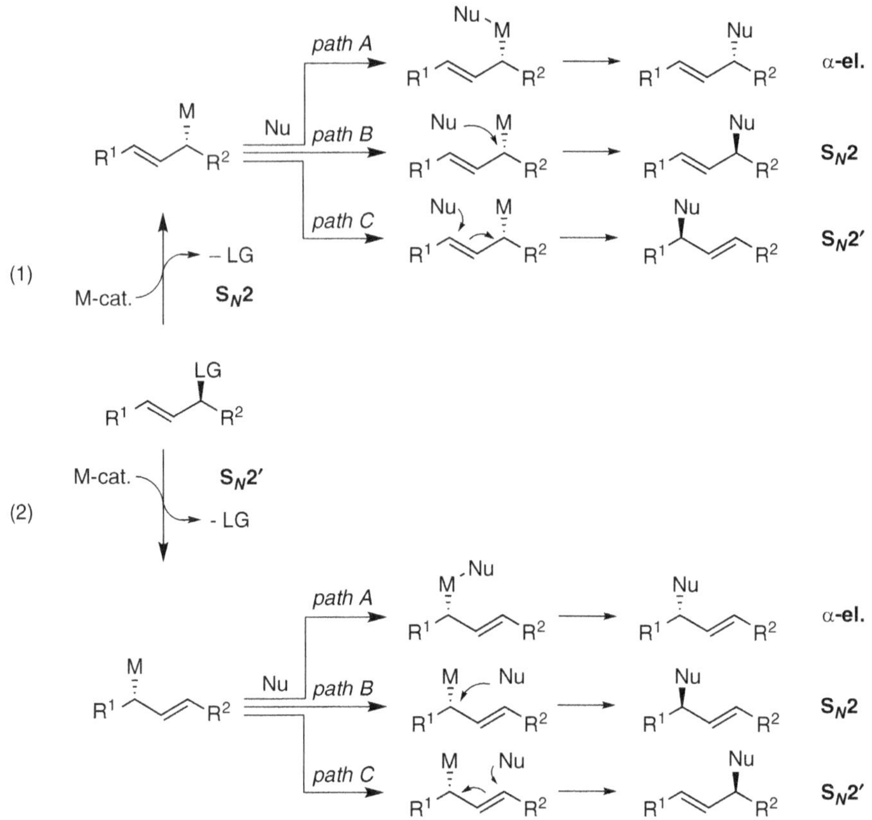

Scheme 2 Regioselectivities in allylic substitutions catalyzed by low-valent metal complexes

allylic cations might be generated leading to a mixture of regio- and stereoisomeric products (Scheme 1).

Substitutions involving low-valent metal complexes also encounter regio- and stereoselectivity problems (Scheme 2). The nucleophilic substitution of the leaving group can occur either in an S_N2- [(1) in Scheme 2] or S_N2'-type mechanism [(2) in Scheme 2]. Depending on the nature of the nucleophile and catalyst employed, the subsequent nucleophilic substitution of the metal can follow either via α-elimination [*path A*, (1) and (2) in Scheme 2], via S_N2-reaction [*path B*, (1) and (2) in Scheme 2] or via S_N2'-type reaction [*path C*, (1) and (2) in Scheme 2]. For reasons of clarity, only strictly concerted and stereospecific S_N2- or S_N2'-antitype mechanistic scenarios are shown in Scheme 2. The situation might, however, be complicated if e.g. the initial S_N2'-anti ionization event is competing with an S_N2'-syn reaction. Erosion in stereo- and regioselectivity can be the result of these competing reactions.

Furthermore, fluxional intermediates such as π-allyl metal-complexes are not shown for reasons of clarity. However, such behavior is known for a variety of late transition metal allyl complexes. This field of allylic substitutions has for long time been a testing ground for low-valent Pd-complexes. Due to tremendous efforts by chemists with an organometallic and synthetic background, this type of catalytic transformation was elaborated into one of the most potent transformation within the field of organometallic catalysis. However, recently other transition metals attracted significant interest within the catalytic community in this field. Different reasons might account for this development with the fact that Pd-complexes do not allow for a general and strict control of the regioselective course of a given transformation might be the most prominent problem one.

In this chapter, the field of allylic substitutions by miscellaneous metals will be covered. We concentrated on the use of group 8–10 metals of periods 4 and 5 since most of the catalytic transformations employing these metals display an unusual high degree of regioselectivity control.

2 Allylic Substitutions Catalyzed by Group 8 Metals (Without Osmium)

2.1 *Iron Catalyzed Allylic Substitutions*

Iron catalyzed substitution reactions can be mainly divided into two categories depending on the catalytic character of the employed iron source. Several formal substitutions are catalyzed by Fe^{3+} salts. In these reactions, the iron salt acts as a Lewis acid, which coordinates the leaving group in the starting material. Due to this activation of the substrate, the attack of the nucleophile is facilitated.

Anionic iron complexes (i.e., ferrate complexes) have found increasing application as nucleophilic catalysts in substitution chemistry. In these reactions, the leaving group is replaced by nucleophilic attack of the ferrate complex in the first step. External attack of a nucleophile or transfer of a ligand from the iron atom via reductive elimination finally generates the substitutions product.

2.1.1 Reactions Catalyzed by Lewis-Acidic Fe-Salts

A complete overview of the field of substitution reactions in which iron salts act as Lewis acidic catalysts is given by a recent review [1]. Here, we like to show some representative examples.

Jana and coworkers reported a direct alkylation of active methylene compounds with benzylic and allylic alcohols catalyzed by $FeCl_3$ (Scheme 3) [2]. In the reaction of cinnamyl alcohol with acetylacetone, the formation of the *ipso*-substitution product was favored. This type of direct substitution of a leaving group was also extended toward propargylic compounds. A number of *O*-, *N*-, or *S*-nucleophiles as well as alkylsilanes were able to react with substituted propargylic alcohols or acetates in the presence of $FeCl_3$ [3, 4].

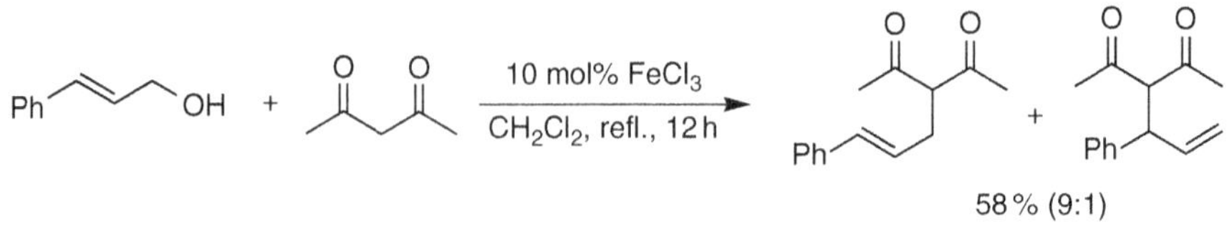

10 mol% $FeCl_3$, CH_2Cl_2, refl., 12 h

58 % (9:1)

Scheme 3 $FeCl_3$ catalyzed direct alkylation with allylic alcohols

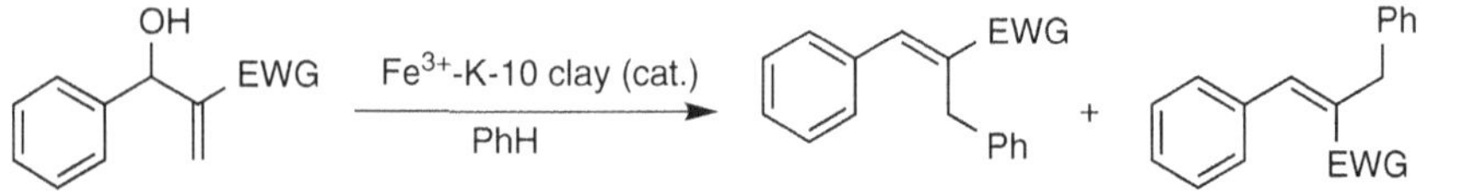

Fe^{3+}-K-10 clay (cat.), PhH

	E-olefin	*Z*-olefin	yield
(EWG = CO_2Me)	96	4	93 %
(EWG = CN)	1	99	89 %

Scheme 4 Lewis acid catalyzed conjugate addition of allylic alcohols

Although iron salts have been widely used as Lewis acidic catalyst in direct S_N reactions, there has been reported only one example for the activation of an allylic leaving group by an iron salt and subsequent conjugate substitution by an incoming nucleophile [5]. Various Baylis–Hillman adducts were coupled to several aromatic compounds in a Friedel–Crafts-type reaction (Scheme 4). Using an immobilized Fe^{3+} source (Fe^{3+} -K10 montmorillonite) as Catalyst, trisubstituted *E*- or *Z*-configured alkenes were obtained. The configuration of the alkenes was strongly dependent on the nature of the electron-withdrawing group in the 2-Position of the Baylis–Hillman adducts.

2.1.2 Reactions Catalyzed by Low-Valent Fe-Complexes

One possible way to generate ferrate complexes is the reaction of an iron salt with an excess of a Grignard or an alkyl zinc reagent. In many cases, these active iron species are formed in situ. Two structurally characterized low-valent Fe complexes are shown in Scheme 5. They are both catalytically low active, while the latter complex is most likely being involved in reactions catalyzed by an iron salt in combination with an alkylmetal reagent [6, 7].

Yamamoto and coworkers reported an iron catalyzed Kharasch-type reaction of Grignard reagents and allylic phosphates [8, 9]. The reaction proceeded in the presence of catalytic amounts of $Fe(acac)_3$ at low temperatures and with high selectivity in favor of the *ipso*-substitution products (Scheme 6). The double bond geometry of the applied allylic phosphonate had no influence on the selectivity of the reaction. This was taken as an experimental indication of a direct S_N-type reaction without intermediate formation of a π-allyl-Fe complex. Interestingly,

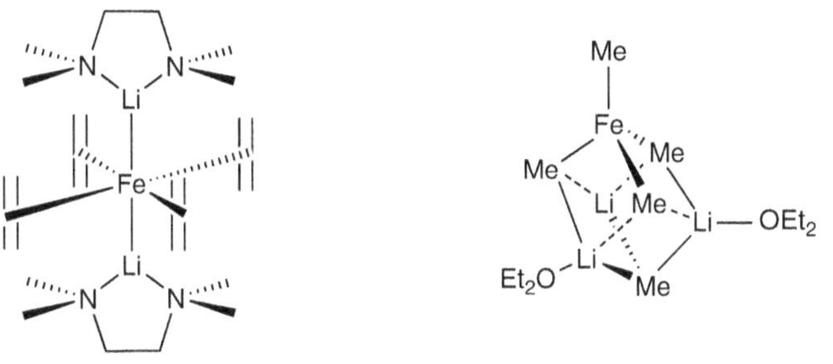

$[(C_2H_4)_4Fe][Li(TMEDA)]_2$

Jonas (1979)

$[((CH_3)_4Fe)(MeLi)][Li(OEt_2)]_2$

Fürstner (2006)

Scheme 5 Structures of isolated catalytically active low-valent Fe complexes

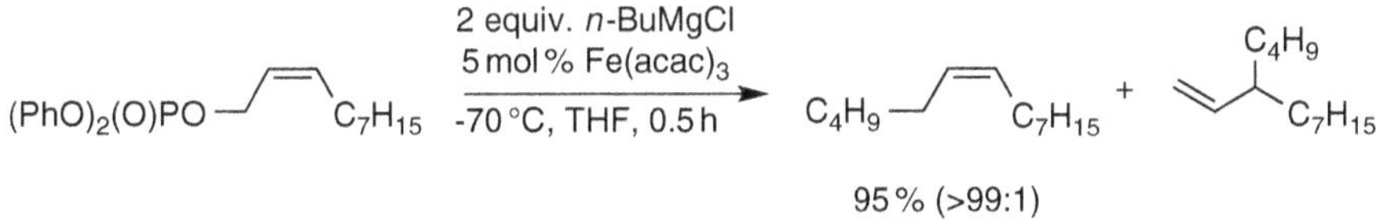

$(PhO)_2(O)PO$–CH₂CH=CH–C_7H_{15} → (2 equiv. *n*-BuMgCl, 5 mol% $Fe(acac)_3$, -70 °C, THF, 0.5 h) → C_4H_9–CH₂CH=CH–C_7H_{15} + branched product; 95% (>99:1)

Scheme 6 Fe-catalyzed substitution of primary allyl phosphates by Grignard reagents

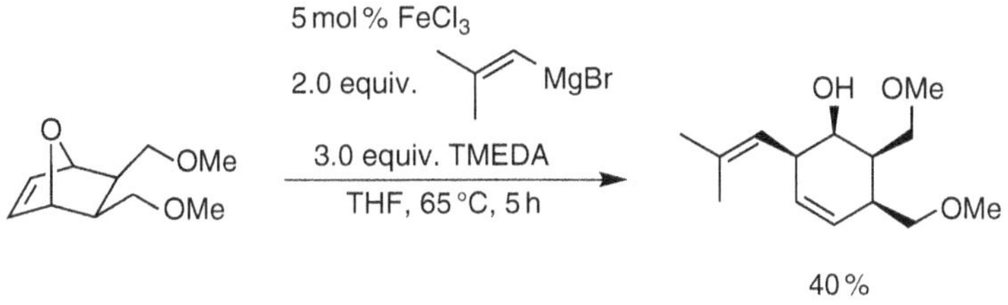

5 mol% $FeCl_3$, 2.0 equiv. MgBr (alkenyl Grignard), 3.0 equiv. TMEDA, THF, 65 °C, 5 h; OMe, OMe, OH, OMe; 40%

Scheme 7 Fe-catalyzed conjugate addition-ring opening of oxanorbornenes

the use of CuCN as catalyst led to a complete inversion of the regioselectivity, while $NiBr_2$ formed the same *ipso*-substitution products such as the iron catalyst. In contrast to these results, the regioselectivity of the analogous iron catalyzed reaction of a propargylic chloride with a Grignard reagent was shown to be highly dependent on the substitution pattern of the propargylic chloride [10].

Another interesting example of this type of reaction is an iron catalyzed conjugate addition-ring opening of oxanorbornenes, which was described by Nakamura and coworkers [11, 12]. The reaction proceeded by treatment of a *meso*-oxanorbornene with the catalyst, which was prepared in situ by addition of a Grignard reagent to $FeCl_3$ in the presence of TMEDA (Scheme 7). The four stereocenters of the ring-opening product were formed both regio- and stereoselective. The proposed mechanism involves carbometallation of the double bond, followed by reductive ring-opening and β-hydride elimination. Interestingly, the ferrate catalyst attacks the double bond from the exo face pointing into the direction of a probable precoordination between metal complex and the oxygen bridge. Based on the earlier results from Pasto and coworkers [13, 14], Fürstner and Mendez reported the closely related synthesis of optically allenol derivatives by addition of Grignard reagents to propargyl epoxides in the presence of $Fe(acac)_3$ [15].

Ladoulis and Nicholas used $Fe_2(CO)_9$ as catalyst for the allylic alkylation of allylic acetates by sodium dimethylmalonate [16, 17]. Mechanistic studies indicated that the Fe(0) complex acts as a precatalyst, which is activated by reaction with the nucleophile (Scheme 8). This ferrate complex as activated species undergoes a ligand exchange with the allylic acetate, which results in the formation of a π-allyl–Fe complex. Attack of the nucleophile liberates the substitution product and regenerates the catalytically active species. However, after these early results no further investigation on this system has been published since then.

Low valent iron complexes such as the Hieber anion $[Fe(CO)_3(NO)]^-$ in the oxidation state of –II are isoelectronic with Pd(0), which is a well-known catalyst for allylic substitution reactions [18–20]. Like Pd, preformed low valent iron complexes can form reactive allyl–Fe complexes as well. Depending on the nature of the iron complex used in the reaction, either σ- or π-allyl–Fe complexes are generated. After some early reports of π-allyl–Fe complexes [21–23], these types of complexes have found application in numerous noncatalytic transformations [24–26].

It was as early as 1979 when Roustan reported the use of catalytic amounts of an iron complex to perform an allylic substitution reaction [27, 28]. He showed that the

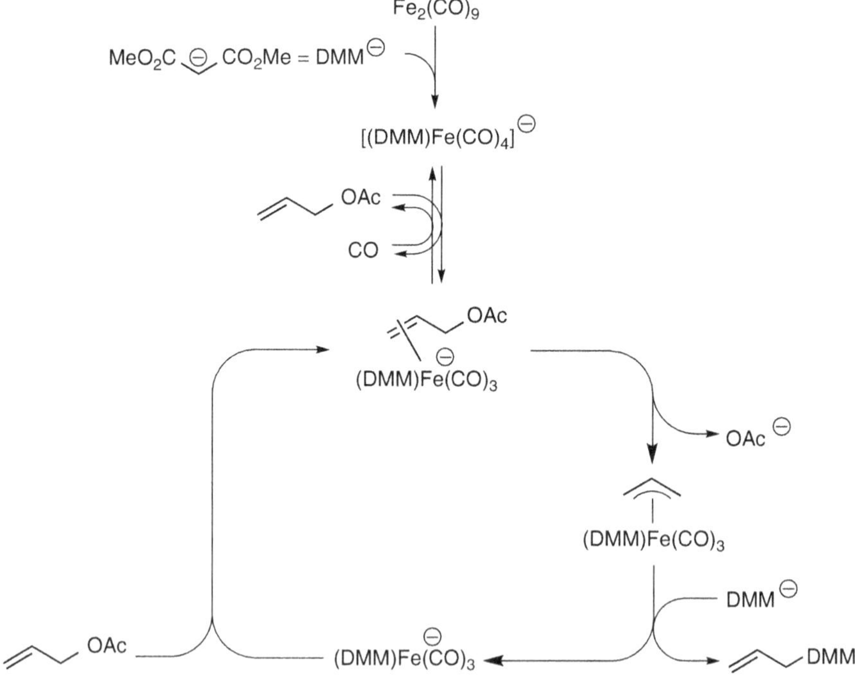

Scheme 8 Allylic alkylation using $Fe_2(CO)_9$ as catalyst – in situ generation of the ferrate

OLG → $NaCH(COOMe)_2$, 25 mol % cat. → MeO_2C / CO_2Me (branched) + MeO_2C / CO_2Me (linear)

80 : 20

		yield
Roustan (1979) :	$Na[Fe(CO)_3(NO)]$ **C1**, THF, reflux	85 %
Xu (1987) :	$[Bu_4N][Fe(CO)_3(NO)]$ **C2**, CO(g), THF, reflux	35 %

Scheme 9 Fe-catalyzed allylic substitution

low valent nitrosyl ferrate $Na[Fe(CO)_3(NO)]$ **C1** forms in the presence of allyl chloride or acetate an iron allyl complex, which reacts with dimethyl malonate to the substitution products (Scheme 9). By using the secondary branched allylic substrate, he observed that the formation of the *ipso*-substitution product was favored. But the reaction was slow and large catalyst loadings were necessary.

Later Xu introduced the more stable $[(Bu)_4N][Fe(CO)_3(NO)]$ **C2**, which is available from $Fe(CO)_5$, $NaNO_2$ and $(Bu)_4NBr$ [29, 30]. They also obtained a preference for the *ipso*-product, but in significantly lower yield. To maintain the catalytic activity of the complex, reactions were performed under CO-gas atmosphere.

Based on these early results, our group developed an improved version using a monodentate σ-donor ligand such as PPh_3 for the stabilization of the intermediate π-allyl Fe-complex [31]. Replacement of one CO-ligand by the phosphin was supposed to prevent the formation of a catalytically inactive allyl Fe-complex like **C4** (Scheme 10). By following this approach, the catalyst stability was improved and the toxic CO gas atmosphere was avoided. DMF as a coordinating solvent additionally increased the nucleophilicity of the ferrate complex.

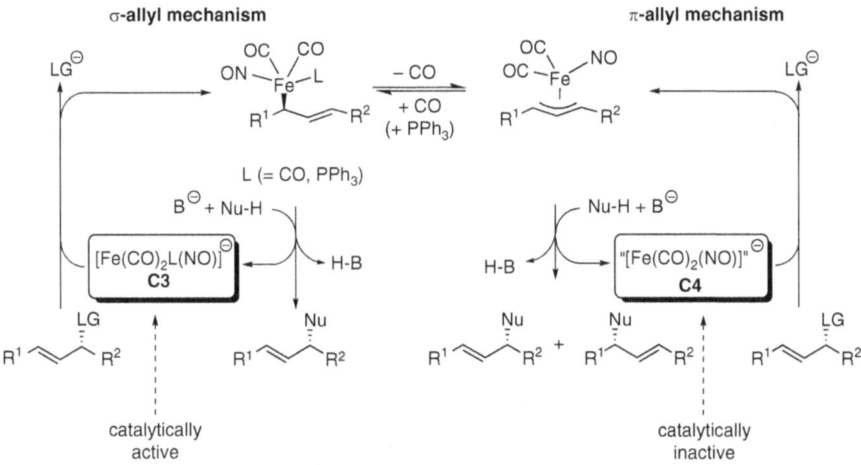

Scheme 10 σ- vs. π-allyl mechanism in Fe-catalyzed allylic substitutions

Various allyl carbonates were transferred under these reaction conditions into the substitution products in high yields and high regioselectivities in favor of the *ipso*-substitution products (Scheme 11). No preformation of the nucleophile was necessary, since the carbonate acted both as leaving group and as in situ base.

					yield
100	: 0		98	: 2	81 %
0	: 100		7	: 93	61 %

Conditions: $CH_2(COOMe)_2$, 2.5 mol % **C2**, 3 mol % PPh_3, DMF, 80 °C, 24 h

Scheme 11 The Fe-catalyzed allylic alkylation

The scope of this reaction was further extended to allylic aminations [32] and sulfonations (Scheme 12) [33]. Due to their basicity, the amination products tend to decompose the catalyst. Therefore, addition of catalytic amounts of piperidine hydrochloride (Pip·HCl) was necessary. According to the results of the alkylation reactions, several aromatic amines were allylated under these optimized reaction conditions regioselectively to the *ipso*-substitution products. The sulfonation reactions were carried out in a polar solvent mixture of DMF and 2-methoxyethanol to dissolve both the nucleophile and the carbonate. This choice of the solvent limits the range of ligands to phosphin ligands. With $P(C_6H_4OCH_3)_3$, which proved to be the ligand of choice, various aromatic sulfinates were converted in combination

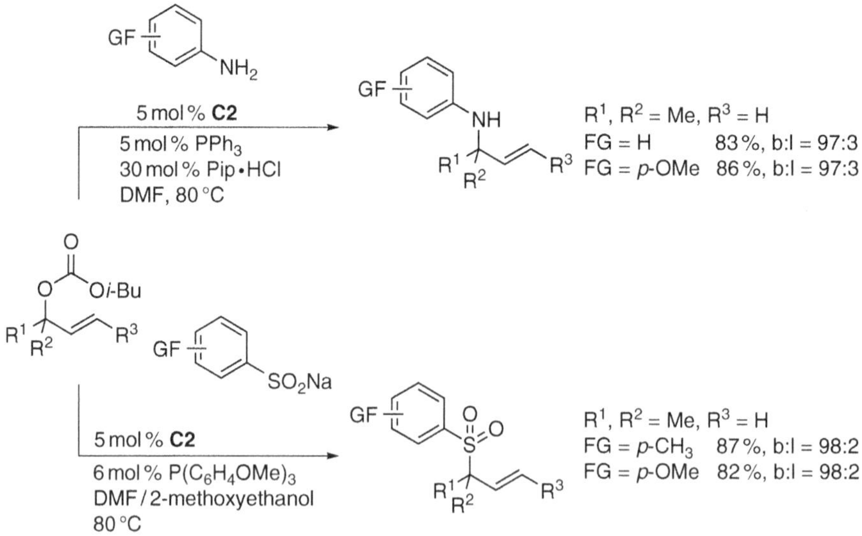

Amination: 5 mol % **C2**, 5 mol % PPh_3, 30 mol % Pip•HCl, DMF, 80 °C

R^1, R^2 = Me, R^3 = H
FG = H 83 %, b:l = 97:3
FG = *p*-OMe 86 %, b:l = 97:3

Sulfonation: 5 mol % **C2**, 6 mol % $P(C_6H_4OMe)_3$, DMF / 2-methoxyethanol, 80 °C

R^1, R^2 = Me, R^3 = H
FG = *p*-CH_3 87 %, b:l = 98:2
FG = *p*-OMe 82 %, b:l = 98:2

Scheme 12 Fe-catalyzed allylic amination and sulfonation with several functionalized carbonates in good to excellent yields and regioselectivities. If chiral allyl carbonates were used, complete retention of the stereochemistry was observed.

Further improvement of the allylic alkylation was achieved by exchange of the PPh_3 with an *N*-heterocyclic carbene ligand (NHC) [34]. In the presence of a *tert*-butyl-substituted NHC ligand **L1**, various *C*-nucleophiles were allylated with good to excellent regioselectivities and complete conversion. The reaction is supposed to proceed through an σ-allyl–Fe intermediate, and complete conservation of the double-bond geometrie of enantiomerically enriched (*E*)- or (*Z*)-carbonates was observed (Scheme 13) [35].

By changing the ligand's topology, a significant change in the regioselective course of the reaction was observed. Whereas the *tert*-butyl substituted ligand **L1** allowed a regio- and stereoselective *ipso*-substitution of branched allyl substrates, use of a sterically less hindered aryl substituted ligand **L2** led to selective formation of the linear product (Scheme 14). This change in regioselectivity suggested an alternative reaction path. We proposed that the aryl substituent forces the reaction to follow a π-allyl mechanism: This ligand-dependent mechanistic dichotomy resembles a promising starting point for the development of an asymmetric Fe-catalyzed allylic substitution. Later on, it was shown that even isolated π-allyl–Fe complexes, which were previously reported to be catalytically inactive, are potent precatalysts in the presence of ligand **L2** [36].

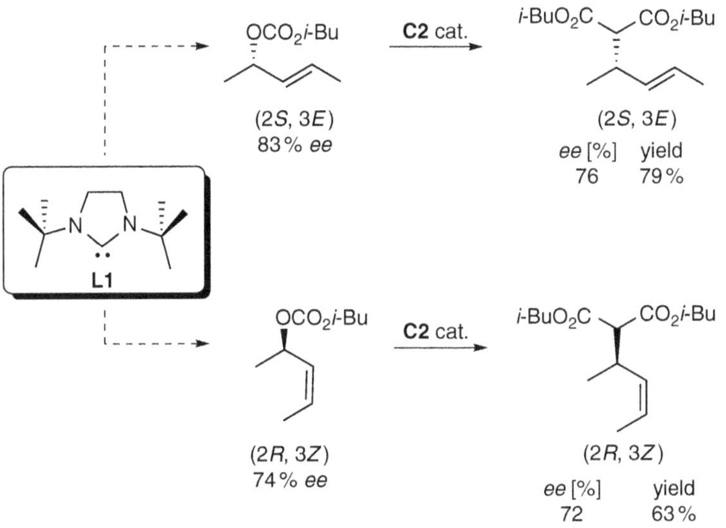

OCO2*i*-Bu (2*S*, 3*E*) 83 % *ee* —**C2** cat.→ *i*-BuO2C CO2*i*-Bu (2*S*, 3*E*) *ee* [%] 76, yield 79 %

L1

OCO2*i*-Bu (2*R*, 3*Z*) 74 % *ee* —**C2** cat.→ *i*-BuO2C CO2*i*-Bu (2*R*, 3*Z*) *ee* [%] 72, yield 63 %

Scheme 13 Fe-catalyzed regio- and stereoselective allylic substitution in the presence of an NHC-ligand

Recently, Trivedi and Tunge reported a decarboxylative allylic etherification catalyzed by $[(Bu)_4N][Fe(CO)_3(NO)]$ **C2**, in which several aromatic allyl

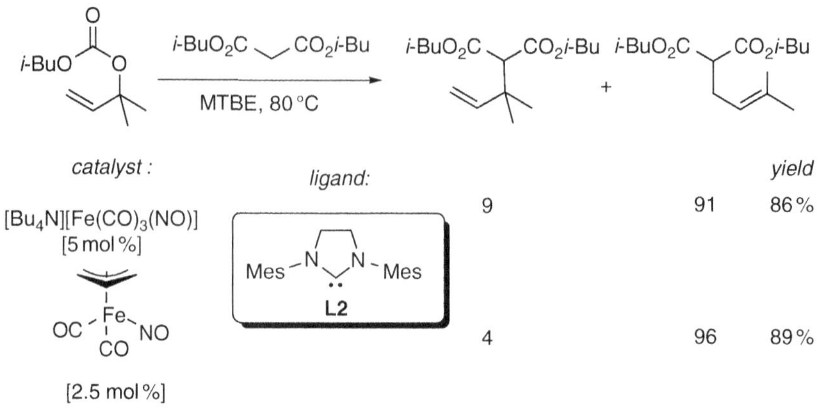

Scheme 14 Fe-catalyzed allylic substitution via π-allyl mechanism

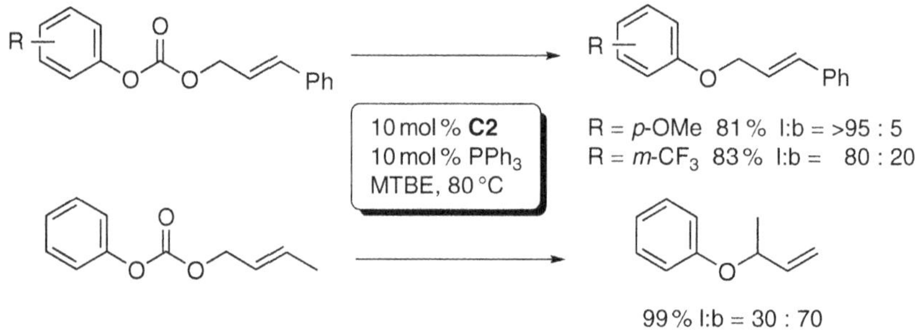

Scheme 15 Decarboxylative allylic etherification

carbonates were converted into their corresponding allyl ethers in high yields (Scheme 15) [37]. Regioselectivities were dependent on the substituent located at the allyl group. Cinnamyl carbonate led to linear products, while crotyl carbonate favored the formation of the branched product. In contrast to the Fe-catalyzed allylic alkylation, the authors assumed here an intermediate π-allyl–Fe complex.

2.2 Ruthenium Catalyzed Allylic Substitutions

The potential of Ruthenium complexes used in allylic substitution reactions was first shown by Tsuji and coworkers in 1985, who applied $RuH_2(PPh_3)_4$ in the allylic alkylation of methyl carbonates with β-ketoesters but only obtained linear products [38]. The use of Ru(cod)(cot) led to the branched substitution product in case of β-ketoesters, but the reaction was not regioselective for malonate nucleophiles [39]. Later contributions revealed that the presence of a Cp* ligand was in most cases

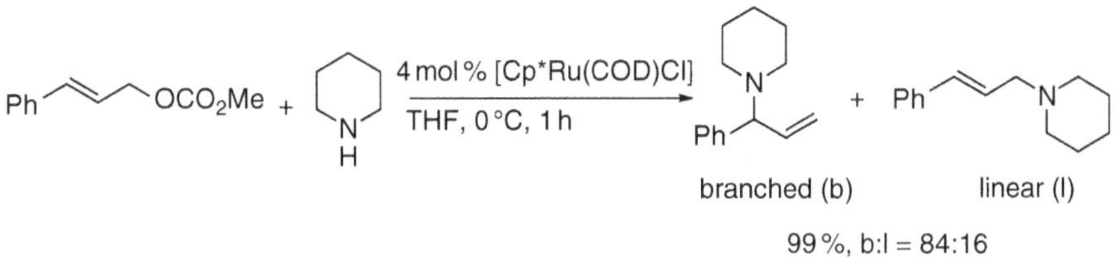

Scheme 16 Regioselective ruthenium-catalyzed allylic substitution

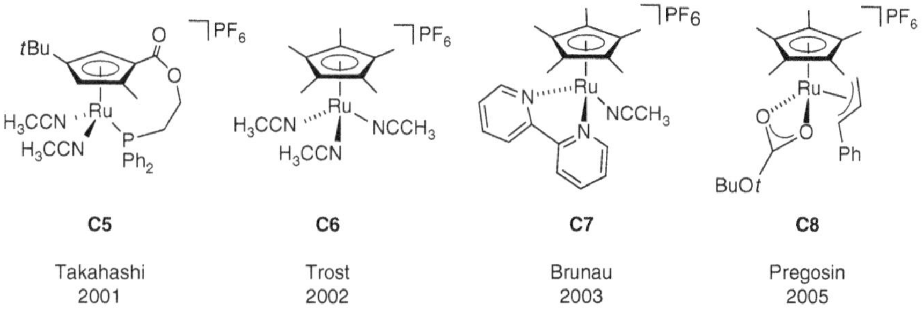

Scheme 17 Ruthenium complexes for regioselective allylic substitution reactions

crucial for obtaining regioselectivity in favor of the branched product (Scheme 16) [40, 41].

In recent years, several catalyst systems were developed for regioselective allylic alkylations, aminations, etherifications, and sulfenylations (Scheme 17) [42–45]. However, there have been only very few examples of enantioselective Ruthenium catalyzed allylic substitutions. Besides asymmetric formation of allylic ethers with $[Cp^*Ru(CH_3CN)_3]PF_6$ **C6** in combination with bisoxazolin ligands, the only example of asymmetric allylic alkylations and aminations was described by Takahashi et al. using planar–chiral cyclopentadienyl Ruthenium complexes **C5** [46, 47]. All these developments are covered by recent reviews [48, 49].

2.2.1 Regioselective Ruthenium Catalyzed Allylic Substitutions

In contrast to the branched-selective allylic substitution, Itoh and coworkers developed a linear-selective allylic alkylation by using $Ru_3(CO)_{12}$/diphenylphosphinebenzoic acid [50]. This concept could be applied to allylic aminations as well [51]. Furthermore, they found that a combination of $[Ru(p\text{-cymene})Cl_2]_2/PPh_3$ leads to retention of the regiochemistry of allylic acetates in the alkylation with $CH_2(COOMe)_2$ [52].

Recently, Bayer and Kazmaier used this system for the regioselective allylic alkylation of chelated enolates (Scheme 18) [53]. Optically active substrates were converted with perfect chirality transfer. Furthermore, the use of the Ruthenium

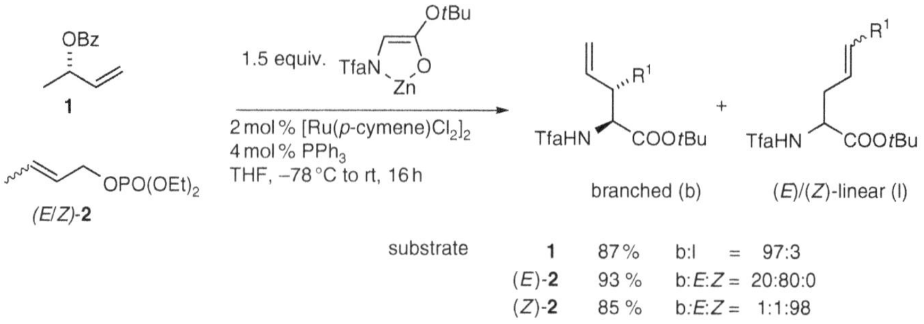

Scheme 18 Ru-catalyzed regioselective allylic alkylation of chelated enolates

catalyst showed a high degree of regioretention. Reaction of a chelated enolate with a secondary carbonate **1** led to formation of the branched substitution product as major product. Even with aryl substituents at the allyl system, branched products were obtained selectively. If the linear substrates (*E*)/(*Z*)-**2** were alkylated with a chelated enolate, the configuration of the double bond was conserved. In case of (*E*)-**2**, a certain amount of the branched product was formed.

Ru(IV)–solvate complexes such as **C9** or **C10** are formed, if carbonate-complex **C8** is decomposed under loss of CO_2 and *t*-butoxide. The two complexes were actually prepared by Pregosin and coworkers by reaction of complex **C6** with cinnamyl chloride and subsequent treatment with $AgPF_6$ in the corresponding solvent (Scheme 19) [54].

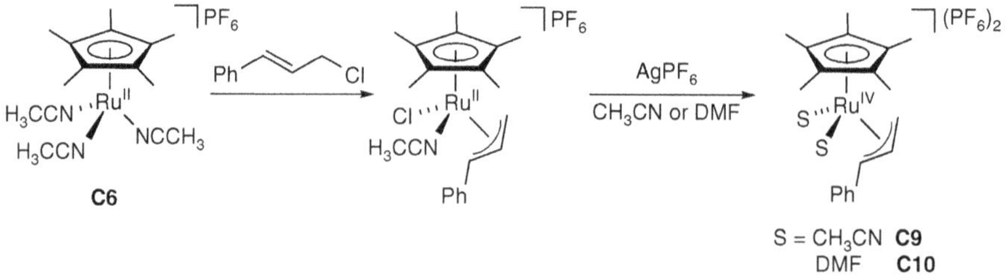

Scheme 19 Synthesis of dicationic Ru(IV)-complexes

In the reaction of phenol with allyl carbonates, these dicationic Ru(IV)-complexes show an unexpected C–C coupling reaction, which can be considered as a Friedel–Crafts-type allylation. By reacting Phenol with a secondary allyl carbonate in the presence of 3 mol% of the dicationic complex **C10**, the linear product of the C–C-coupling reaction was obtained with 100% conversion in a *o*:*p*:*m* ratio of 10:6:84 (Scheme 20). The scope of the reaction was extended to several phenol derivatives, thioanisol as well as some xylole derivatives.

Besides allyl carbonates, free allyl alcohols were converted to the desired C–C-coupling products under similar reaction conditions using complex **C9** [55]. The most important issue of the proposed mechanism was the controlled release of a proton during the reductive attack of the phenol on the allyl system forming a

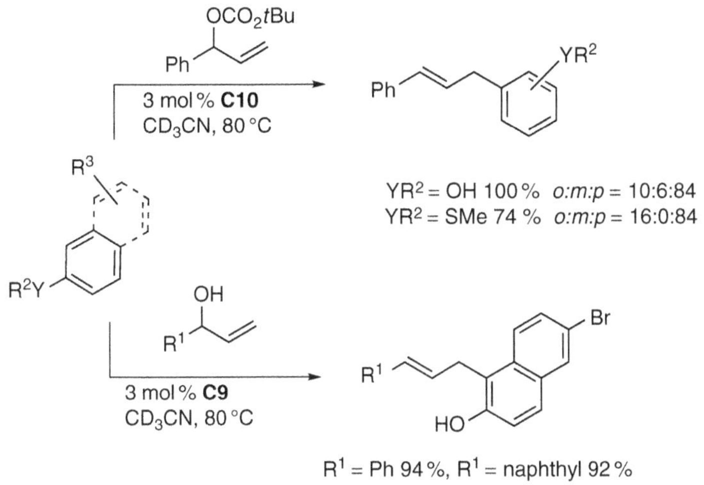

Scheme 20 Ru-catalyzed Friedel–Crafts type C–C-coupling of Phenol with allyl carbonates and allyl alcohols

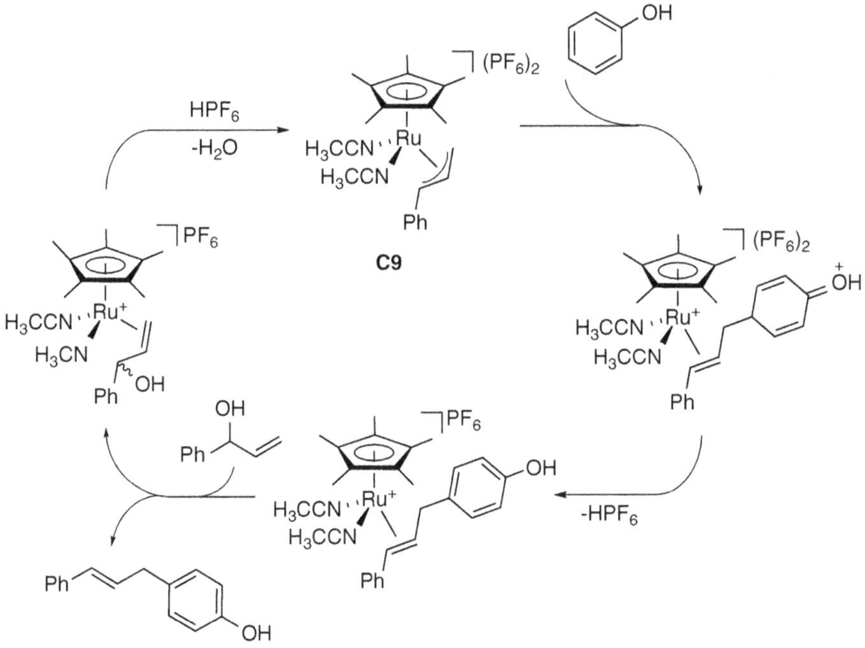

Scheme 21 Proposed mechanism of the Friedel-Crafts type C–C-coupling of Phenol with allyl alcohols

Ruthenium(II) complex (Scheme 21). After dissociation of the product and coordination of the allyl alcohol, the hydroxyl group of the allyl alcohol was converted into a leaving group by protonation. Loss of H_2O reformed the Ru(IV) allyl

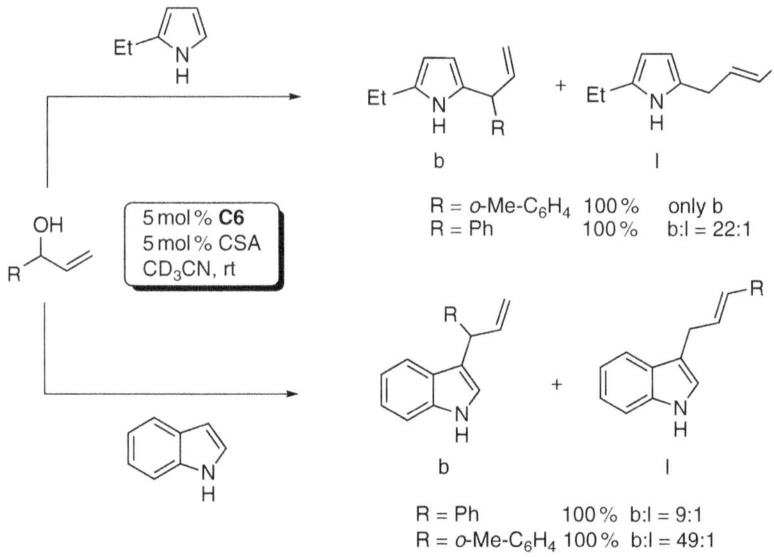

Scheme 22 Regioselective Ru-catalyzed allylation of indole and pyrrol derivatives

complex. They assumed that the use of a catalyst in a higher oxidation state facilitates the controlled release of a proton with the consequence that only a small amount of free acid was formed during the reaction.

Furthermore, they extended this chemistry to the regioselective allylation of Indoles [56]. By changing the cinnamyl moiety into an unsubstituted allyl moiety in complex **C9**, they obtained the linear C–C-coupling product selectively after reaction of indole with allyl alcohol, without *N*-allylation. Only after long reaction times with an excess of allyl alcohol, they obtained the *N*-allylated product exclusively.

Regioselective allylation of indoles and pyrroles to the branched substitution product was achieved by use of complex **C6** in the presence of a sulfonate source such as CSA or *p*-TSA (Scheme 22) [57]. Under these reaction conditions, several pyrrol derivatives such as 2-Ethylpyrrol were allylated with several aromatic allyl alcohols at C-5. Allover the branched products were obtained in high yield. Allylation of indole with aromatic allyl alcohols gave the branched 3-allyl indoles in 100% conversion and good to excellent b:l selectivities.

The role of the sulfonate was further investigated and complex **C11** was proposed as a probable catalytic intermediate. If **C11** was allowed to react with indole, only a modest amount of the desired allylation product was formed. But addition of CH_3CN led to fast formation of the allylation product. NMR-studies showed that formation of complex **C12** is fast, while the bis-nitrile complex **C13** is formed slowly (Scheme 23). Due to this fact, they propose complex **C12** as catalytic active species.

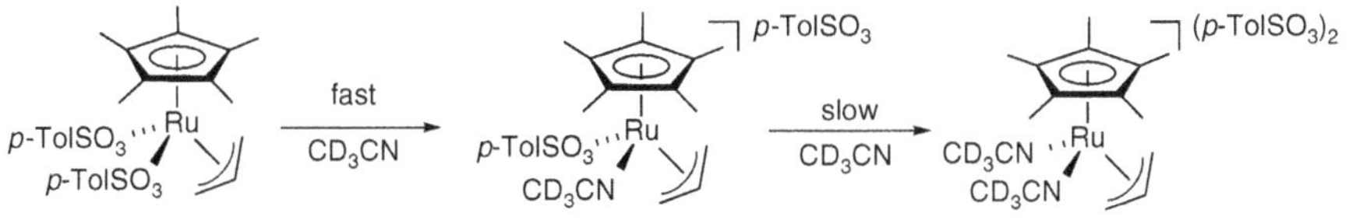

C11 C12 C13

Scheme 23 Formation of the catalytic active species by replacement of a p-TolSO$_3^-$- with a CD$_3$CN ligand

After Kitamura and coworkers reported the use of **C14** (Scheme 24) for the catalytic formation and cleavage of allyl ethers, the group of Bruneau used similar complexes for the etherification, alkylation and amination of allylic carbonates [58–60].

Scheme 24 CpRu(2-quinolinecarboxylato) allyl complexes for allylic substitution reactions

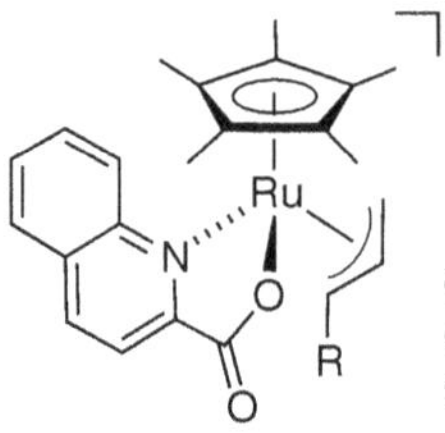

C14 R = H, X = PF$_6$
C15 R = Me, X = PF$_6$
C16 R = nPr, X = BF$_4$

They further extended the scope of this catalyst system to the substitution of 1,3-dienic carbonates [61]. Nucleophiles such as aromatic alcohols, dimethyl malonate, and pyrrolidin were used. Alkylation of a dienic carbonate in the presence of **C15** formed the branched product almost exclusively. By changing the catalyst to **C16**, β-silylated carbonates were converted with several C-, O-, and N-nucleophiles to the corresponding substituted branched α-vinylsilanes [62] (Scheme 25).

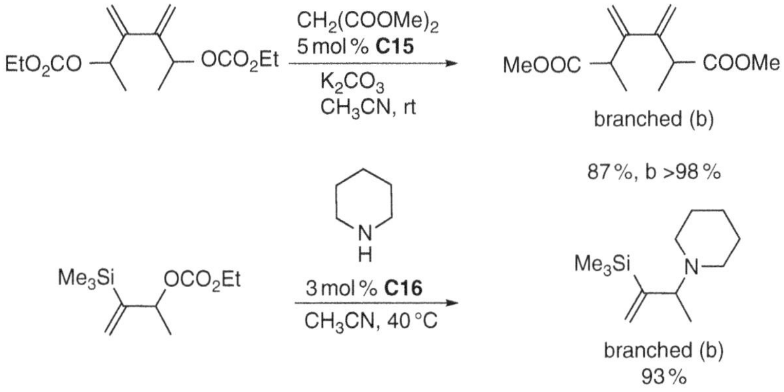

Scheme 25 Allylic substitutions reactions with 1,3-dienic- and β-silylated carbonates

Very recently, they reported the synthesis of a Cp*Ru complexes **C17** containing a diphenylphosphinobenzene sulfonate (DPPBS) ligand (Scheme 26) and its application in the *O*-allylation of *p*-methoxyphenol with allyl chlorides and in the branched-selective allylation of indoles with allylic alcohols [63].

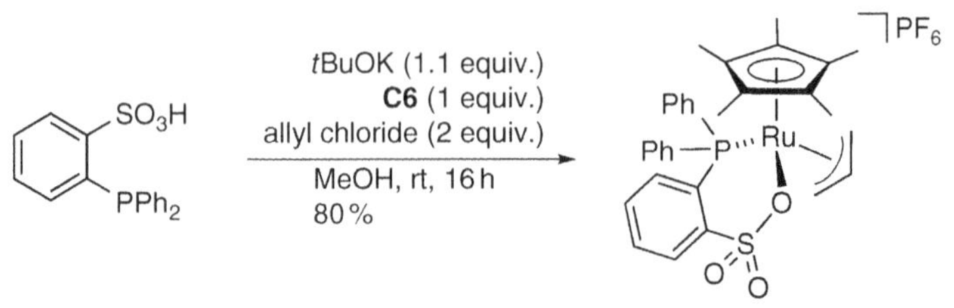

Scheme 26 Synthesis of a Cp*Ru(DPPBS)-complex **C17**

2.2.2 Asymmetric Ruthenium Catalyzed Allylic Substitutions

As mentioned earlier, there are only a few examples of asymmetric Ruthenium-catalyzed allylic substitutions [46, 47]. After they reported some early examples of enatioselective allylic alkylations, aminations and kinetic resolution of racemic allyl carbonates, Onitsuka and coworkers recently reported the enantioselective *O*-allylation of alcohols and esters [64, 65]. In their early reports, they showed already that the selectivity and the direction of chiral induction using the planar–chiral cyclopentadienyl Ruthenium complexes **C5** depend on the substitution pattern on the Cp ring and the tether length of the diarylphosphin group. Using 3 mol% of **C5**, phenol was allylated with cinnamyl chloride in high yield and 95% *ee* (Scheme 27). Under similar reaction conditions, sodium benzoate was well allylated in excellent yield and high *ee*.

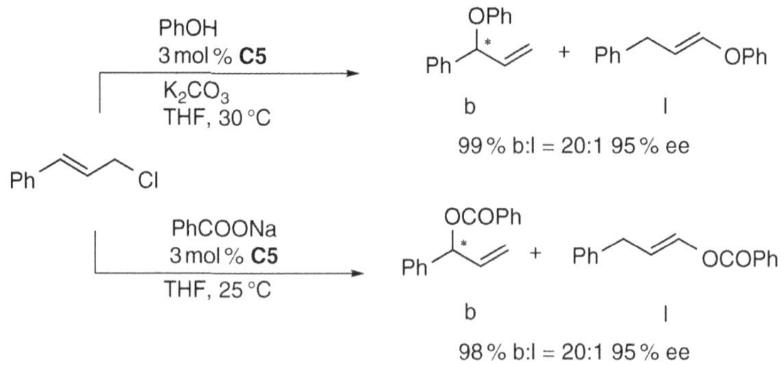

Scheme 27 Asymmetric etherification and esterification with planar-chiral cyclopentadienyl Ruthenium complex **C5**

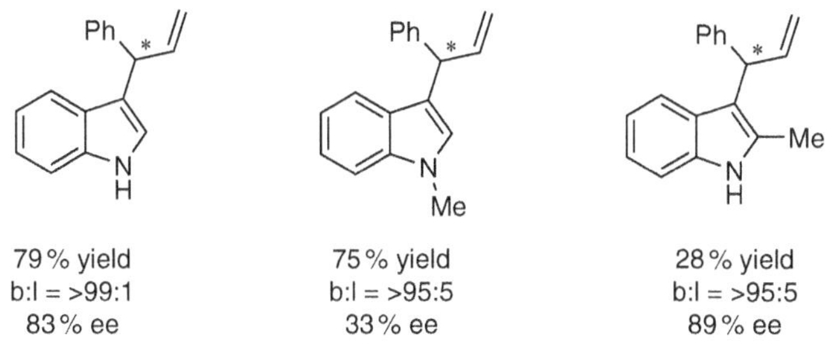

79 % yield	75 % yield	28 % yield
b:l = >99:1	b:l = >95:5	b:l = >95:5
83 % ee	33 % ee	89 % ee

Scheme 28 Asymmetric allylation of indoles with planar–chiral cyclopentadienyl Ruthenium complex **C5**

Scheme 29 Ligands for the asymmetric Carroll-rearrangement of β-ketoesters

L3 L4

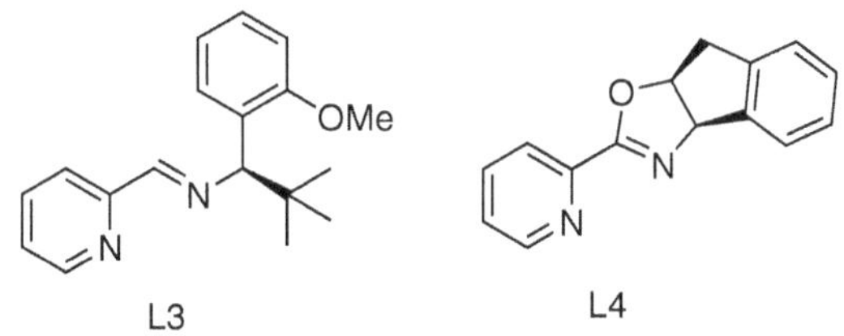

This type of catalyst was further extended to the regio- and enantioselective allylation of indoles with cinnamyl chloride [66]. Indole gave the branched substitution product with excellent regioselectivity and good *ee* (Scheme 28). *N*-methylation led to almost complete loss of enantiomeric purity of the favored branched product, while methylation in 2-position led to good regio- and enantioselectivity, but the product was only isolated in poor yield.

As an intramolecular decarboxylative allylic alkylation, Burger and Tunge described the Ruthenium-catalyzed Carroll rearrangement of allyl β-ketoesters to γ,δ-unsaturated ketones [67]. This [3,3]-rearrangement proceeded in the presence of $[(Cp^*RuCl)_4]$ and bpy-ligand regioselectively to the branched substitution products. If chiral allyl β-ketoesters were used, the reaction proceeded highly stereospecific [68]. By using chiral pyridine–diimine ligands **L3** or **L4** (Scheme 29) in combination with $[CpRu(CH_3CN)_3]PF_6$, Lacour and coworkers reported an asymmetric version of this Carroll rearrangement [69].

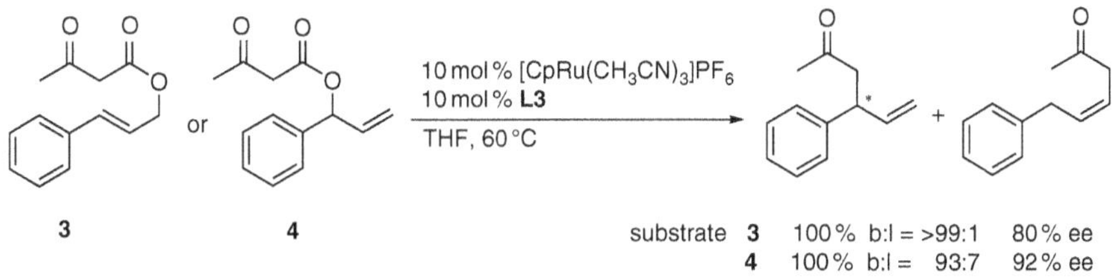

3 or **4**

10 mol % $[CpRu(CH_3CN)_3]PF_6$
10 mol % **L3**
THF, 60 °C

substrate			
3	100 %	b:l = >99:1	80 % ee
4	100 %	b:l = 93:7	92 % ee

Scheme 30 Asymmetric Carroll-rearrangement of β-ketoesters

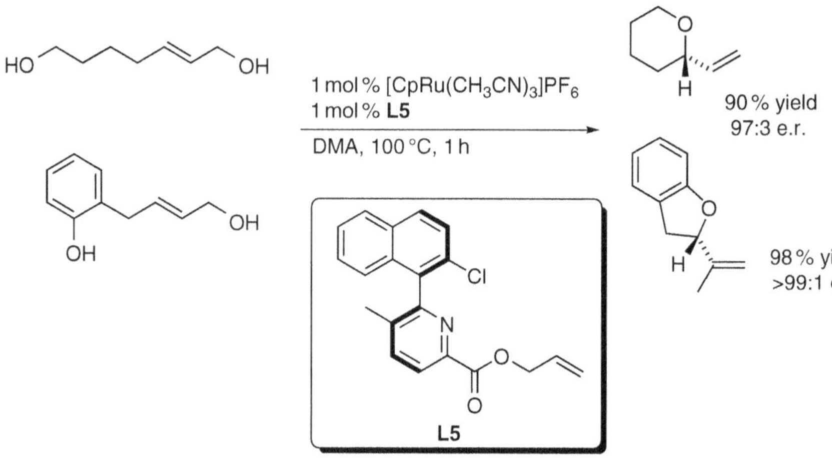

Scheme 31 Asymmetric dehydrative cyclization of ω-hydroxy allyl alcohols

Reaction of the linear allyl β-ketoester **3** led to formation of the chiral branched product in a b:l ratio of >99:1 with 80% *ee*. If the chiral secondary allyl β-ketoester **4** was used, the reaction proceeded highly regio- and stereospecific (Scheme 30). Reduction of the catalyst loading to 2.5 mol% was achieved by changing the metal source to the airstable [CpRu(η^6-naphthalene)]PF_6 [70]. Addition of $Mg(OTf)_2$ as Lewis acid allowed further reduction of the catalyst loading to 2 mol% and a decrease of the temperature to 25°C [71]. They further applied a combination of [CpRu($CH_3CN)_3$]PF_6 and Ligand **L4** in the enantio- and regioselective decarboxylative etherification of allyl aryl carbonates [72].

As a chiral extension of the above-mentioned complex **C14**, the group of Kitamura developed **L5** as a new chiral ligand for asymmetric dehydrative cyclization of ω-hydroxy allyl alcohols (Scheme 31) [73]. Reaction of (2*E*)-heptene-1,7-diol in the presence of [CpRu($CH_3CN)_3$]PF_6 in combination with ligand (*R*)-**L5** afforded the cyclization product quantitatively with 6-exo-trig selectivity and with an enantiomeric ratio of $S{:}R = 97{:}3$. Besides aliphatic, aromatic alcohols were used as nucleophiles at the ω-hydroxy end of the substrates, which led to the cyclization products in high yields and high enantiomeric ratios as well.

3 Allylic Substitutions Catalyzed by Group 9 Metals (Without Iridium)

3.1 *Cobalt Catalyzed Allylic Substitution*

Cobalt catalyzed allylation reactions have not been extensively studied when compared to analogous reactions using other transition-metals. Although Cobalt catalysis is still in an exploratory phase, the low cost of cobalt complexes and their interesting mode of action make them an attractive alternative for use in C–C-bond formation reactions as evidenced by the growing number of studies that have been recently reported in this field [74–77]. Among other reasons, Cobalt catalysts allow for a broad functional group tolerance as well as for mild reaction conditions and simple experimental procedures.

3.1.1 Direct Cobalt Catalyzed Allylic Substitution

Allylic Alkylation of 1,3-Dicarbonyl Compounds

In the early 1990s, Iqbal and coworkers reported an allylation reaction of 1,3-dicarbonyl compounds with allylacetates using catalytic amounts of cobalt(II)-chloride in 1,2-dichloroethane [78, 79]. In contrast to metal-catalyzed allylic substitution reactions that are performed in basic conditions using the stabilized anions derived from 1,3-dicarbonyl compounds, the cobalt catalyzed reaction was performed employing neutral 1,3-dicarbonyl compounds (Scheme 32). The yields were modest to good. The regioselectivity as well as the stereochemistry (*Z*:*E*) were strongly influenced by the nature of the 1,3-dicarbonyl compound. While the reaction of 2,4-pentanedione afforded allylated products in good yields and regioselectivities (via allylic transposition or without any rearrangement depending on the allylic substrate), methyl acetoacetate exhibited only poor regioselectivities and lower yields. Furthermore, the reaction of 2,4-pentanedione proved stereoselective as it gave the (*E*)-isomer as the major product.

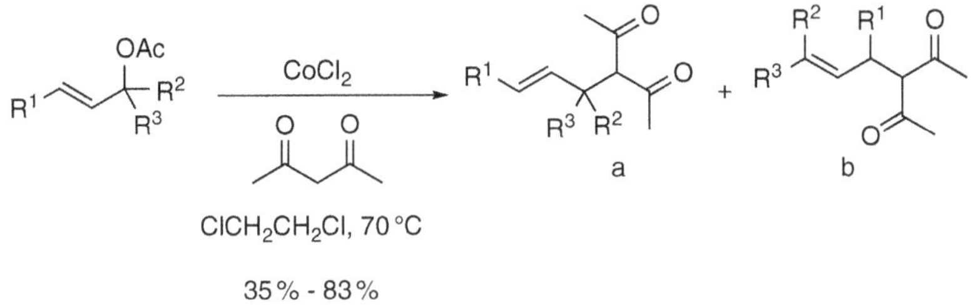

OAc, R^1, R^2, R^3 — $CoCl_2$, $ClCH_2CH_2Cl$, 70 °C, 35 % - 83 % — a + b

Scheme 32 Co-catalyzed allylic alkylation with 1,3-dicarbonyl compounds

Cobalt Catalyzed Allylation of Arylhalides

In 2003, Gosmini and Périchon examined the cobalt-catalyzed direct electrochemical cross-coupling between aryl or heteroaryl bromides or chlorides and allylacetate or -carbonate [80, 81]. The reaction was performed with a variety of substituted arylhalides providing the allylation product in moderate to good yields. However, electrochemical reactions are generally considered as being more difficult to handle than conventional chemical methods and are seldom applied on larger scale. Therefore, Gosmini and Périchon investigated the transposition of this electrochemical process into a chemical reaction. They reported a chemical coupling reaction of allylacetate with arylchloride in using $CoBr_2$ as catalyst in a mixture acetonitrile/pyridine in the presence of Mn powder as reducing agent and $FeBr_2$ (Scheme 33) [82]. Various arylchlorides bearing electron-withdrawing substituents such as CN or CO_2Et were coupled with allylacetate affording the arylallyl products in good yields. However, the mechanism of the reaction is not clear as the exact role of the additive $FeBr_2$ was not elucidated.

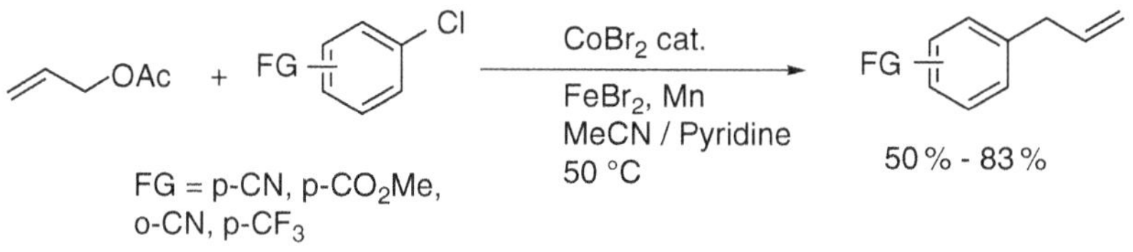

OAc + FG–C₆H₄–Cl → (CoBr₂ cat.; FeBr₂, Mn; MeCN / Pyridine; 50 °C) FG–C₆H₄–allyl, 50 % - 83 %

FG = p-CN, p-CO$_2$Me, o-CN, p-CF$_3$

Scheme 33 Direct chemical coupling of allylacetate with functionalized arylchlorides

Dehydrative Coupling of Allylic Alcohols to Allylamides

Iqbal and coworkers described the conversion of allylic alcohols to allyl amides with acetonitrile and acetic anhydride using catalytic amounts of $CoCl_2$ [83]. The reaction with tertiary alcohols afforded selectively the transposed allylic amides in modest to good yields while secondary alcohols gave a mixture of regioisomers (Scheme 34). A mechanism derived from the Ritter reaction was postulated for this process based on the nucleophilic attack of the solvent acetonitrile on a cobalt allyl intermediate arising from the reaction of allyl acetate and $CoCl_2$. The allylic compound obtained might react with acid acetic to give an imidate ester hydrolyzed under basic conditions to lead to the allylamide.

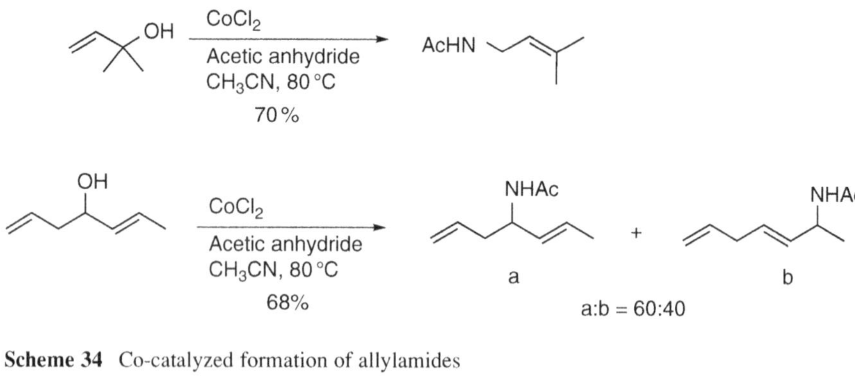

OH → ($CoCl_2$; Acetic anhydride; CH_3CN, 80 °C; 70 %) AcHN

OH → ($CoCl_2$; Acetic anhydride; CH_3CN, 80 °C; 68%) NHAc (a) + NHAc (b)

a:b = 60:40

Scheme 34 Co-catalyzed formation of allylamides

3.1.2 Cobalt Catalyzed Allylation Reactions with Organometallic Reagents

Co Catalyzed Allylation with Organozinc Reagents

Knochel and Reddy investigated the utilization of organozinc reagents in Cobalt and Iron catalyzed reactions. In the course of these studies, they reported a Cobalt catalyzed allylic alkylation reaction with dialkylzinc reagents using $CoBr_2$ as catalyst [84]. This reaction led selectively to the S_N2-substitution products with complete retention of the stereochemistry at the double bond. Thus, while the reaction of dipentylzinc ($pent_2Zn$) with geranyl chloride gave the (E)-diene almost exclusively (E:Z (%) = 98:2), the reaction of dipentylzinc with neryl chloride afforded the (Z)-diene (E:Z (%) = 2:98) (Scheme 35). Interestingly, the same reaction performed in presence of the soluble copper salt CuCN·2LiCl provided selectively the S_N2'-substitution product. The utilization of alkylzinc halides instead of dialkylzinc compounds provided the same reaction yields but required longer reaction times and it was shown that both R-groups in R_2Zn were transferred. Allylic phosphates or allylic chlorides react under these conditions affording the S_N2-product with the same regio- and stereoselectivity.

Recently, Knochel and coworkers reexamined this reaction and reported an extension of the method to functionalized diarylzinc reagents obtained from the corresponding aryliodides through an iodine–zinc exchange (Scheme 36) [85]. Thus geranyl-, neryl- or cinnamyl- chlorides or phosphates were coupled to functionalized diarylzinc reagents to afford selectively the S_N2-products with good yields and complete retention of the double bond configuration. Among the catalysts tested out in these reaction, $Co(acac)_2$ gave the highest conversion.

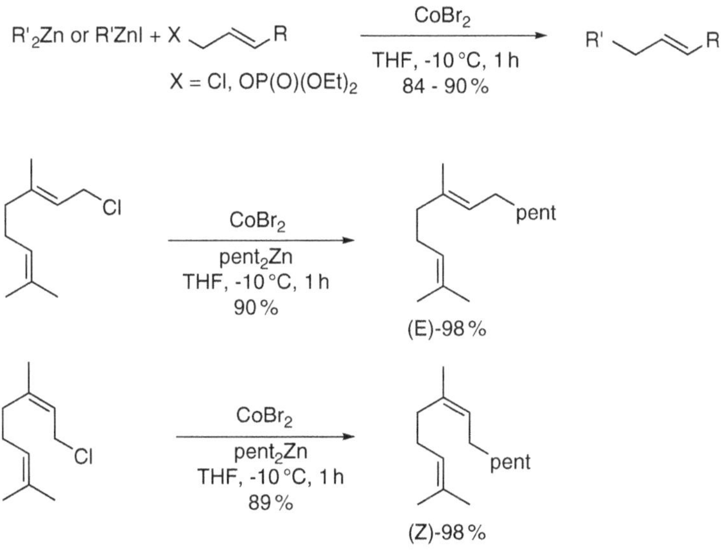

R'$_2$Zn or R'ZnI + X⌒R —[$CoBr_2$; THF, -10 °C, 1 h; 84 - 90 %]→ R'⌒R
X = Cl, OP(O)(OEt)$_2$

Geranyl chloride —[$CoBr_2$; $pent_2Zn$; THF, -10 °C, 1 h; 90 %]→ pent product (E)-98 %

Neryl chloride —[$CoBr_2$; $pent_2Zn$; THF, -10 °C, 1 h; 89 %]→ pent product (Z)-98 %

Scheme 35 Co-catalyzed allylic alkylation with dialkylzinc compounds and alkylzinc iodides

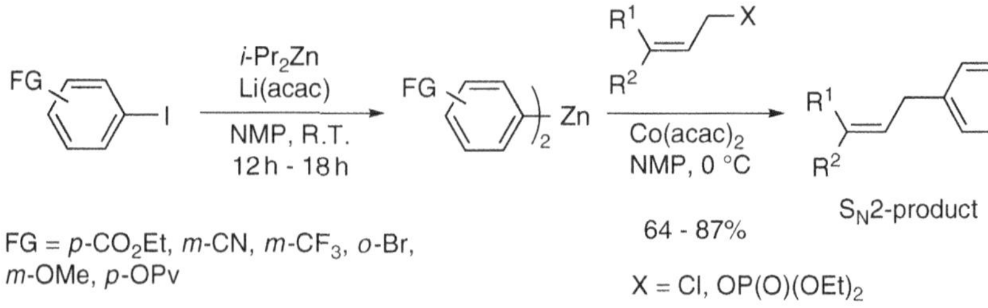

Scheme 36 Co-catalyzed allylic arylation with functionalized diarylzinc reagents

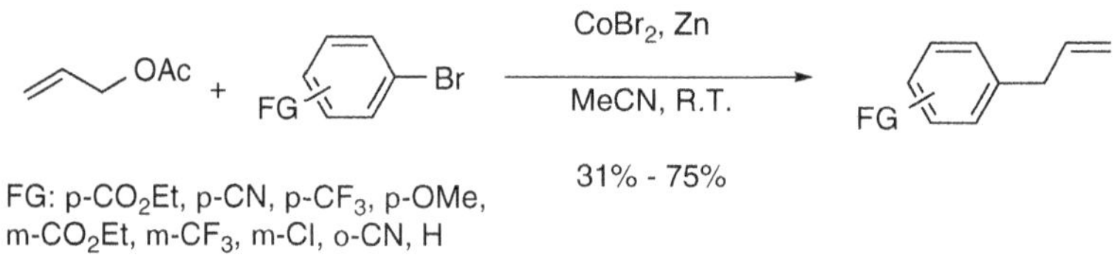

Scheme 37 Co-catalyzed allylic arylation with functionalized arylzinc reagents formed in-situ

It should be noted that the presence of a nitrile or an ester group on the aryl moiety is tolerated. However, this method requires the preparation of the organozinc reagents from aryliodides in a preliminary step before performing the coupling reaction with the allylic derivative.

Gosmini and Périchon developed a Co-catalyzed allylic arylation reaction with allylacetates and arylbromides via the formation of arylzinc bromide reagents [82]. The arylzinc bromides were formed in situ by $CoBr_2$-catalyzed insertion of readily available zinc powder at R.T. in acetonitrile (Scheme 37). Thus, in this reaction the Cobalt bromide in presence of zinc powder as reducing agent proved to catalyze both the formation of the arylzinc reagents from the corresponding arylbromides and the coupling reaction of the arylzinc compounds with allylacetate. The presence of various electron-withdrawing or electron-donating functional groups such as ester, nitrile, and methoxy in *ortho-*, *meta-* or *para-*positions is tolerated. Modest to good yields were obtained depending on the aromatic substrate, and it was shown that the utilization of PhBr and $ZnBr_2$ as additives led to higher yields by restraining the formation of the by-products ArH and ArAr. However, the reaction of substituted allylacetates such as crotylacetate or cinnamylacetate afforded the expected products in poor yields.

Co-Catalyzed Allylation with Organomagnesium Reagents

Oshima investigated the allylic substitution reaction of allylic ethers with phenyl and trimethylsilylmethyl Grignard reagents in the presence of cobalt-phospine complexes [86, 87]. The success of the cobalt-catalyzed reaction with phenyl grignard reagents proved to be strongly dependent on the allylic substrate, the ligand and the solvent.

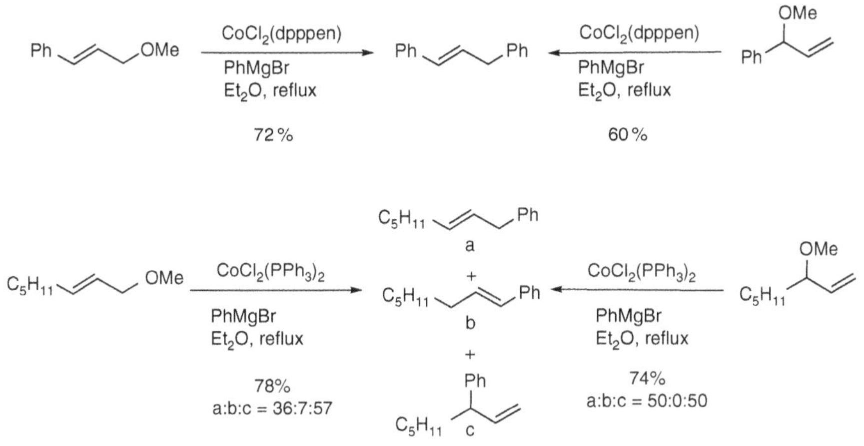

Scheme 38 Co-catalyzed allylic arylation of allylic ethers with PhMgBr

Thus, the reaction of cinnamyl ether with PhMgBr required the utilization of 1,5-diphenylphosphinopentane (DPPPEN) as ligand and ether as solvent at reflux. Under these conditions, the reactions with linear or branched allylic ether substrates afforded exclusively the linear product with good yield (Scheme 38). It should be noted that the reaction of cinnamyl bromide under similar conditions gave a mixture of dimeric products. The formation of dimeric products might suggest that single electron transfer from the cobalt complex lead to a cinnamyl radical intermediate. The reaction of (E)-2-octenyl ether required triphenylphosphine as ligand (the reaction with DPPPEN gave only a poor yield) and afforded a mixture of linear and branched product (Scheme 38).

In contrast, the reaction with Me_3SiCH_2MgCl was only slightly determined by the reaction conditions [86, 87]. Although the best ligand proved to be 1,6-bis (diphenylphosphino)hexane (dpph), the reaction could also be performed with ligandless $CoCl_2$ affording the product with excellent yield and good regio-selectivity (Scheme 39).

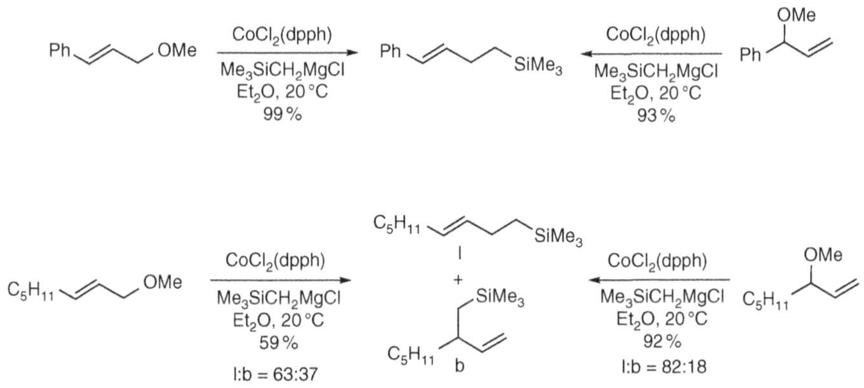

Scheme 39 Co-catalyzed allylic arylation of allylic ether with PhMgBr

3.2 *Rhodium Catalyzed Allylic Substitution Reactions*

3.2.1 Regiospecificity of the Rh-Catalyzed Allylic Alkylation

In 1984, Tsuji et al. described the first regiospecific C-allylation reaction of carbonucleophiles with unsymmetrical acyclic allylic carbonates under neutral conditions [88]. Interestingly, the Wilkinson catalyst [$RhCl(PPh_3)_3$] itself showed almost no activity, while good results were obtained by the addition of phosphine ligands as $P(^nBu)_3$ or $P(OEt)_3$. PPh_3 or dppe proved to be ineffective. The highest activity for this reaction was observed by using $RhH(PPh_3)_4$ in the presence of P $(nBu)_3$ (Scheme 40).

Additional studies comparing the Rh-catalyzed allylation with the analogous palladium-catalyzed version highlight the unusual regiospecificity of this reaction [38]. Indeed, the rhodium-catalyzed reaction of the branched allylic carbonate lead to predominant formation of the corresponding branched product, while the linear products were favored in the reaction with the linear allylic carbonate. This is in contrast to the palladium-catalyzed reaction which led to nearly the identical branched and linear products ratios and yields independent of the substrate used (Table 1). From these results, Tsuji et al. postulated that the Rh-catalyzed reaction

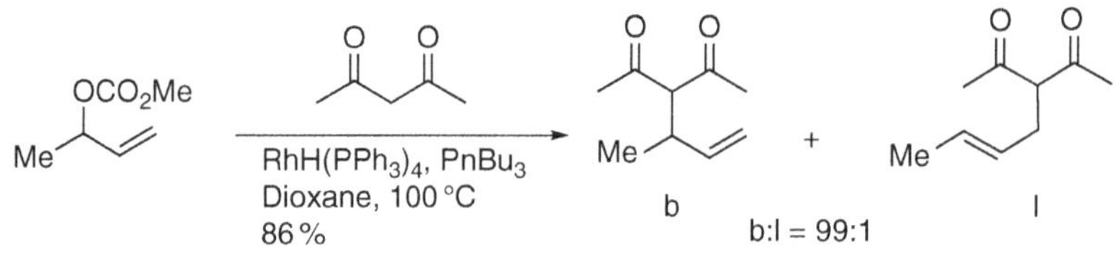

$RhH(PPh_3)_4$, $PnBu_3$
Dioxane, 100 °C
86 %

b b:l = 99:1 l

Scheme 40 Regiospecific Rh-catalyzed allylic alkylation

Table 1 Rh- and Pd-catalyzed allylic alkylations

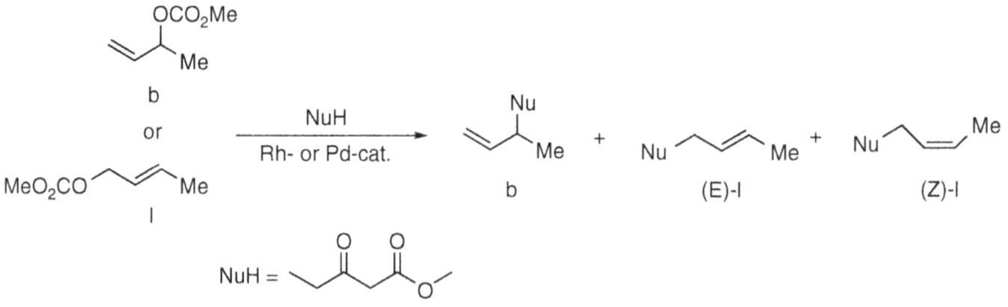

b or l

NuH
Rh- or Pd-cat.

b + (E)-l + (Z)-l

NuH =

Entry	Substrate	Catalyst	Reaction conditions	Product b:(E)-l:(Z)-l	Yields (%)
1	b	$RhH(PPh_3)_4$, $PnBu_3$	Dioxane,100°C	86:12:2	81
2	b	$Pd_2(dba)_3$, $CHCl_3$, PPh_3	THF, R.T.	27:65:8	89
3	l	$RhH(PPh_3)_4$, $PnBu_3$	Dioxane,100°C	28:63:9	97
4	l	$Pd_2(dba)_3$, $CHCl_3$, PPh_3	THF, R.T.	29:63:8	93

might proceed through an σ-allyl organorhodium intermediate and not through a π-allyl-Rhodium complex, although the identical product ratio obtained with both Rh- and Pd-catalysts in the reaction with the linear substrates might indicate a similar catalytic intermediate.

Fourteen years later, Evans and Nelson demonstrated that the use of a triorganophosphite-modified Wilkinson's catalyst increased the reaction rates and enhanced the regioselectivity toward the branched product [89]. Moreover, the reaction proceeded under much milder conditions (Scheme 41). These results were explained by the higher π-acceptor character of the triorganophosphite ligand which might promote an SN-1-type allylic substitution favoring the formation of the branched product in increasing the electrophilic nature of the allyl-Rhodium intermediate. The reaction with tertiary allylic substrates gave the highest regioselectivities with $P(OPh)_3$ whereas the best ligand for the reaction with secondary allyl carbonates was $P(OCH_2CF_3)_3$. Takeuchi and Kitamura obtained similar results in using $[Rh(COD)Cl]_2$ as catalyst for the allylic alkylation of secondary allylic acetates [90] and more recently, Martin et al. reported the use $[Rh(CO)_2Cl]_2$ as catalyst [91]. In these conditions, the substitution product arising from the nucleophilic substitution at the carbon atom bearing the carbonate group was obtained with good to excellent yields as well as excellent regioselectivity. Furthermore, the reaction with enantioenriched allylic carbonate proved to proceed with net retention of configuration (Scheme 42) [91]. This result is consistent with

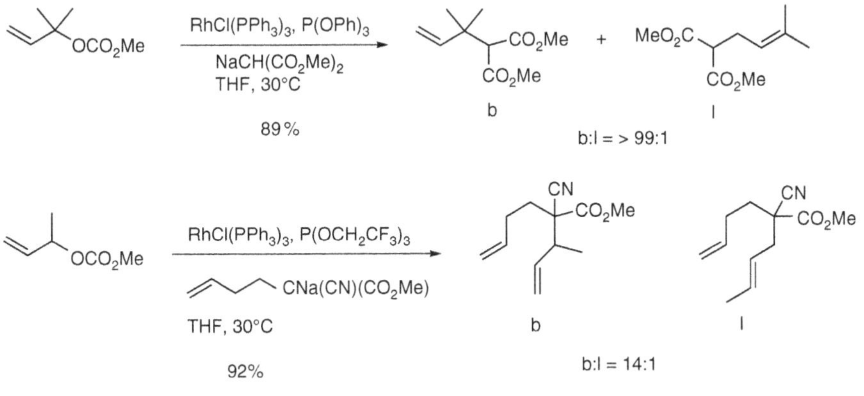

OCO2Me
RhCl(PPh3)3, P(OPh)3
NaCH(CO2Me)2
THF, 30°C
89%
CO2Me
CO2Me
b
+
MeO2C
CO2Me
l
b:l = > 99:1

OCO2Me
RhCl(PPh3)3, P(OCH2CF3)3
CNa(CN)(CO2Me)
THF, 30°C
92%
CN
CO2Me
b
CN
CO2Me
l
b:l = 14:1

Scheme 41 Rh-catalyzed allylic alkylation with triorganophosphite-modified Wilkinson's catalyst

OCO2Me
99 % ee
[Rh(CO)2Cl]2
NaCH(CO2Me)2
DMF, 0 °C
93 %
MeOOC COOMe
98 % ee
Regioselectivity: 93:7

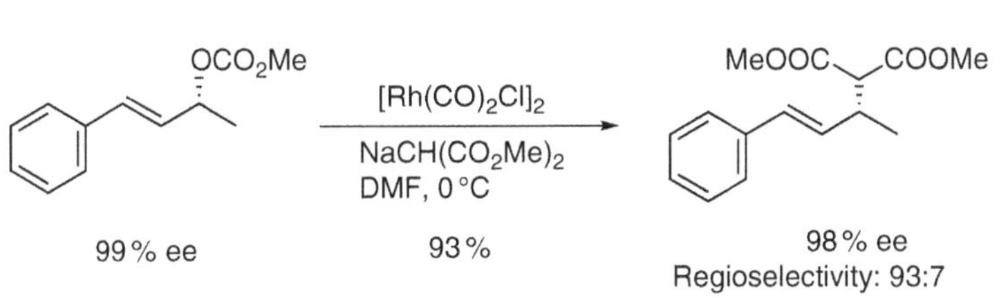

Scheme 42 Enantiospecific Rh-catalyzed allylic alkylation

the mechanism for the Rh-catalyzed allylic alkylation postulated before by Evans and Nelson [92].

3.2.2 Stereospecificity of the Rh-Catalyzed Allylic Alkylation

The investigation of Evans and Nelson on the stereochemical outcome of the Rh-catalyzed allylic alkylation showed that the reaction of enantiomerically enriched acyclic allylic carbonates with soft C-nucleophiles proceeds with net retention of absolute configuration [92]. This result is in strong contrast to what is observed in Pd-catalyzed reaction. The formation of an enyl($\sigma + \pi$) complex as intermediate was postulated and such complexes could be characterized by IR, NMR, and X-ray analysis [93, 94]. In such intermediates, the rotation around the σ-bond is hindered by the coordination of the double bond toward the Rhodium. Hence if the nucleophilic attack is faster than the isomerization ($k_2 > k_1$), there is retention of the regio- and the stereochemical information.

On the other hand according to this model, the reaction of nonracemic allylic carbonates with hard nucleophiles should proceed with complete inversion of the absolute configuration due to the tendency of hard nucleophiles to add directly to the rhodium and thereby lead to the alkylation product via reductive elimination ($k_3 > k_2$) (Scheme 43).

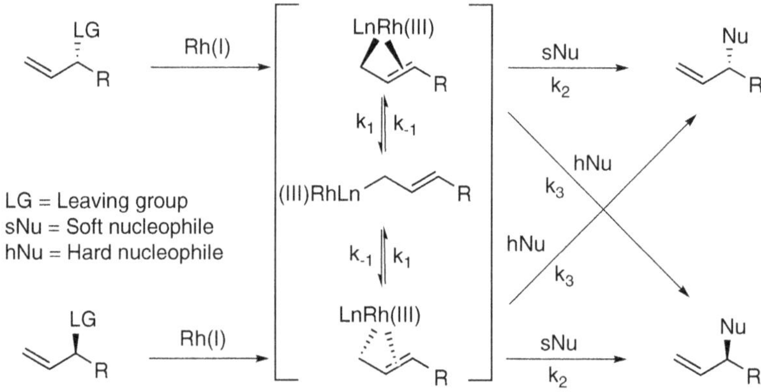

Scheme 43 Postulated mechanism for the Rh-catalyzed allylic alkylation

3.2.3 Reaction with C-Nucleophiles

Malonates and Sulfonylacetates as Nucleophiles

Based on this mechanism, asymmetric allylic alkylation reactions can be easily achieved by the use of chiral allyl substrates and symmetric C-nucleophiles such as malonate derivatives. However, only the stereogenic center of the allyl moiety can be controlled this way and the use of unsymmetric nucleophiles generally leads to mixtures of diastereoisomers in addition to regioisomers. Evans and Kennedy investigated the allylic alkylation reaction with the sodium salt of methylphenylsulfonylacetate [95]. The reaction proceeds with good yields and excellent chirality transfer to the branched product. However, this product was obtained as a 1:1 mixture of diastereoisomers due to the epimerization of the sulfonylacetates (Scheme 44).

Kazmaier and Stolz reported in 2006 a regio- and stereoselective rhodium-catalyzed allylic alkylation using chelated amino acid ester enolates [96]. Due to the chelation, the allylic amino acids are obtained with very high diastereoselectivities. Very surprising was the excellent chirality transfer observed for all linear products while the same reaction performed with a Pd-catalyst afforded only racemic products (linear as well as branched). This result was explained by the very slow isomerization of the allyl-rhodium intermediates formed. The regioselectivity of the reaction was shown to be strongly dependent of the allylic substrate (Scheme 45).

Hayashi and coworkers investigated a ligand-controlled enantioselective Rh-catalyzed allylic alkylation of unsymmetrical substrates in using $Rh(dpm)(C_2H_4)_2$ as catalyst [97]. The use of the P,N-ligand **L6** (Scheme 46) for an intermolecular enantioselective Rh-catalyzed allylic alkylation reaction had been previously reported by Pregosin and coworkers [98]. Hayashi et al. postulated that in order to obtain high enantioselectivity starting from a racemic secondary allylic acetate, the rhodium-allyl intermediate must have a longer lifetime, to allow equilibration between the isomeric intermediates which is in contrast to enantiospecific allylic

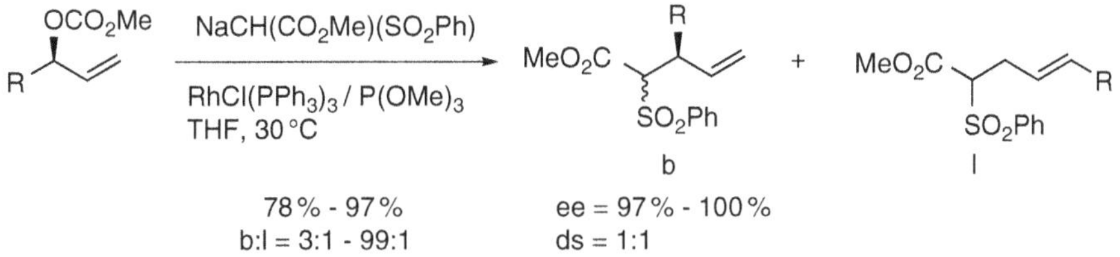

OCO2Me
R
NaCH(CO2Me)(SO2Ph)
RhCl(PPh3)3 / P(OMe)3
THF, 30 °C
MeO2C
R
SO2Ph
b
+
MeO2C
R
SO2Ph
l
78 % - 97 %
b:l = 3:1 - 99:1
ee = 97 % - 100 %
ds = 1:1

Scheme 44 Rh-catalyzed allylic alkylation reaction with the sodium salt of methylphenylsulfonylacetate

OPO(OEt)2
R
TFA–N
OtBu
Zn O
RhCl(PPh3)3 / P(OR)3
THF, –78 °C - R.T.
TFAHN
CO2tBu
R
b
+
R
CO2tBu
l
NHTFA
R = n-Pr, Y = 97 %, b:l = 71:29, ds = 92 %, ee (b) = 98 %, ee (l) = 83 %
R = i-Pr, Y = 95 %, b:l = 7:93, ds = 88 %, ee (b) = 98 %, ee (l) = 89 %

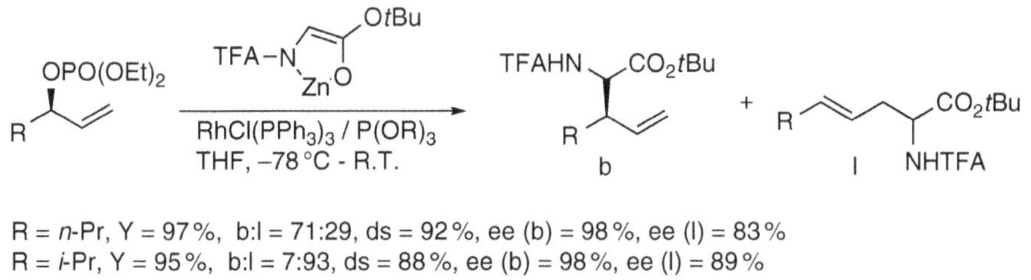

Scheme 45 Stereoselective Rh-catalyzed allylic alkylation with chelated enolate

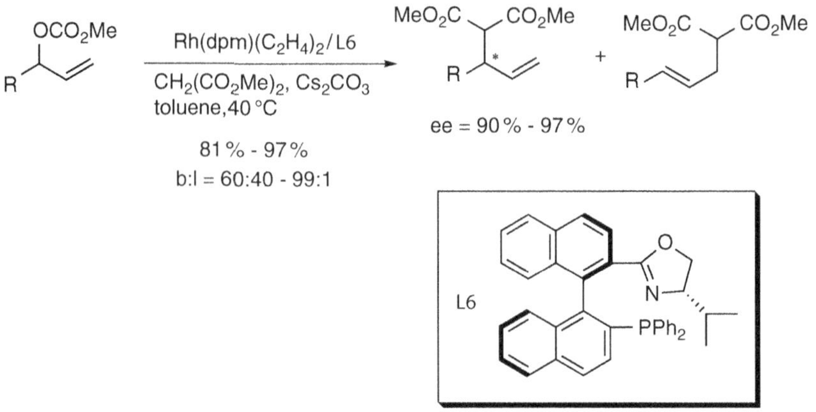

Scheme 46 Enantioselective Rh-catalyzed allylic alkylation

alkylation, where the rhodium-allyl intermediate must have a short lifetime in order to avoid equilibration and thus racemization as described by Evans and coworkers. Thus, they could show that the low concentration of the malonate nucleophile increased the enantioselectivity of the reaction by increasing the lifetime of the allyl-rhodium intermediate. The use of Cs_2CO_3 as a base in place of NaH gave the best enantioselectivity. This result was explained by the weaker basicity of Cs_2CO_3 that keeps the concentration of the malonate nucleophile lower (Scheme 46).

Ketones and Ester Enolates as Nucleophiles in Rh-Catalyzed Allylic Alkylation

Matsuda and coworkers reported the use of trimethylsilylenol ethers as ketone equivalent for allylic alkylation reactions with a wide range of allylic carbonates in using $[Rh(COD)(MePh_2P)_2]OTf$ as catalyst [99, 100]. However, only poor regio- and diastereoselectivities were obtained. The Wilkinson catalyst and $RhH(PPh_3)_4$ classically used in rhodium-catalyzed allylic alkylation reactions proved to be ineffective in this reaction. Evans and Leahy re-examined this reaction and showed that the use of copper(I) enolates as softer and less basic nucleophiles increased dramatically the regioselectivity [101] (Scheme 47). This reaction could be extended to allylic alkylation of α-substituted enolates.

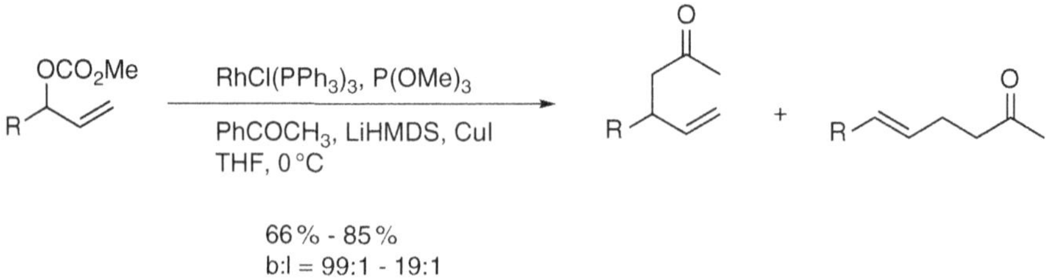

Scheme 47 Rh-catalyzed allylic alkylation with copper(I) enolates

Organozinc, Organomagnesium, and Organoboronic Reagents as Nucleophiles in Rh-Catalyzed Allylic Alkylation

Rhodium-catalyzed allylic alkylation could also be achieved with unstabilized nucleophiles such as organozinc or organoboronic reagents. The main difficulty of these reactions relied on the basic nature of these nucleophiles which tend to promote elimination reactions of the metal–allyl intermediates or hydrolysis of the leaving group in the allyl alcohol. Furthermore, hard nucleophiles proved to undergo allylic alkylation with net inversion of absolute configuration as previously mentioned [92].

Evans and Uragushi reported a regio- and enantiospecific rhodium-catalyzed allylic arylation with arylzinc reagents [102]. Surprisingly, the classically used trimethylphosphite modified Wilkinson's catalyst was not an effective catalyst with arylzinc reagents and the best catalyst for this reaction proved to be a hydrotris (pyrazolyl)borate rhodium complex. The regioselectivity could be dramatically improved by using allylic substrates bearing fluorinated leaving groups and by adding additional LiBr to the catalyst prior to the introduction of the arylzinc specie and the allylic carbonate (Scheme 48). Martin and coworkers obtained similar results in using $[Rh(CO)_2Cl]_2$ as catalyst and an enantioenriched allylic methyl carbonate [91]. Oshima et al. described a rhodium-catalyzed coupling reaction of an allylchloride derivative and allylzinc bromide [103]. The expected diene was obtained with good yield and regioselectivity. The best yield and selectivity were observed with $[Rh(COD)Cl]_2$ associated with the ligand TMEDA (Scheme 49).

$[Rh(COD)Cl]_2$ could also be used as a catalyst for rhodium-catalyzed coupling reaction of allylic methyl ethers and allylmagnesium reagents [103]. The reaction showed a good regioselectivity; however, only modest yields were obtained.

Kabalka and coworkers reported the Rh-catalyzed coupling reaction of a variety of electron-rich and electron-deficient substituted aryl-, vinyl-, and

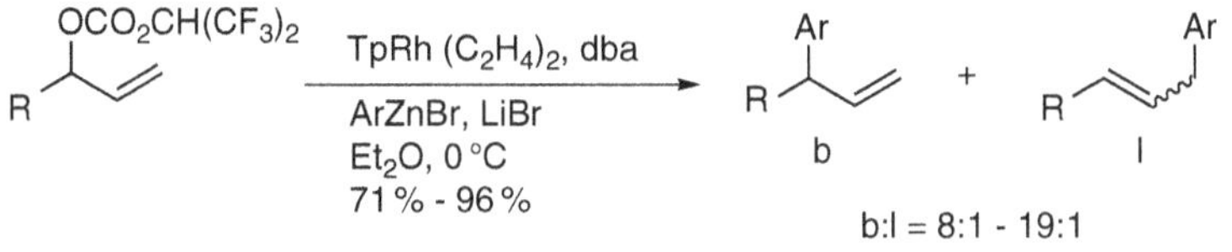

$OCO_2CH(CF_3)_2$ R — $TpRh(C_2H_4)_2$, dba; ArZnBr, LiBr; Et_2O, 0 °C; 71 % - 96 % → Ar R b + Ar R l

b:l = 8:1 - 19:1

Scheme 48 Utilization of $TpRh(C_2H_4)_2$ as catalyst for the allylic alkylation with arylzinc reagents

Ph Cl — ZnBr; $[RhCl(COD)]_2$, TMEDA; THF; 87 % → Ph b + Ph l

b:l = 83:17

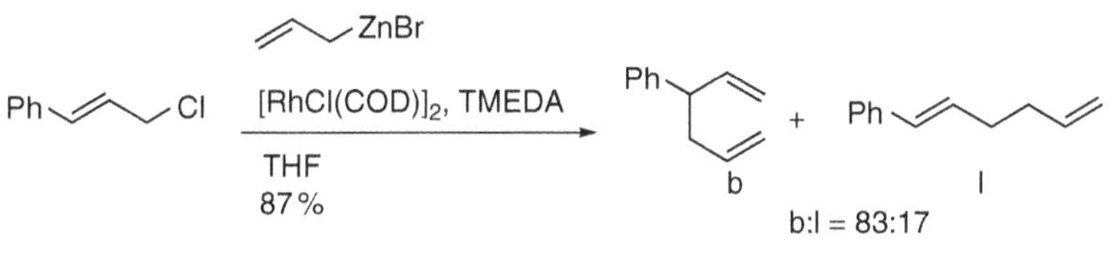

Scheme 49 Rh-catalyzed allylic allylation reaction with allylchloride derivatives and allylzinc reagents

heteroarylboronic acids with cinnamyl alcohols in ionic liquid medium [104]. Among the catalysts tested, hydrated rhodium (III) chloride [$RhCl_3.xH_2O$] proved the most efficient. The allylic alkylation products were obtained with modest to good yields. The presence of $Cu(OAc)_2$ as an additive increased the reaction yields (Scheme 50). Considering the poor leaving group ability of hydroxide, the use of allylic alcohols without activation is a significant achievement especially in term of atom-economy. However, no information was provided about the regio- and stereochemical features of the reaction.

In 2006, Gong and coworkers investigated the rhodium-catalyzed asymmetric nitroallylation of arylboronic acids and arylzinc chloride with cyclic nitroallyl acetates [105]. A catalyst screening indicated that the optimal catalytic system for the nitroallylation of arylzinc chlorides with 2-nitrocyclohex-2-enyl acetate was the rhodium complex prepared from $Rh(acac)(C_2H_4)_2$ and (R)-BINAP. The reaction was carried out at 0°C. High yields of up to 93% were obtained as well as high enantioselectivities of up to 96% *ee*. For the nitroallylation reaction of arylboronic acids with 2-nitrocyclohex-2-enyl acetate, the most efficient catalyst proved to be $[Rh(OH)(COD)]_2$ with (S)-BINAP. Modest to good yields were obtained and excellent enantioselectivities of up to 99% *ee* were observed (Scheme 51). Chiral synthetic intermediates could be prepared from the substituted nitroalkenes obtained by this asymmetric nitroallylation. The first total synthesis of optically pure (+)-β-lycorane was completed starting from the asymmetric nitroallylation of organozinc reagents with nitroallylacetates [105].

The same year, Lautens et al. investigated the enantio-, regio-, and diastereoselective rhodium(I)-catalyzed desymmetrization of meso-cyclic dicarbonate substrates by using a rhodium(I) catalyst formed in situ from $[Rh(OH)(COD)]_2$

Ph⁄⁄OH → Ph⁄⁄Ar
$RhCl_3.xH_2O$
$ArB(OH)_2$, $Cu(OAc)_2$
Ionic liquid, 50 °C
33 - 78%

Scheme 50 Direct Rh-catalyzed cross-coupling of cinnamyl alcohols with aryl boronic acids

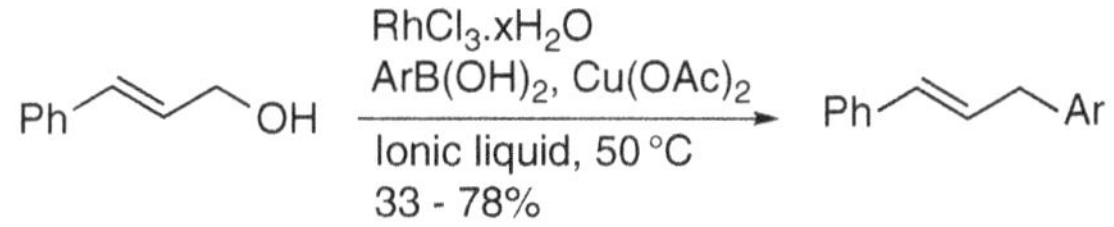

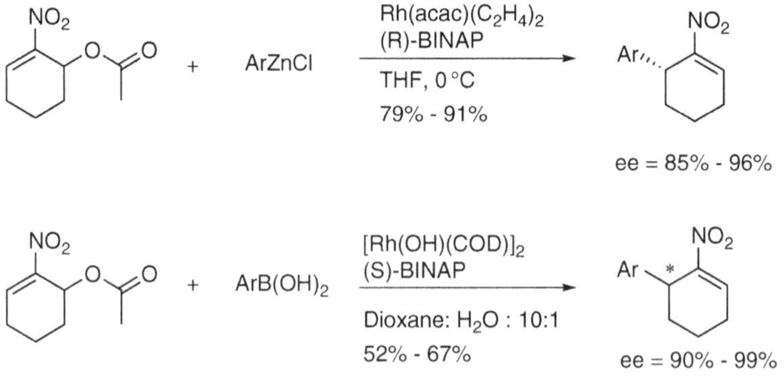

NO_2 (2-nitrocyclohex-2-enyl acetate) + ArZnCl → Ar (product with NO_2)
$Rh(acac)(C_2H_4)_2$
(R)-BINAP
THF, 0 °C
79% - 91%
ee = 85% - 96%

NO_2 (2-nitrocyclohex-2-enyl acetate) + $ArB(OH)_2$ → Ar * (product with NO_2)
$[Rh(OH)(COD)]_2$
(S)-BINAP
Dioxane: H_2O : 10:1
52% - 67%
ee = 90% - 99%

Scheme 51 Enantioselective Rh-catalyzed nitroallylation reaction of arylzinc reagents and arylboronic acids with 2-nitrocyclohex-2-enyl acetate

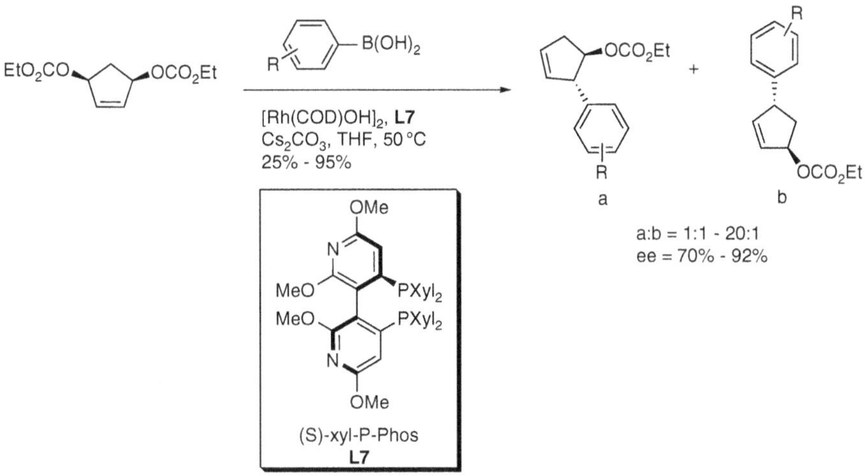

Scheme 52 Enantio-, regio- and diastereoselective Rh(I)-catalyzed desymmetrization of meso-cyclic dicarbonate

and Xyl-P-PHOS **L7** [106]. The allylic substitution products were obtained under mild conditions in high yields, with regioselectivities of up to 20:1 and excellent enantioselecivities of up to 92% (Scheme 52). Best conversions, regioselectivities, and enantioselectivities were obtained with arylboronic reagents bearing a strongly electron-withdrawing group in the *para*-position. *Ortho*-substituted arylboronics gave low conversions and regioselectivities. Substrates containing a nitrogen atom did not react probably due to the coordination of nitrogen to the rhodium complex.

3.2.4 Reaction with N- and O-Nucleophiles

Rhodium-Catalyzed Allylic Amination

Transition-metal catalyzed allylic amination reactions are particularly interesting as allylic amines are ubiquitous in a variety of biologically important natural and unnatural products [107].

Evans and coworkers reported a regio- and enantiospecific allylic amination reaction using the trimethylphosphite modified Wilkinson's catalyst $RhCl(PPh_3)_3$, $P(OMe)_3$ [108]. The amination reaction proceeds with net retention of absolute configuration. With the lithium anion *N*-toluenesulfonyl-*N*-alkylamines, the reaction afforded the allylic amination products with high yields and excellent stereospecificity, the regioselectivities were strongly dependent of the allylic substrates. This reaction could be extended to the allylic amination reaction of anilines derivatives (Scheme 53) [109].

Martin et al. examined the ability of $[Rh(CO)_2Cl]_2$ to catalyze allylic aminations [91]. The first reactions carried out with allyl carbonate and the salts of secondary sulfonamides were unsuccessful, the allylic substrates were almost entirely recovered at the end of the reactions. Inspired by the work of Lautens about halide effects in transition-metal catalysis [110, 111], they examined the use of iodide ion as an additive and they discovered that allylic amination reactions with pyrrolidine and methylbenzylamine as nucleophiles proceeded readily and efficiently at room temperature in the presence of tetra-*n*-butylammonium iodide TBAI (20 mol%). This result was explained by the increased stability of the Rh(I)–I complex obtained by exchanging the bridging chloride ion of $[Rh(CO)_2Cl]_2$ with an iodide anion that might be less prone to react with amines to form an unreactive complex. Yields are generally high and the regiospecificities are good to excellent dependent upon the allylic substrates (Scheme 54). The stereospecificity was not investigated.

Recently, Evans and Clizbe developed a regio- and stereospecific Rhodium-catalyzed allylic amination reaction with aza-ylide obtained from 1-amino-pyridinium iodide as stabilized N-nucleophile [112]. They could show that the nature of the stabilizing group was critical for this reaction. Thus, ammonium and pyridinium ylides are more reactive than phosphonium and sulfonium ylides as they are less stabilized therefore more nucleophilic. Among the bases tested out in this reaction, LiHMDS proved to give the best yield. The nature of the base had no influence on the regioselectivity. The yield could be increased by quenching the reaction with acetic acid and lithium iodide to circumvent problems arising from the formation of methoxyde by fragmentation of the methylcarbonate leaving group. Various racemic secondary allylic carbonates could be used, the yields were good and the variation of the allylic substrates had no influence on the regio-selectivities (>19:1 in all cases) (Scheme 55). Additional studies demonstrated that

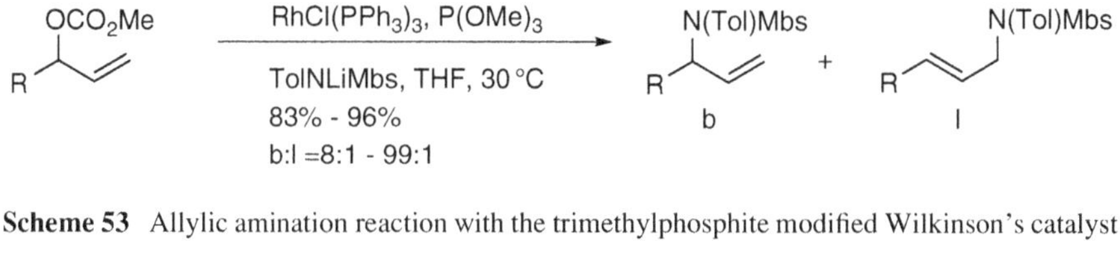

OCO_2Me — $RhCl(PPh_3)_3$, $P(OMe)_3$ / BnNLiTs, THF, 30 °C / 86% - 94% / b:l = 9:1 - 99:1 → N(Bn)Ts (b, ee = 100%) + N(Bn)Ts (l)

OCO_2Me — $RhCl(PPh_3)_3$, $P(OMe)_3$ / TolNLiMbs, THF, 30 °C / 83% - 96% / b:l =8:1 - 99:1 → N(Tol)Mbs (b) + N(Tol)Mbs (l)

Scheme 53 Allylic amination reaction with the trimethylphosphite modified Wilkinson's catalyst

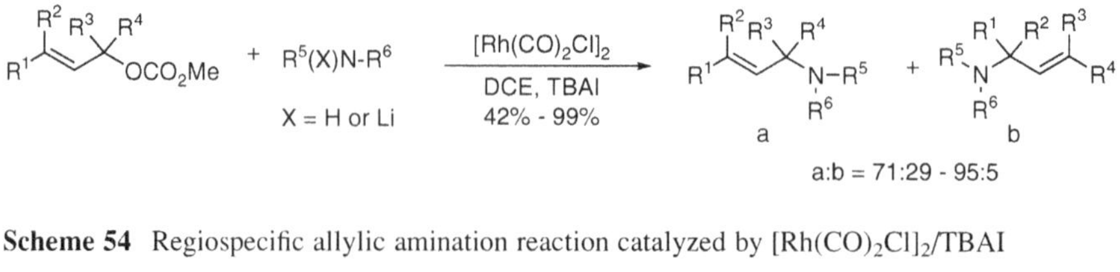

R^1, R^2, R^3, R^4, OCO_2Me + $R^5(X)N\text{-}R^6$ (X = H or Li) — $[Rh(CO)_2Cl]_2$ / DCE, TBAI / 42% - 99% → a + b; a:b = 71:29 - 95:5

Scheme 54 Regiospecific allylic amination reaction catalyzed by $[Rh(CO)_2Cl]_2$/TBAI

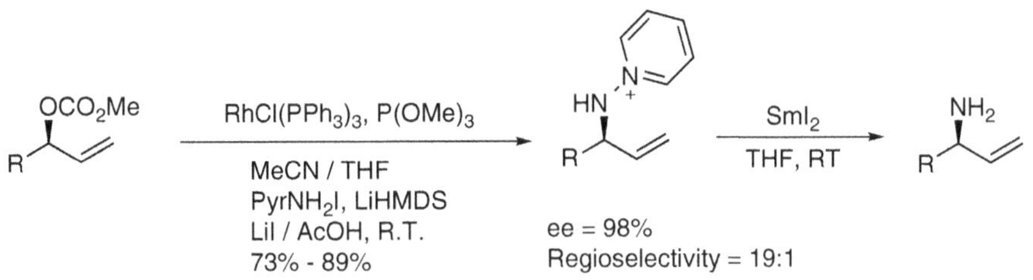

Scheme 55 Regio- and stereospecific Rh-catalyzed allylic amination reaction with aza-ylide

aza-ylides reacted with nonracemic allylic carbonate to provide chiral allylic pyridinum salts with retention of absolute configuration. The reductive cleavage of the pyridinium salt with samarium diiodide gave the expected enantiomerically enriched primary allylic amine [112]. A one pot protocol could be developed based on the utilization of 1-aminopyridinium iodide as a new ammonia equivalent for the construction of primary allylic amines.

An enantioselective version of rhodium-catalyzed allylic amination was reported by Vrieze and coworkers [113]. They developed a new method for kinetic resolution of unsymmetrical acyclic allylic carbonates and the concurrent synthesis of enantioenriched secondary amines. Interestingly, among the catalysts screened in this reaction, the three best catalysts were (S)-Binapine(**L10**)-Rh, (R,R)-Ph-BPE (**L9**)-Rh, and (S,S,R,R)-Tangphos(**L8**)-Rh, all three are commercially available and were unprecedented for use in allylic substitution reaction. The carbonates could be isolated in >99.9% *ee* and the amines were isolated in 91–99.9% *ee* (Scheme 56). Furthermore, the process proved regioselective as no formation of linear product was observed.

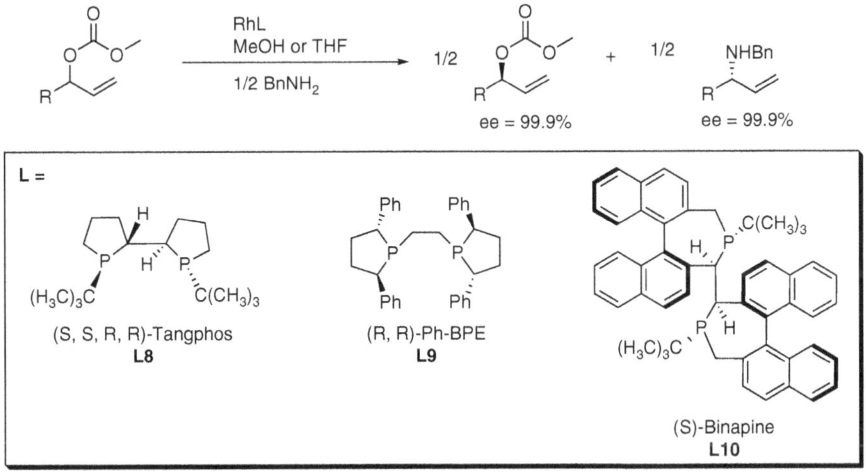

Scheme 56 Rh-catalyzed kinetic resolution of unsymmetrical acyclic allylic carbonates and concurrent synthesis of enantioenriched secondary amines

Rhodium-Catalyzed Allylic Etherification

Evans and Leahy described the first Rhodium-catalyzed allylic etherification with *ortho*-substituted phenols in using the trimethylphosphite modified Wilkinson's catalyst $RhCl(PPh_3)_3$, $P(OMe)_3$ [114]. A temperature of 0°C was crucial for obtaining good yields and selectivities (Scheme 57). It was shown that the reaction proceeds with net retention of absolute configuration. The ability to use the more challenging secondary and tertiary alcohols in rhodium-catalyzed allylic etherification reactions was also examined.

The rhodium-catalyzed allylic etherification was unsuccessful with hard alkali metal alkoxides, this problem could be circumvent in using copper alkoxides obtained by transmetallation of the alkali derivative with a copper halide salt in order to softens the basic character of the metal alkoxyde [115]. Furthermore, yields could be increased by utilizing *tert*-butyloxycarbonate as leaving group which restrains the competitive transacylation reaction of the leaving group by the nucleophile (Scheme 58).

Investigations on the stereospecificity of the Rhodium-catalyzed etherification reaction showed a dramatic halide effect. The examination of other copper(I) halide salts indicated that utilization of copper(I) chloride or bromide improved the chirality transfer. This result was explained by the halide's influence on the π–σ–π isomerization supported by the fact that the rate of racemization observed varies according to the *trans*-effect of each halide ligand (I > Br > Cl) [116]. Furthermore, the best stereospecificities were obtained with lower temperatures (Scheme 59). Additional studies highlighted the fact that the copper alkoxide and the lithium iodide were both necessary for optimal catalytic activity [115].

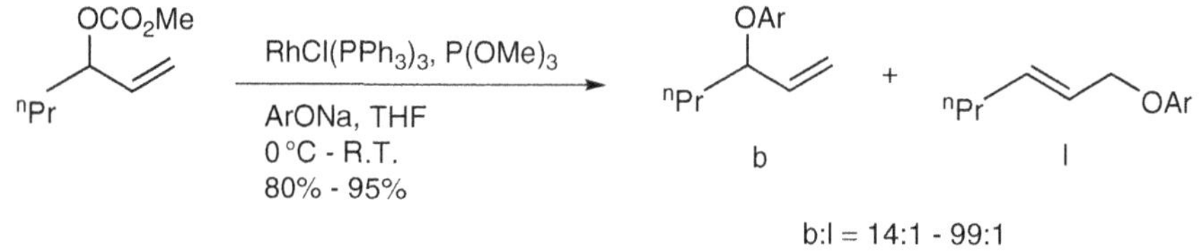

OCO$_2$Me, ^{n}Pr → [$RhCl(PPh_3)_3$, $P(OMe)_3$; ArONa, THF; 0 °C - R.T.; 80% - 95%] → OAr, ^{n}Pr (b) + ^{n}Pr, OAr (l)

b:l = 14:1 - 99:1

Scheme 57 Rh-catalyzed allylic etherification with *ortho*-substituted phenols in using the trimethylphosphite modified Wilkinson's catalyst

OCO$_2$^{t}Bu, R → [$RhCl(PPh_3)_3$, $P(OMe)_3$; iPr$_2$CHOLi, CuI; THF, 0 °C - R.T.; 46% - 73%] → O, ^{i}Pr, ^{i}Pr, R (b) + R, O, Pr^i, ^{i}Pr (l)

b:l = 18:1 - 99:1

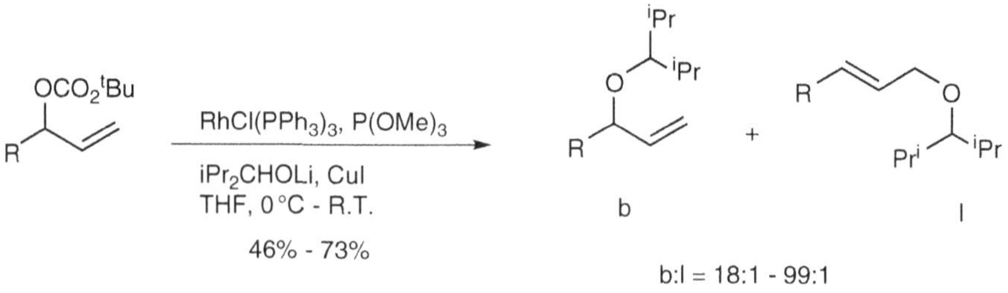

Scheme 58 Rh-catalyzed allylic etherification with copper alkoxides

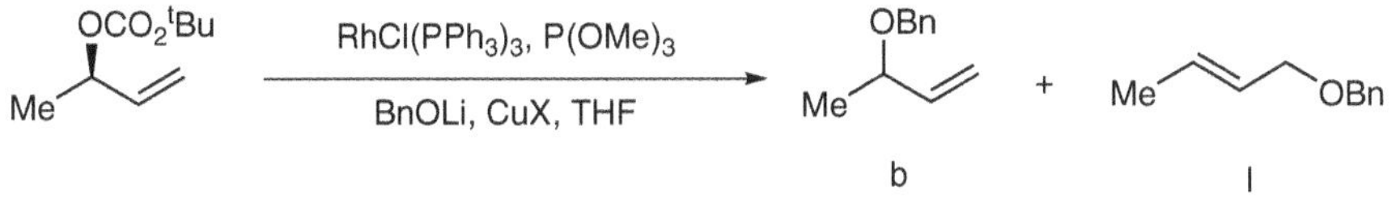

$X = I$, Yield = 84%, b:l = 99:1, ee = 41%
$X = Br$, Yield = 86%, b:l = 91:9, ee = 85%
$X = Cl$, $T = 0\,°C$, Yield = 81%, b:l = 99:1, ee = 88%
$X = Cl$, $T = -10\,°C$, Yield = 81%, b:l = 99:1, ee = 96%

Scheme 59 Halide effect in Rh-catalyzed allylic etherification with copper alkoxides

4 Allylation Reactions Catalyzed by Group 10 Metals (Without Palladium)

Nickel has been considered a cheap and readily available substitute for Palladium in catalytic reactions. Although a direct comparison is inevitable, we would like to emphasize the reactivity of allylation reactions deviating from the Palladium-catalyzed reactions.

4.1 Nickel Catalyzed Allylic Allylation Reactions

4.1.1 Nickel-Catalyzed Allylic Alkylations

Nickel-catalyzed allylic alkylations have recently been described in a monograph on organonickel chemistry. Therefore, only a representative overview will be given [117, 118].

Allylic alkylations using Nickel have mainly been described for hard nucleophiles such as Grignard-reagents. The inversed l to b ratio (linear to branched) for Nickel catalysis compared to the identical Palladium source is noteworthy. Examples of leaving groups used for this transformation are: $-OSiEt_3$, –OTHP, –Cl, and –OH [119–121] (Scheme 60).

Although this trend in selectivity for Nickel-catalyzed reactions is similar for a manifold of reactions (vide infra), a severe dependence on substrate and/or the catalyst system is usually observed. For example, the preservation of conjugation may reverse the above selectivity when a phenyl-substituent is used (Scheme 61) [122].

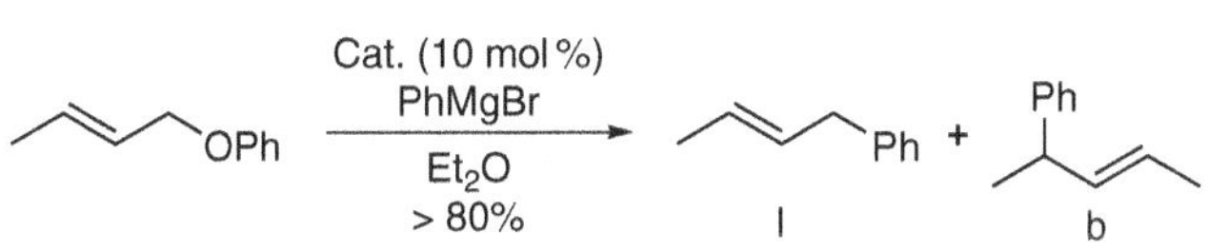

$NiCl_2(dppf)$: l:b = 14:86
$PdCl_2(dppf)$: l:b = 90:10

Scheme 60 Allylic alkylation using Grignard-reagents

Reactions of this type (employing *hard* nucleophiles) result in inversion of stereochemistry when appropriate substrates are used [123]. This is in agreement with an inner-sphere mechanism.

Furthermore, was the use of various organoboronate reagents as the nucleophilic partner successfully carried out [117, 118].

An example is the allylic alkylation of cyclic-carbonate **5** with organoboronate **6** using $NiCl_2$(dppf) as a catalyst [124]. This reaction proceeds with inversion of the stereochemistry, as observed for Grignard reagents. The use of this type of nucleophile with related Palladium complexes results in exclusive methoxylation and no carbon–carbon-bond formation (Scheme 62).

In contrast, the use of soft carbonnucleophiles, such as malonates, is only of limited synthetic utility [125–127].

Only few asymmetric allylic alkylations using Nickel catalysis have been described. First examples relied on the use of cyclic substrates for high enantioselectivities (Table 2).

Although the use of acyclic systems is somewhat limited, there are examples of moderate enantioselectivity using a catalyst system consisting of $Ni(cod)_2$ and (*S*,*S*)-chiraphos [132].

Scheme 61 Substrate controlled formation of linear product [122]

Scheme 62 Allylic alkylation with organoboron reagent [124]

Table 2 Nickel catalyzed asymmetric allylic alkylation

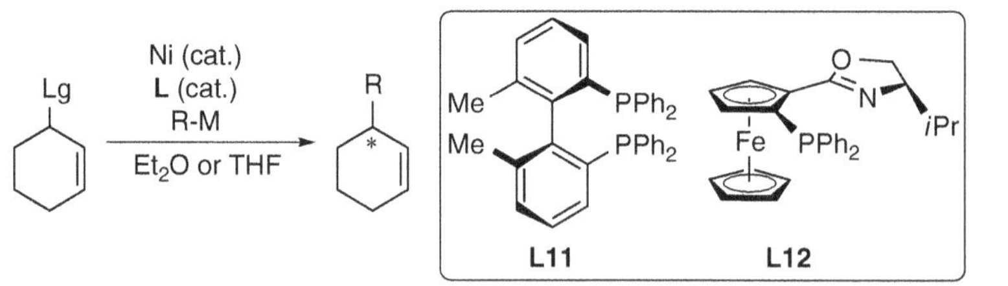

Entry	Lg	R–M	Ni-source	Ligand	Yield (%)	*ee* (%)	Reference
1	OPh	EtMgBr	$NiBr_2$	**L11**	84	83	[128, 129]
2	OPh	PhMgBr	$Ni(acac)_2$	**L12**	98	82	[130]
3[a]	OAc	$PhB(OH)_2$	$Ni(acac)_2$	**L13**	81	50	[131]

[a] $HAl(i Bu)_2$ (16 mol%) and KOH (3 equiv.) was added

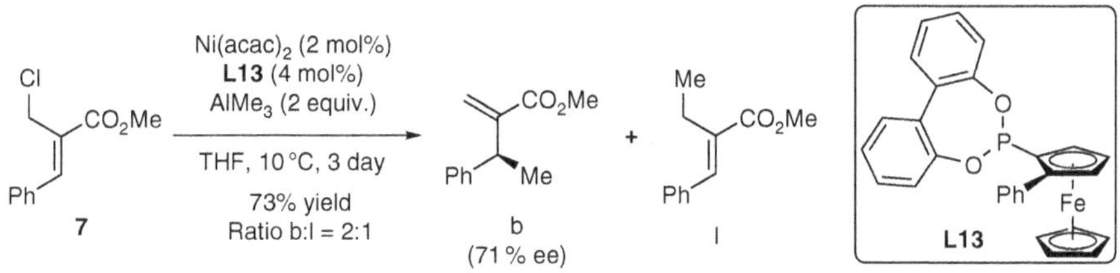

Scheme 63 Nickel-catalyzed allylic aubstitution using Ferrophite-Ligand **L13** [133]

Table 3 Homoallylic alcohols as nucleophile-surrogates [135]

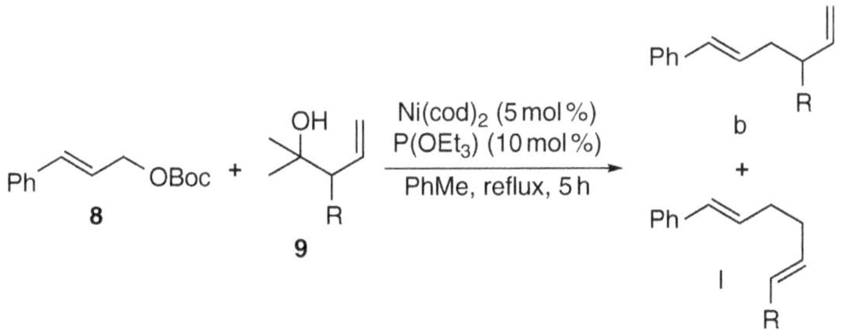

Entry	R	Yield	b:l
1	nC_7H_{15}	60	26:74
2	cC_6H_{11}	77	9:91
3	*t*Bu	79	3:97
4	Ph	100	0:100
5	4-MeC_6H_4	96	0:100
6	1-Naphthyl	99	0:100
7	$SiMe_3$	89	0:100

Novak et al. found that asymmetric allylic alkylation of allylic compound **7** with $AlMe_3$ is promoted in the presence of $Ni(acac)_2$ and a Ferrophite-Ligand **L13** [133] (Scheme 63). This reaction could also be carried out using DABAL-Me_3 (DABAL-Me_3 is an adduct of DABCO and $AlMe_3$, which is solid and airstable, unlike the pyrophoric $AlMe_3$ itself) [134].

As described above, the nucleophiles employed in allylic substitutions are mostly of organometallic nature representing a carbanion-equivalent. Less intuitively have homoallylic alcohols been employed as nucleophiles in Nickel catalyzed allylation reactions by Sumida et al. which generate a carbon-nucleophile via retro-allylation [135]. Treating allyl-carbonate **8** with homoallylic alcohol **9** and catalytic amounts of $Ni(cod)_2$ and $P(OEt)_3$ results in the formation of 1,6-hexadienes in good yields (Table 3).

However, this reaction only provides moderate branched to linear ratio of products when primary alkylsubstituents are used in place of R (nC_7H_{15}, Me also delivered poor selectivity). Changing to secondary and tertiary alkyl-substituent increased selectivity drastically, and finally changing for aromatic or silyl substituents resulted in selective formation of the linear products.

This reaction has been proposed to proceed according to the depicted catalytic cycle (Scheme 64).

In recent years, carbon–carbon-bond cleavage catalyzed by transition metals has emerged as a new challenge in catalysis. Among the successfully employed metals is Nickel [136, 137]. The group around Oshima has been able to develop an allylic substitution reaction which employs allylic malonates **10** as substrates and aryl-zinc compounds as nucleophiles in the presence of a catalytic amount $NiBr_2(PPh_3)_2$ and 2 equiv. of $MgBr_2$ [138] (Scheme 65). The inclusion of $MgBr_2$ proved to be

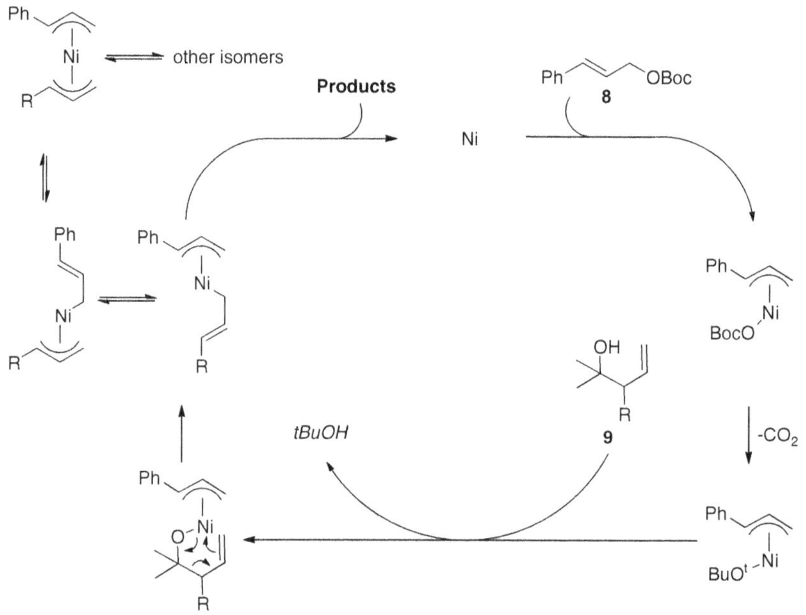

Scheme 64 Proposed catalytic cycle for the coupling of allylcarbonates and homoallylic alcohols [135]

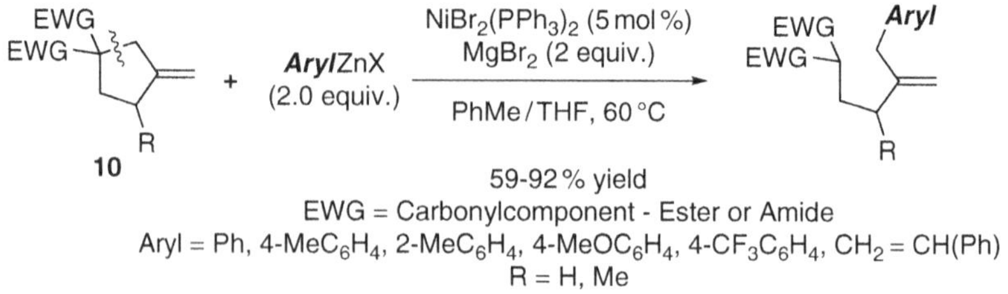

Scheme 65 Nucleophilic allylic substitution of malonates [138]

essential, which was attributed to the activation of the dicarbonylcomponent (EWG). This reaction was also applicable to acyclic systems.

4.1.2 Nickel-Catalyzed Allylic Amination Reactions

Although Nickel-catalyzed allylic substitutions are most known for their ability to tolerate hard nucleophiles (vide supra), some examples of allylic amination have been described.

An early example describes the allylic substitution of allylalcohol by morpholine in the presence of a catalytic amount of $Ni(nBu_3P)_2Br_2$ and K*Ot*Bu [139, 140] (Scheme 66).

Bricout et al. were able to expand on the utility of soft nucleophiles in Nickel-catalyzed reactions and showed that amination of allylalcohol and its derivatives was readily catalyzed by Nickel-bisphosphine complexes [127, 141, 142]. They observed rates for the reaction depending on the ligand: dppb > dppp > dppe. Exponential plots for the conversion were observed for this type of reaction, which was attributed to the built-up of diethylammonium acetate. The inclusion of ammonium salts, such as diethylammonium acetate, diethylammonium phenoxide, or tetrabutylammonium hexafluorophosphate, had a rate enhancing effect leading to linear plots (Table 4).

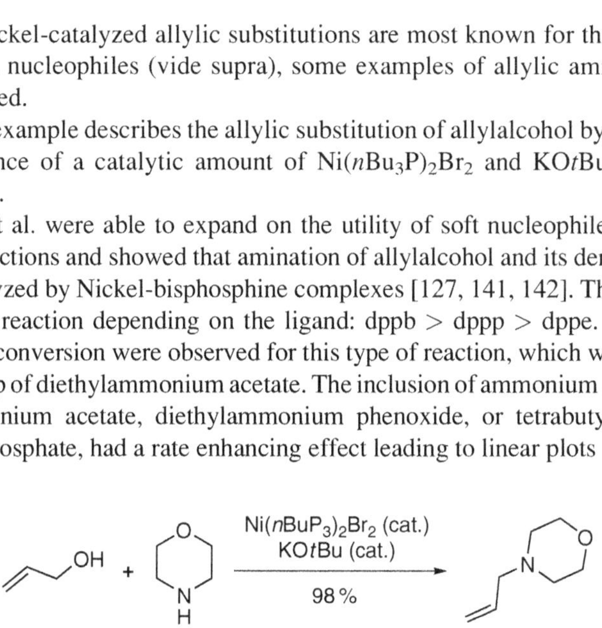

OH + morpholine (O, N–H) —[$Ni(nBuP_3)_2Br_2$ (cat.), K*Ot*Bu (cat.); 98 %]→ N-allylmorpholine

Scheme 66 First Nickel catalyzed allylic amination [139]

Table 4 Ni catalyzed allylic amination [127, 141, 142]

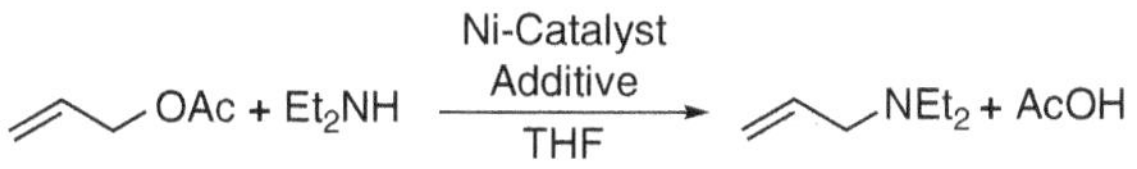

OAc + Et_2NH —[Ni-Catalyst, Additive; THF]→ NEt_2 + AcOH

Entry	Catalyst[a]	Additive	T (°C)	t[b] (min)
1	$Ni(dppe)_2$	–	80	270
2	$Ni(dppp)_2$	–	80	165
3	$Ni(dppb)_2$	–	80	90
4	$Ni(dppb)_2$	$[H_2NEt_2][OAc]$[c]	50	36
5	$Ni(dppb)_2$	$[H_2NEt_2][OPh]$[c]	50	9
6	$Ni(dppb)_2$	$[NBu_4][PF_6]$[d]	50	14

[a]Catalyst prepared in situ from $Ni(cod)_2$ and the appropriate ligand
[b]Time required for full conversion
[c]100 eq. With respect to Ni-catalyst
[d]5 eq. With respect to Ni-catalyst

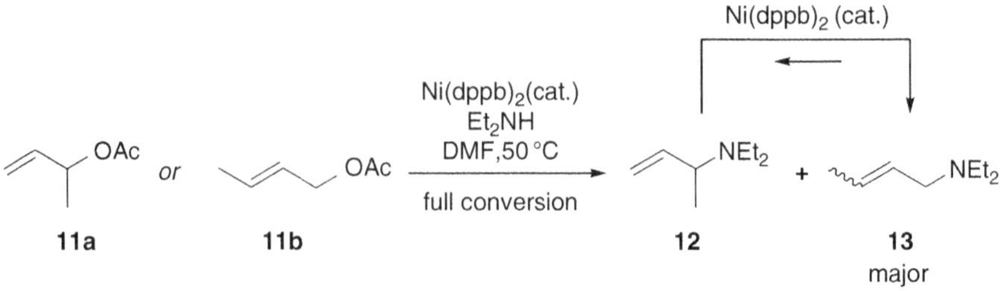

Scheme 67 Possible intermediates in asymmetric allylic alkylation [127, 141, 142]

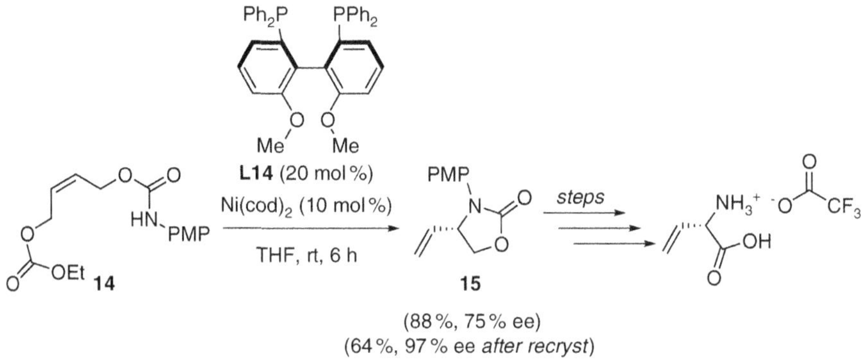

Scheme 68 Preparation of L-vinylglycine using Nickel catalyzed asymmetric allylic amination [144]

When substituted allylic acetate **11a** or **11b** was used preferential formation of the linear product **13** was observed. However, when the reaction was monitored accumulation of the branched product **12** was observed which rearranged into the linear product in the presence of the Nickel catalyst (Scheme 67).

More recently in situ *enzymatic screening* (ISES) was employed as a tool for reaction discovery leading to the finding of Nickel-catalyzed intramolecular allylic amination [143–145]. This could further be used to develop an asymmetric version of this reaction. The utility of this process was demonstrated by the synthesis of L-vinylglycine. Treatment of cyclisation precursor **14** with $Ni(cod)_2$ in the presence of MeO-BIPHEP **L14** gave compound **15** in 88% yield and 75% *ee*. This could be increased to 97% *ee* upon recrystalization. Standard functional group interconversions gave L-vinylglycine (Scheme 68).

4.1.3 Nickel-Catalyzed Allylation Reactions of Miscellaneous Nucleophiles

Within this section, we would like to give a few examples of nucleophiles which are neither noncarbon nor nitrogen based.

An early example employing sulfinates as nucleophiles has been described by Cuvigny and Julia [125]. Depending on the catalytic system, they were able to selectively prepare either the linear sulfone **16a** or the branched sulfone **16b**. This resembles a case of reagent control (Scheme 69).

Recently, Yatsumonji et al. found that $Ni[P(OEt)_3]_4$ catalyzed the reaction between allylic acetates and either thiols or alcohols [146]. This reaction proceeds with retention of both regio- and stereochemistry (Scheme 70).

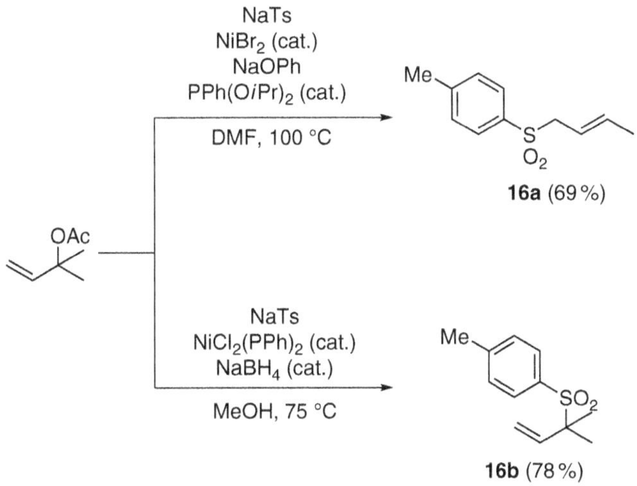

NaTs
$NiBr_2$ (cat.)
NaOPh
$PPh(O\textit{i}Pr)_2$ (cat.)
DMF, 100 °C
Me
S
O_2
16a (69 %)
OAc
NaTs
$NiCl_2(PPh)_2$ (cat.)
$NaBH_4$ (cat.)
MeOH, 75 °C
Me
SO_2
16b (78 %)

Scheme 69 Reagent controlled allylic sulfonylation [125]

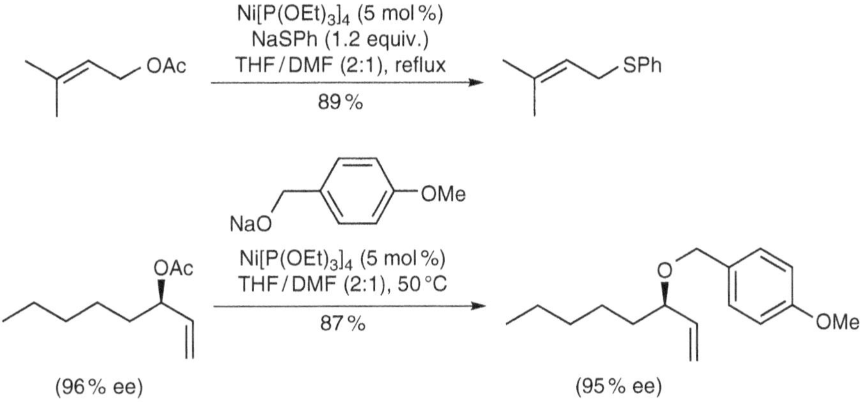

OAc
$Ni[P(OEt)_3]_4$ (5 mol %)
NaSPh (1.2 equiv.)
THF / DMF (2:1), reflux
89 %
SPh
NaO
OMe
OAc
$Ni[P(OEt)_3]_4$ (5 mol %)
THF / DMF (2:1), 50 °C
87 %
O
OMe
(96 % ee)
(95 % ee)

Scheme 70 Allylic etherification and sulfenylation [146]

A rare example of carbon–phosphorus bond formation was described by Lu et al. [147]. Treatment of allylic acetates **17a** or **17b** with $HP(O)(OEt)_2$ in the presence of

N,O-Bis(trimethylsilyl)acetamide (BSA) and a catalytic amount of $Ni(cod)_2$ results in the selective formation of **18** regardless of the starting material (Scheme 71).

Organoboron reagents have turned out to be among the most versatile synthetic reagents. Successful Nickel-catalyzed borylation has been described for allylic epoxides using $Ni(cod)_2$ and *racemic*-BINAP (Scheme 72) [148]. The allylic boronic ester **19** could be reacted further in situ, i.e. with aldehydes.

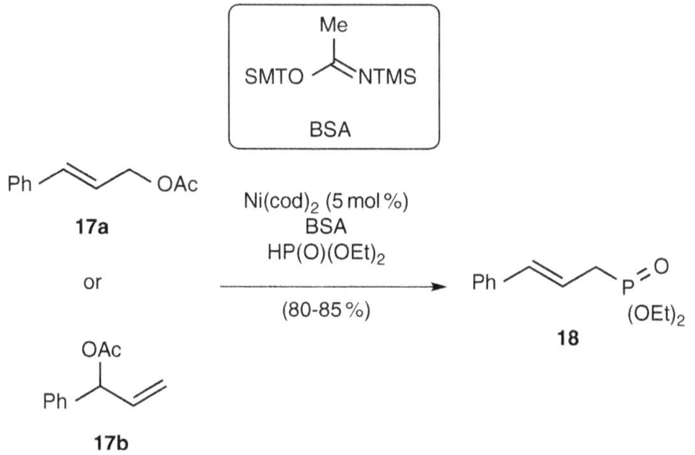

Me
SMTO NTMS
BSA

Ph OAc
17a
or
OAc
Ph
17b

$Ni(cod)_2$ (5 mol %)
BSA
$HP(O)(OEt)_2$
(80-85 %)

Ph P=O
$(OEt)_2$
18

Scheme 71 Allylic phosphorylation [147]

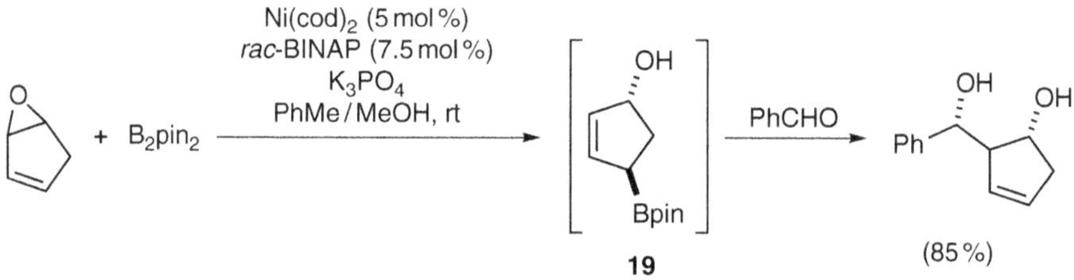

O
+ B_2pin_2

$Ni(cod)_2$ (5 mol %)
rac-BINAP (7.5 mol %)
K_3PO_4
PhMe / MeOH, rt

OH
Bpin
19

PhCHO

OH
OH
Ph
(85 %)

Scheme 72 Allylic borylation and subsequent reaction with aldehyde [148]

4.2 *Platinum Catalyzed Allylation Reactions*

In the past decades, Palladium has been one of the workhorses for allylic substitution reactions. While a manifold of transition metals (vide supra) has been used for this transformation, only few applications remain investigated for Platinum. Usually, a lower reactivity of Platinum allows for mechanistic investigations of Palladium-catalyzed reactions including the characterization of intermediates, which are usually too reactive in the case of Palladium [149, 150]. Despite this, a few transformations have been found and developed for Platinum-catalyzed allylic substitution reactions in which reactivity is observed differing from Palladium [151].

4.2.1 Platinum-Catalyzed Allylic Alkylation Reactions

The early success of Palladium in allylic substitution reactions [152], including the early moderate success in enantioselective variants of this transformation [153], resulted in the attempt to utilize organo-Platinum-complexes as mechanistic probes for this reaction. Kurosawa observed inversed selectivity for the reaction between butenyl acetates and malonates, the preferential formation of the branched products, when otherwise identical Platinum complexes were compared to their Palladium congeners [154]. In analogy to this study, Brown and MacIntyre observed the same effect when using chiral DIOP-Platinum-(η^3-allyl)BF_4 complexes. Complexes of this type only delivered poor enantioselectivities when applied in the previously mentioned reaction (up to 23% *ee* for Platinum-DIPAMP-complex) [155].

More than a decade later were Blacker et al. able to describe a highly enantioselective allylic alkylation using platinum complexes [156, 157]. Key to obtain enantioselectivity was the use of a bidentate-phosphino-oxazoline ligand (see complex **C18**). Interestingly was this complex less reactive than for example simple $Pt(PPh_3)_4$. The reaction only proceeded well at elevated temperatures.

An interesting effect observed was the erosion of *ee* when excess ligand was added. It is believed that this reaction can proceed via the two different intermediates **C19** and **C20** out of which the bis-phosphine-complex is more reactive and prone to form due to hemilability of the phosphino-oxazoline ligand (Scheme 73). This also explains why the addition of PPh_3 resulted in a more reactive Platinum-species resulting in almost racemic product (Table 5).

Murai and coworkers were able to develop an allylic alkylation featuring an additional nucleophilic substitution at the central allyl carbon [158]. Treatment of allylic acetate **20** with malonate **21** in the presence of a catalytic amount of Pt $(C_2H_4)(PPh_3)_2$ resulted in the selective formation of **23**. This product is formed via π-allyl complex **C21**,which forms a Platina-cyclobutane **C22** upon reaction with a nucleophile, following elimination to give **C23** which upon reaction with a further equivalent of nucleophile results in the formation of the product **23**. The analogous

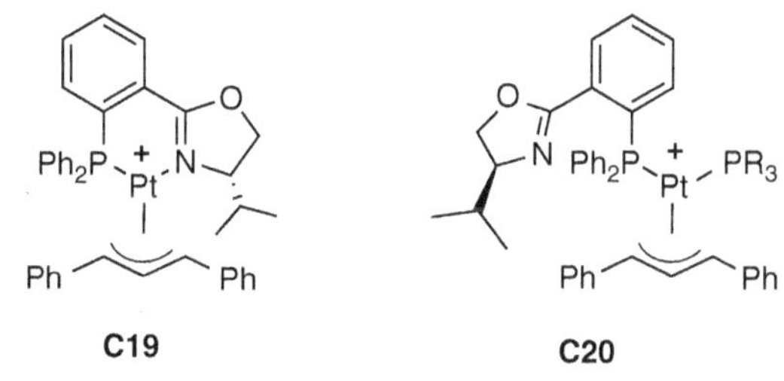

Scheme 73 Possible intermediates in asymmetric allylic alkylation [156, 157]

Table 5 Pt catalyzed enantioselective allylic alkylation [156, 157]

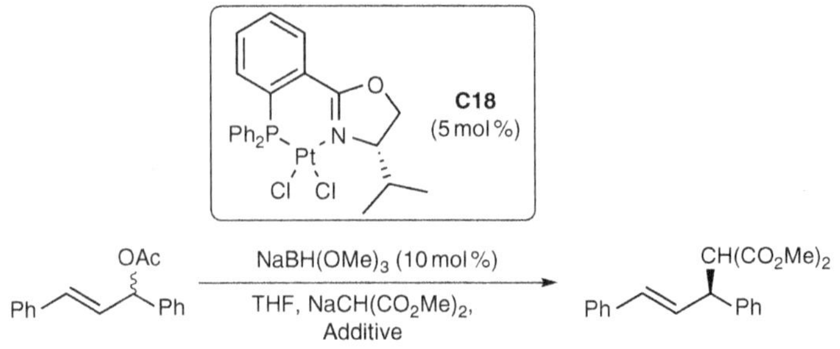

C18 (5 mol%)

NaBH(OMe)$_3$ (10 mol%), THF, NaCH(CO$_2$Me)$_2$, Additive

Entry	Additive	T [°C]	t [h]	Conversion (yield)	*ee*
1	None	20	20	–	–
2	None	65	44	65 (–)	77 (*S*)
3	5% ligand	65	35	100 (93)	83 (*S*)
4	10% ligand	65	44	100 (–)	61 (*S*)
5	5% PPh$_3$	20	16	100 (91)	2 (*S*)

reaction only resulted in the regular allylic alkylation product **22** when Pd(PPh$_3$)$_4$ was used as a catalyst. This methodology could further be developed for the synthesis of furans and hydrofurans [159–161].

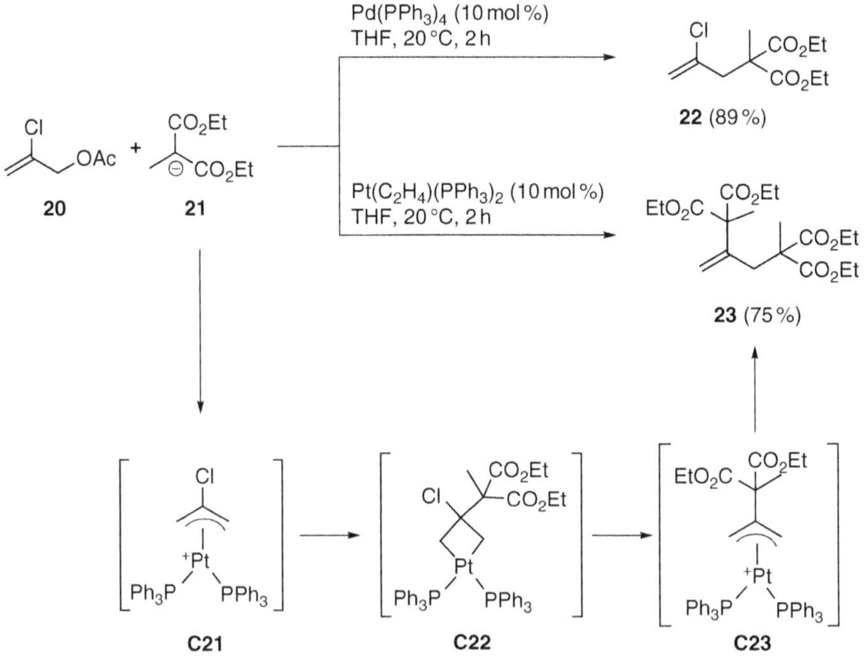

Pd(PPh$_3$)$_4$ (10 mol%), THF, 20 °C, 2 h → **22** (89%)

Pt(C$_2$H$_4$)(PPh$_3$)$_2$ (10 mol%), THF, 20 °C, 2 h → **23** (75%)

20 + **21** → **C21** → **C22** → **C23**

Scheme 74 Platinum-catalyzed double alkylation [158]

4.2.2 Platinum-Catalyzed Allylic Amination Reactions

Allylic amination reactions using Platinum are scarce despite the widespread application of Palladium in this field. Examples of Platinum-catalyzed allylations of aminonaphthalenes using allyacetates in the presence of catalytic amounts of $Pt(acac)_2$ and a phosphine have been described recently [162]. It may be noticed that this transformation could be carried out in water as the sole solvent [163]. A field of special interest is the direct application of allylic alcohols (which have not been preactivated as i.e. the acetate, carbonate, phosphate, etc.) in order to minimize waste generation. This type of reaction was successfully developed for Palladium [164]. The reactions usually require the use of a promoter (i.e., $SnCl_2{\cdot}H_2O$, $Ti(OiPr)_4$ or BEt_3). This general concept was readily transferred to Platinum catalysis. The use of 1 mol% $Pt(acac)_2$ in the presence of PPh_3 and 0.25 equiv. of $Ti(OiPr)_4$ allowed for the preparation of *N*-allyl-anilines (Scheme 75) [165]. The use of substituted alcohols resulted under preferential formation of the linear products.

Although this type of procedure represents a significant advancement in the direct use of allylic alcohols in substitution reactions, the use of a promoter free protocol is desirable. Promoter-free protocols have been developed for Palladium,

OH
$Pt(acac)_2$ (cat.)
PPh_3 (cat.)
$Ti(O\textit{i}Pr)_4$
PhH, reflux
R^2
NHR^1
NR^1
N
+
major
minor
R^1 = H, Me, Et, Ally l
R^2 = OMe, Me, CO_2Et, Cl, Br, I, CN, NO_2
50-99 % yield

Cl
NH_2
Me
OH
24a
or
OH
24b
$Pt(acac)_2$ (cat.)
PPh_3 (cat.)
$Ti(O\textit{i}Pr)_4$
PhH, reflux
Cl
Me
HN
+
from **24a** 55 % (E:Z = 83:17) 37 %
from **24b** 51 % (E:Z = 83:17) 39 %

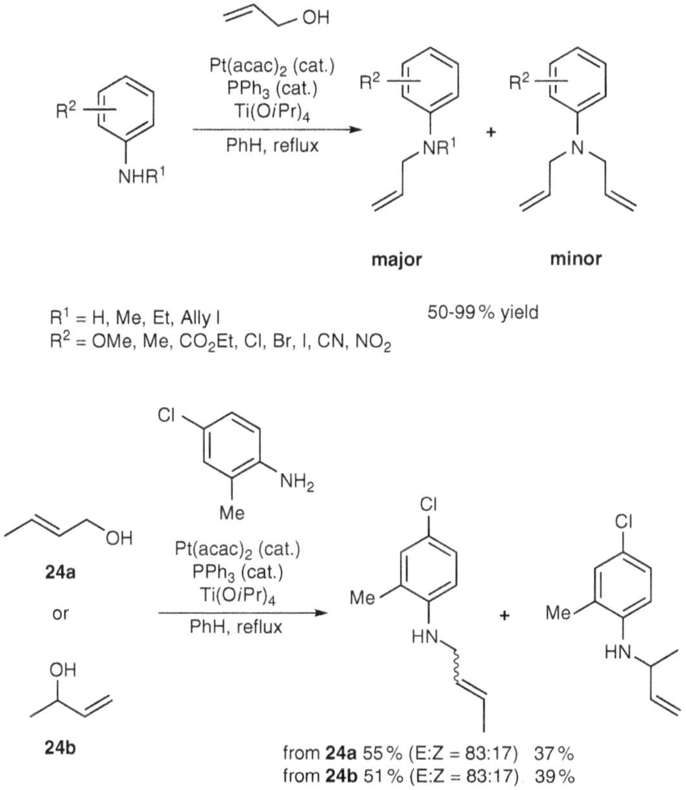

Scheme 75 Pt catalyzed N-allylation in the presence of the promoter $Ti(O\textit{i}Pr)_4$ [165]

suffering from preferential formation of diallylated products [166, 167]. The groups of Oshima [168, 169] and le Floch [170] were able to circumvent this by the application of Platinum catalysis employing large-bite angle phosphine ligands ($\beta > 100°$) (Scheme 76).

Both aromatic and aliphatic amines were tolerated under the reaction conditions. Oshima et al. further observed that microwave irradiation had a beneficial effect (see Table 6). It is notable that reactions of substituted allylic alcohols (not shown) resulted in the selective formation of linear products.

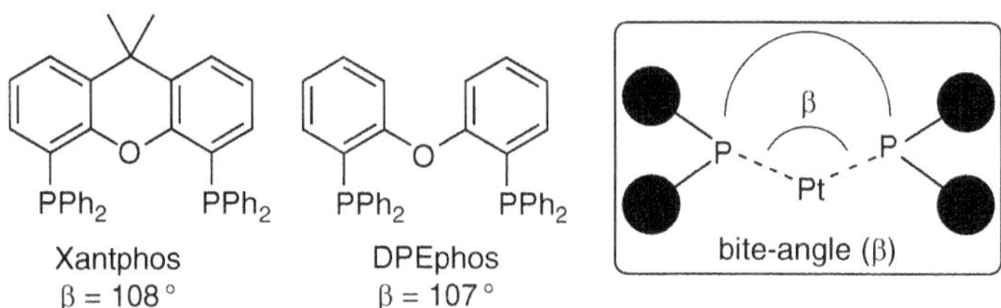

Scheme 76 Large bite-angle ligands used in Pt-catalyzed allylic amination [168, 169]

Table 6 Pt catalyzed *N*-allylation under microwave conditions [168, 169]

Ph⁄⁄OH + Amine —conditions→ Ph⁄⁄NHR **25** + Ph⁄⁄NR$_2$ **26**

Entry (conditions)	Amine	Yield of **25** (%)	Yield of **26** (%)
1 (a)	H, NH$_2$	88	6
2 (a)	MeO, NH$_2$	85	7.5
3 (a)	F$_3$C, NH$_2$	86	5
4 (b)	NH$_2$	80	10
5 (b)	NH$_2$	85	7
6 (b)	O, NH	96	–

Conditions: (a) Pt(cod)Cl$_2$ (0.5 mol%), Xantphos (0.5 mol%), DMF, 50°C, 1 h, microwave; (b) Pt(cod)Cl$_2$ (1.0 mol%), DPEphos (2.0 mol%), PhMe, 60°C, 2–4 h, microwave

The mechanism of this reaction was investigated both experimentally and theoretically via DFT calculations.

References

1. Plietker B (ed) (2008) Iron catalysis in organic chemistry. Wiley-VCH, Weinheim
2. Jana U, Biswas S, Maiti S (2007) Tetrahedron Lett 48:4065–4069
3. Zhan Z-P, Liu H-J (2006) Synlett 2278–2280
4. Zhan Z-P, Yu J, Liu H, Cui Y, Yang R, Yang W (2006) J Org Chem 71:8298–8301
5. Das B, Maijhi A, Banerjee J, Chowdhury N, Venkateswarlu K (2005) Chem Lett 34:1492–1493
6. Jonas K, Schiefenstein L, Krüger C, Tsay Y-H (1979) Angew Chem 91:590–591, Angew Chem Int Ed (1979) 18:550–551
7. Fürstner A, Krause H, Lehmann C W (2006) Angew Chem 118:454–458, (2006) Angew Chem Int Ed 45:440–444
8. Yanagisawa A, Nomura N, Yamamoto H (1991) Synlett 513–514
9. Yanagisawa A, Nomura N, Yamamoto H (1994) Tetrahedron 50:6017–6028
10. Hashmi ASK, Szeimies G (1994) Chem Ber 127:1075–1089
11. Nakamura M, Hirai A, Nakamura E (2000) J Am Chem Soc 122:978–979
12. Nakamura M, Matsuo K, Inoue T, Nakamura E (2003) Org Lett 5:1373–1375
13. Pasto DJ, Hennion GF, Shults RH, Waterhouse A, Chou S-K (1976) J Org Chem 41:3496
14. Pasto DJ, Chou S-K, Waterhouse A, Shults RH, Hennion GF (1978) J Org Chem 43:1385–1388
15. Fürstner A, Mendez M (2003) Angew Chem 115:5513–5515, (2003) Angew Chem Int Ed 42:5355–5357
16. Ladoulis SJ, Nicholas KM (1985) J Organomet Chem 285:C13–C16
17. Silverman GS, Strickland S, Nicholas KM (1986) Organometallics 5:2124–2127
18. Hieber W, Beutner H (1960) Z Naturforsch Teil B 15:323–324
19. Hieber W, Beutner H (1963) Z Anorg Allg Chem 320:101–111
20. Trost BM, Van Vranken DL (1996) Chem Rev 96:395–422
21. Emerson GF, Pettit R (1962) J Am Chem Soc 84:4591–4592
22. Whitesides TH, Ahrhardt RW (1971) J Am Chem Soc 93:5296–5298
23. Pearson AJ (1975) Tetrahedron Lett 16:3617–3618
24. Ley SV, Liam R, Meek G (1996) Chem Rev 96:423–442
25. Enders D, Jandeleit B, von Berg S (1997) Synlett 421–431
26. Enders D, Jandeleit B, von Berg S, Raabe, Runsink J (2001) Organometallics 20:4312–4332 and references therein
27. Roustan J-L, Abedini M, Baer HH (1979) Tetrahedron Lett 3721–3724
28. Roustan J-L, Abedini M, Baer H (1989) J Organomet Chem 376:C20–C22
29. Xu Y, Zhou B (1987) J Org Chem 52:974–977
30. Zhou B, Xu Y (1988) J Org Chem 53:4419–4421
31. Plietker B (2006) Angew Chem 118:1497–1501, Angew Chem Int Ed 45:1469–1473
32. Plietker B (2006) Angew Chem 118:6200–6203, Angew Chem Int Ed 45:6053–6056
33. Jegelka M, Plietker B (2009) Org Lett 11:3462–3465
34. Glorius F (ed) (2007) N-heterocyclic carbenes in transition metal catalysis. Springer, Hamburg
35. Plietker B, Dieskau A, Möws K, Jatsch A (2007) Angew Chem 120:204–207, (2008) Angew Chem Int Ed 47:198–201
36. Holzwarth M, Dieskau A, Tabassam M, Plietker B (2009) Angew Chem 121:7387–7391, Angew Chem Int Ed 48:7251–7255
37. Trivedi R, Tunge JA (2009) Org Lett 11:5650–6652

38. Minami I, Shimizu I, Tsuji J (1985) J Organomet Chem 296:269–280
39. Zhang S-W, Mitsudo T, Kondo T, Watanabe Y (1993) J Organomet Chem 450:197–207
40. Kondo T, Ono H, Satake N, Mitsudo T, Watanabe Y (1995) Organometallics 14:1945–1953
41. Trost BM, Fraisse PL, Ball ZT (2002) Angew Chem 114:1101–1103, (2002) Angew Chem Int Ed 41:1059–1061
42. Kondo T, Morisaki Y, Uenoyama S, Wada K, Mitsudo T (1999) J Am Chem Soc 121:8657–8658
43. Mbaye M D, Demersemann B, Renaud J-L, Toupet L, Bruneau C (2003) Angew Chem 115:5220–5222, (2003) Angew Chem Int Ed 42:5066–5068
44. Mbaye MD, Demersemann B, Renaud J-L, Toupet L, Bruneau C (2004) Adv Synth Catal 346:835–841
45. Hermatschweiler R, Fernández I, Breher F, Pregosin P S, Veiros L-F, Calhorda M J (2005) Angew Chem 117:4471–4474, (2005) Angew Chem Int Ed 44:4397–4400
46. Mbaye MD, Renaud J-L, Demersemann B, Bruneau C (2004) Chem Commun 1870–1871
47. Matsushima Y, Onitsuka K, Kondo T, Mitsudo T, Takahashi S (2001) J Am Chem Soc 123:10405–10406
48. Bruneau C, Renaud J-L, Demersemann B (2006) Chem Eur J 12:5178–5187
49. Murahashi S-I (ed) (2004) Ruthenium in organic synthesis. Wiley-VCH, Weinheim
50. Kawatsura M, Ata F, Wada S, Hayase S, Uno H, Itoh T (2007) Chem Commun 298–300
51. Kawatsura M, Ata F, Hirakawa T, Hayase S, Itoh T (2008) Tetrahedron Lett 49:4873–4875
52. Kawatsura M, Ata F, Hayase S, Itoh T (2007) Chem Commun 4283–4285
53. Bayer A, Kazmaier U (2010) Org Lett 12:4960–4963
54. Fernández I, Hermatschweiler R, Breher F, Pregosin PS, Veiros L-F, Calhorda MJ (2006) Angew Chem 118:6535–6540, (2006) Angew Chem Int Ed 45:6386–6391
55. Fernández Nieves I, Scott D, Gruber S, Pregosin P S (2007) Helv Chim Acta 90: 271–276.
56. Zaistev AB, Gruber S, Pregosin PS (2007) Chem Commun 4692–4693
57. Zaistev AB, Gruber S, Plüss PA, Pregosin PS, Veiros LF, Wörle M (2008) J Am Chem Soc 130:11604–11605
58. Saburi H, Tanaka S, Kitamura M (2005) Angew Chem 117:1758–1760, (2005) Angew Chem Int Ed 44:1730–1732
59. Tanaka S, Saburi H, Kitamura M (2006) Adv Synth Catal 348:375–378
60. Zhang H-J, Demerseman B, Toupet L, Xi Z, Bruneau C (2008) Adv Synth Catal 350:1601–1609
61. Achard M, Derrien N, Zhang H-J, Demerseman B, Bruneau C (2009) Org Lett 11:185–188
62. Zhang H-J, Demerseman B, Toupet L, Xi Z, Bruneau C (2009) Organometallics 28:5173–5182
63. Sundararaju B, Achard M, Demerseman B, Toupet L, Sharma GVM, Bruneau C (2010) Angew Chem 122:2842–2845, (2010) Angew Chem Int Ed 49:2782–2785
64. Onitsuka K, Okuda H, Sasai H (2008) Angew Chem 120:1476–1479, (2008) Angew Chem Int Ed 47:1454–1457
65. Kanbayashi N, Onitsuka K (2010) J Am Chem Soc 132:1206–1207
66. Onitsuka K, Kameyama C, Sasai H (2009) Chem Lett 38:444–445
67. Burger EC, Tunge JA (2004) Org Lett 6:2603–2605
68. Burger EC, Tunge JA (2005) Chem Commun 2835–2837
69. Constant S, Tortoioli, S, Müller J, Lacour J (2007) Angew Chem 119:2128–2131, (2007) Angew Chem Int Ed 46:2082–2085
70. Linder D, Buron F, Constant S, Lacour J (2008) Eur J Org Chem 5778–5785
71. Linder D, Austeri M, Lacour J (2009) Org Biomol Chem 7:4057–4061
72. Austeri M, Linder D, Lacour J (2008) Chem Eur J 14:5737–5741
73. Tanaka S, Seki T, Kitamura M (2009) Angew Chem 121:9110–9113, (2009) Angew Chem Int Ed 48:8948–8951
74. Cahiez G, Moyeux A (2010) Chem Rev 1435
75. Gosmini C, Begouin J-M, Moncomble A (2008) Chem Commun 3221

76. Hess W, Treutwein J, Hilt G (2008) Synthesis 22:3537
77. Jeganmohan M, Cheng C-H (2008) Chem Eur J 14:10876
78. Bhatia B, Reddy MM, Iqbal J (1993) Tetrahedron Lett 34:6301–6304
79. Maikap GC, Reddy MM, Mukhopadhyay M, Bhatia B, Iqbal J (1994) Tetrahedron 50:9145–9156
80. Gomes P, Gosmini C, Périchon J (2003) J Org Chem 68:1142–1145
81. Gomes P, Buriez O, Labbé E, Gosmini C, Périchon J (2004) J Electroanal Chem 562:255–260
82. Gomes P, Gosmini C, Périchon J (2003) Org Lett 5:1043–1045
83. Nayyar NK, Reddy MM, Iqbal J (1991) Tetrahedron Lett 32:6965–6968
84. Reddy CK, Knochel P (1996) Angew Chem Int Ed 35:1700–1701
85. Dunet G, Knochel P (2007) Synlett 1383–1386
86. Mizunami K, Yorimitsu H, Oshima K (2004) Chem Lett 33:832–833
87. Yasui H, Mizunami K, Yorimitsu H, Oshima K (2006) Tetrahedron 62:1410–1415
88. Tsuji J, Minami I, Shimizu I (1984) Tetrahedron Lett 25:5157–5160
89. Evans PA, Nelson JD (1998) Tetrahedron Lett 39:1725–1728
90. Takeuchi R, Kitamura N (1998) New J Chem 22:659–660
91. Ashfeld BL, Miller KA, Smith AJ, Tran K, Martin ST (2007) J Org Chemn 72:9018–9031
92. Evans PA, Nelson JD (1998) J Am Chem Soc 120:5581–5582
93. Lawson DN, Osborn JA, Wilkinson G (1966) J Chem Soc A 1733–1736
94. Tanaka I, Jin-no N, Kushida T, Tsutsui N, Ashida T, Suzuki H, Sakurai H, Moro-oka Y, Ikawa T (1983) Bull Chem Soc Jpn 56:657–661
95. Evans PA, Kennedy LJ (2000) Org Lett 2:2213–2215
96. Kazmaier U, Stolz D (2006) Angew Chem Int Ed 45:3072–3075
97. Hayashi T, Okada A, Suzuka T, Kawatsura M (2003) Org Lett 5:1713–1715
98. Selvakumar K, Valentini M, Pregosin PK (1999) Organometallics 18:4591–4597
99. Murakoa T, Matsuda I, Itoh K (2000) Tetrahedron Lett 41:8807–8811
100. Murakoa T, Matsuda I, Itoh K (2000) J Am Chem Soc 122:9552–9553
101. Evans PA, Leahy DK (2003) J Am Chem Soc 125:8974–8975
102. Evans PA, Uraguchi D (2003) J Am Chem Soc 125:7158–7159
103. Yasui H, Mizutani K, Yorimitsu H, Oshima K (2006) Tetrahedron 62:1410–1415
104. Kabalka GW, Dong G, Venkataiah B (2003) Org Lett 5:893–895
105. Dong L, Xu Y-J, Yuan W-C, Cui X, Cun L-F, Gong L-Z (2006) Eur J Org Chem 4093–4105
106. Menard F, Chapman TM, Dockendorff C, Lautens M (2006) Org Lett 8:4569–4572
107. Johannsen M, Jorgensen KA (1998) Chem Rev 98:1689–1708
108. Evans PA, Robinson JE, Nelson JD (1999) J Am Chem Soc 121:6761–6762
109. Evans PA, Robinson JE, Moffett KK (2001) Org Lett 3:3269–3271
110. Lautens M, Fagnou K, Yang D (2003) J Am Chem Soc 125:14884–14892
111. Fagnou K, Lautens M (2002) Angew Chem Int Ed 41:26–47
112. Evans PA, Clizbe EA (2009) J Am Chem Soc 131:8722–8723
113. Vrieze DC, Hoge GS, Hoerter PZ, Van Haitsma JT, Samas BM (2009) Org Lett 11:3140–3142
114. Evans PA, Leahy DK (2000) J Am Chem Soc 122:5012–5013
115. Evans PA, Leahy DK (2002) J Am Chem Soc 124:7882–7883
116. Basolo F, Pearson RG (2007) The *trans* effect in metal complexes. In: Cotton FA (ed) Progress in inorganic chemistry, vol 4. Wiley, Hoboken
117. Kobayashi Y (2005) Reactions of alkenes and allyl alcohol derivatives. In: Tamaru Y (ed) Modern organonickel chemistry. Wiley-VCH, Weinheim
118. Shintani R, Hayashi T (2005) Asymmetric synthesis. In: Tamaru Y (ed) Modern organonickel chemistry. Wiley-VCH, Weinheim
119. Chuit C, Felkin H, Frajerman C, Roussi G, Swierczewski G (1968) J Chem Soc Chem Commun 1604–1605
120. Hayashi T, Konishi M, Yokota K, Kumada M (1981) J Chem Soc Chem Commun 313–314
121. Hayashi T, Konishi M, Yokota K, Kumada M (1985) J Organomet Chem 285:359–373

122. Wenkert E, Fernandes JB, Michelotti EL, Swindell CS (1983) Synthesis 701–703
123. Consiglio G, Morandini F, Piccolo O (1981) J Am Chem Soc 103:1846–1847
124. Kobayashi Y, Ikeda E (1994) J Chem Soc Chem Commun 1789–1790
125. Cuvigny T, Julia M (1983) J Organomet Chem 250:c21–c24
126. Alvarez E, Cuvigny T, Julia M (1988) J Organomet Chem 339:199–212
127. Bricout H, Carpentier J-F, Mortreux A (1995) J Chem Soc Chem Commun 1863–1864
128. Consiglio G, Indolese A (1991) Organometallics 10:3425–3427
129. Indolese AF, Consiglio G (1994) Organometallics 13:2230–2234
130. Chung K-G, Miyake Y, Uemura S (2000) J Chem Soc Perkin Trans 1:2725–2729
131. Chung K-G, Miyake Y, Uemura S (2000) J Chem Soc Perkin Trans 1:15–18
132. Nomura N, RajanBabu TV (1997) Tetrahedron Lett 38:1713–1716
133. Novak A, Fryatt R, Woodward S (2007) C R Chimie 10:206–212
134. Biswas K, Prieto O, Goldsmith PJ, Woodward S (2005) Angew Chem Int Ed 44:2232–2234
135. Sumida Y, Hayashi S, Hirano K, Yorimitsu H, Oshima K (2008) Org Lett 10:1629–1632
136. Nakao Y, Yada A, Ebata S, Hiyama T (2007) J Am Chem Soc 129:2428–2429
137. Nájera C, Sansano JM (2009) Angew Chem Int Ed 48:2452–2456
138. Sumida Y, Yorimitsu H, Oshima K (2010) Org Lett 12:2254–2257
139. Furukawa J, Kiji J, Yamamoto K, Tojo T (1973) Tetrahedron 29:3149–3151
140. Chiusoli GP, Costa M, Pallini L, Terenghi G (1981) Transit Met Chem 6:317–318
141. Bricout H, Carpentier J-F, Mortreux A (1998) Tetrahedron 54:1073–1084
142. Bricout H, Carpentier J-F, Morteux A (1998) J Mol Catal A Chem 136:243–251
143. Berkowitz DB, Bose M, Choi S (2002) Angew Chem Int Ed 41:1603–1607
144. Berkowitz DB, Maiti G (2004) Org Lett 6:2661–2664
145. Berkowitz DB, Shen W, Maiti G (2004) Tetrahedron Asymm 15:2845–2851
146. Yatsumonji Y, Ishida Y, Tsubouchi A, Takada T (2007) Org Lett 9:4603–4606
147. Lu X, Zhu J, Huang J, Tao X (1987) J Mol Catal A Chem 41:235–243
148. Crotti S, Bertolini F, Macchia F, Pineschi M (2009) Org Lett 11:3762–3765
149. Huang T-M, Chen J-T, Lee G-H, Wang Y (1991) Organometallics 10:175–179
150. Dick AR, Kampf JW, Sanford MS (2005) Organometallics 24:482–485
151. Clarke ML (2001) Polyhedron 20:151–164
152. Trost BM (1977) Tetrahedron 33:2615–2649
153. Trost BM, Van Vranken DL (1996) Chem Rev 96:395–422
154. Kurosawa H (1979) J Chem Soc Dalton Trans 939–943
155. Brown JM, Macintyre JE (1985) J Chem Soc Perkin Trans 2:961–970
156. Blacker AJ, Clark ML, Loft MS, Williams JMJ (1999) Chem Commun 913–914
157. Blacker AJ, Clarke ML, Loft MS, Mahon MF, Humphries ME, Williams JMJ (2000) Chem Eur J 6:353–360
158. Ohe K, Matsuda H, Morimoto T, Ogoshi S, Chatani N, Murai S (1994) J Am Chem Soc 116:4125–4126
159. Kadota J, Komori S, Fukumoto Y, Murai S (1999) J Org Chem 64:7523–7527
160. Kadota J, Chatani N, Murai S (2000) Tetrahedron 56:2231–2237
161. Kadota J, Katsuragi H, Fukumoto Y, Murai S (2000) Organometallics 19:979–983
162. Yang S-C, Feng W-H, Gan K-H (2006) Tetrahedron 62:3752–3760
163. Gan K-H, Jhong C-J, Shue Y-J, Yang S-C (2008) Tetrahedron 64:9625–9629
164. Muzart J (2007) Eur J Org Chem 2007:3077–3089
165. Yang S-C, Tsai Y-C, Shue Y-J (2001) Organometallics 20:5326–5330
166. Kayaki Y, Koda T, Ikariya T (2004) J Org Chem 69:2595–2597
167. Kinoshita H, Shinokubo H, Oshima K (2004) Org Lett 6:4085–4088
168. Utsunomiya M, Miyamoto Y, Ipposhi J, Oshima T, Mashima K (2007) Org Lett 9:3371–3374
169. Ohshima T, Miyamoto Y, Ipposhi J, Nakahara Y, Utsunomiya M, Mashima K (2009) J Am Chem Soc 131:14317–14328
170. Mora G, Piechaczyk O, Houdard R, Mézailles N, Le Goff X-F, le Floch P (2008) Chem Eur J 14:10047–10057

Top Organomet Chem (2012) 38: 321–340
DOI: 10.1007/3418_2011_13

Published online: 26 July 2011

Enantioselective Allylic Substitutions in Natural Product Synthesis

Barry M. Trost and Matthew L. Crawley

Abstract Enantioselective transition metal-cataylzed allylic substitution reactions constitute a broad and deep methodology that has expanded to incorporate ever more metals and ligands. The utility of these processes is reflected in the wide range of natural product and drug target total syntheses that incorporate asymmetric allylic alkylations as the key step or steps.

Keywords Allylic alkylation · Catalysis · Enantioselective · Natural products · Palladium

Contents

B.M. Trost (✉)
Department of Chemistry, Stanford University, Stanford, CA 94306, USA
e-mail: bmtrost@stanford.edu

M.L. Crawley
Main Line Health, 1160 W. Swedesford Rd., Southpoint One, Berwyn, PA 19312, USA
e-mail: crawleym@mlhs.org

1 Introduction

The evolution of the transition metal-catalyzed allylic substitution reaction has been a remarkable process. Initially, a limited reaction of mainly mechanistic interest, it has exploded into one of the most powerful tools available for the asymmetric formation of a variety of bond types on highly complex substrates [1]. As described in the previous chapters, its scope now extends to palladium-, iridium-, molybdenum-, tungsten-, and copper-catalyzed processes with additional metals making headway as well [2]. In sharp contrast to most transition metal-catalyzed enantioselective processes which overwhelmingly utilize only one mechanism for enantiodiscrimination, the asymmetric allylic alkylation (AAA) has numerous mechanisms, lending it greater power in total synthesis of biologically important natural products.

This chapter focuses on a key validation of the methodologies discussed to this point: their application in the total synthesis of complex natural products or drug targets. As the mechanism has already been discussed in detail and the ligands and selectivity thoroughly reviewed, this chapter will directly address the pivotal and representative total syntheses that highlight the power of this transformation. The selection criteria for included examples requires both high selectivity and yield in allylic substitution reaction, a target of significant complexity or biological relevance, and a strategy with potential broad applicability to other synthetic targets. While a large part of the focus is on syntheses completed in the last decade, several older representative syntheses are included both as landmarks of achievement and for purposes of comparison. This is not intended to be a comprehensive review of all syntheses containing AAA reactions, rather a compilation of the most instructive examples.

The chapter is divided into sections based on the type of nucleophile employed, focusing first on carbon–nitrogen bond forming reactions, then carbon–carbon bond formation, and finally carbon–oxygen processes. It is important to note that while remarkable progress has been made with respect to enantioselectivity, scope and yield with many transition-metal catalyzed AAA reactions, palladium-catalyzed processes still dominate the literature for total synthesis in part because of their robust nature and versatility and in part as the methods have been available longer than some of the more recent developments with iridium and copper. In fact, all of the critical examples discussed herein for C–N and C–O bond forming processes in total synthesis are for palladium-catalyzed reactions, whereas C–C bond forming processes extend to a range of transition-metals. Furthermore, the palladium-catalyzed AAA processes have been the most effective among the available metals at inducing stereochemistry of prochiral nucleophiles as well as electrophiles, although there are an increasing number of such reactions using molybdenum.

2 Total Syntheses Utilizing Nitrogen Nucleophiles

Asymmetric formation of carbon–nitrogen bonds is one of the most significant challenges in the preparation of biologically important natural products and drug targets. While a variety of transition-metal catalyzed approaches have been developed to address this necessity, few are as versatile or selective as the palladium-catalyzed AAA reaction with nitrogen nucleophiles. While basic amines do participate in allylic substitution reactions, stabilized nitrogen nucleophiles, such as imides and sulfonamides, have been more frequently employed.

2.1 (R)-Vigabatrin

One of the most simple but yet most elegant examples of a nitrogen nucleophile allylic substitution in a total synthesis is in the asymmetric synthesis of (*R*)-vigabatrin, an important marketed drug for the treatment of epilepsy [3]. In the first and key step of the total synthesis, phthalimide was reacted with butadiene monoepoxide under weakly basic conditions in the presence of chiral ligand **L1** to afford adduct **1** in 98% yield with 96% ee (Scheme 1). This is particularly remarkable as the starting epoxide is a racemate, showing a true dynamic kinetic asymmetric transformation process. With this key building block **1** in hand, three subsequent steps afforded the anticonvulsant (*R*)-vigabatrin in four total steps with 59% overall yield from butadiene monoepoxide.

2.2 (+)-Polyoxamic Acid

The versatility of the phthalimide nucleophile was further demonstrated in another concurrent synthesis of (+)-polyoxamic acid, a novel amino acid substituent of the

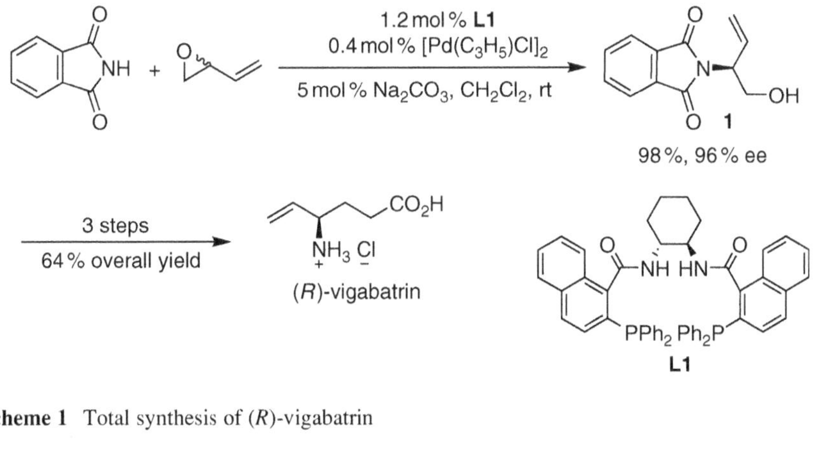

1.2 mol % **L1**
0.4 mol % $[Pd(C_3H_5)Cl]_2$
5 mol % Na_2CO_3, CH_2Cl_2, rt
1
98 %, 96 % ee
3 steps
64 % overall yield
(*R*)-vigabatrin
L1

Scheme 1 Total synthesis of (*R*)-vigabatrin

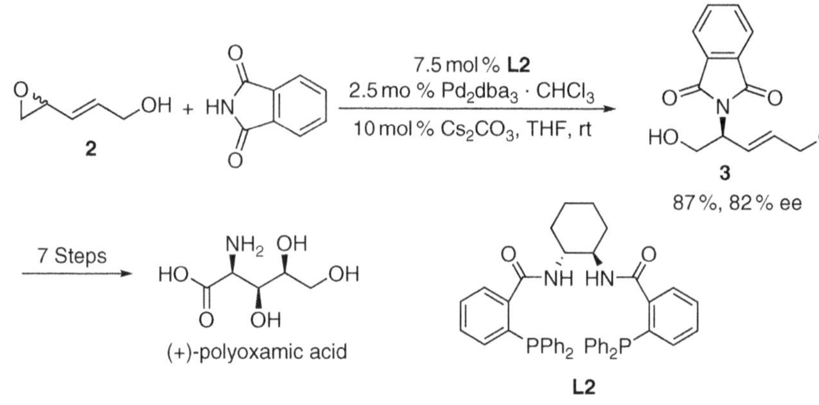

Scheme 2 Total synthesis of (+)-polyoxamic acid

polyoxins and nikkomycins [4]. In this case the vinyl epoxide **2** generated a symmetric π–allyl complex in the presence of ligand **L2** which gave rise to amino diol **3** in excellent yield with 82% ee (Scheme 2). This intermediate **3** was subsequently transformed in seven steps to (+)-polyoxamic acid.

2.3 (−)-Tubifoline

The sulfonamides are another highly effective class of nitrogen nucleophiles that have been employed in asymmetric allylic substitution reactions. One effort in particular, the total synthesis of (−)-tubifoline, stands out [5, 6]. This natural product is a member of the strychnos indole alkaloid family and has been of target of interest due to its significant biological activity. The key reaction is that of cyclic allylic phosphonate **4** with substituted *N*-tosylaniline **5** via a meso π–allyl intermediate (Scheme 3). The reaction is guided by a novel BINAPO ligand **L3** and affords substituted product **6** in good yield with moderately high enantioselectivity. This highly functionalized building block **6** was readily converted to the natural product (−)-tubifoline in a series of steps.

2.4 Agelastatin A

A powerful example of nitrogen nucleophiles in allylic substitution reaction was reported as the core process for the total synthesis of agelastatin A, a bioactive natural product that may be useful in the treatment of Alzheimer's disease [7]. A previously unused class of stabilized nucleophiles, electron deficient pyrroles, were successfully tested as nucleophiles. This result enabled the reaction of

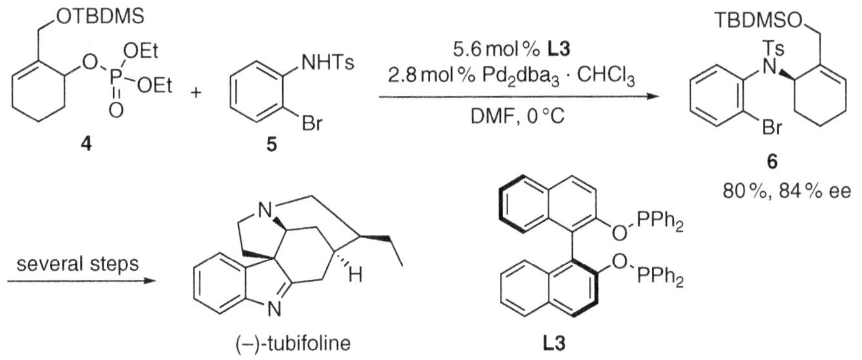

Scheme 3 An enantioselective synthesis of (−)-tubifoline

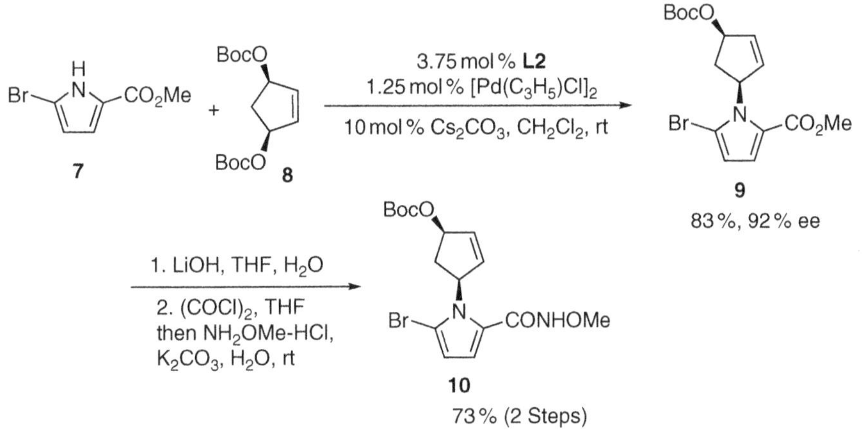

Scheme 4 Enantioselective synthesis of (−)-agelastatin part I

bromopyrrole **7** to undergo a palladium-catalyzed amination with symmetric dicarbonate **8** to afford building block **9** in 83% yield and 92% ee (Scheme 4). With a slight reduction in enantioselectivity, it was possible to further increase the yield to 93%. This ester of adduct **9** was hydrolyzed and the hydroxamic amide **10** formed in two steps to setup for the next allylic substitution reaction.

This time a diastereoselective palladium catalyzed process enabled the transformation of chiral carbonate **10** to tricyclic core **11** in 91% yield as a single diastereomer (Scheme 5). Of note, the opposite piperazinone regioisomer could be obtained by a minor modification of the conditions, namely replacing the stoichiometric cesium carbonate with catalytic acetic acid. The highly functionalized core **11** could then be transformed in a highly efficient four-step effort to afford (+)-agelastatin A. The dual palladium-catalyzed allylic substitutions, both with nitrogen nucleophiles, demonstrate the utility of these processes in total synthesis.

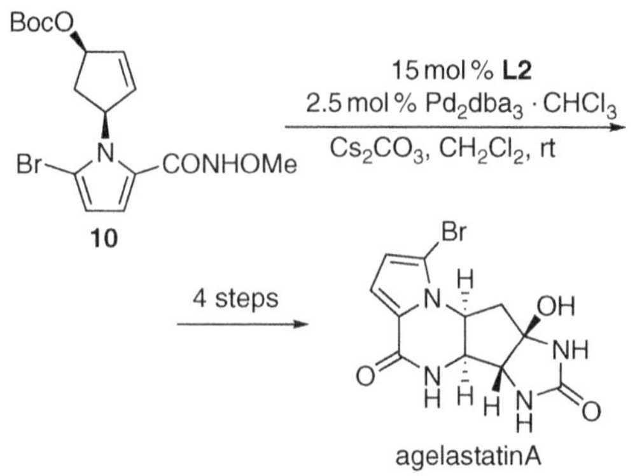

Scheme 5 Enantioselective synthesis of (−)-agelastatin part II

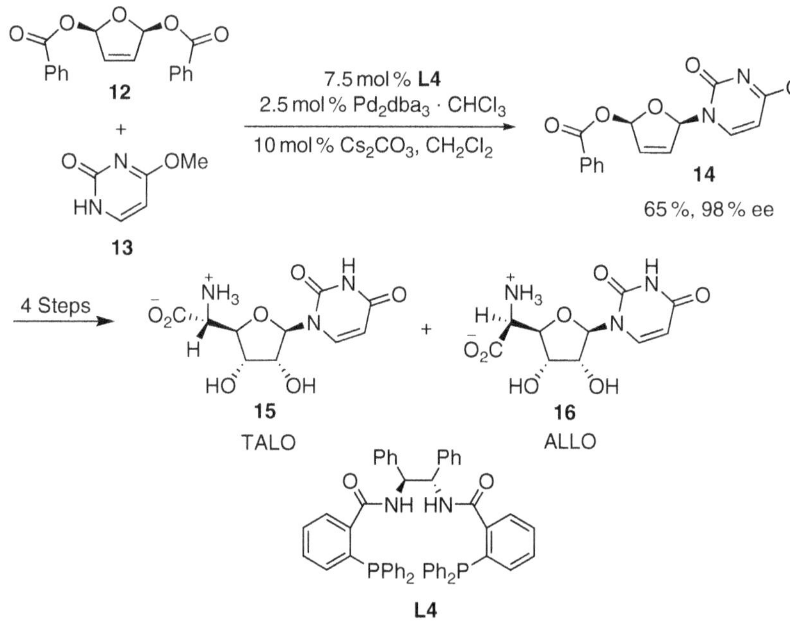

Scheme 6 Synthesis of the polyoxin–nikkomycin nucleoside core

2.5 Polyoxin/Nikkomycin Nucleosides

As with sulfonamide nucleophiles, heterocyclic amines have been employed in several total syntheses, notably those of carbonucleosides. In one representative example, reaction of the meso bis-carbonate **12** with uracil equivalent **13** afforded chiral core **14** in a very high 98% ee and 65% (Scheme 6). In this case, the diphenyl ligand **L4** was employed. The building block **14** was then transformed in four subsequent steps to the polyoxin–nikkomycin nucleoside cores **15/16** in an overall yield of 39%. It is notable that this approach is in sharp contrast to the chiral-pool methods which have limited access to the enantiomers based on the availability of the requisite chiral starting material.

3 Total Syntheses Utilizing Carbon Nucleophiles

The most widespread use of transition metal-catalysis in carbon–carbon bond forming reactions involves linking two sp^2 centers together or a sp^2 to a sp^3 center. The fact that allylic substitution reactions occur at sp^3 centers and additionally enable the formation of two contiguous chiral centers is what has made this process so vital in a range of complex natural product and clinical drug targets. This part of the chapter first examines developments of the application of copper- and molybdenum-catalyzed carbon–carbon bond forming processes to total synthesis, and then closes with palladium-catalysis driven syntheses.

3.1 Copper-Catalyzed Reactions in Total Syntheses

In the past decade, impressive results have been obtained for copper-catalyzed allylic substitution reactions, particularly those employing organozinc and organomagnesium reagents [24]. The ability to employ hard nucleophiles in these processes nicely compliments the soft nucleophiles utilized in palladium-catalyzed chemistry. Despite the ability to introduce unstabilized organometallics with remarkable functional group and reaction condition compatibility, the recent development of these processes has thus far limited reports of application in natural product total synthesis. However, two total syntheses effectively demonstrate the power of this type of approach.

3.1.1 (*R*)-(−)-Elenic Acid

In the total synthesis of (*R*)-(−)-elenic acid, a naturally occurring topoisomerase II inhibitor, the key step is the asymmetric introduction of a methyl substituent to an ester-substituted allylic phosphonate **17** (Scheme 7) [8]. In this reaction, dimethylzinc was employed as the nucleophile and the process is catalyzed by modified peptidic scaffold **L5** to give methylated allylic ester **18** with complete control of regioselectivitiy (>20:1 branched/linear) in good yield with high enantiomeric excess (90%). This key building block **18** was efficiently transformed in just two steps (a cross-metathesis reaction followed by conversion of the ester to the acid) to afford the natural product.

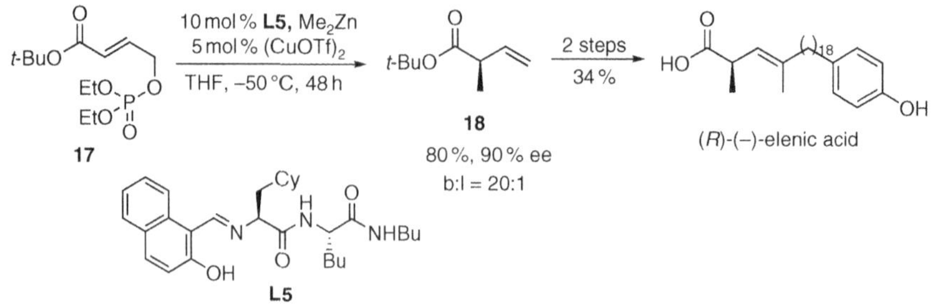

Scheme 7 Asymmetric synthesis of (*R*)-(−)-elenic acid

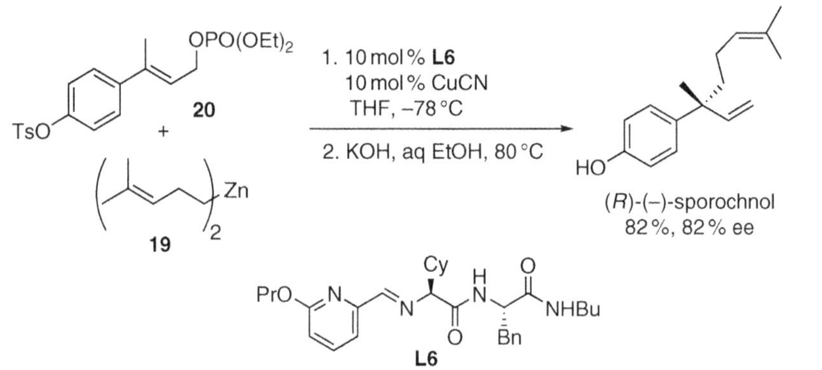

Scheme 8 Enantioselective total synthesis of (*R*)-(−)-sporochnol

3.1.2 (*R*)-(−)-Sporochnol

While the introduction of a chiral tertiary center has significant value in total synthesis, there remain a variety of methods, namely asymmetric hydrogenation, which give nice alternative access to these types of tertiary substrates. A truly differentiating property of copper-catalyzed allylic substitution reactions is the ability to generate chiral quaternary centers efficiently. In the total synthesis of (*R*)-(−)-sporochnol, a naturally occurring fish deterrent, dialkylzinc reagent **19** was reacted with substituted-allylic phosphonate **20** in the presence of peptidic ligand **L6** to afford the direct precursor to the natural product target (Scheme 8) [9]. This transformation proceeds with 82% two step yield (after liberation of the phenol) and 82% ee, and notably forms a single regioisomeric adduct.

3.2 *Molybdenum-Catalyzed Reactions in Total Syntheses*

The inherent preference for branched versus linear alkylation makes the molybdemum-catalyzed allylic substitution reaction the perfect complement to the palladium-catalyzed reaction for soft nucleophiles (see also Chap. 5). For certain substrates where it is difficult if not impossible to overcome the regioselectivity issues with palladium, asymmetric molybdenum processes provide a solution. While the literature has a significantly smaller number of examples for molybdenum in total synthesis than palladium, there are several reports that highlight both the scope and depth of this methodology as well as their excellent performance when employed. Three examples have been selected for this section.

3.2.1 (−)-Δ^9-*trans*-Tetrahydrocannabinol

The first example comes from the synthesis of (−)-Δ^9-*trans*-tetrahydrocannabinol, the primary pyschomimetic component of marijuana [10]. The strategy for this approach was to set the key chiral center with the molybdenum catalyzed-reaction of achiral carbonate **20** with dimethylmalonate in the presence of chiral ligand **L7** (Scheme 9). This reaction proceeded with very high (97%) enantioselectivity and a good yield of 84%. It is of note that a single regioisomer of product formed even with the steric congestion caused by the flanking *o*,*o*-dimethoxy groups, though that is presumably what slowed the rate of the reaction. The core building block **21** allowed for the completion of the total synthesis in a series of steps to give the natural product **38** in 30% overall yield.

3.2.2 Clinical Candidate Intermediate

The synthesis of an important clinical candidate intermediate was selected because of the demonstrated robustness of this process [11]. Starting with racemic carbonate **22**, the sodium salt of dimethyl malonate afforded functionalized product **23** in 84–91% yield (batch dependent) with 19:1 branched to linear selectivity in 97% ee (Scheme 10). Furthermore, this dynamic kinetic asymmetric transformation was run industrially on multi-kilogram scale, proving the scalability of this type of molybdenum-catalyzed process. The adduct **23** was then cyclized to **24**, which

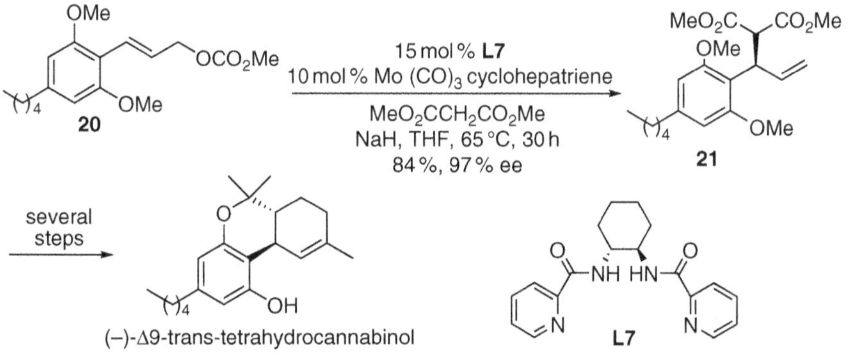

OMe
OCO$_2$Me
OMe
20
15 mol % **L7**
10 mol % Mo (CO)$_3$ cyclohepatriene
MeO$_2$CCH$_2$CO$_2$Me
NaH, THF, 65 °C, 30 h
84 %, 97 % ee
MeO$_2$C CO$_2$Me
OMe
OMe
21
several steps
OH
(−)-Δ9-trans-tetrahydrocannabinol
NH HN
L7

Scheme 9 An asymmetric synthesis of (−)-Δ^9-*trans*-tetrahydrocannabinol

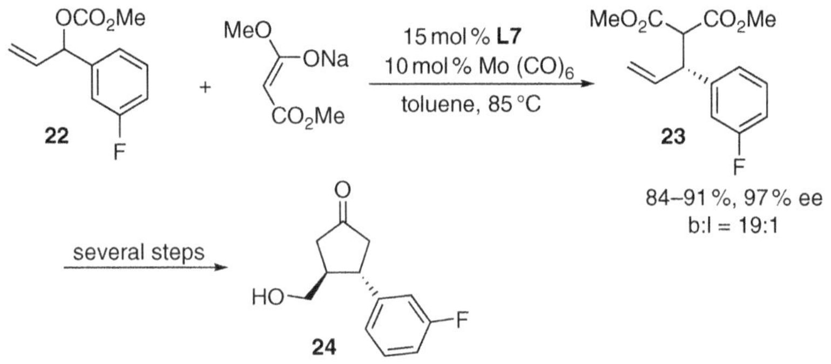

Scheme 10 Mo-catalyzed allylation in the synthesis of A clinical candidate building block

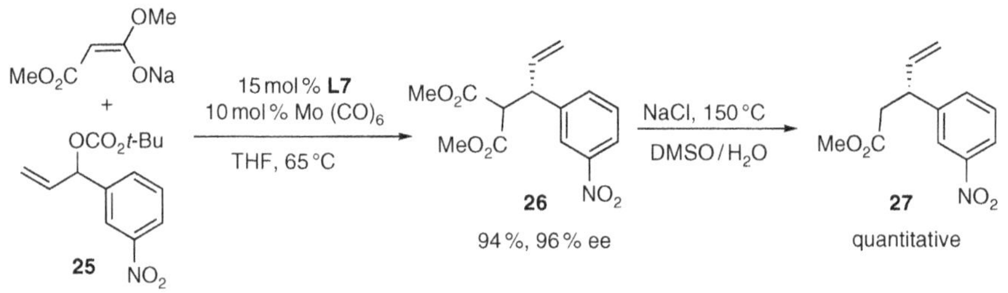

Scheme 11 Synthesis of western half of tipranavir

was used as the core building block for several preclinical and clinical candidate studies.

3.2.3 Tipranavir

Another synthesis of a clinically important target that employs molybdenum-catalyzed process as one of the key steps was reported in the total synthesis of tipranavir, a non-peptidic protease inhibitor that went through clinical trials as a therapeutic agent against human immunodeficiency virus [12]. This synthesis is truly a show case for allylic substitution reactions as it involved both molybdenum- and palladium-catalyzed dynamic kinetic asymmetric reactions to form the two key halves of the molecule. In a scheme nearly identical to the one described above in the synthesis of a key drug intermediate (both were published at the same time) reaction with racemic carbonate **25** in the presence of the sodium salt of dimethyl malonate and ligand **L7** afforded branched regioisomer **26** as a single isomer in 94% yield with 96% ee (Scheme 11). This substrate **26** was readily decarboxylated quantitatively under classic thermal conditions to give adduct **27**.

The other large fragment of the molecule was assembled in a palladium-catalyzed DYKAT using an oxygen nucleophile. To complete the tipranavir story, this reaction

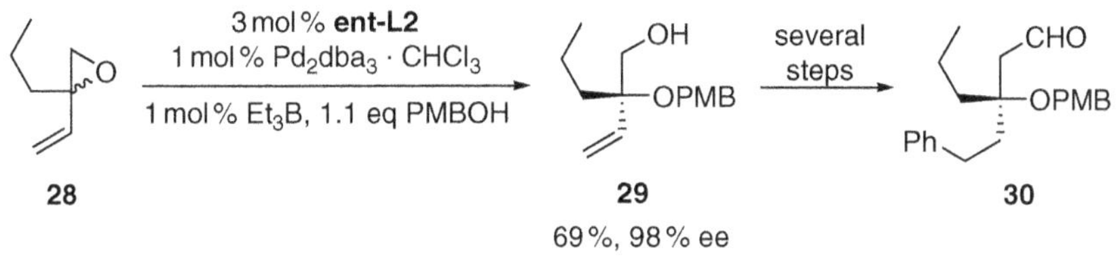

Scheme 12 Synthesis of the eastern half of tipranavir

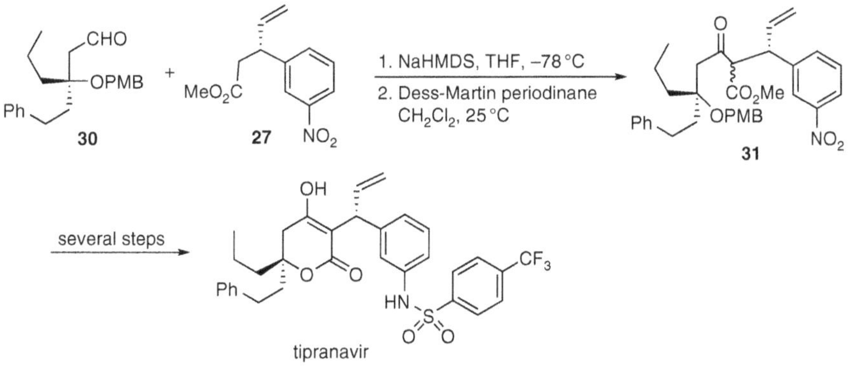

Scheme 13 Completion of the total synthesis of tipranavir

is covered in this section rather than later. In this sequence epoxide **28** was reacted with PMB alcohol, co-catalyzed by triethylboron, in the presence of chiral ligand **ent-L2**. This afforded the chiral quaternary diol **29** in 69% yield with 98% ee (Scheme 12). This fragment was subsequently transformed over several steps to key intermediate **30**.

The two key fragments **27** and **30** were coupled under basic conditions with subsequent oxidation to afford key intermediate **31** (Scheme 13). This highly functionalized intermediate is then transformed in a number of steps to tipranavir.

3.3 *Palladium-Catalyzed Reactions in Total Syntheses*

Palladium-catalyzed allylic substitution reactions remain one of the most robust, selective, and efficient methods to asymmetrically form carbon–carbon bonds (see also Chap. 3). In particular, the ability to generate two contiguous tertiary or even quaternary chiral centers has caused this methodology to be incorporated in a wide range of natural product and drug target total syntheses. A handful of these syntheses were selected to be discussed herein, with attention given to capturing different types of nucleophiles, substrates and types of mechanisms, all of which

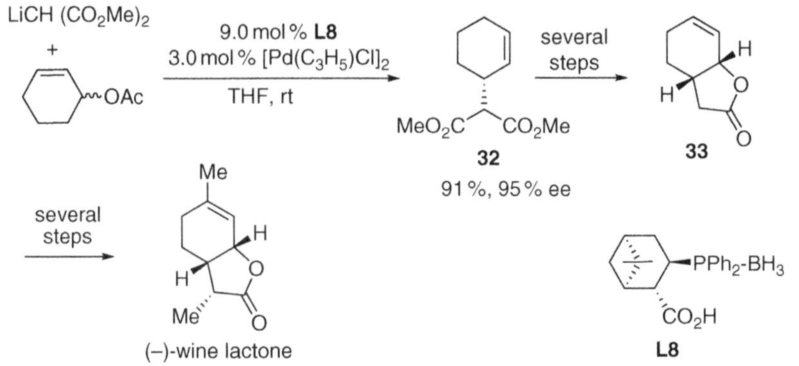

Scheme 14 An enantioselective synthesis of (−)-wine lactone

highlight the power of this catalytic process in the synthesis of complex biologically relevant targets.

3.3.1 (−)-Wine Lactone

A typical palladium-catalyzed allylic alkylation utilized to evaluate new ligands involves the reaction of cyclohex-2-enyl acetate with dimethyl malonates (Scheme 14). This acetate substrate is useful as it generates an accessible, symmetric, cyclic π–allyl intermediate. One total synthesis of the natural product (−)-wine lactone incorporated this classic reaction utilizing a more recent mixed phosphine–borane acid complex ligand **L8** to afford product **32** in a robust 91% yield with 95% ee [13]. It is worth noting that comparable yields and enantioselectivities can be obtained with other ligands as well. The key intermediate **32** was efficiently transformed in four steps to a bicyclic core **33**, which after some functional group manipulation was further reacted to afford the natural product.

3.3.2 (+)-Valienamine

Though reported over a decade ago, the total synthesis of (+)-valienamine remains a classic and powerful example of the utility of palladium-catalyzed allylic substitution reactions with a carbon nucleophile [14]. The strategy involved the desymmetrization of the meso allylic dicarbonate **34** with phenylsulfonyl-substituted nitromethane nucleophile **35** in the presence of the standard Trost ligand **ent-L2** (Scheme 15). What is unique about this reaction is an initial substitution reaction leads to intermediate **36**, which subsequently undergoes a second allylation to afford bicyclic adduct **37** in 87% yield with complete control of both enantio- and diastereoselectivity. This core building block **37** was transformed into the glycosidase inhibitor natural product (+)-valienamine in 13 steps.

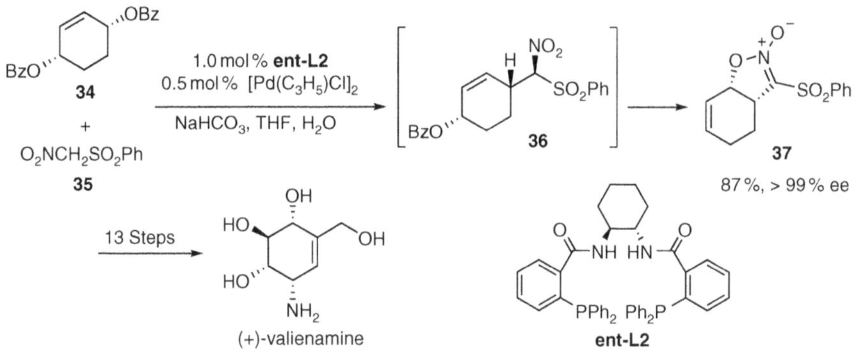

Scheme 15 Asymmetric total synthesis of (+)-valienamine

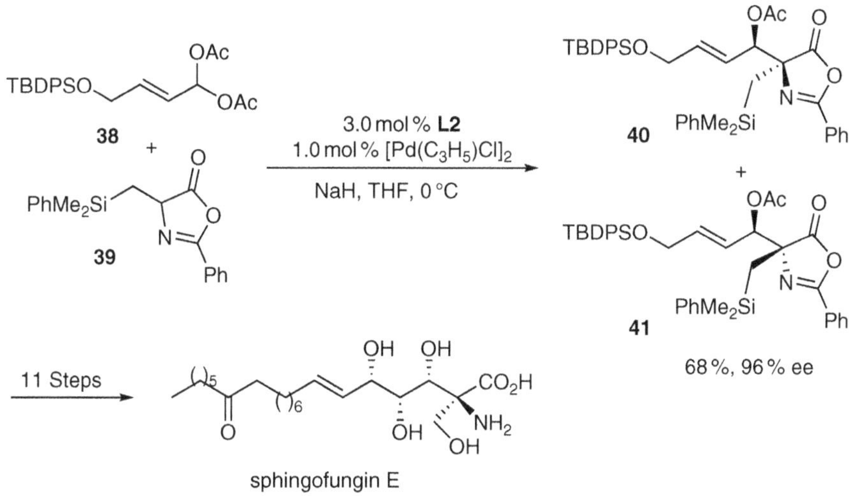

Scheme 16 The total synthesis of sphingofungin E

3.3.3 Sphingofungin E

Sphingofungin E is a natural product that has a role in a diverse array of biological processes, and as such is a target of interest for total synthesis. The strategy to construct it involved a novel type of leaving group: a prochiral allylic gem-diacetate whereby the selective ionization of the acetate would drive the enantioselectivity [15]. Reaction of the gem-diacetate **38** with an azlactone nucleophile **39** in the presence of ligand **L2** afforded a 2.4:1.0 diastereomeric mixture of products (**40**:**41**) in 68% yield, with the adducts possessing 96% ee (Scheme 16). The major diastereomer **40** was subsequently transformed into sphingofungin E in an additional 11 steps. It is of note that another isomer (sphingofungin F) was also synthesized utilizing this methodology.

3.3.4 Horsefiline

Indole related nucleophiles, such as oxindole, represent an important yet difficult class of carbon-nucleophiles for allylic alkylation reactions. Despite the challenges associated with this motif, it was core to the synthetic strategy in the total synthesis of horsfiline, a natural product leaf extract with structural similarities to the active indole alkaloid natural products [16]. The key step of the synthetic sequence was the reaction of oxindole **42** with allyl acetate catalyzed by palladium and ligand **L2**. Under optimized conditions, this resulted in 84% ee of product **43** (one recrystallization to 98%) with quantitative yield (Scheme 17). This core intermediate **43** was readily transformed to the natural product in a series of five steps. It is of note that only 0.25% catalyst loading was required for the allylic substitution reaction.

3.3.5 Viridenomycin

Another synthesis which highlights the utility of the palladium-catalyzed allylic substitution reaction with carbon nucleophiles was reported in the synthesis of the cyclopentyl core **46** of the natural product viridenomycin, a potent antitumor antibiotic [17]. This series created a chiral quaternary center by the reaction of oxo ester **44** with isoprene monoepoxide in the presence of ligand **ent-L2** (Scheme 18). The reaction proceeded with 71% yield with tandem cyclization to the hemiacetal **45** with remarkable 94% ee. That key building block **45** was then transformed to the cyclopentyl core **46** in ten overall steps. This core **46** is a known synthetic precursor to the fully elaborate viridenomycin natural product.

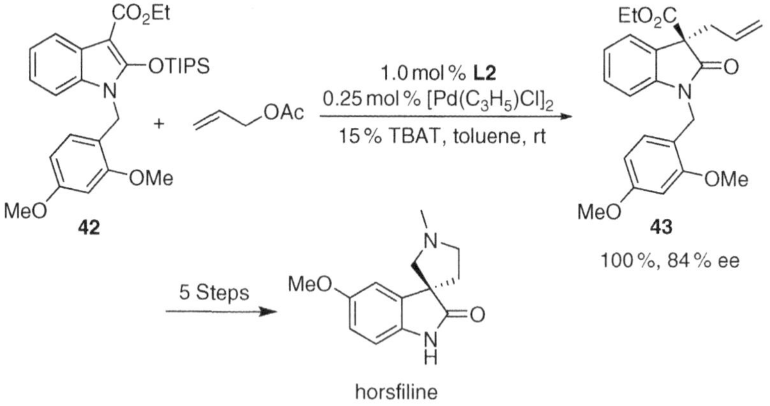

CO_2Et OTIPS N OMe MeO **42** + OAc

1.0 mol% **L2**
0.25 mol% $[Pd(C_3H_5)Cl]_2$
15% TBAT, toluene, rt

EtO_2C O N OMe MeO **43**
100%, 84% ee

5 Steps

MeO N O N H
horsfiline

Scheme 17 An efficient enantioselective total synthesis of horsfiline

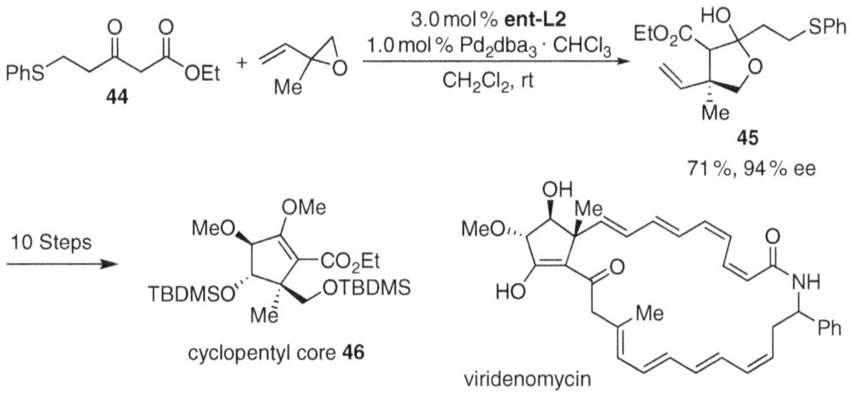

Scheme 18 Asymmetric total synthesis of viridenomycin

4 Total Syntheses Utilizing Oxygen Nucleophiles

The palladium-catalyzed allylic substitution reaction with oxygen nucleophiles has become one of the most efficient and selective ways to asymmetrically form a carbon–oxygen bond. There are examples with a range of nucleophiles, including alcohols as in the synthesis of tipranavir (see Sect. 3.2.3). However, by far the most commonly utilized nucleophiles are stabilized phenols, which are both highly effective nucleophiles and versatile building blocks in total synthesis.

4.1 Calanolides

The enantioselective total syntheses of calanolide A and calanolide B, both natural products that demonstrate HIV-1 specific reverse transcriptase inhibition, nicely demonstrate the power of the Pd-catalyzed allylic substitution reaction [18]. Starting with highly functionalized phenol **47**, allylic carbonate **48** was reacted in the presence of highly constrained ligand **L9** to afford chiral tricycle **49** (Scheme 19). The reaction proceeded in good yield (85%) with outstanding regioselectivity (92:8) and near complete control of enantioselectivity (98%). It is of interest to note that the ligand employed was designed to help reverse the intrinsic regioselectivity of attack of the phenol **47** on the unsymmetric π–allyl intermediate, favoring the more substituted product. With this chiral fragment **49** in hand, a series of steps readily transformed **49** to (−)-calanolide A (R_1 = H; R_2 = OH) and (−)-calanolide B (R_1 = OH; R_2 = H).

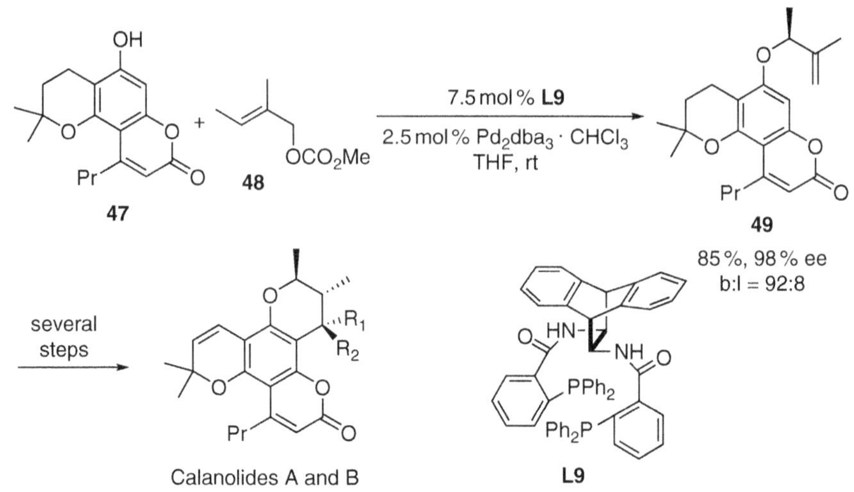

Scheme 19 An enantioselective approach to the calanolides

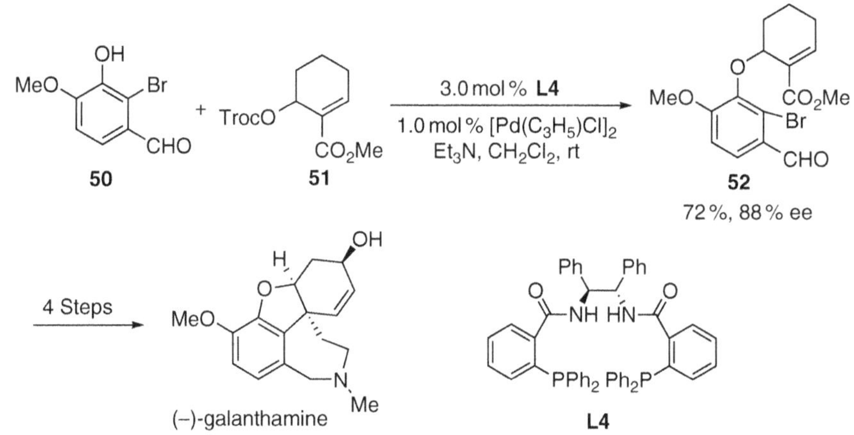

Scheme 20 Total synthesis of (−)-galanthamine

4.2 (−)-Galanthamine

Another representative and robust example of this transformation was reported in the total synthesis of (−)-galanthamine, a potent acetylcholine esterase inhibitor [19, 20]. Employing cyclic carbonate **51** and sterically encumbered phenol **50**, the allylic substitution reaction in the presence of ligand **L4** afforded chiral adduct **52** in 72% yield with 88% ee (Scheme 20). It should be noted that the open diphenyl ligand **L4** was employed to accommodate the steric congestion of both the nucleophile **50** and electrophile **51** in this reaction. This functionalized core **52** was

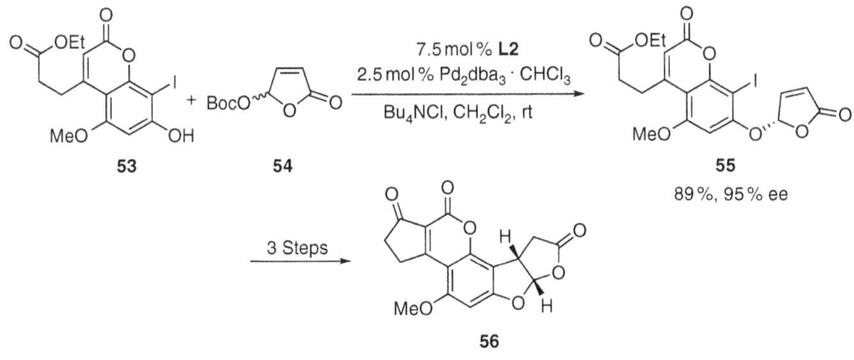

Scheme 21 An asymmetric approach to (−)-aflatoxin B lactone **56**

subsequently cyclized and transformed in a series of four steps to the natural product (−)-galanthamine.

4.3 (−)-Aflatoxin B

Another compelling application of an oxygen nucleophile utilized in an allylic substitution reaction that drives a total synthesis was reported in the enantioselective approach to (−)-aflatoxin B lactone **56** [21] (Scheme 21). The process in this total synthesis differs from the last two in that phenol **53** is reacted in a dynamic kinetic asymmetric transformation with racemic butenolide **54**. This transformed the racemic starting material to substituted product **55** in 89% yield with 95% ee. The pivotal intermediate **55** was further reacted in a series of three steps to afford (−)-aflatoxin B lactone **56**, completing the formal synthesis of the natural product aflatoxin B.

4.4 (+)-Brefeldin A

As in the last synthesis, the ability to complete a dynamic kinetic asymmetric transformation has significant appeal for early use in a synthesis, as 100% of the racemic starting material can in theory be converted to the desired chiral adduct. Following the same strategy for the core as in the aflatoxin B lactone synthesis, 2-naphthol was reacted with butenolide **54** in the presence of ligand **L2** to afford ether **57** in 87% yield with 97% ee [22, 23] (Scheme 22). This fragment **57** was utilized as the core in the total synthesis (+)-brefeldin A, a biologically active macrolactone with good in vitro potency against a variety of cancer cell lines. It is

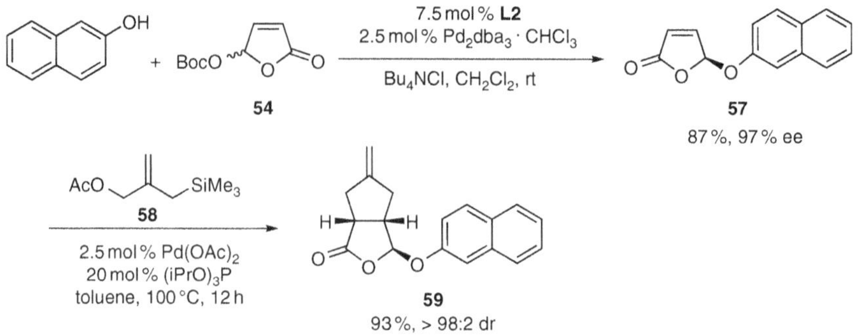

Scheme 22 Enantioselective synthesis of (+)-brefeldin A (part I)

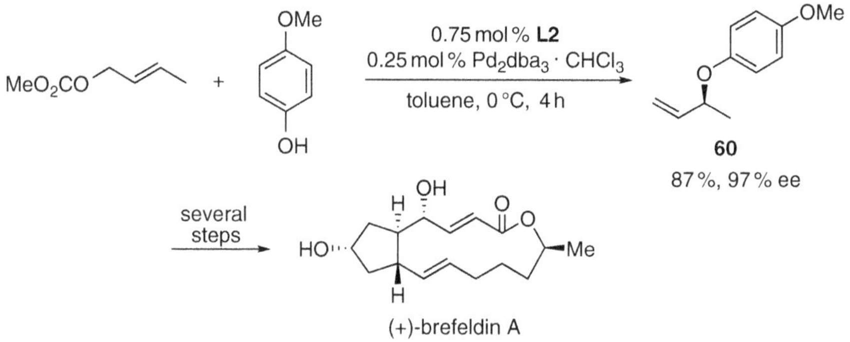

Scheme 23 Enantioselective synthesis of (+)-brefeldin B part II

of note that this was the first of three palladium-catalyzed reactions that generate a π–allyl complex in this total synthesis. The second followed in the next step, whereby a 1,3-dipole fragment **58**, which generates trimethylenemethane (TMM) in situ from an initially formed palladium complex. That complex subsequently undergoes a cycloaddition with the core **57** to give bicyclic adduct **59** in 93% yield with complete control of diastereoselectivity.

Separately, another regio- and enantioselective allylic substitution reaction with paramethoxy phenol and crotyl carbonate generated chiral ether **60** in 95% yield with 90% ee and a compelling 96:4 ratio of regioisomers (Scheme 23). It is notable that only 0.25% catalyst loading was required and in lower catalyst loading increased the enantioselectivity and regioselectivity of the reaction, suggesting recognition of enantiotopic faces of the π–allyl complex was the mechanism for selectivity. After elongation of ether fragment **60** and further transformation of core **59**, the total synthesis of (+)-brefeldin A was completed. This synthesis truly served as a showcase for palladium-catalyzed allylic substitution reactions.

5 Conclusion

The scope and depth of enantioselective transition-metal catalyzed allylic substitutions has grown substantially over the past several decades giving rise to what is arguably one of the most extensive toolboxes for chiral transformations in existence. While research on new methodologies with a range of metals continues, the ultimate test for the utility of any new allylic alkylation process is its applicability to and incorporation in synthesis of complex natural products. The examples that were described show a powerful but limited set of syntheses that highlight the range but not the extent of transition metal-catalyzed allylic substitution reactions in the preparation of targets of biological interest. There is no doubt with the pace and volume of new discoveries that an update to this chapter will be needed in short order.

References

1. Trost BM, Crawley ML (2003) Asymmetric transition metal-catalyzed allylic alkylations: applications in total synthesis. Chem Rev 103:2921–2943
2. Crawley ML (2011) Allylic substitution reactions. In: Evans PA (ed) Stereoselective synthesis, Section 3.9, Georg Thieme Verlag KG, Berlin, ASAP
3. Trost BM, Lemoine RC (1996) An asymmetric synthesis of vigabatrin. Tetrahedron Lett 37:9161–9164
4. Trost BM, Krueger AC, Bunt RC, Zambrano J (1996) On the question of asymmetric induction with acyclic allylic substrates. An asymmetric synthesis of (+)-polyoxamic acid. J Am Chem Soc 118:6520–6521
5. Mori M, Nakanishi M, Kajishima D, Sato Y (2001) A new and general synthetic pathway to stychnos indole alkaloids: total syntheses of (−)-dehydrotubifoline and (−)-tubifoline by palladium-catalyzed asymmetric allylic substitution. Org Lett 3:1913–1916
6. Mori M, Nakanishi M, Kajishima D, Sato Y (2003) A novel and general synthetic pathway to strychnos indole alkaloids: total syntheses of (−)-tubifoline, (−)-dehydrotubifoline, and (−)-strychnine using palladium-catalyzed asymmetric allylic substitution. J Am Chem Soc 125:9801–9807
7. Trost BM, Dong G (2006) New class of nucleophiles for palladium-catalyzed asymmetric allylic alkylation total synthesis of agelastatin A. J Am Chem Soc 128:6054–6055
8. Murphy KE, Hoveyda AH (2003) Enantioselective synthesis of α-alkyl–β, γ-unsaturated esters through efficient Cu-catalyzed allylic alkylations. J Am Chem Soc 125:4690–4691
9. Luchaco-Cullis CA, Mizutani H, Murphy KE, Hoveyda AH (2001) Modular pyridinyl peptide ligands in asymmetric catalysis: enantioselective synthesis of quaternary carbon atoms through copper-catalyzed allylic substitutions. Angew Chem Int Ed 40:1456–1460
10. Trost BM, Dogra K (2007) Synthesis (−)-Δ9-*trans*-tetrahydrocannabinol: stereocontrol via Mo-catalyzed asymmetric allylic alkylation reaction. Org Lett 9:861–863
11. Palucki M, Um JM, Yasuda N, Conlon DA, Tsay FR, Hartner FW, Hsiao Y, Marcune B, Karady S, Hughes DL, Dormer PG, Reider PJ (2002) Development of a new and practical route to chiral 3,4-disubstituted cyclopentanones: asymmetric alkylation and intramolecular cyclopropanation as key C-C bond-forming steps. J Org Chem 67:5508–5516

12. Trost BM, Andersen NG (2002) Utilization of molybdenum- and palladium-catalyzed dynamic kinetic asymmetric transformations for the preparation of tertiary and quaternary stereogenic centers: a concise synthesis of tipranavir. J Am Chem Soc 124:14320–14321
13. Bergner EJ, Helmchen G (2000) Synthesis of enantiomerically pure (−)-wine lactone based on a palladium-catalyzed enantioselective allylic substitution. Eur J Org Chem 419–423
14. Trost BM, Chupak LS, Lübbers T (1998) Total synthesis of (+/−)- and (+)-valienamine via a strategy derived from new palladium-catalyzed reactions. J Am Chem Soc 120:1732–1740
15. Trost BM, Lee CB (2001) Gem-diacetates as carbonyl surrogates for asymmetric synthesis. Total synthesis of sphingofungin E and F. J Am Chem Soc 123:12191–12201
16. Trost BM, Brennan MK (2006) Palladium asymmetric allylic alkylation of prochiral nucleophiles: horsfiline. Org Lett 8:2027–2030
17. Trost BM, Jiang C (2003) Pd-catalyzed asymmetric allylic alkylation. A short route to the cyclopentyl core of viridenomycin. Org Lett 5:1563–1565
18. Trost BM, Toste FD (1998) A catalytic enantioselective approach to chromans and chromanols. A total synthesis of (−)-calanolides A and B and the vitamin E nucleus. J Am Chem Soc 120:9074–9075
19. Trost BM, Toste FD (2000) Enantioselective total synthesis of (−)-galanthamine. J Am Chem Soc 122:11262–11263
20. Trost BM, Tang W (2002) An efficient enantioselective synthesis of (−)-galanthamine. Angew Chem Int Ed 41:2795–2797
21. Trost BM, Toste FD (1999) Palladium-catalyzed kinetic and dynamic kinetic asymmetric transformation of 5-acyloxy-2-(5H)-furanone. enantioselective synthesis of (−)-aflatoxin B lactone. J Am Chem Soc 121:3543–3544
22. Trost BM, Crawley ML (2002) 4-aryloxybutenolides as "Chiral Aldehyde" equivalents: an efficient enantioselective synthesis of (+)-brefeldin A. J Am Chem Soc 124:9328–9329
23. Trost BM, Crawley ML (2004) A "Chiral Aldehyde" equivalent as a building block towards biologically active targets. Chem Eur J 10:2237–2252
24. Langlois J-B, Alexakis A (2011) Copper-catalyzed Enantioselective Allylic Substitution. Top Organomet Chem, DOI: 10.1007/3418_2011_12

Index

Lightning Source UK Ltd.
Milton Keynes UK
UKOW05n1327071016

284730UK00001B/98/P